Synthetic Biology Conference 2016 (SEED 2016)

Engineering, Evolution, and Design

Chicago, Illinois, USA
18-21 July 2016

ISBN: 978-1-5108-3202-2

Printed from e-media with permission by:

Curran Associates, Inc.
57 Morehouse Lane
Red Hook, NY 12571

Some format issues inherent in the e-media version may also appear in this print version.

Printed by Curran Associates, Inc. (2017)

For permission requests, please contact AIChE
at the address below.

AIChE
120 Wall Street, FL 23
New York, NY 10005-4020

Phone: (800) 242-4363
Fax: (203) 775-5177

www.aiche.org

Additional copies of this publication are available from:

Curran Associates, Inc.
57 Morehouse Lane
Red Hook, NY 12571 USA
Phone: 845-758-0400
Fax: 845-758-2633
Email: curran@proceedings.com
Web: www.proceedings.com

TABLE OF CONTENTS

KEYNOTE 2

SESSION 7: CELL-FREE SYSTEMS

SESSION 8: PLANTS

SESSION 9: BIOLOGICAL CIRCUITS IN NATURAL AND ENGINEERED SYSTEMS

ACS SYNTHETIC BIOLOGY YOUNG INVESTIGATOR AWARD LECTURE

SESSION 10: BIO-DESIGN AUTOMATION

Programmable On-Chip DNA Compartments as 'Artificial Cells'

Bar-Ziv, R., Weizmann Institute of Science

Originally presented on: 7/18/2016 9:30:00 - 9:55:00

The assembly of artificial cells capable of executing DNA programs has been an important goal for basic research and technology. We assemble 2D DNA compartments fabricated in silicon as â??artificial cellsâ?? capable of metabolism, programmable protein synthesis, and communication. We programmed gene expression cycles in separate compartments, as well as protein synthesis fronts propagating in a coupled 1D system of compartments. Gene expression in the DNA compartments reveals a rich, dynamic system that is controlled by geometry. The organization of matter in the compartment suggested conditions for controlled assembly of biological machines. This puts forth a man-made biological system with programmable information processing from the gene to a â??cellâ??, and up to the â??multicellularâ?? scale.

References

A. Tayar, E. Karzbrun, V. Noireaux, R.H. Bar-Ziv, Propagating gene expression fronts in a one-dimensional coupled system of artificial cells. *Nature Phys.* **11**, 1037â??1041 (2015).

E. Karzbrun, A. M. Tayar, V. Noireaux, R.H. Bar-Ziv, Programmable on-chip DNA compartments as artificial cells. *Science.* **345**, 829â??832 (2014).

D. Bracha, E. Karzbrun, G. Shemer, P. A. Pincus, R.H. Bar-Ziv, Entropy-driven collective interactions in DNA brushes on a biochip. *Proc. Natl. Acad. Sci. U. S. A.* **110**, 4534â??8 (2013).

Y. Heyman, A. Buxboim, S. G. Wolf, S. S. Daube, R.H. Bar-Ziv, Cell-free protein synthesis and assembly on a biochip. *Nature Nanotech.* **7**, 374â??378 (2012).

Ribosome Drafting

Next-generation Biophysical Models & Algorithms for Engineering Genetic Systems

Salis Lab

Penn State University

http://salislab.net

PENNSTATE

SEED 2016, Chicago

The Design Challenge in Synthetic Biology

A genetic system's function is encoded within its DNA sequence

We pick a DNA sequence, and (we hope) it has the desired function

Desired Regulation

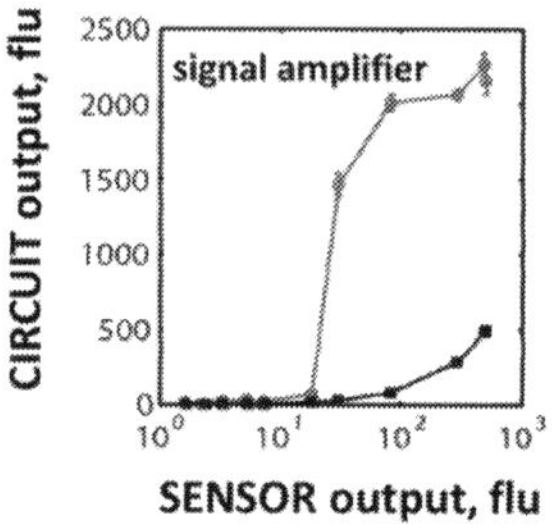

Optimally balanced expression levels

1000X

2X

200X

And No Surprises!

efficient transcriptional termination

no repetitive sequences

no internal promoters/RBSs/terminators

few RNAse binding sites

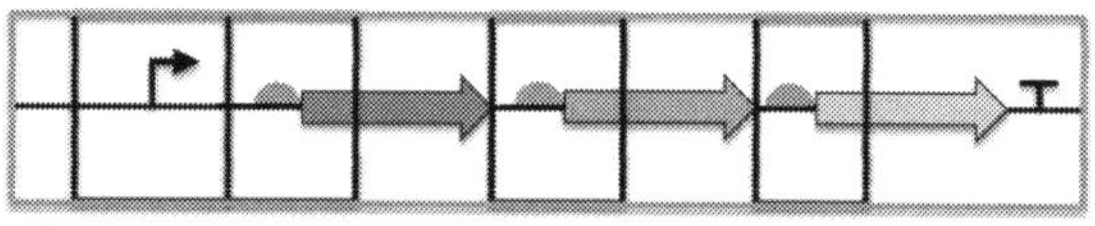

The Design Challenge in Synthetic Biology

A genetic system's function is encoded within its DNA sequence

Out of an astronomical number of DNA sequences, some are much better than others

60 bp promoter = $\mathbf{10^{36}}$ ways to regulate transcription

70 nt RBS/CDS = $\mathbf{10^{42}}$ ways to regulate translation

100 AA protein = $\mathbf{10^{51}}$ ways to regulate protein folding

1 operon = 10^{351} sequences

vary transcription rate by >1000-fold

vary translation rate by >100,000-fold

1D expression space = 10^{8}-fold

an *N*-enzyme pathway = an *N*-dimensional enzyme expression space

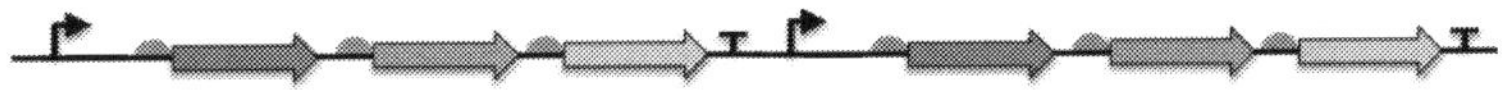

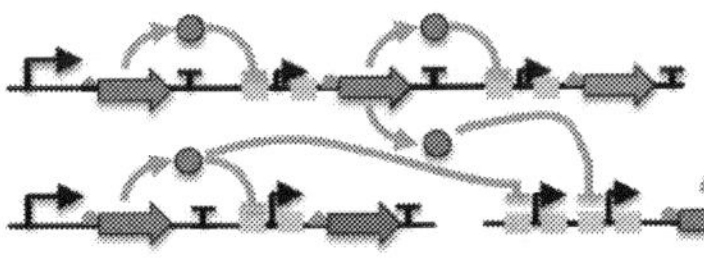

an N-regulator genetic circuit = an N-D *Pt* design space

Sequence-to-Function Models

Biophysical models can predict the sequence-function relationship for parts & systems

genetic system sequence → biophysical model prediction → **genetic system function**

RBS Calculator

RBS sequence **CDS sequence**

AUCCGAUAACUAGUCACACAGUAAAAAUUUAGUUAUGACCACCUUUCACAAAGCAAGCGGA
GUGCUUCUUGUGCCGACUUACGAGCAUCUAGCGAGCAUCUAGCGACUACUGAC . . . **UAA**

↓

a ribosome-mRNA free energy model

$$\Delta G_{total} = \Delta G_{mRNA:rRNA} + \Delta G_{spacing} + \Delta G_{start} + \Delta G_{standby} - \Delta G_{mRNA}$$

↓

translation rate predictions using statistical thermodynamics

$$r_{translation} \propto \exp(-\beta \Delta G_{tot})$$

↓

experimental validation on large/diverse sequence data-sets

Our Collection of Sequence-to-Function Biophysical Models

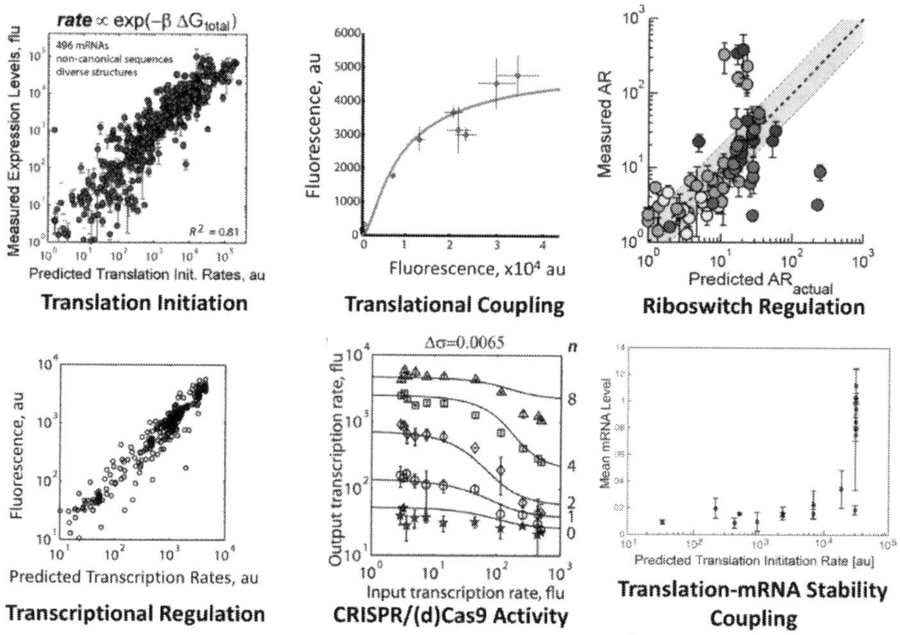

Solving the Design Challenge

Biophysical models enable automated, predictive design of genetic systems

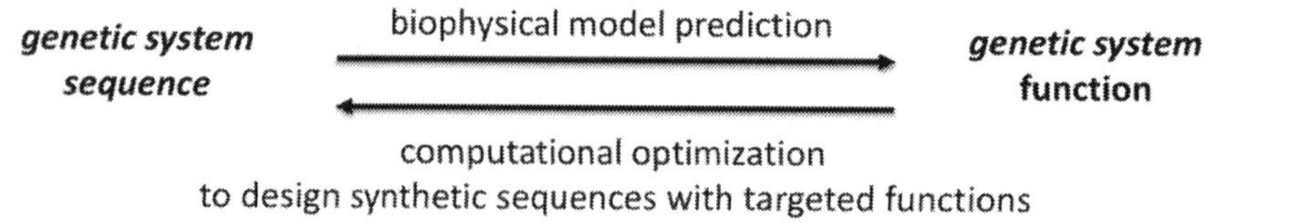

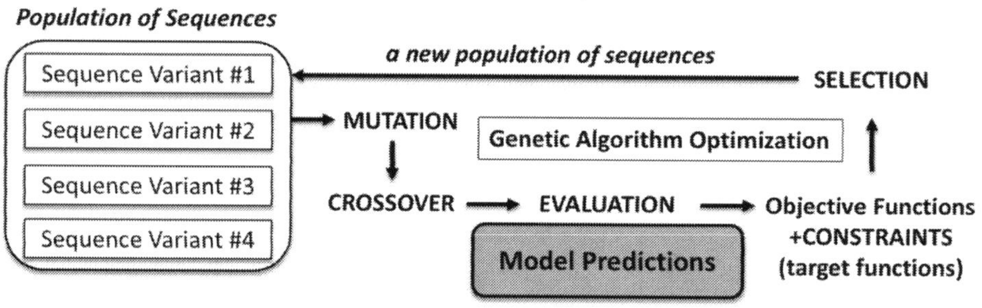

Operon Design

online @ http://salislab.net/software

Pick your Models + Design Rules

improve **Evolutionary Robustness**
improve mRNA stability
eliminate undesired genetic elements

Pick your Promoter

range of transcription rates
stress responsive (heat, toxins, load)
metabolite responsive (feedstock types)

Pick your Protein Coding Sequences

targeted translation rates
synonymous codon usage optimization
eliminate ribosomal pausing (optional)

Pick your Transcriptional Terminator
non-repetitive, high efficiency

Pick your Host Organism & click ***GO***
Design searches 10^{350} possible solutions
in 4-24 hours on a 16-core node

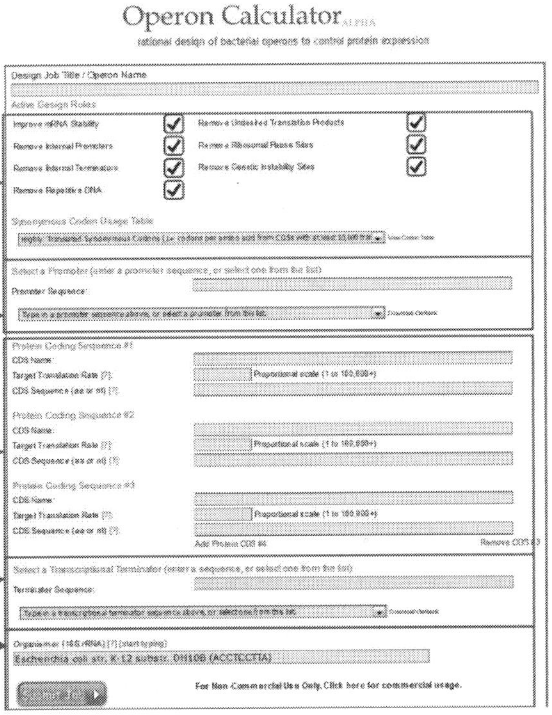

Operon Design

online @ http://salislab.net/software

designed RBS
for target translation rates

no internal promoters
no internal terminators
(intrinsic or rho-dependent)
no repetitive sequences

3 RNAse E binding site
No RNAse III binding sites
1 x Tn10 site
1 x IS10 site
3 x IS231 sites

Pareto optimal trade-offs
= multiple solutions

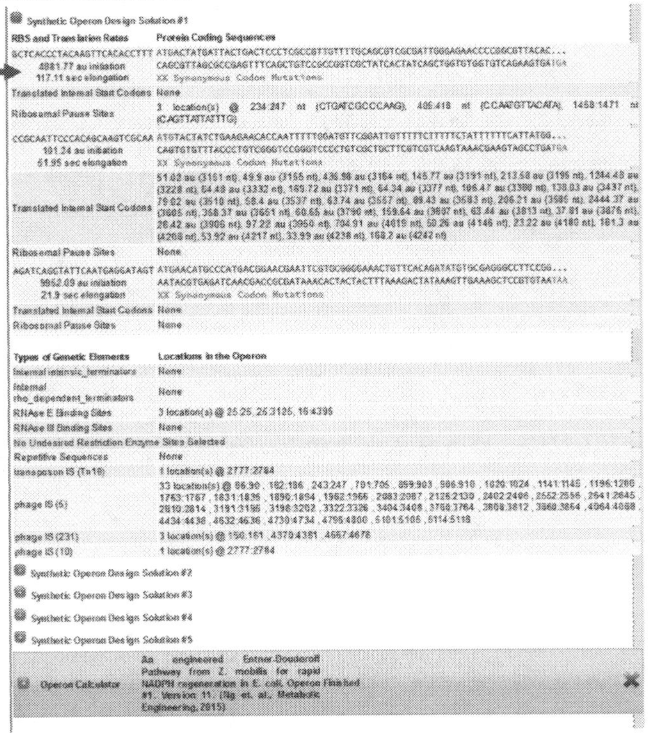

Sensor, Circuit, and Pathway Applications

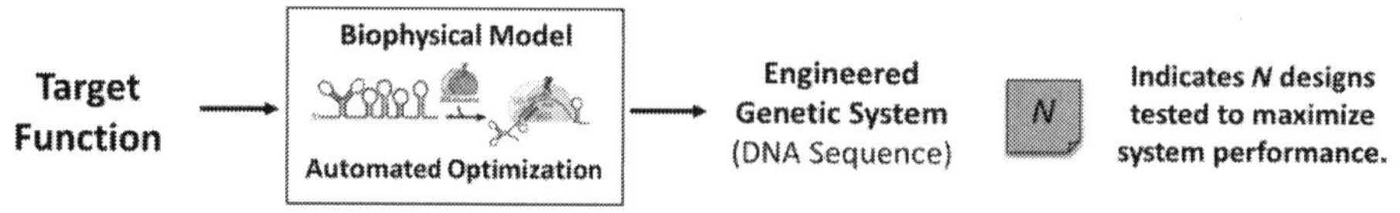

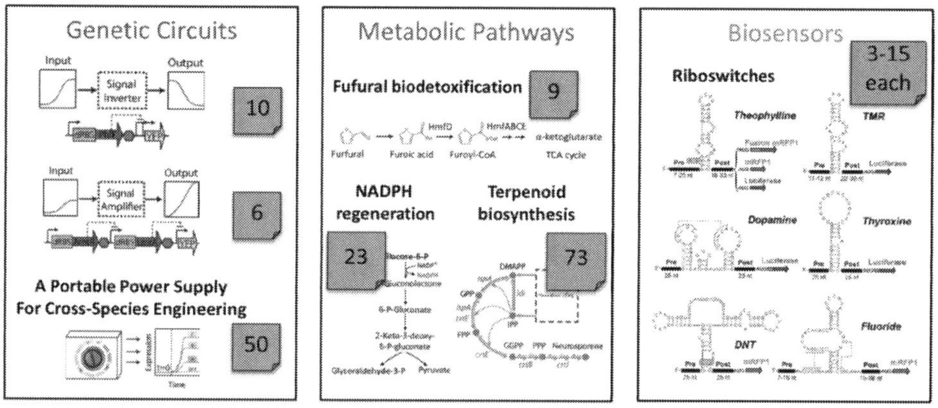

Farasat et. al. *Mol. Sys. Bio.*, 2014
Ng. et. al. *Metabolic Eng.*, 2015
Kushwaha & Salis, *Nat. Comm.*, 2015
Espah Borujeni et. al., NAR Breakthrough Article, 2016

Automated Design Methods for Genetic Systems

Design/Optimize RBSs & RBS libraries, Operons, Riboswitches, ...

Optimizing Recombinant Protein Expression

Antibody expression (Makino et. al., 2011)
P450 cytochrome expression (Chang et al., 2014)
Salmonella protein expression (Medina et. al., 2011)
Xylanase and glucanase expression (Liu et. al., 2012)
Nitrile hydratase expression (Kang et. al., 2014)
Secreted protein expression (Heggeset et. al. 2013)

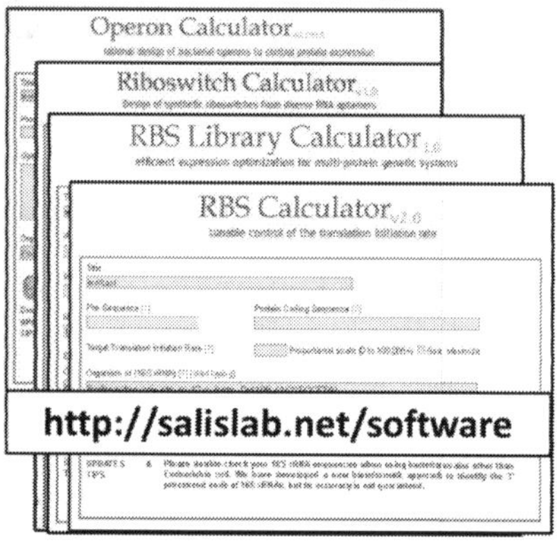

Pathway Optimization & Circuit Engineering

Xylanase and glucanase expression (Liu et. al., 2012)
Nitrile hydratase expression (Kang et. al., 2014)
Secreted protein expression (Heggeset et. al. 2013)

Third–party validation (Biggs et. al., 2014)

FR900098 biosynthesis (Freestone et. al., 2015)
6-ACA biosynthesis for nylon (Zhou et. al., 2015)
NADPH regeneration (Ng et. al., 2015)
Xylitol biosynthesis (Su et. al., 2015)
Riboflavin biosynthesis (Lin et. al., 2014)
Terpenoid biosynthesis (Zhou et. al., 2014)
Amorphadiene biosynthesis (Nowroozi et. al., 2014)
2,3-butanediol biosynthesis (Oliver et. al., 2014)
Engineering memory genetic devices (Yang et. al., 2014)
Butanol biosynthesis (Lim et. al., 2013)
Fatty acid biosynthesis (Lennen et. al., 2013)
Spectinabilin biosynthesis (Shao et. al., 2013)
Nitrogen fixation (Temme at. al., 2012)
Isoprene biosynthesis (Zurbriggen et. al. 2012)

Biophysical Models: Predicting Sequence-to-Function

Our models quantify several biophysical interactions (colors)

But not all models use all colors

How do we <u>systematically</u> develop & test improved models?
Better Models = Better Designs

the rich color palette of physical chemistry & biophysics | **only thermodynamics** | **next-generation models**

Predicted Transcription Rates, au

Transcriptional Regulation

Input transcription rate, flu

CRISPR/(d)Cas9 Activity

Translation-mRNA Stability Coupling

The Road to Design Perfection: One Example Question

which happens first?

and how does the mRNA sequence control the answer?

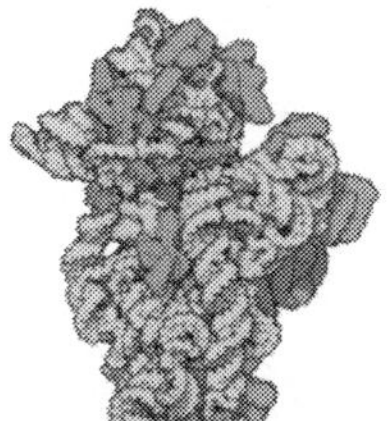

unfolded mRNA

Shine-Dalgarno start codon

. . . CACACAAGGAGACCAUAUGAUCAUUAAUUAUAUCGUUUUCAAA . . .

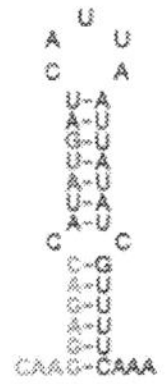

the ribosome binds the mRNA & initiates translation

OR

the mRNA folds into a structure

The Road to Design Perfection

What do we need to develop ever more-accurate sequence-to-function models?

1. a Model Test System with a database of characterized & DIVERSE sequences

750 individually characterized sequences
RBSs, riboswitches, & operons from *E. coli*, *B. subtilis*, *C. glutanicum*, *P. fluorescens*, *B. fragilis*

$$H_{seq} = \sum_j \sum_{i=1}^{4} p_i log_2 p_i$$

Higher Sequence Entropy → More Diverse Data-set
a quantitative metric to prevent cherry-picking!

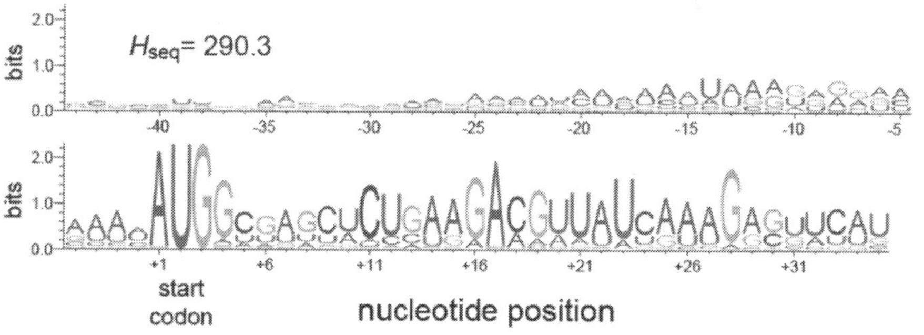

Traveling the Road to Design Perfection

What do we need to develop ever more-accurate sequence-to-function models?

1. a Model Test System with a database of characterized & DIVERSE sequences

RBS Calculator v2.0
53% of sequences within 2-fold model error
69% of sequences within 3-fold model error

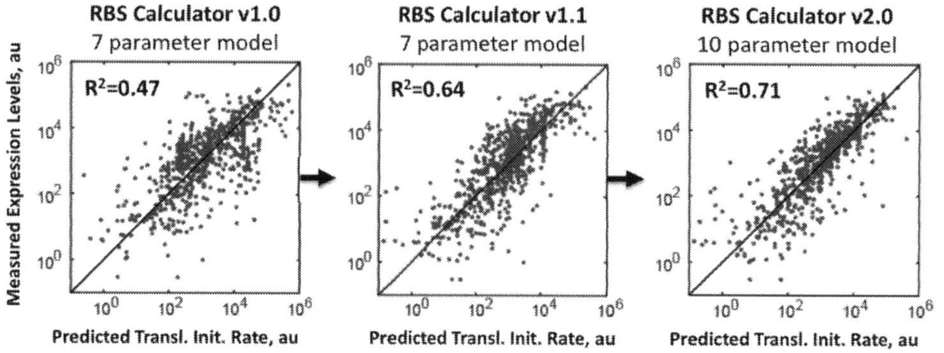

v1.0: Salis, Mirsky, & Voigt, Nat. Biotech. 2009; v1.1: Salis, 2010; v2.0: Espah Borujeni et. al., Nucl Acid. Res., 2013

The Road is Not Straight

What do we need to develop ever more-accurate sequence-to-function models?

1. a Model Test System with a database of characterized & DIVERSE sequences

750 individually characterized sequences

RBSs, riboswitches, & operons from *E. coli*, *B. subtilis*, *C. glutanicum*, *P. fluorescens*, *B. fragilis*

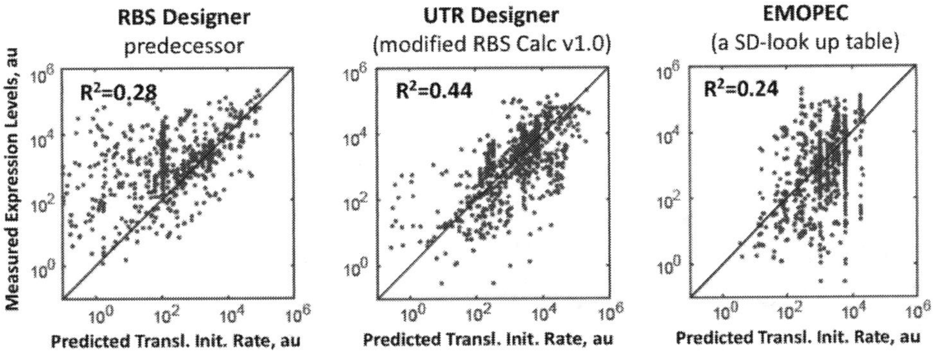

RBS Designer: Na, Lee, & Lee, BMC Sys. Bio., 2009

UTR Designer: Seo et. al. Met. Eng., 2013

EMOPEC: Bonde et. al. Nat. Methods., 2016

There Might be Some Dead Ends

What do we need to develop ever more-accurate sequence-to-function models?

1. a Model Test System with a database of characterized & DIVERSE sequences

750 individually characterized sequences

RBSs, riboswitches, & operons from *E. coli*, *B. subtilis*, *C. glutanicum*, *P. fluorescens*, *B. fragilis*

Claims of superior accuracy need to be tested on the same data-set

and that data-set should be sufficiently large & diverse to reflect actual usage

We ♥ Suppl. Data

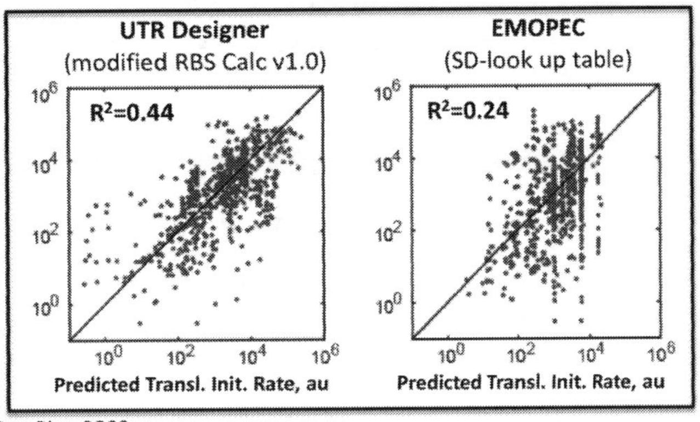

RBS Designer: Na, Lee, & Lee, BMC Sys. Bio., 2009

UTR Designer: Seo et. al. Met. Eng., 2013

EMOPEC: Bonde et. al. Nat. Methods., 2016

And Sometimes You Need a Bridge

What do we need to develop ever more-accurate sequence-to-function models?

1. a Model Test System with a database of characterized & DIVERSE sequences

15933 sfGFP-expressing mRNAs characterized by Flow-Seq in *E. coli*
(Kosuri et. al. 2013; Goodman et. al. 2013)

But <u>none of the models</u> can (accurately) predict their translation rates!
Why?!

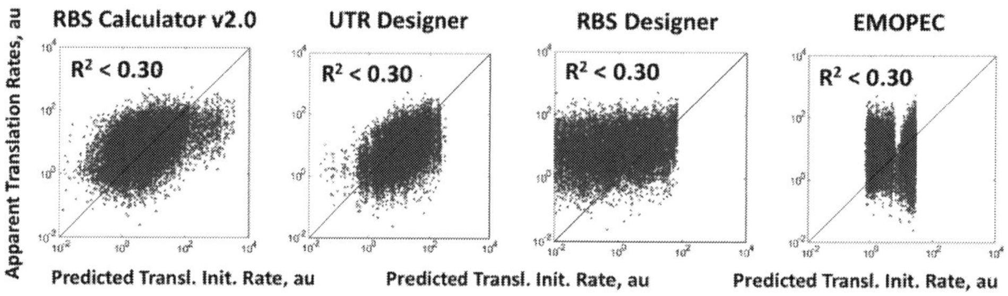

Non-Equilibrium Models of Translation Initiation

Using the Model Test system
we noticed that mRNAs with **long structures** had higher model error

Q: How does the rate of RNA folding influence a mRNA's translation rate?

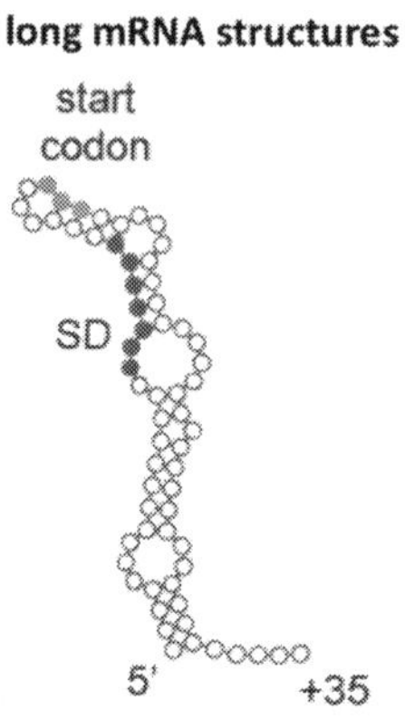

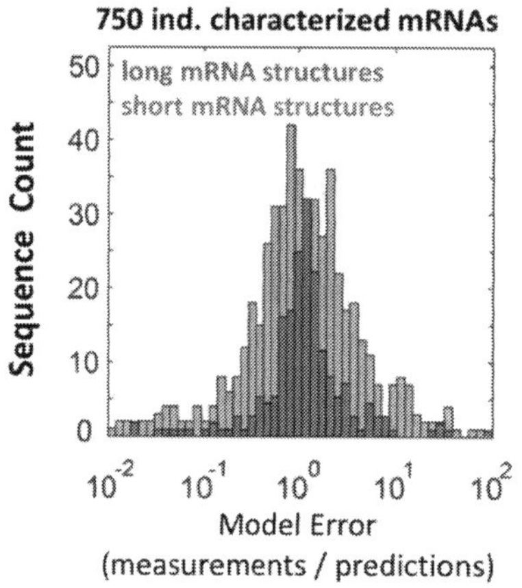

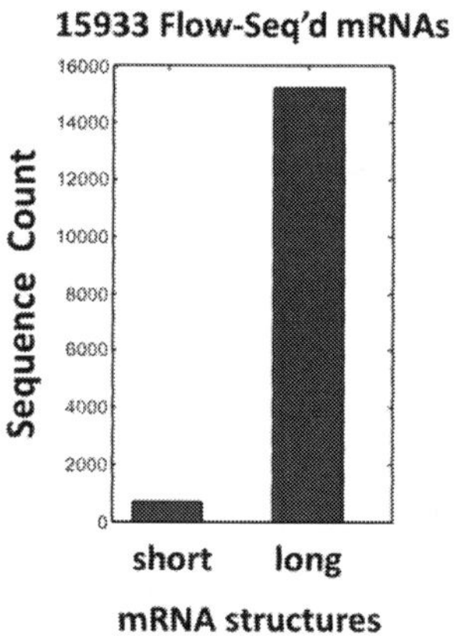

Design of RNA with Desired Folding Kinetics

What do we need to develop ever more-accurate sequence-to-function models?

2. new approaches to **predict**, design, and measure the *dynamics* of RNA folding

RNA dynamics affects transcriptional termination, translation, and mRNA stability

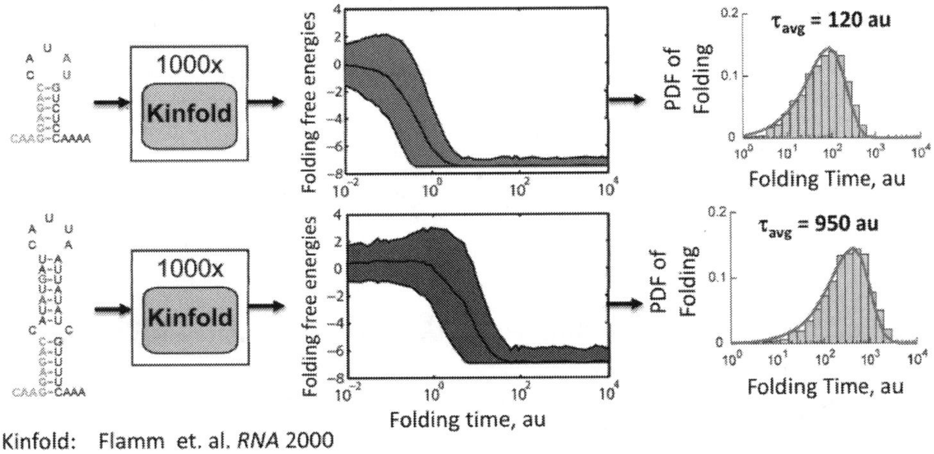

Kinfold: Flamm et. al. *RNA* 2000

Design of RNA with Desired Folding Kinetics

What do we need to develop ever more-accurate sequence-to-function models?

2. new approaches to predict, **design**, and measure the *dynamics* of RNA folding

We computationally designed pairs of RNA hairpins to have: the same folding energetics (thermodynamics) but very different folding rates (kinetics)

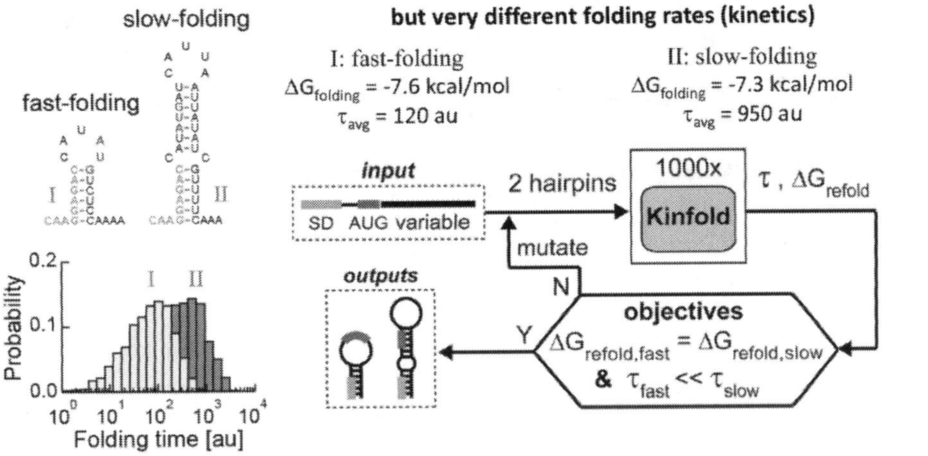

Non-Equilibrium Measurements of RNA Folding

What do we need to develop ever more-accurate sequence-to-function models?

2. new approaches to predict, design, and **measure** the *dynamics* of RNA folding

We applied time-correlated single photon counting (TCSPC) to measure the *dynamics* of RNA folding

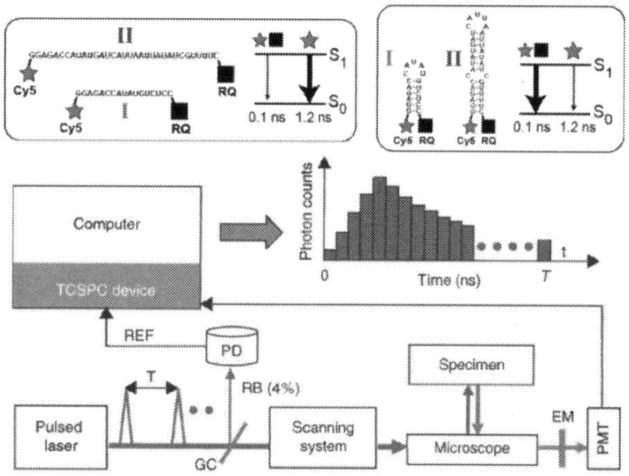

FREE fluorophore: **1.2** ns lifetime
QUENCHED fluorophore: **0.1** ns lifetime

TCSPC measures the photon lifetime distribution in a sample

From these distribution, we can determine the fraction of fluorophore that was
FREE vs. QUENCHED
UNFOLDED vs. FOLDED RNA

Non-Equilibrium Measurements of RNA Folding

Q: What were the relative folding rates of the fast-folding and slow-folding RNAs?
A: The slow-folding RNA folded after 5 min, much slower than the fast-folding RNA

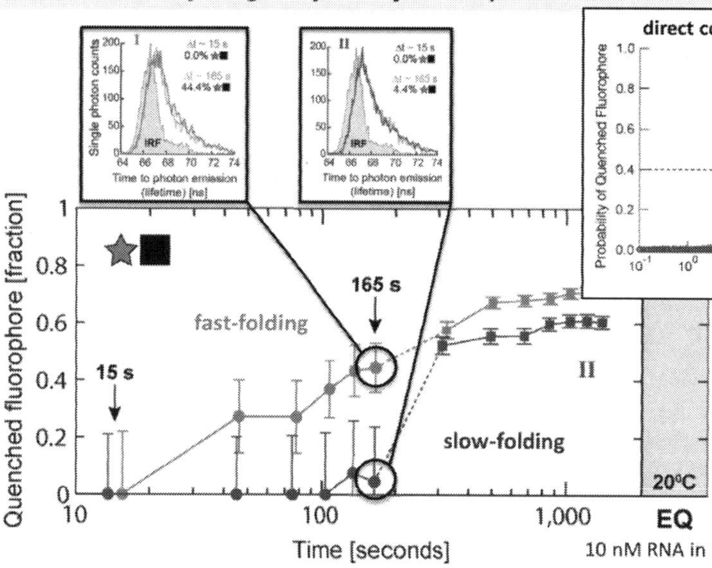

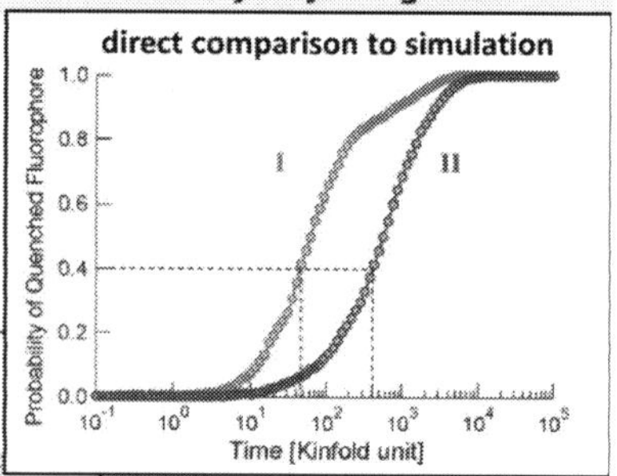

***in vitro* flash-cool RNA folding @ 10 nM was very slow**

this was a surprising result!

10 nM RNA in 100 mM NaCl buffer, N = 2

Non-Equilibrium Models of Translation Initiation

What do we need to develop ever more-accurate sequence-to-function models?

3. advanced theory to *predict* gene expression levels in non-equilibrium systems

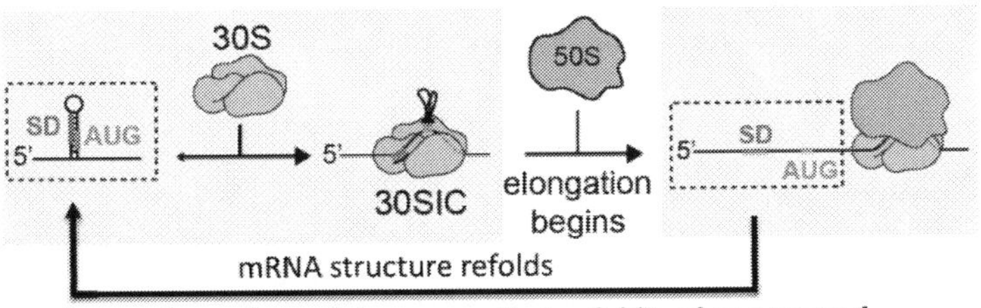

a thermodynamic model assumes that refolding has occurred

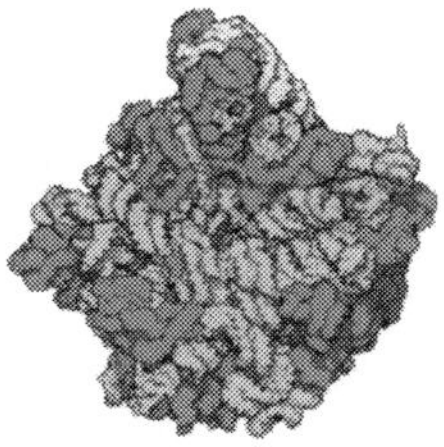

which happens first?
RNA refolding or ribosome binding?

. . . CACACAAGGAGACCATATGATCATTAATTATATCGTTTTCAAA . . .

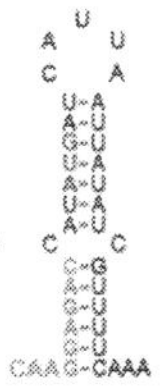

Non-Equilibrium Models of Translation Initiation

What do we need to develop ever more-accurate sequence-to-function models?

3. advanced theory to *predict* gene expression levels in non-equilibrium systems

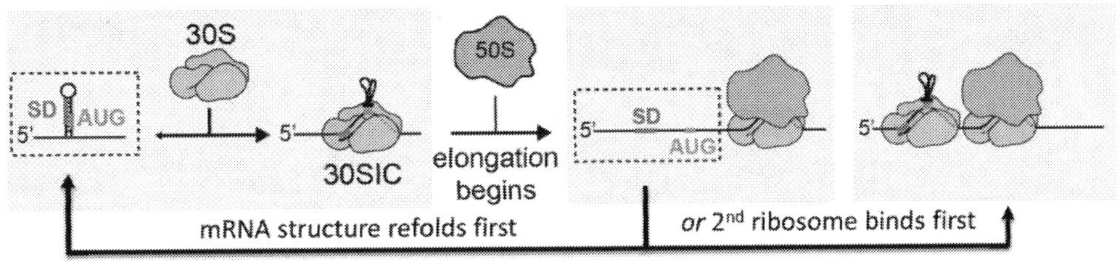

Ribosome Drafting:

during cycles of translation

successive ribosomes bind *faster* than the mRNA's refolding rate

Espah Borujeni & Salis, *JACS*, 2016

Ribosome Drafting on mRNAs

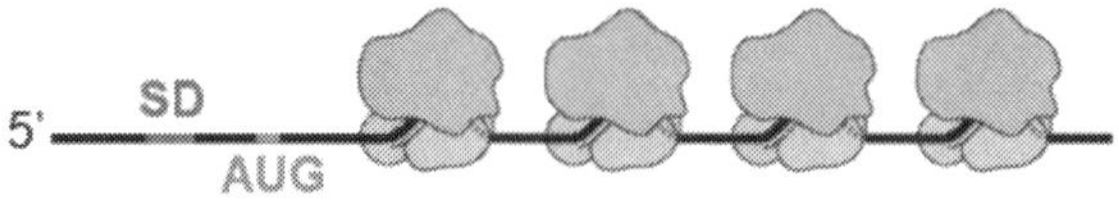

Ribosome Drafting takes place when:

1. Inhibitory RNA structures have slow refolding kinetics
2. The ribosome's binding rate is fast

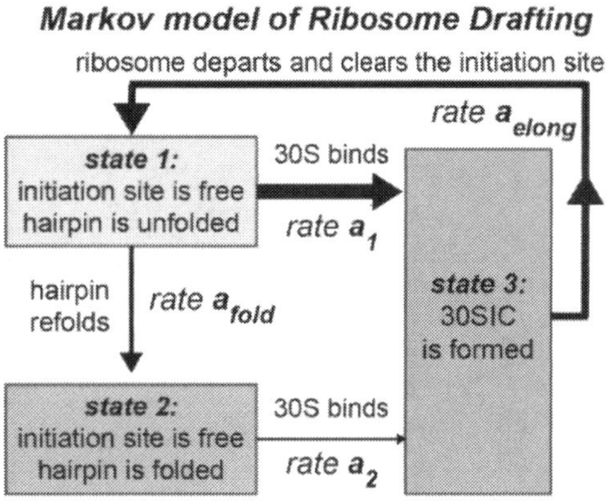

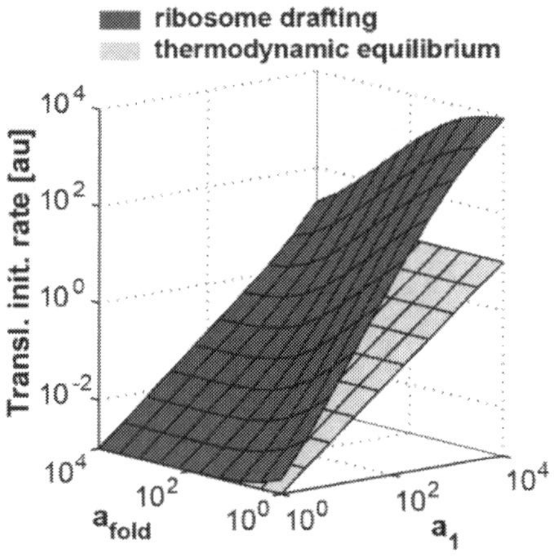

Ribosome Drafting on mRNAs

Q: Does Ribosome Drafting take place inside cells?
A: YES, a slow-folding mRNA had a 70-fold higher translation rate, compared to a fast-folding mRNA. Both mRNAs have the same folding energetics.

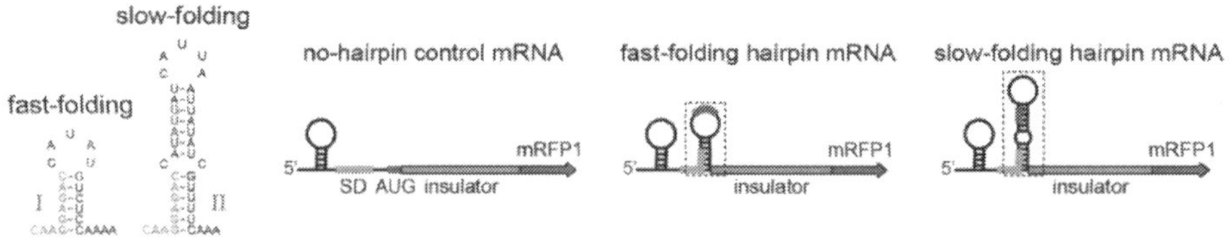

mRNA level measurements

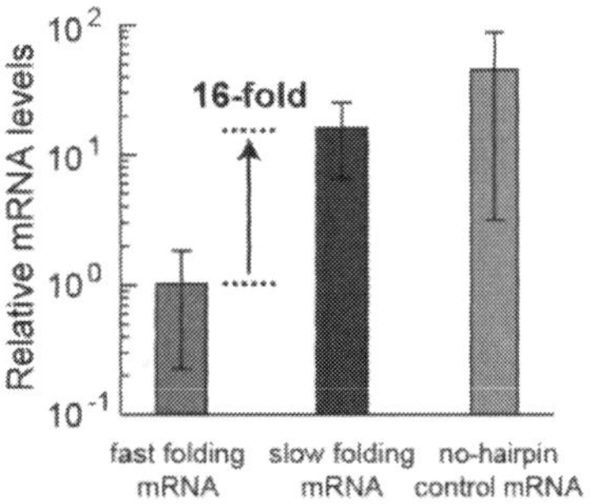

protein expression measurements

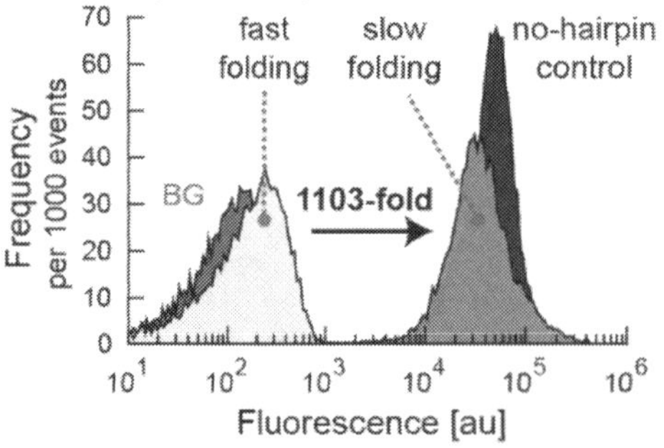

Ribosome Drafting on mRNAs

Q: How does Ribosome Drafting depend on the mRNA sequence?

We tested the Ribosome Drafting mechanism by characterizing mRNAs with fast-folding versus slow-folding inhibitory RNA hairpins (same energetics) while varying the ribosome's binding rate

Markov Model

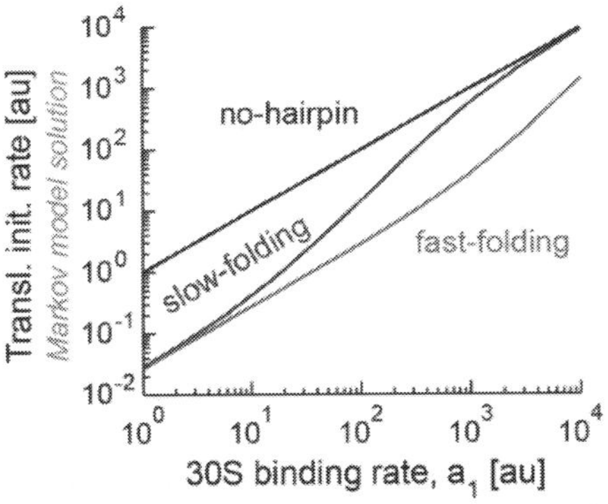

Experimental Design

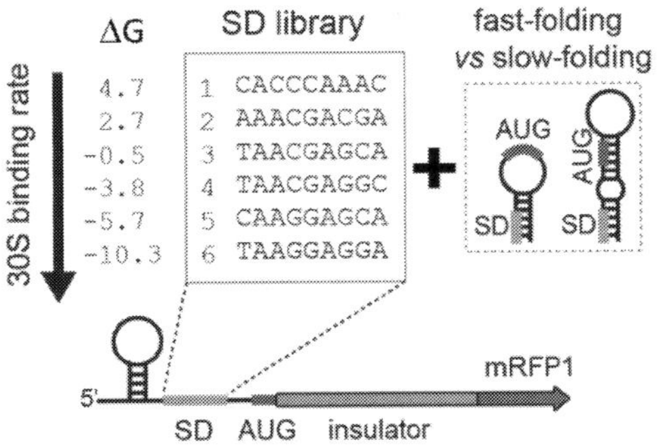

Ribosome Drafting on mRNAs

Q: How does Ribosome Drafting depend on the mRNA sequence?
A: Refolding rate, ribosome binding rate, & folding energetics are important

If the mRNA does not have any inhibitory mRNA structures then the equilibrium RBS Calculator v2.0 model works well

Markov Model

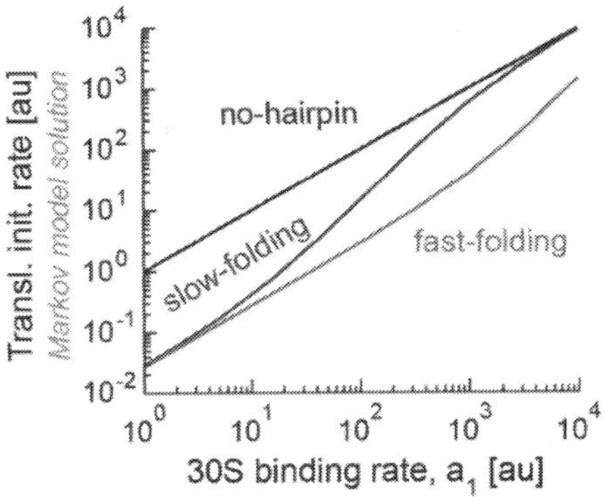

Experimental Measurements

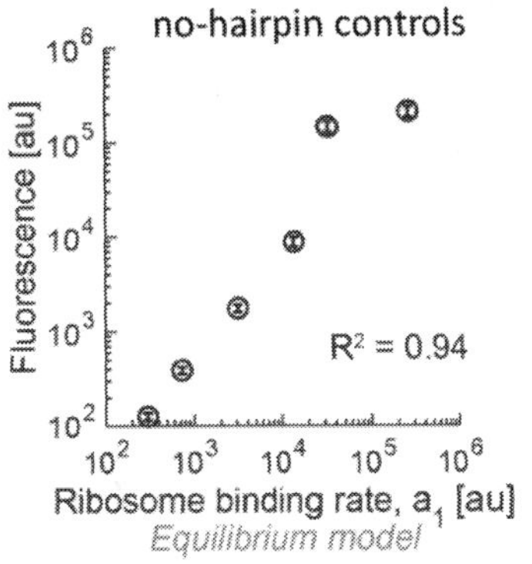

Ribosome Drafting on mRNAs

Q: How does Ribosome Drafting depend on the mRNA sequence?
A: Refolding rate, ribosome binding rate, & folding energetics are important

But if the mRNA has slow-folding mRNA structures
then we must add the Ribosome Drafting mechanism to the model

Markov Model

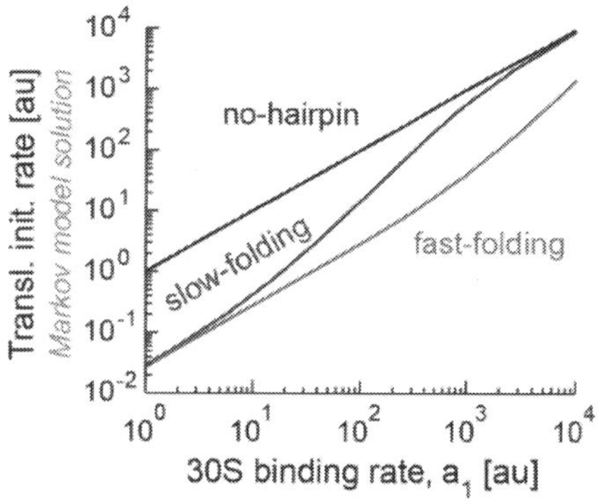

Experimental Measurements

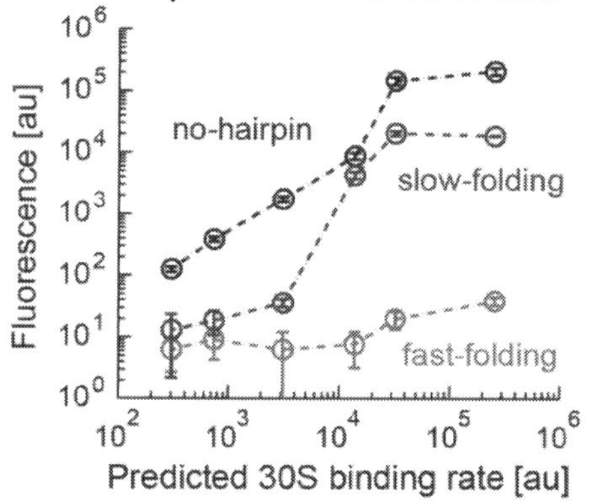

Ribosome Drafting on mRNAs

Q: How does Ribosome Drafting depend on the mRNA sequence?
A: Refolding rate, ribosome binding rate, & folding energetics are important

We introduced fast-folding & slow-folding mRNA hairpins
with **high** ribosome binding rates & ***varying folding energetics***

Ribosome Drafting can take place with strong or weak mRNA structures

Markov Model

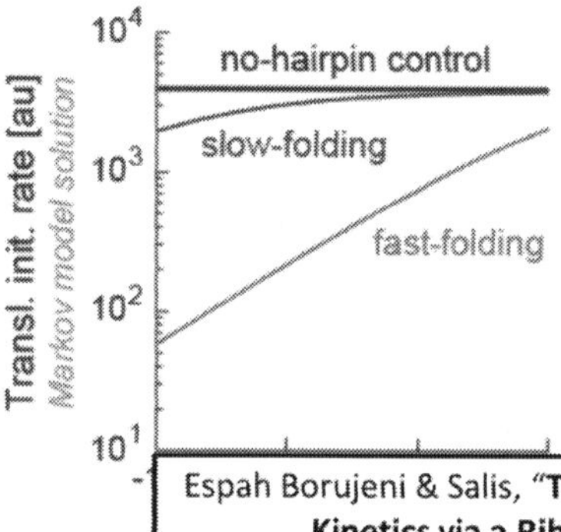

Experimental Measurements

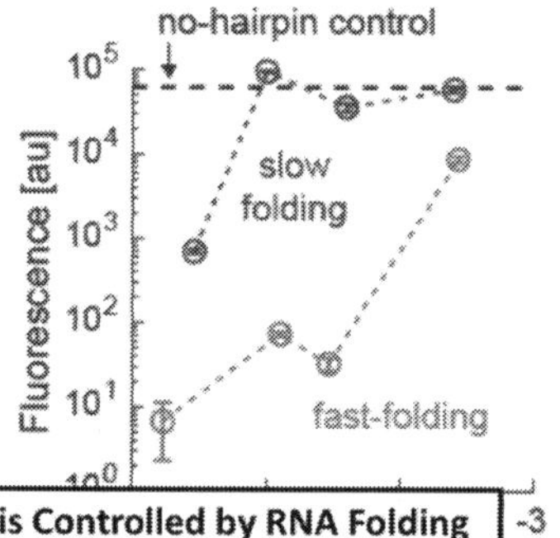

Espah Borujeni & Salis, "**Translation Initiation is Controlled by RNA Folding Kinetics via a Ribosome Drafting Mechanism**", *JACS*, 2016

Next-generation Non-equilibrium Biophysical Models

What do we need to develop ever more-accurate sequence-to-function models?

1. a Model Test System with a database of characterized & DIVERSE sequences
2. new approaches to predict, design, and measure the *dynamics* of RNA folding
3. advanced theory to *predict* gene expression levels in non-equilibrium systems

The 6000+ registered researchers @ http://salislab.net/software
have designed over 100,000 synthetic DNA sequences using *our equilibrium models*

Next-generation biophysical models will be essential
to achieving the next-level design of mRNAs, riboswitches, operons, and genomes

the rich color palette of physical chemistry & biophysics

equilibrium models

non-equilibrium models

Acknowledgements

Graduate students
Sean Halper
Grace Vezeau
Alexander Reis
Daniel Cetnar

Undergraduate students
Phillip Clauer, Danielle Hand,
Kavya Vaseekaran, Lipika Gadila

Alumni
Tian Tian
Manish Kushwaha
Amin Espah Borujeni
Iman Farasat
Long Chen
Chiam Yu Ng

Current Funding

ONR Revolutionary Research . . . Relevant Results
AFOSR AIR FORCE OFFICE OF SCIENTIFIC RESEARCH
NSF

psiee penn state institutes of energy and the environment

Operon Design

Q: How can we optimize multi-protein genetic systems with the fewest experiments?

Step 1: Apply the Operon Calculator to design a GOOD "v1.0" genetic system

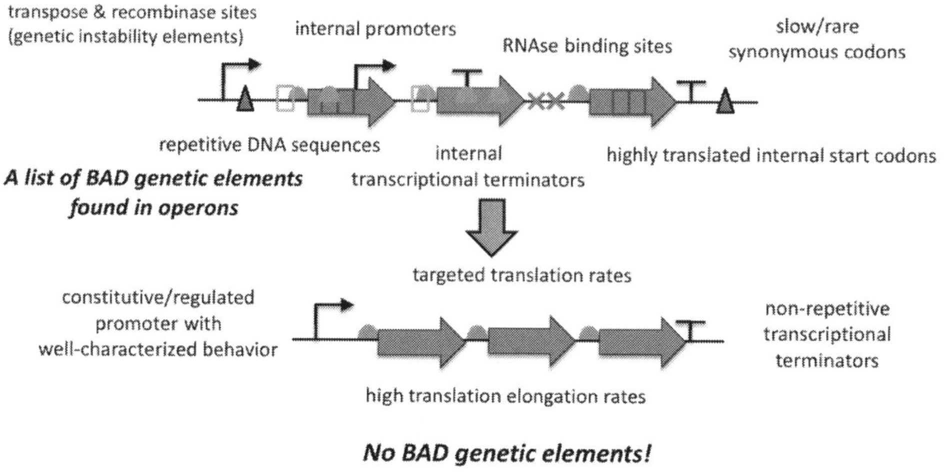

Shadows on the Wall

How can we understand complex sequence-function relationships inside cells?

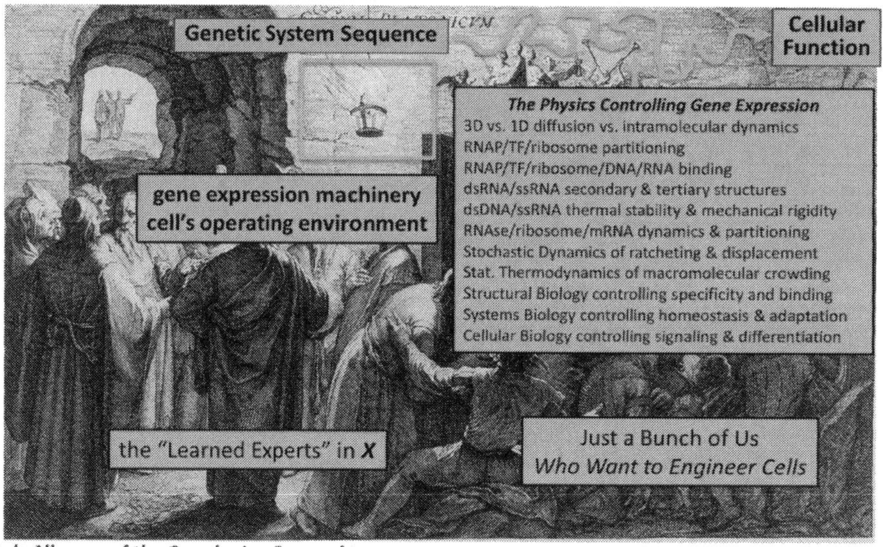

Plato's Allegory of the Cave by Jan Saenredam

Shadows on the Wall

How can we understand complex sequence-function relationships inside cells?

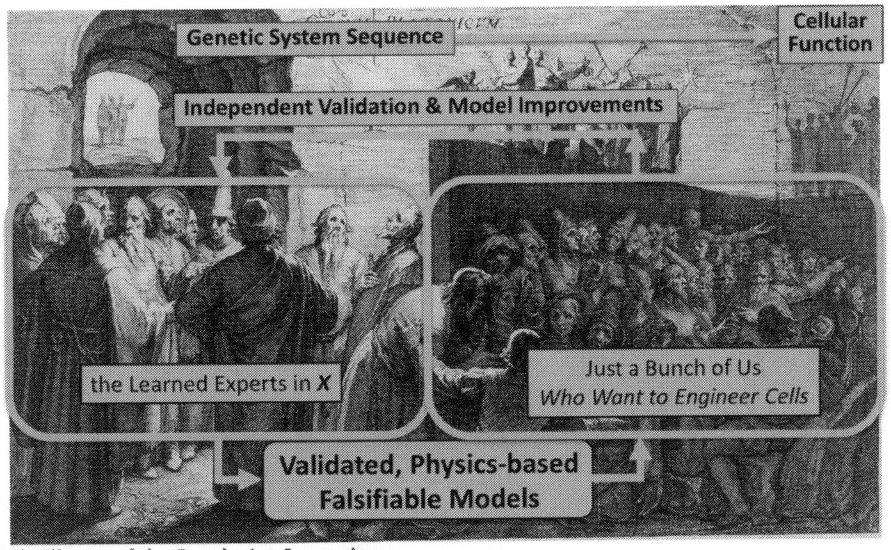

Plato's Allegory of the Cave by Jan Saenredam

Ribosome Drafting **on mRNAs**

Q: How does Ribosome Drafting depend on the mRNA sequence?
A: Refolding rate, ribosome binding rate, & folding energetics are important

We introduced fast-folding & slow-folding mRNA hairpins
with **low** ribosome binding rates & ***varying folding energetics***

There are stable, slow-folding structures and weak, fast-folding structures

Markov Model

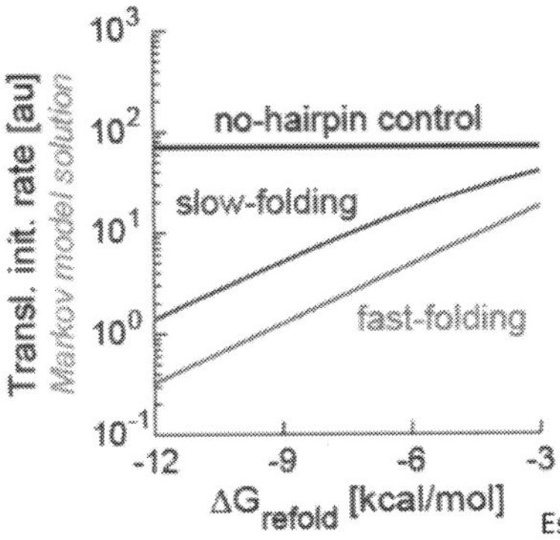

Experimental Measurements

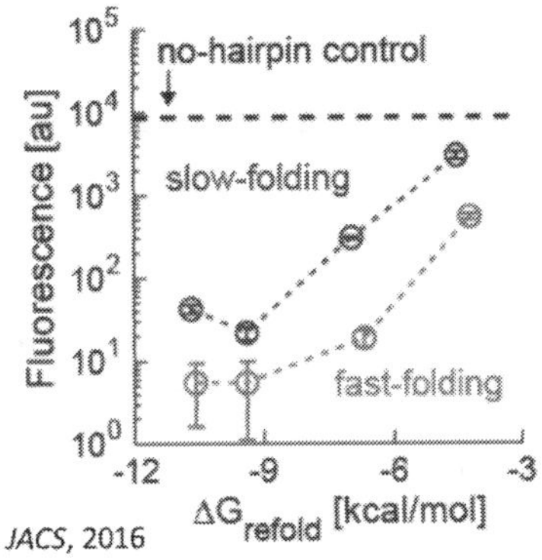

Espah Borujeni & Salis, *JACS*, 2016

Recoding Genomes to Expand Life's Functional Repertoire

Isaacs, F.,

Originally presented on: 7/18/2016 11:30:00 - 12:00:00

The conservation of the genetic code permits organisms to share beneficial traits via horizontal gene transfer, and enables the accurate expression of heterologous genes in non-native organisms. Fundamental changes to the genetic code could significantly enhance our understanding of the origins of the canonical code and reveal new subtleties of how genetic information is encoded and exchanged. I will describe the construction of Genomically Recoded Organisms (GROs), in which all 321 UAG stop codons were replaced with the synonymous UAA stop codon in *E. coli*. This GRO exhibited improved properties for incorporation of nonstandard amino acids that expand the chemical diversity of proteins *in vivo* and demonstrated multi-virus resistance. Finally, I will describe the engineering of the GRO to depend on synthetic amino acids aimed at construction of safe GMOs unable to grow in the wild. This works demonstrates the application of genome-scale engineering toward the fundamental redesign of genomes and construction of orthogonal biological systems with new properties.

Organs-on-Chips Technology: Micro-Engineered System to Emulate Living Human Biology

Hamilton, G. A.,

Originally presented on: 7/18/2016 13:30:00 - 14:00:00

Dr. Geraldine A. Hamilton, PhD will discuss how the Organs-on-Chips technology is being used to better understand human biology and disease – offering new applications to advance drug development, precision medicine and consumer product design. Organ-Chips - such as the lung, liver, brain or skin - are micro-engineered environments lined with living human cells and tissues. Each chip recreates the natural physiology and mechanical forces that cells experience within the human body.

Probing Stem Cell Differentiation with Synthetic Biology

Deans, T. L., University of Utah

Originally presented on: 7/18/2016 14:00:00 - 14:30:00

The demand for stem cells is anticipated to continue to rise due to their expected ability to treat disease, use in novel diagnostic technologies, and for pharmaceutical screening. However, before these goals can be realized, a better understanding of the mechanisms regulating their fate is required. Cells have the remarkable ability to continuously sense, integrate, and store relevant physiological and biological information throughout their lives. They integrate the many signals that surround them and execute cellular behaviors based on these inputs. Both intrinsic (transcription factor expression), as well as extrinsic (environmental) mechanisms are thought to be involved in the regulation of stem cell self-renewal, and their commitment to differentiate into more specialized cell types. This interplay between intrinsic and extrinsic cues in differentiation poses challenges to studying the mechanisms involved in their proliferation and terminal differentiation. These challenges can be addressed using genetic circuits to tightly control gene expression in dynamic patterns, in addition to programming cells to sense and record changes in their microenvironment as cells undergo differentiation.

Advanced Biocomputers in Mammalian Cells

Wong, W.,

Originally presented on: 7/18/2016 14:30:00 - 15:00:00

Synthetic genetic circuits can revolutionize biotechnology, but engineering complex decision-making circuits remains challenging due to unintended interactions between cascading biological components. We have previously developed a single-layered biocomputeration platform in mammalian cells called Boolean Logic and Arithmetic through DNA Excision (BLADE). Here we will describe our recent efforts extending the BLADE platform to control guide RNA expression, thus allowing logic control of endogneous mammalian gene expression throught CRISPR/Cas. We have also developed temporal logic circuits based on the BLADE platform. These temporal logic circuits can generate at least five unique states from two inputs, and discriminate the order at which the two signals appear. Together, these circuit designs will provide a new set of tools to integrate multiple features for enhancing tissue-specific trans- and endogenous gene regulation in animal models.

Teaching Synthetic Transcription Factors to Read an Epigenetic Code

Haynes, K., Arizona State University
Nyer, D. B.,
Vargas, D.,
Originally presented on: 7/18/2016 15:00:00 - 15:30:00

We have developed a novel engineered protein to manipulate aberrant DNA-protein packaging (chromatin) in cancer cells. "PcTF" (Polycomb-based transcription factor) contains an N-terminal peptide motif "PCD" from the effector protein CBX8, which recognizes the histone modification signal H3K27me3. The C-terminal VP64 activation domain of PcTF reactivates epigenetically-silenced genes. This work represents a critical advance in the emerging field of epigenetic engineering by determining the function of a peptide module that reads chromatin marks. Other work that has analyzed formaldehyde-fixed chromatin suggests that CBX8 recognizes only a subset of H3K27me3 marks. In order to determine PCD activity in live cells, we used PcTF to activate H3K27me3-marked genes. We expressed PcTF from transfected plasmid DNA in three different human cancer-derived cell types (ATCC: U-2 OS/ bone, SK-N-SH/ brain, and K562/ blood). Quantitative sequencing of total messenger RNA (RNA-seq) was used to compare target-gene expression in PcTF-treated and untreated cancer-derived cells. In human cancer-derived cell lines, silenced H3K27me3-enriched genes (verified by chromatin immunoprecipitation) show higher levels of mRNA in PcTF-expressing cells compared to mock-transfected controls. We identified direct targets, genes with PcTF signals near their promoters, as well as hundreds of secondary targets. PcTF co-regulates clusters of genes within a subset of H3K27me3-associated regions, demonstrating that the CBX8 motif is selective. PcTF-activated genes verified by qRT-PCR include master transcriptional regulators such as HOX, RUNX3, CASZ1, and IRF8. These regulators and their targets are inhibitors of cancer cell growth in many reported studies. Thus, PcTF represents a potentially powerful new fusion-protein-based method for cancer treatment.

Developing Synthetic Biotics for the Treatment of Human Metabolic Disease

Kotula, J. W., Synlogic, Inc.
Miller, P.,
Originally presented on: 7/18/2016 16:00:00 - 16:15:00

The concurrent maturation and convergence of the fields of microbiome research and synthetic biology have made it possible to impact the host's metabolic processes, immune system, and disease states by engineering bacteria that operate from within the gut. At Synlogic, we aim to use the tools of synthetic biology to program probiotic organisms to perform a pharmacologically predictable and quantifiable function in order to treat metabolic diseases from within the existing host/microbiota communities. We have begun to engineer bacterial strains that absorb and convert toxic metabolites, produce therapeutic compounds, and interact with the immune system. These synthetic biotics have proven effective in genetic mouse models of disease and are exciting prospects for clinical development.

We have designed our synthetic biotics to be non-pathogenic, express a single or series of therapeutic genes, be contained within the gut and survive long enough in the competitive gut ecosystem to be effective. To achieve these design goals, we have had to understand and engineer genetic solutions for metabolic disease pathways, clinically suitable induction mechanisms, dynamics of *E. coli* transit through and residence within the mammalian gut, and the effects of containment strategies on viability and productivity.

Traditionally, synthetic biology circuits have been characterized on plasmids, in a controlled lab setting, and activated by well-characterized promoters such as tet, lac, and ara. However, plasmid-based synthetic organisms may not perform reliably during scale-up or within the gut, synthetic circuits may face intense selective pressures in competitive environments, and the inducers of the tet, lac, and ara promoters may be not be suitable for human therapeutic use. Therefore, we have developed robust and reusable genomic integration systems, evaluated the genetic stability of our circuits, and identified promoter schemes for reliable induction in a clinical setting.

Specifically, this approach has proven effective at reducing blood ammonia in genetic mouse models of urea cycle disorders, reducing blood phenylalanine in genetic mouse models of phenylketonuria, and increasing gut short-chain fatty acid concentrations in mouse models of gut dysfunction. Our work demonstrates that the power of synthetic biology can be applied toward generating therapeutic organisms that perform from within the mammalian gut.

Enabling Next Generation Biological Research through Synthetic Biology Tools

Trefzer, A.,

Originally presented on: 7/18/2016 16:15:00 - 16:30:00

Over the past decade Synthetic Biology has emerged as a new discipline in biotechnology that enables an engineering approach to biology. This approach is beginning to transform how biological research is conducted. It is expanding into many application areas, including the integrated analysis of complex pathways, the construction of new biological parts, enzyme engineering and the re-design of existing, natural biological systems. These new application areas and approaches are enabled by new tools emerging from Synthetic Biology.

ThermoFisher Scientific is a leader in serving science and development of tools for Synthetic Biology. We present here our integrated approach to provide tools to the community with a focus on efficient engineering of DNA via DNA-synthesis and genome editing tools. Overall this toolbox enables researchers to carry out engineering of biological systems at all scales faster and more efficient than before in development of biologic solutions. Recent advances in tool development and their applications will be shared.

The Development of Platform-Based Technologies for the Optimization of Sustainably Produced Chemicals

Culler, S. J., Genomatica, Inc.
Originally presented on: 7/18/2016 16:30:00 - 16:45:00

Genomatica has established an integrated computational/experimental bioengineering platform to design, create, and optimize novel high-producing organisms and bioprocesses. Here we present the use of our platform to develop *E. coli* strains for the production of the industrial chemical 1,4-butanediol (BDO) from carbohydrate feedstocks. These efforts have culminated in commercial scale production of over 5M lbs of BDO from dextrose. BDO is a chemical intermediate (>3B lbs/yr) produced commercially through energy-intensive petrochemical processes from hydrocarbon feedstocks and is used in a variety of products including automotive, electronics and apparel. Therefore, this product represents an opportunity to make a significant impact on the replacement of traditional petrochemical processes with benign bioprocesses using renewable feedstocks.

We are continuing to enhance our technology platform, enabling faster development of new processes with increased chance of success. Here we describe the application of cell-free transcription-translation (TX-TL), towards improving the design-build-test-learn cycle for the construction of a high-performing microorganism capable of producing BDO and other industrial chemicals from carbohydrates. We used systems biology approaches including ^{13}C-flux analysis, proteomics, and metabolomics to characterize TX-TL applied to different engineered metabolic pathways, including BDO, and compared the results to the analogous *in vivo* processes. These studies provide critical insight into the active metabolism of TX-TL and mechanisms of redox and energy regeneration. Learnings from these studies were applied towards improving the TX-TL platform for the production of intermediate and specialty chemicals. The presentation will highlight the successful integration of the TX-TL technology into our rational and high-throughput strain engineering platform, and will show the potential advantages for using cell-free systems towards the optimization of sustainably produced chemicals.

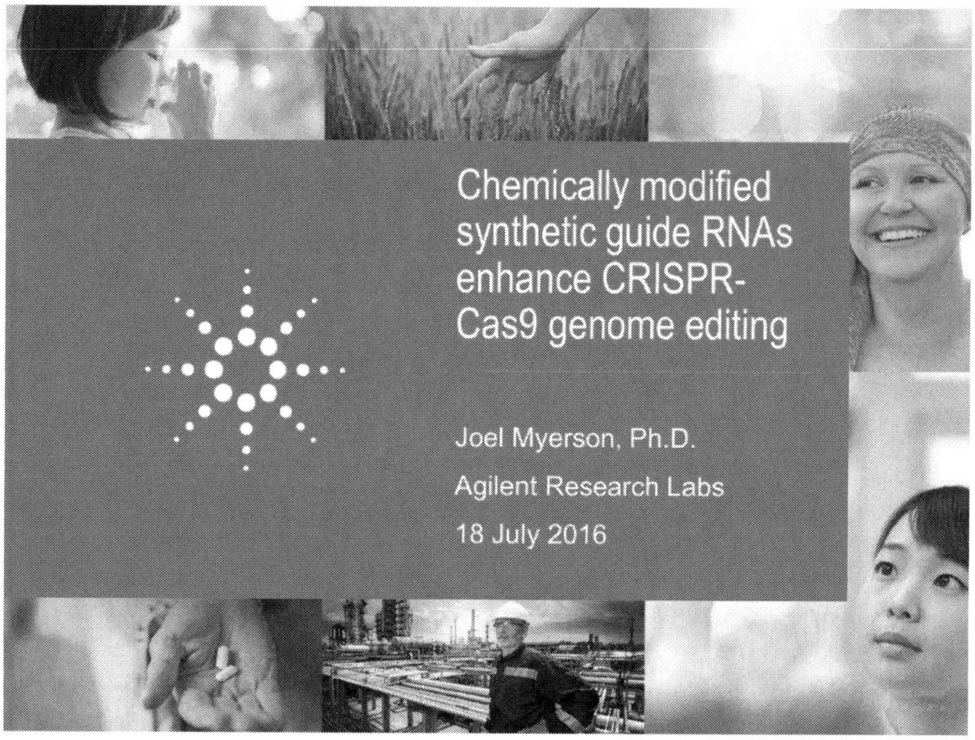

Why chemically synthesized RNA?

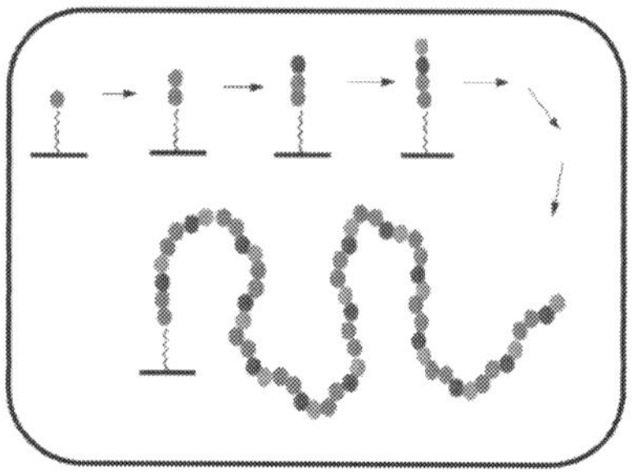

Chemical synthesis allows us to create RNAs with:

- User defined sequences containing defined ends
- Consistent purity in well-defined amounts at different scales
- Combinations of natural, modified natural or unnatural nucleotides at defined position that can impact:
  - Stability
  - Specificity
  - Activity
  - Structure
  - Persistence
  - Conjugation
  - Labeling
  - Visualization

Agilent Technologies

TC-RNA Chemistry

Making RNA synthesis as robust and easy as DNA synthesis

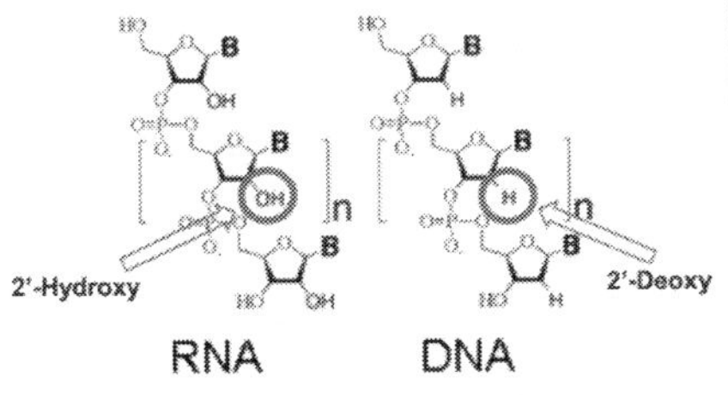

- Breakthrough chemistry
  → **High quality, biologically active RNA**
- High efficiency per cycle
  → **Routine synthesis of 120-mers**
- Cost-effective monomers
  → **Incorporation of modifications**

J|A|C|S

JOURNAL OF THE AMERICAN CHEMICAL SOCIETY

Streamlined Process for the Chemical Synthesis of RNA Using 2′-*O*-Thionocarbamate-Protected Nucleoside Phosphoramidites in the Solid Phase

Douglas J. Dellinger, Zoltán Timár, Joel Myerson, Agnieszka B. Sierzchala, John Turner, Fernando Ferreira, Zoltán Kupihár, Geraldine Dellinger, Kenneth W. Hill, James A. Powell, Jeffrey R. Sampson, and Marvin H. Caruthers

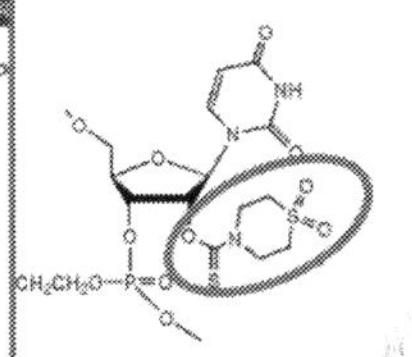

Agilent Technologies

Robust chemical synthesis of single guide RNAs (sgRNAs)

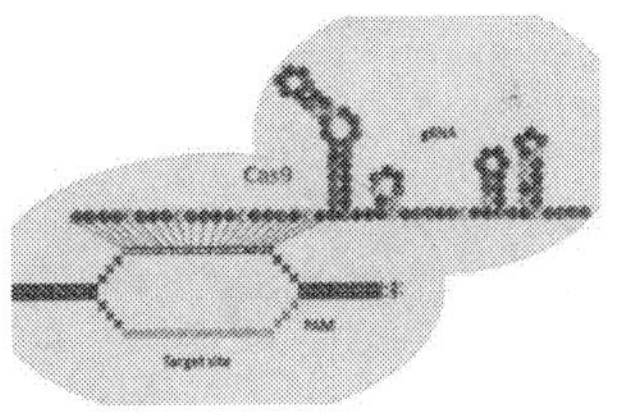

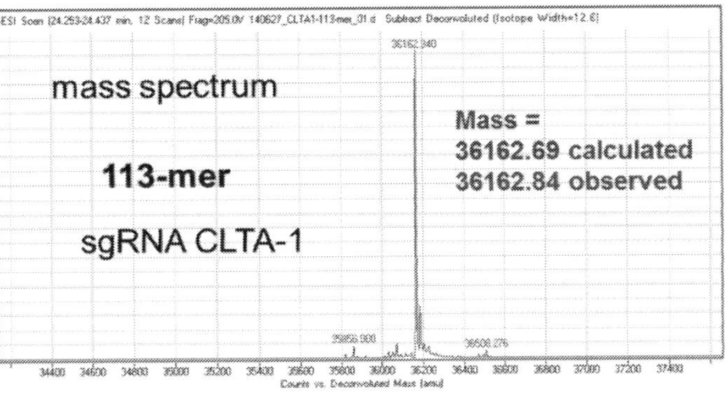

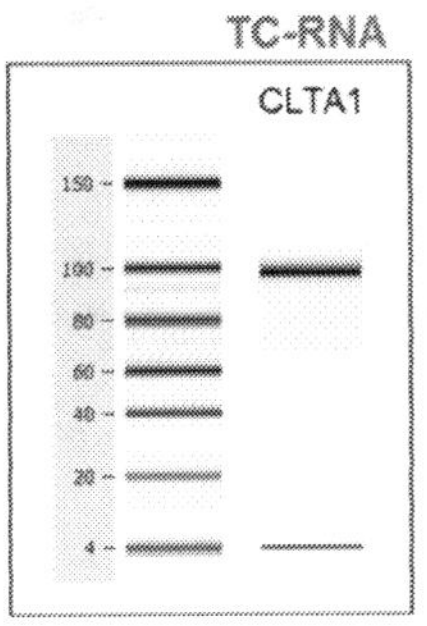

AGUCCUCAUCUCCCUCAAGCGUUUAAGAGCUAUGCUGGUAACAGCAUAGCAAGUUUAAAUAAGG
CUAGUCCGUUAUCAACUUGAAAAAGUGGCACCGAGUCGGUGCUUUUUUU

Agilent Technologies

Chemically modified sgRNAs for increasing Stability

M =	2'-O-methyl
MS =	2'-O-methyl 3'-phosphorothioate
MSP =	2'-O-methyl 3'-thioPACE

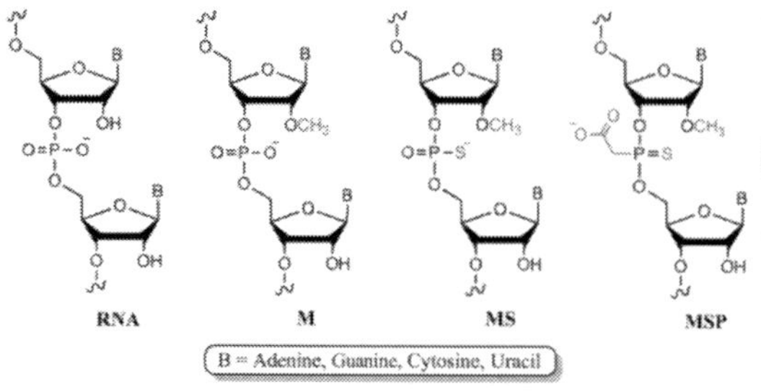

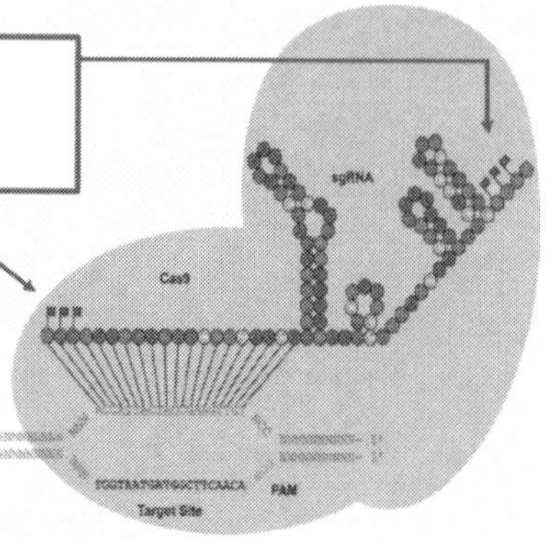

NATURE BIOTECHNOLOGY published online 29 June 2015; doi:10.1038/nbt.3290

Chemically modified guide RNAs enhance CRISPR-Cas genome editing in human primary cells

Ayal Hendel[1,5], Rasmus O Bak[1,5], Joseph T Clark[1], Andrew B Kennedy[2], Daniel E Ryan[2], Subhadeep Roy[3], Israel Steinfeld[4], Benjamin D Lunstad[3], Robert J Kaiser[2], Alec B Wilkens[1], Rosa Bacchetta[1], Anya Tsalenko[2], Douglas Dellinger[3], Laurakay Bruhn[2] & Matthew H Porteus[1]

Chemically modified sgRNAs with Cas9 mRNA mediate high frequencies of gene disruption in CD34+ HSPCs

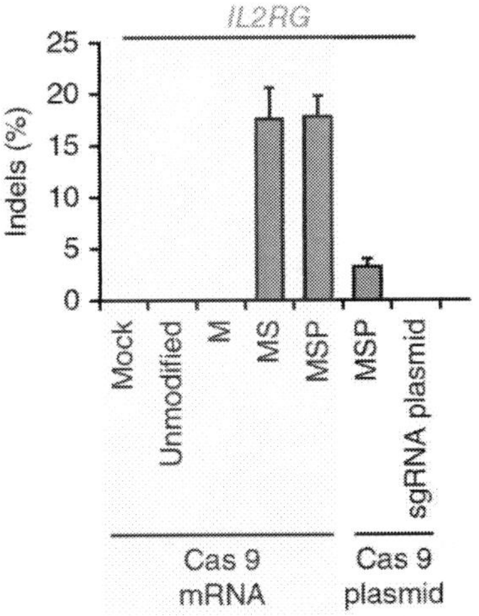

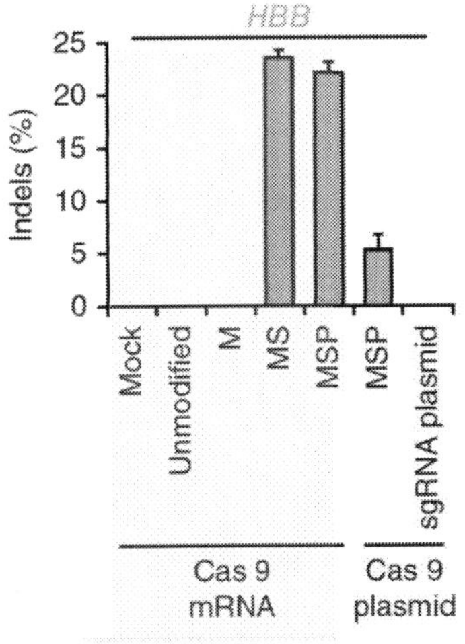

15 ug Cas9 **mRNA** + 10 ug sgRNA
1 ug Cas9 **plasmid** + 10 ug sgRNA

Agilent Technologies

Chemically modified sgRNAs with Cas9 mRNA or protein mediate high frequencies of *CCR5* indels in stimulated primary human T cells

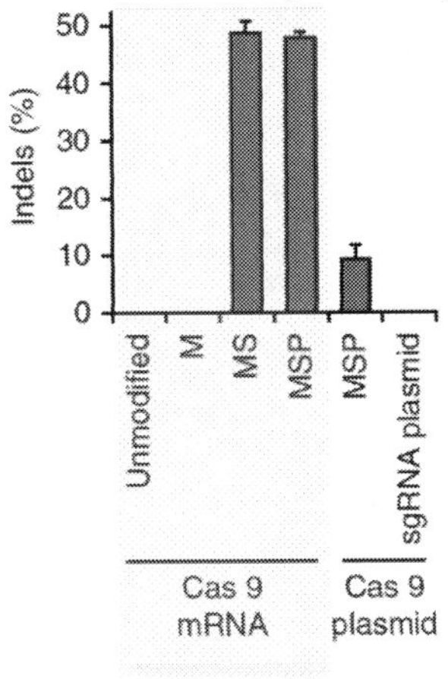

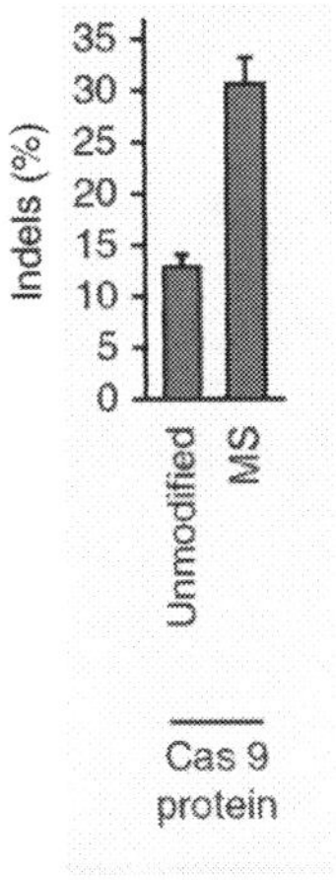

15 ug Cas9 **mRNA** + 10 ug sgRNA
1 ug Cas9 **plasmid** + 10 ug sgRNA

15 ug Cas9 protein + 7.6 ug sgRNA

Agilent Technologies

Chemically modified sgRNAs co-delivered with Cas9 mRNA facilitate high frequencies of indels in cell lines

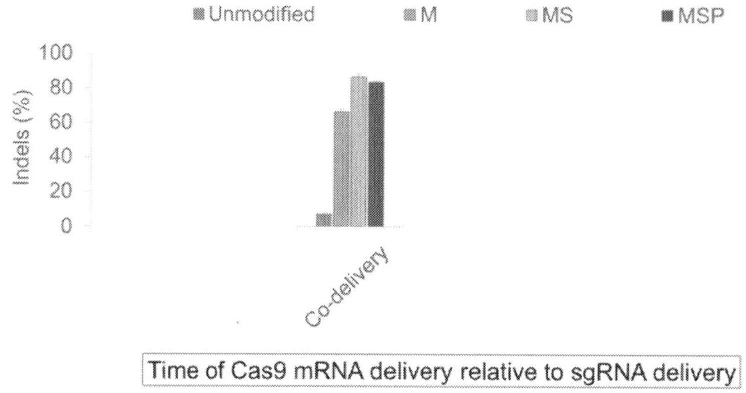

K562 cells: 10 ug IL2RG sgRNA
15 ug Cas9 mRNA

Agilent Technologies

Chemically modified sgRNAs co-delivered with Cas9 mRNA facilitate high frequencies of indels in cell lines

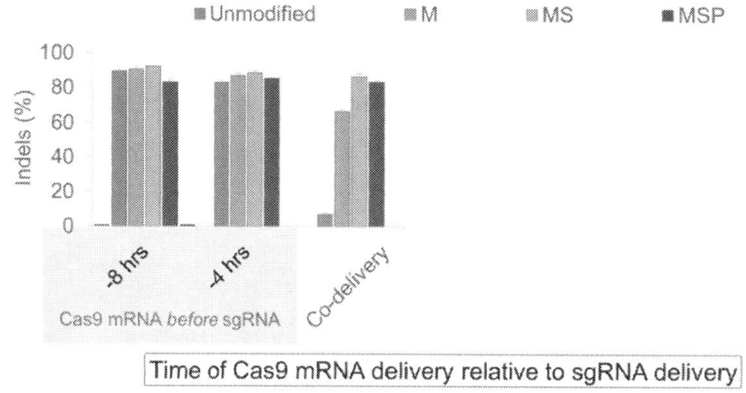

K562 cells: 10 ug IL2RG sgRNA
15 ug Cas9 mRNA

Agilent Technologies

Chemically modified sgRNAs co-delivered with Cas9 mRNA facilitate high frequencies of indels in cell lines

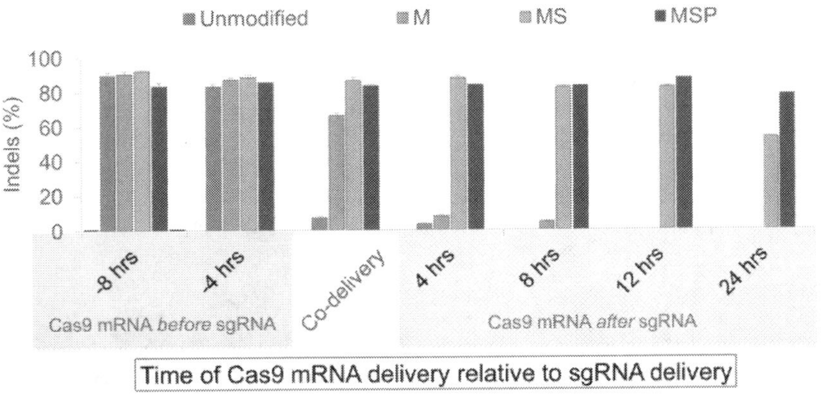

K562 cells: 10 ug IL2RG sgRNA
15 ug Cas9 mRNA

Chemically modified sgRNAs facilitate high frequencies of HR in cell lines

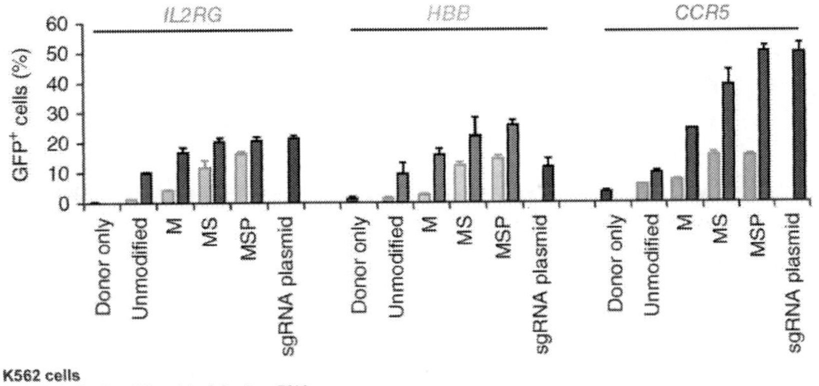

K562 cells
1ug (light bar) or 20ug (dark bar) sgRNA
2ug Cas 9 plasmid
5 ug GFP encoding donor plasmid

Why chemically synthesized guide RNAs?

- Combinations of natural, modified natural or unnatural nucleotides at defined positions that can impact:
  - Stability
  - Activity
  - Specificity
  - Cross-linking
  - Labeling

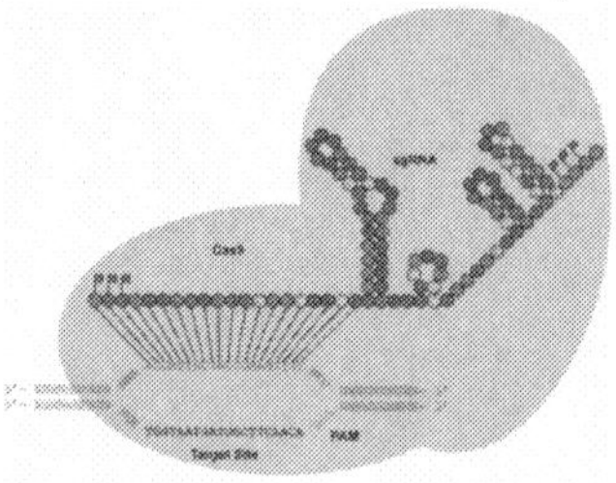

- Consistent purity in well-defined amounts from micrograms → milligrams → grams

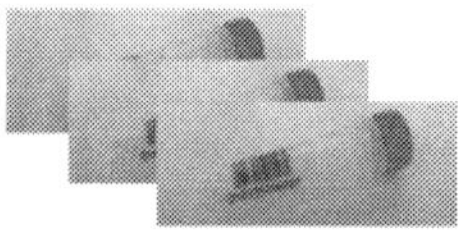

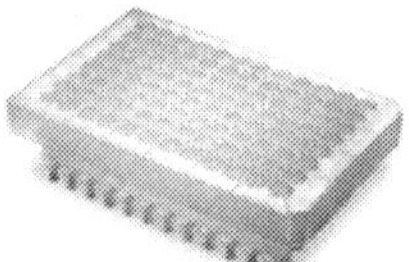

Agilent Technologies

Agilent's Nucleic Acid Solutions Division (NASD)

Nucleic Acid Therapeutics Contract Manufacturing and Associated Services:

- **Development & manufacture of Oligonucleotide APIs**
- **Pre-clinical, Clinical, & Commercial use**
- **DNA , RNA, duplexes, standard and novel chemical modifications, standard and novel conjugation chemistries**
- **GMP and GLP quality systems ICH compliant**
- **Serving customers in Asia, Europe, & North America for over a decade**

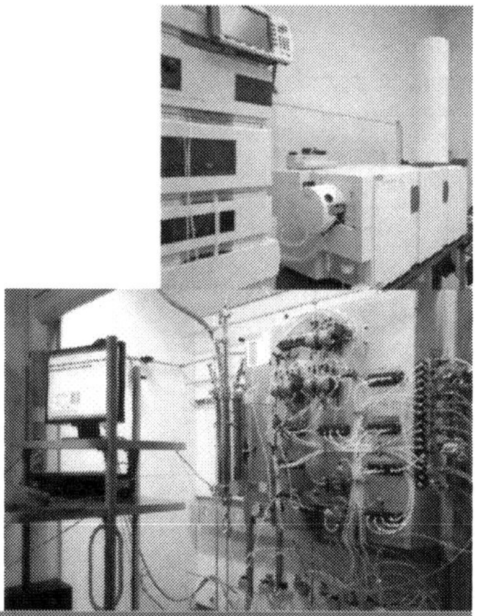

Agilent Technologies

How long can we go with TC-RNA chemistry?

162mer 5'-GAGGGAAACUUUGAGAUGCCUUGCAAAGGGUAUGGUAAUAAGCUGACGGACAUGGUCCUA ACCACGCAGCCAAGUCCUAAGUCAACAGUGGGUGCGAGAUCUUCUGUUGAUAUGGAUGCAGUU CAACCAAAAUCAACCUAAAACUUACACAUGGGUGCGAGU-3'

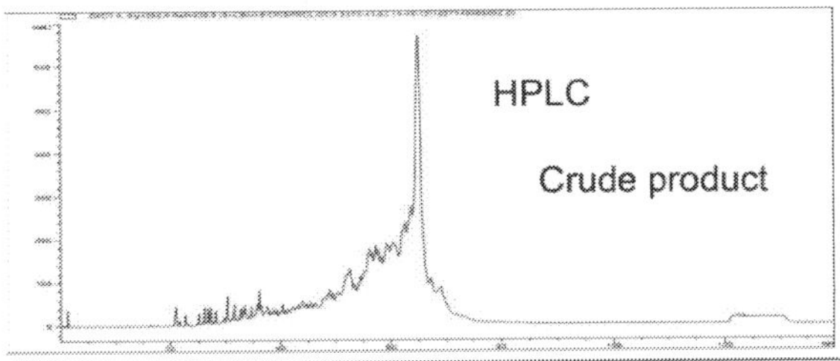

LC/MS on Agilent 1290 Infinity Quaternary HPLC;
6430 QTOF Mass Spectrometer

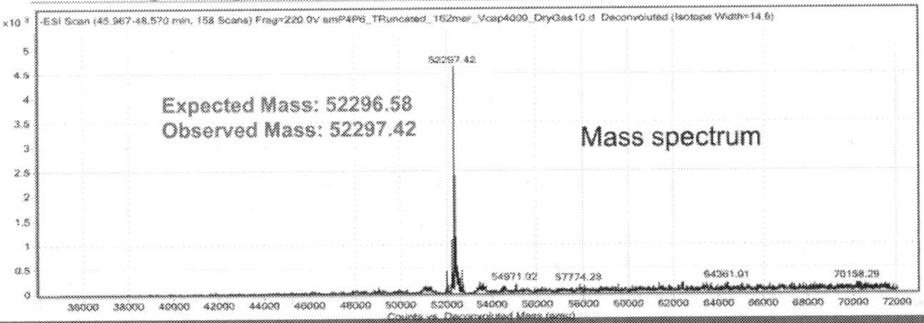

Acknowledgements

Agilent Technologies

Agilent Research Labs

Laurakay Bruhn
Bo Curry
Doug Dellinger
Rob Kaiser
Andrew Kennedy
Ben Lunstad
Smruti Phadnis
Subhadeep Roy
Dan Ryan

Israel Steinfeld
Anya Tsalenko
Peter Tsang
Chong Yung

Gary Carter
Gigi Dellinger
Steve Laderman
Jeff Sampson

joel_myerson@agilent.com

Stanford MEDICINE

Ayal Hendel
Rasmus Bak
Joseph Clark
Alex Wilkens
Rosa Bacchetta
Matthew Porteus

Commercial Scale Production of Fuels and Chemicals from Waste and Low Cost Resources via an Integrated, Multi-Scale Gas-Fermentation Platform

Köpke, M., LanzaTech, Inc.
Juminaga, A.,
Mitchell, W. P., LanzaTech
Simpson, S. D., LanzaTech
Originally presented on: 7/18/2016 17:00:00 - 17:15:00

LanzaTech has developed novel gas fermentation technology that capture and utilize greenhouse gases for production of fuels and chemicals. In contrast to traditional fermentation that use sugars as substrate (and releases CO_2 as byproduct), gas fermentation utilizes C1 substrates carbon monoxide (CO) or CO_2. This enables a diverse range of feedstock options including waste gases from industrial sources (e.g., steel mills and processing plants) or syngas generated from any biomass resource (e.g., agricultural waste, municipal solid waste, or organic industrial waste). Biomass is thence gasified, allowing for maximum yields and complete carbon utilization including the recalcitrant lignocellulosic fraction that cannot be utilized in traditional sugar fermentation.

At the heart of the technology is an acetogenic microbe capable of autotrophic growth. While the potential of acetogens have been recognized for decades, development has been severely limited by the challenging nature of these organisms and the lack of genetic tools. Since its inception in 2005, LanzaTech has evolved a strain of *Clostridium autoethanogenum* for fast autotrophic growth and developed robust strain engineering and Synthetic Biology tools including several genome editing methods, libraries of validated genetic parts, design algorithms, a validated genome scale model, and automated strain engineering workflows.

Using this platform, production of over 25 new molecules have been demonstrated from gas, in several cases exceeding reported production rates and yields of native producers engineered *E. coli* or yeast strains. Proprietary, scalable reactor designs and optimized process chemistry, ensure efficient, continuous, single-pass gas conversion with a high selectivity to the product of interest. The process has been successfully scaled up from the laboratory bench to fully integrated 100,000-gallon/year pre-commercial demonstration plants with a total of over 40,000 hours on stream. Currently, LanzaTech is constructing first commercial plants with partner ArcelorMittal in Belgium, Shougang Steel in China and China Steel in Taiwan.

Automated Construction of Synthetic DNA Fragments Using the BioXp™ 3200 System

Lyons, K.,

Originally presented on: 7/18/2016 17:15:00 - 17:30:00

Whereas basic cloning methods have become a part of the standard laboratory skillset, building synthetic constructs *de novo* remains specialized and is commonly outsourced to service labs. Outsourced DNA synthesis times can be lengthy and delayed and researchers are completely dependent on the service house. Although methods for oligo assembly have been described in the literature, successful assembly of synthetic DNAs often requires a holistic understanding of the characteristics of the sequence. These characteristics can be elucidated using various bioinformatic tools and, when large data sets are available, machine learning algorithms, but only some of these tools are readily available. Additionally, synthesis design, raw materials, and chemistries can strongly affect success of construct building. Here, we describe the capability of a synthetic DNA printer, the BioXp™ 3200 System, which allows benchtop DNA assembly in any lab. Analysis and design are done upstream and custom reagents are shipped, the user loads and starts the machine, and synthetic DNA fragments are produced in an overnight run, either as linear material or cloned into a vector. This automation reduces synthetic molecule construction to a simple, attainable process and is the result of the convergence of bioinformatics, biology, and engineering. We continue to develop the platform for construction of molecules with increased complexity and additional downstream applications.

Genetic Constructor

Outline

- Who is Eli?
- Who is Autodesk?
- What is Genetic Constructor?

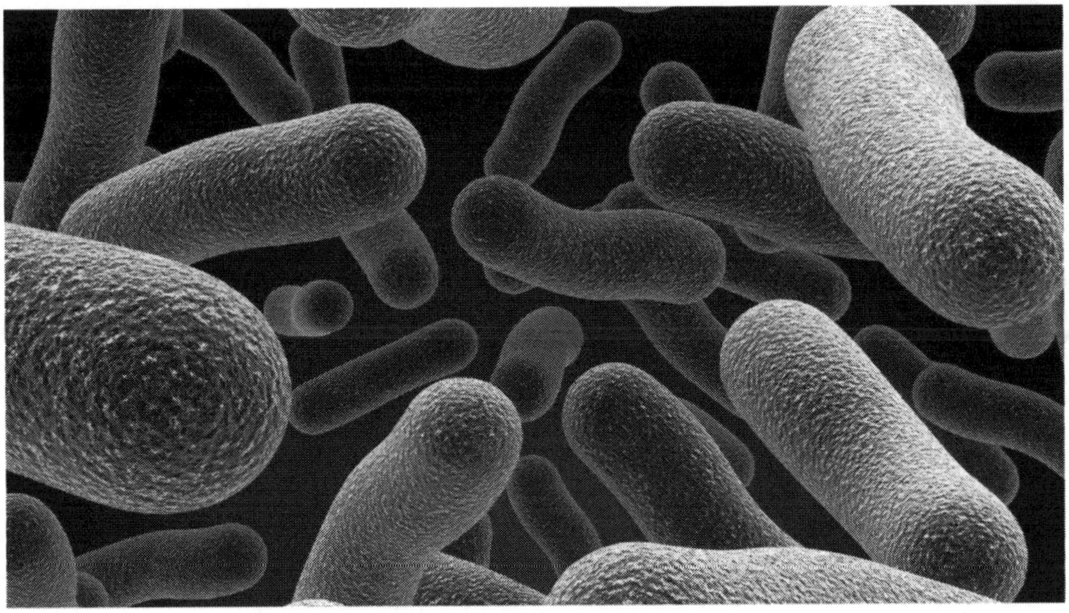

SANTA RAFAELA

Dove
NEW
PANTENE
promosi

Help people imagine, design, and create a better world

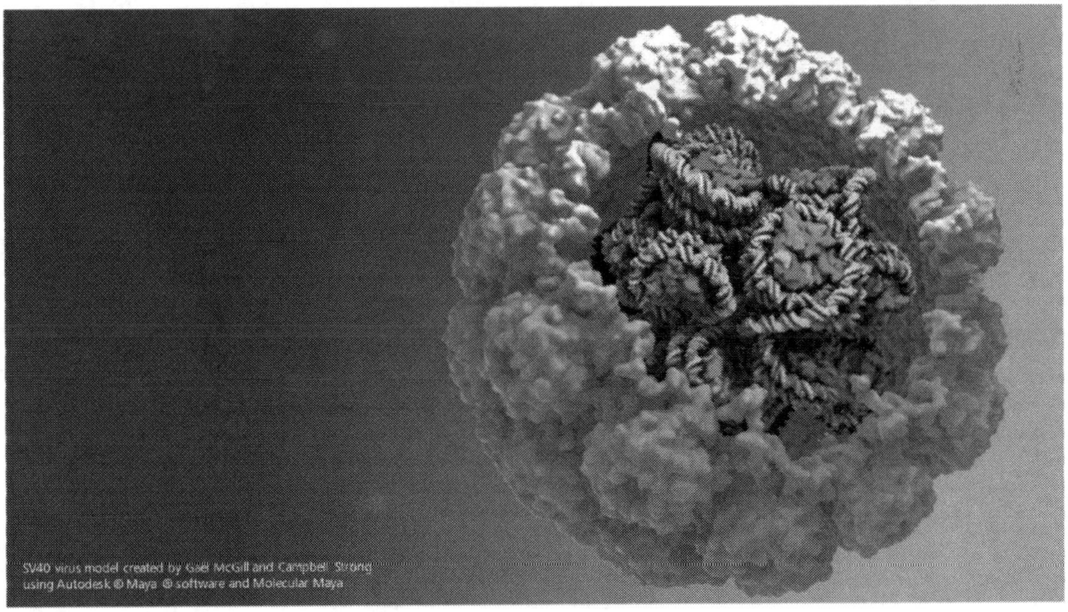
SV40 virus model created by Gaël McGill and Campbell Strong
using Autodesk ® Maya ® software and Molecular Maya

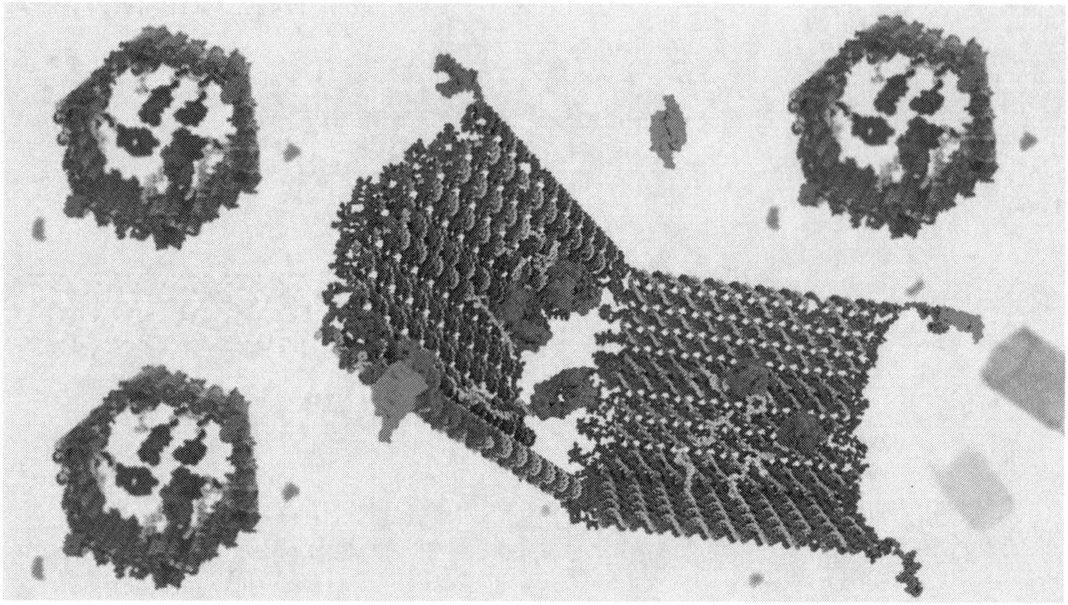

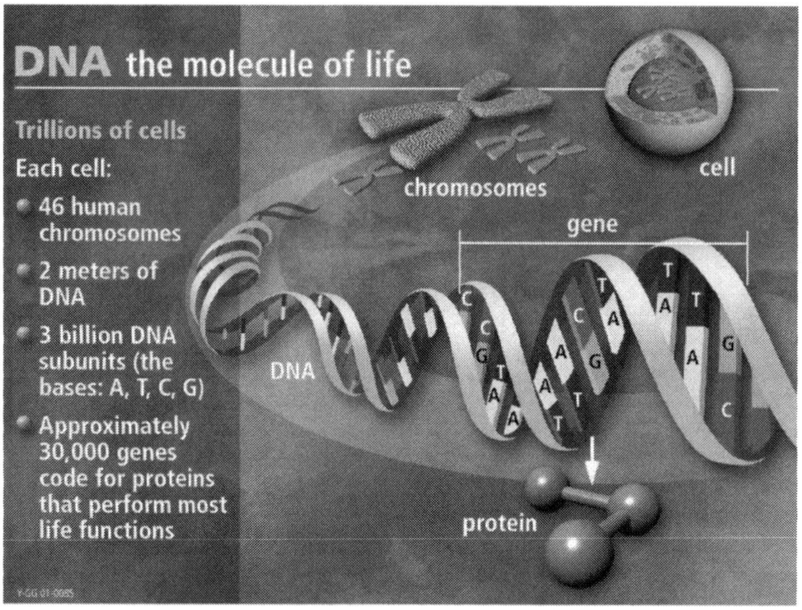
DNA the molecule of life
Trillions of cells
Each cell:
46 human chromosomes
2 meters of DNA
3 billion DNA subunits (the bases: A, T, C, G)
Approximately 30,000 genes code for proteins that perform most life functions
chromosomes
cell
gene
DNA
protein

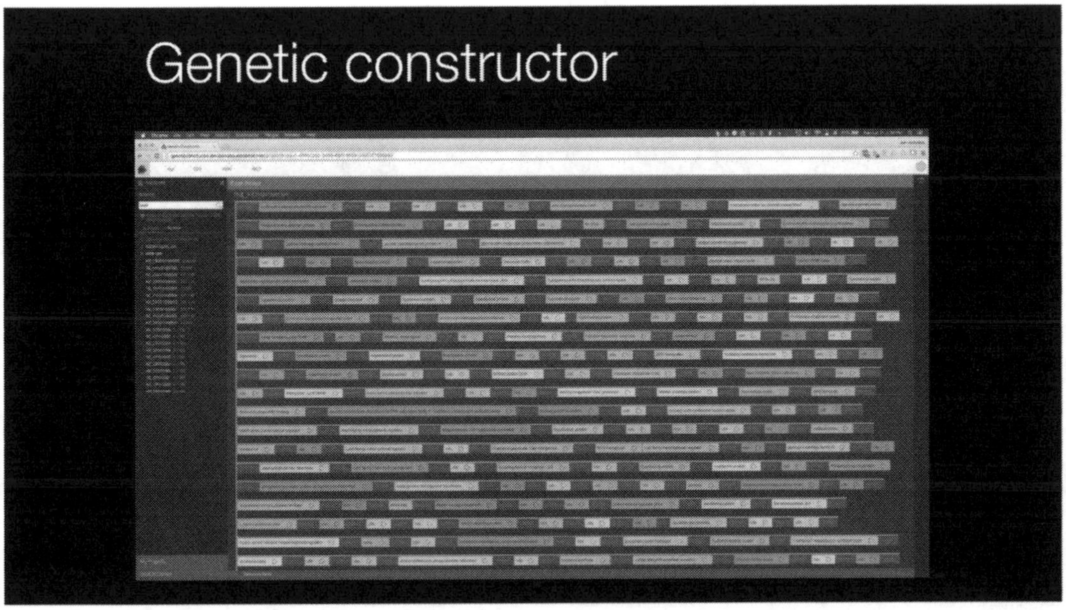
Genetic constructor

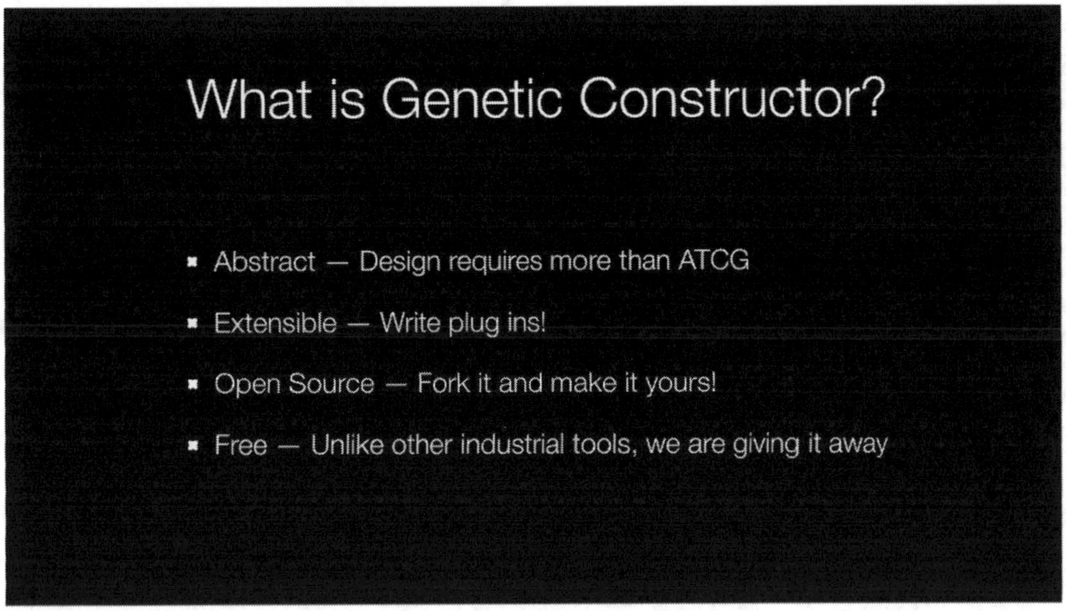
What is Genetic Constructor?
Abstract — Design requires more than ATCG
Extensible — Write plug ins!
Open Source — Fork it and make it yours!
Free — Unlike other industrial tools, we are giving it away

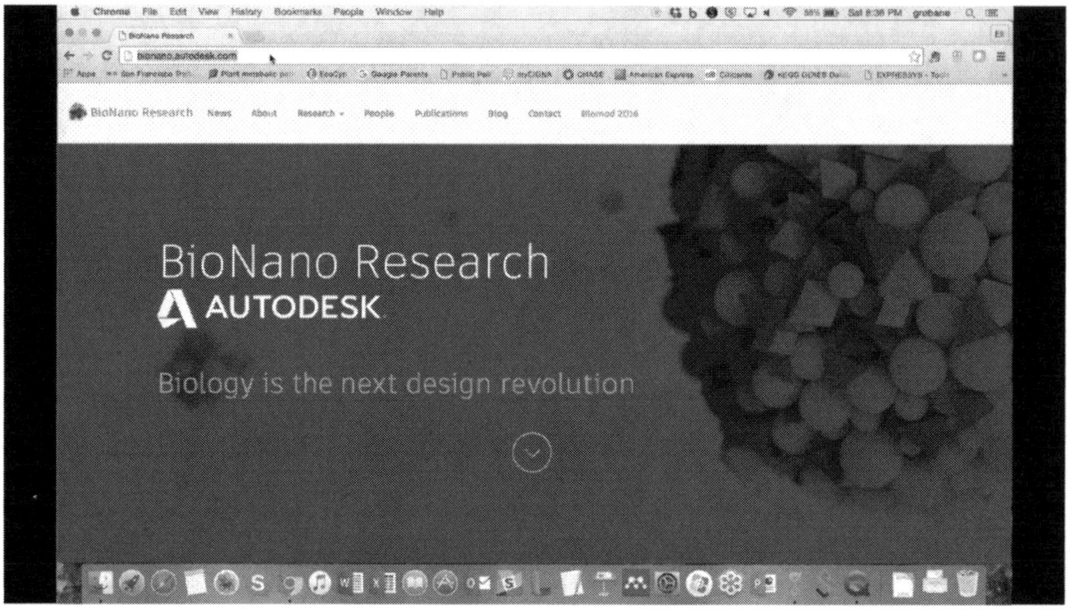

Thank you

www.geneticconstructor.com

bionano.autodesk.com

eli.groban@autodesk.com

A Synthetic DNA Base Pair Enables the Discovery of Novel Protein Therapeutics

Ptacin, J.,

Originally presented on: 7/18/2016 17:45:00 - 18:00:00

Over the past two decades, synthetic biologists have developed elegant methods to add diversity or re-appropriate redundancy in the genetic code for the introduction of a vast array of unnatural amino acids with unique properties. These efforts promise to revolutionize applied science with new molecular tools and strategies for protein-based therapeutics. However, unknown biological constraints have thus far largely precluded the encoding of more than one unnatural amino acid at a time. As an alternative strategy to expand the genetic code without interfering with the naturally-evolved coding structure, the Romesberg lab at the Scripps Research Institute has developed an unnatural DNA base pair that enables addition of multiple orthogonal codons to the natural set. This base pair is fully-compatible with replication and transcription in vitro, and was demonstrated to be faithfully replicated in E. coli cells.

Synthorx Inc. is a drug discovery company that was recently established to commercialize this technology and develop novel protein-based therapeutics in vitro and within semi-synthetic cells. In collaboration with the Romesberg lab, we have developed an in vitro protein expression system that is compatible with the expanded genetic code. Using this system, we have discovered a large variety of unnatural codons that are fully orthogonal and support high-level transcription and translation of unnatural genetic information into unnatural proteins. Importantly, these codons function as orthogonal genetic elements that do not require modification or manipulation of the organism's natural coding structure. We will discuss the features and performance of our system for in vitro protein production as well as our efforts to expand this technology into E. coli. Our ability to genetically-encode protein therapeutics of any size that contain multiple, distinct unnatural amino acids allows for the development of novel therapeutic proteins that have not been accessible using existing technologies.

A CRISPR based approach for Trackable Multiplexed Engineering on a Genome Scale

Andrew Garst
Muse Biotechnology

INSPIRING GENOMIC DESIGN

Building tools to general purpose molecular biology

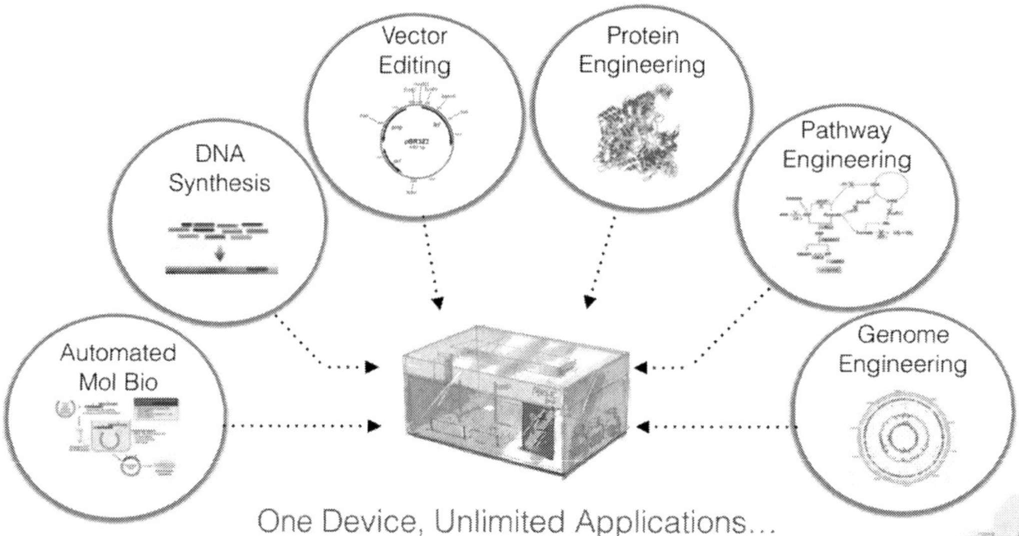

One Device, Unlimited Applications...

The approach:
CRISPR-Enabled Trackable Genome Engineering

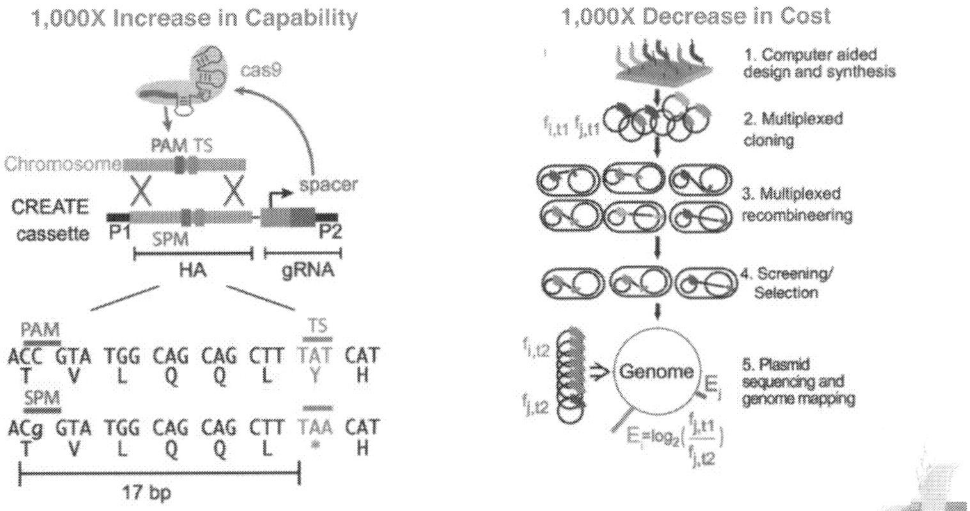

Barcoded libraries for high-throughput, high resolution genomics

Garst et al., in final review

A Wide Range of Designer Modifications

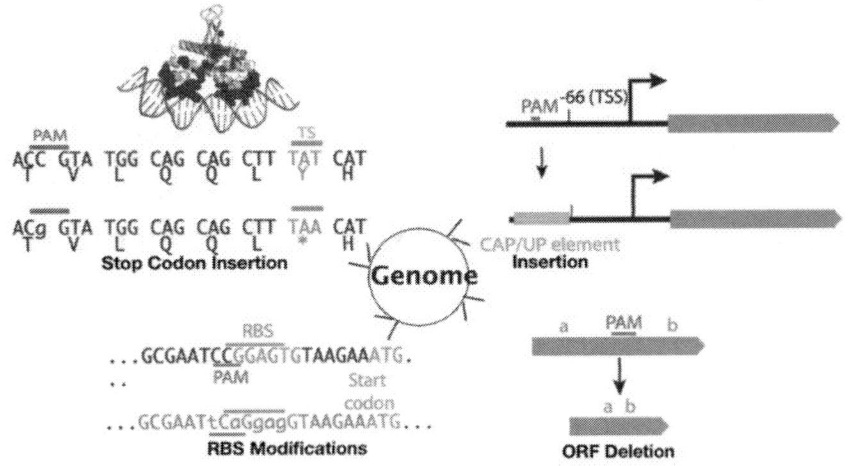

Extensible genomic design and query functionality!

Initial genome scale engineering design

Transcription factors/regulators

Global	Local
crp	*rpoE*
fis	*rpoH*
fnr	*rpoN*
ihfA	*rpoS*
appY	*marA*
rho	*marR*
nusA	*soxR*
	fadR

Transcription machinery

rpoA
rpoB
rpoC
rpoD

Biosynthesis

Isoprenoids	Other
dxs	
idi	*metA*
ispB	*pntA*
ispC	*folA*
ispD	*gapA*
ispE	*galK*
ispF	
ispG	**Transport**
ispH	*acrB*
ispU	
pdxJ	

Promoter libraries

UP element insertion (-66 from TS) RegulonDB
CAP element insertion (-60 from TS) RegulonDB

52,356 total oligos
48,080 protein targeting cassettes/36 proteins/2404 codons
4269 promoter targeting cassettes
Rational targeting of active sites for multiple targets!

Protein Engineering

Saturation mutagenesis of essential *E. coli* dihydrofolate reductase - FolA; complete substitution structure-activity mapping identifying novel mutations

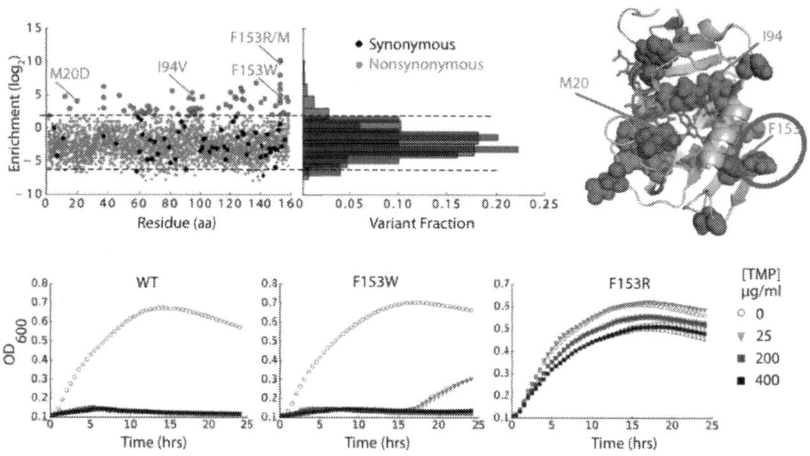

Designed, built, and screened by a single researcher in 3 weeks

Garst, et al. in review (visit Gur Pines, poster #158)

High resolution genomics for antibiotic resistance mapping

Drug binding site identification and discover from >45,000 mutations

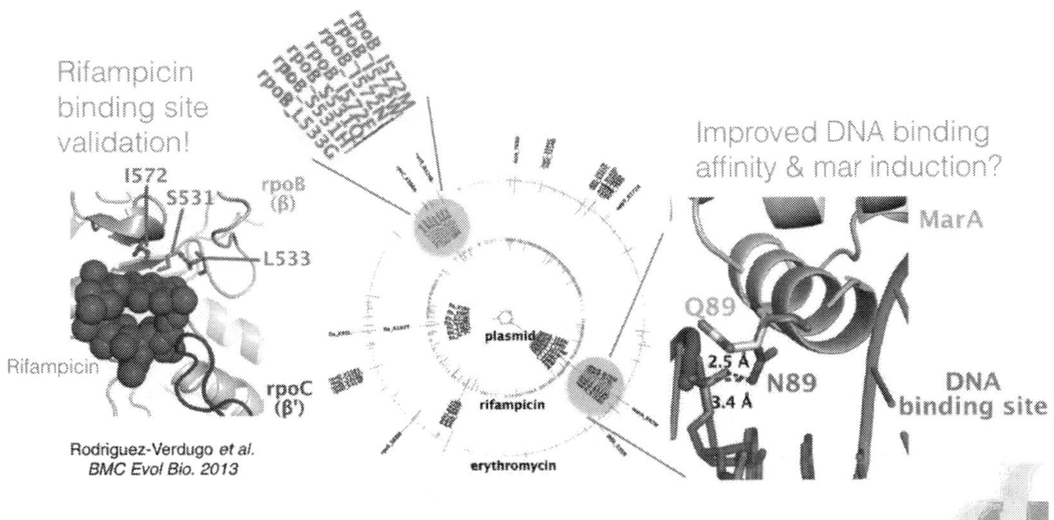

Garst, et al. in review

Pathway Library Construction- Lysine

~ **16,520 variants** of lysine biosynthesis and regulation (18 genes).
Construction and initial selections: **1 researcher, 2 weeks**

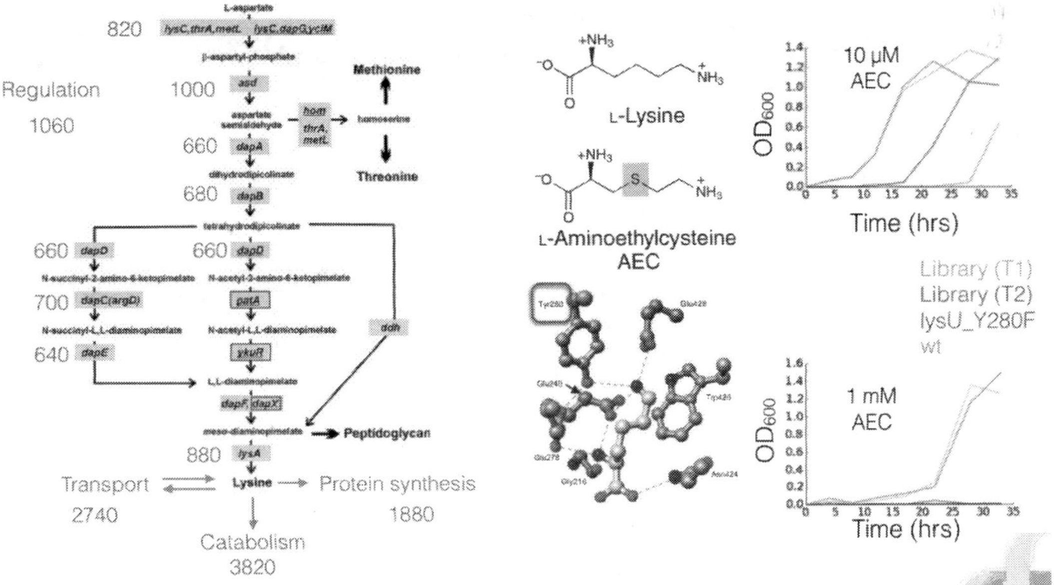

Creating a new paradigm in metabolic engineering

Visit Marcelo Bassalo (poster #8)

Rodinov et al. NAR. 2003
Blount et al. Nat. Biotech. 2006
Ataide et al. ACS Chem. Biol. 2007

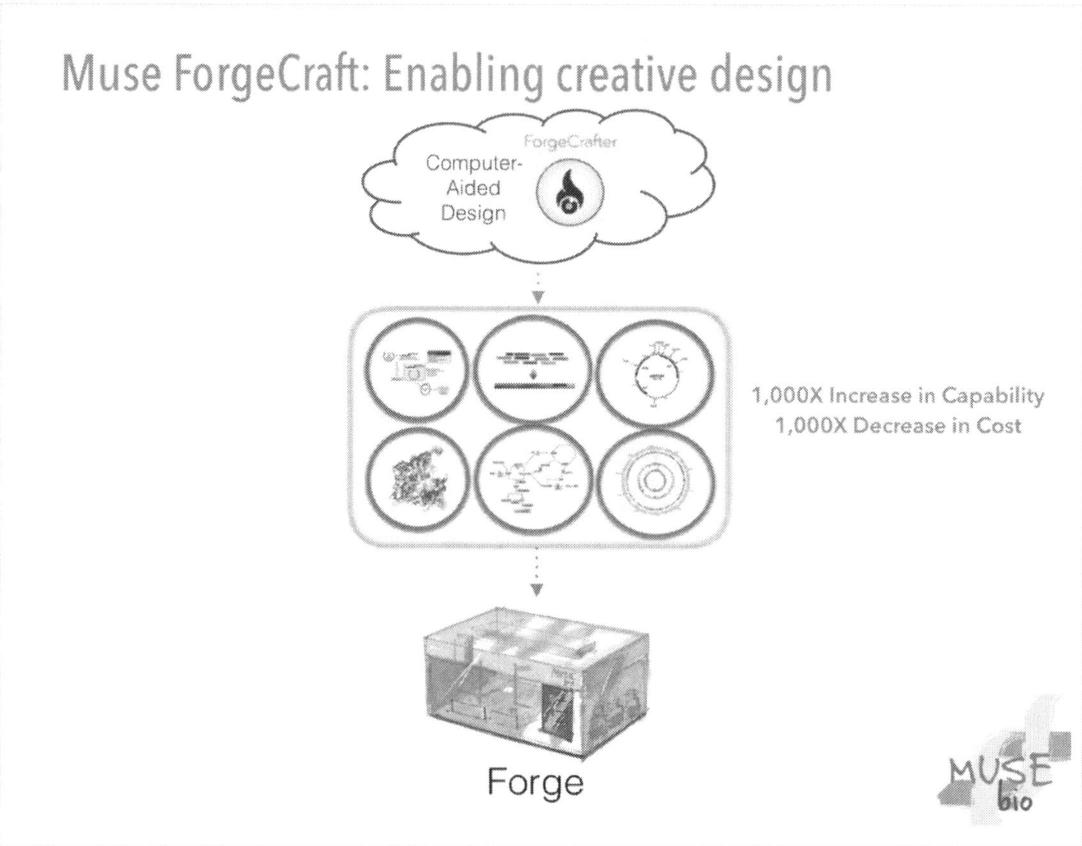
Muse ForgeCraft: Enabling creative design
ForgeCrafter
Computer-
Aided
Design
1,000X Increase in Capability
1,000X Decrease in Cost
Forge
MUSE
bio

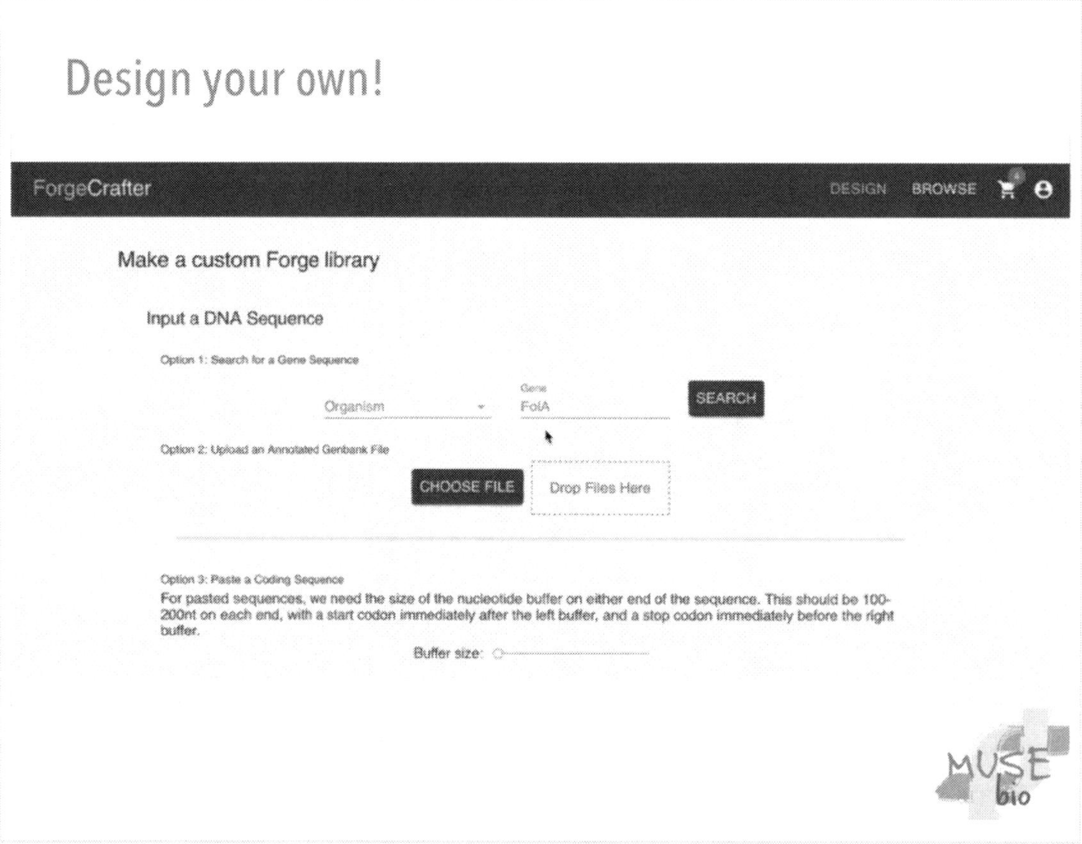
Design your own!
ForgeCrafter
DESIGN
BROWSE
Make a custom Forge library
Input a DNA Sequence
Option 1: Search for a Gene Sequence
Organism
Gene
FolA
SEARCH
Option 2: Upload an Annotated Genbank File
CHOOSE FILE
Drop Files Here
Option 3: Paste a Coding Sequence
For pasted sequences, we need the size of the nucleotide buffer on either end of the sequence. This should be 100-200nt on each end, with a start codon immediately after the left buffer, and a stop codon immediately before the right buffer.
Buffer size:
MUSE
bio

ForgeCrafting with Thought Leaders

UCSD

CEDARS-SINAI

Ucla

Your Name Here

Berkeley
UNIVERSITY OF CALIFORNIA

Duke
UNIVERSITY

Major Chemical Company 1*

Company 2

....

Company X

MUSE bio

Reach out and tell us how we can help!

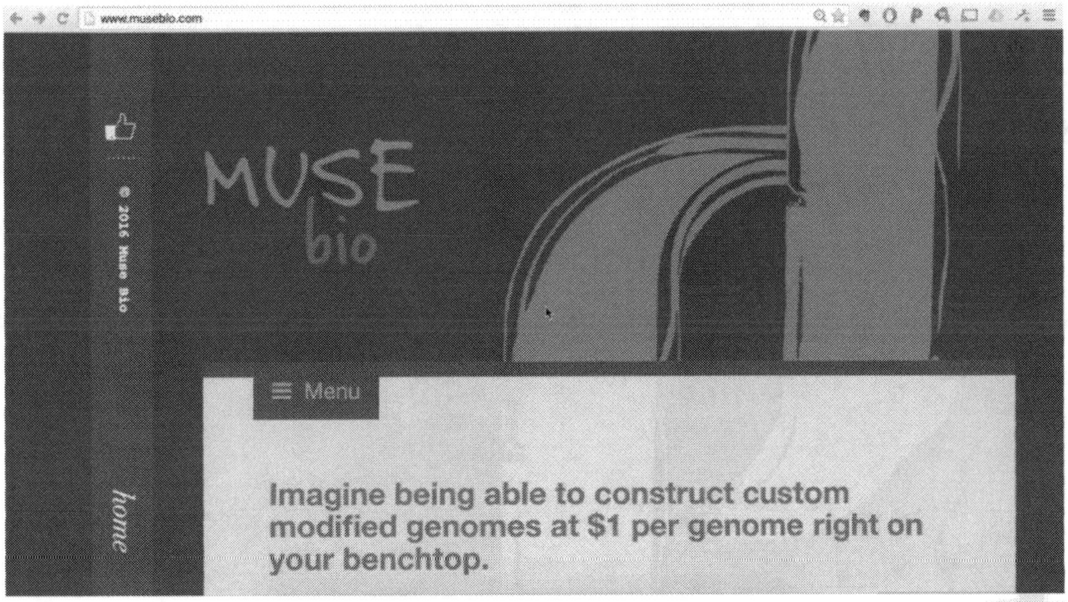

www.musebio.com

MUSE bio

Our Muses

Tanya
Lipscomb
Automation

Andrea
Edwards
Design

Andrew Garst
Build

Eileen Spindler
Test

Questions?

www.musebio.com

Combinatorial Engineering

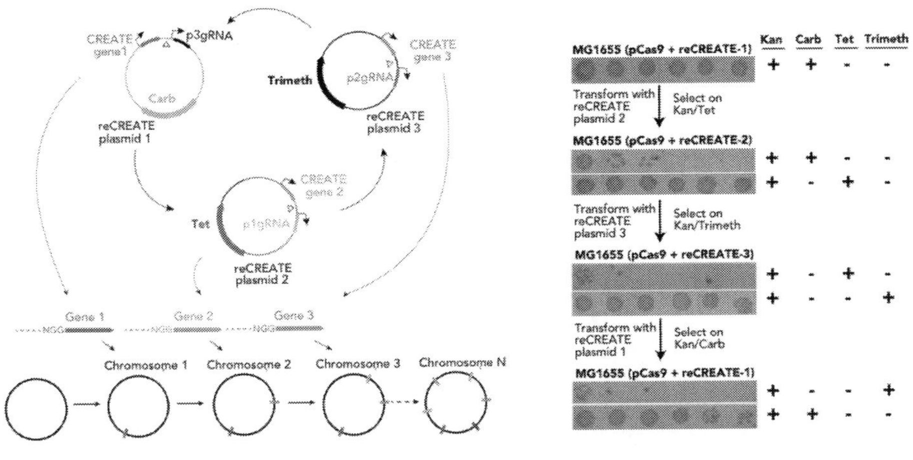

Recursive editing to construct complex, custom genomes

Bassalo, University of Colorado
Ronda, et al. CRMAGE, 2016

The MUSE Vision: end to end user automation....

Reduced cost and time at any scale compared to state of the art approaches

Scale	Edit Type	Average Library Size	Forge Cost	State of the Art Cost
Protein	Complete Substitution	1,000s	$1,000	$100,000+
Pathway	Rational Saturation of ~10 genes	10,000s	$2,000	$1M
Genome	ALE Reconstruction, Promoter Replacement, Complex Phenotypes	100,000	$11,000	Not Offered

Throughput:
State of the Art - 96 rationally designed strains = 2-4 weeks
Muse - 100,000 rationally designed strains= 2-3 weeks

Technology That Scales

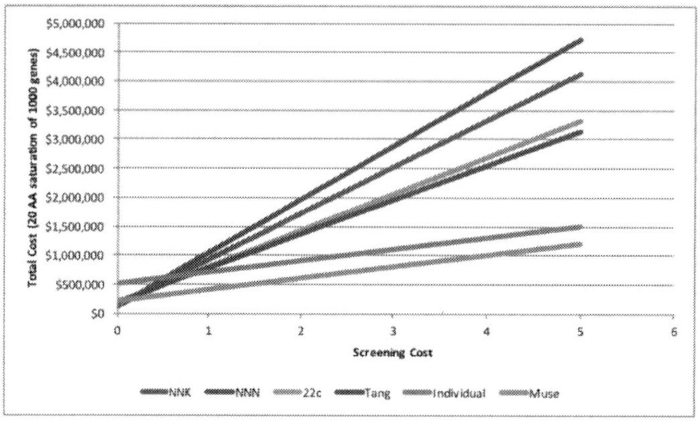

MUSE bio

Design of Reprogrammed and New Molecular and Cellular Functions

Kortemme, T.,

Originally presented on: 7/19/2016 9:00:00 - 9:30:00

There has been exciting progress in the computational design of proteins with new structures and functions, highlighting the potential to advance many applications in biological engineering, as well as to provide insights into the design principles of function. Many significant challenges remain, both in the accuracy of current computational approaches, and in the complexity of functions that can be designed at present. I will discuss our recent progress with computational methods and describe new approaches and their application to engineer sensor/actuators that detect and respond to small molecules in living cells, and to reprogram the function of biological machines.

Rapid Engineering of New Functions in Yeast with Synthetic Chromosomes

Ellis, T., Imperial College London

Originally presented on: 7/19/2016 9:30:00 - 10:00:00

Synthetic DNA and modular assembly has revolutionised cellular engineering, allowing new constructs to be built and optimised in days. Through our involvement with the Sc2.0 synthetic yeast project, we are now able to show how a synthetic modular genome can accelerate synthetic biology further. Alongside construction of a 670 kb synthetic yeast chromosome, we have developed protocols to generate diverse pools of yeast strains in hours via the Sc2.0 SCRaMbLE system, which uses Cre/lox recombination to radically change chromosome content and arrangement. In the presence of synthetic constructs, such as those endowing the biosynthesis of pigments or catabolism of xylose, SCRaMbLE can produce 'specialist' host strains overnight with significant improvements in performance in terms of growth and yield. SCRaMbLE can also be used to directly add new genes to Sc2.0 genomes, simply by adding linear synthetic DNA to cells as their chromosomes recombine. With these new modular genome methods together with combinatorial modular DNA construction we have been able to construct a yeast that produces and secretes the beta-lactam antibiotic penicillin.

Microfluidic Electroporation Techniques to Accelerate Discovery in Synthetic Biology

Garcia, P. A., Massachusetts Institute of Technology
Ge, Z., Massachusetts Institute of Technology
Moran, J. L., Massachusetts Institute of Technology
Buie, C. R., Massachusetts Institute of Technology
Originally presented on: 7/19/2016 10:00:00 - 10:30:00

One major limitation of genetic engineering and synthetic biology is the inability to incorporate genetic material into many bacteria due to the challenge of permeating the cell envelope while maintaining high cell viability. There are millions of species of bacteria on the planet, yet we can only genetically engineer several dozen. New technologies are needed to unlock the bacterial kingdom's potential to impact human life and solve many challenges of interest to mankind. In this presentation we will show recent work utilizing novel microfluidic approaches to enable electroporation-based genetic transformation for applications in synthetic biology.

Electroporation results from exposure of cells to external electric fields of sufficient strength to transiently disrupt the plasma membrane. It is often used for intracellular delivery of molecules such as drugs, proteins, DNA, or RNA. Recently, flow-through microfluidic electroporation platforms with channels of uniform cross-sectional area have demonstrated genetic transformation with uniform electric fields. However, this process has limited utility since in many cases the optimal conditions (e.g., field strength) for genetic transformation are unknown (as in difficult-to-transfect or intractable organisms). A platform to rapidly and quantitatively determine optimal conditions for electroporation-based transformation will accelerate the development of new chassis organisms for synthetic biology while advancing the fundamental science behind electroporation.

We have developed a flow-through electroporation platform that enables experimental characterization and/or genetic transformation by removing the electric field as a variable in the genetic transformation process. The novel microfluidic device generates a linear electric field gradient that allows for spatial evaluation of the electric field required for electroporation, without using discrete experimental steps. Our rapid microfluidic platform is capable of determining electric field thresholds for electroporation in a single experiment. Fluorescence-encoded DNA plasmids or nucleic acid stains permeate cell membranes in regions where the electric field is sufficiently high to induce membrane permeabilization. We then correlate the fluorescent region in the channel with the range of electric fields that results in successful electroporation, effectively analyzing a continuous spectrum of experimental conditions in a single experiment. Both gram-positive (*Corynebacterium glutamicum* and *Mycobacterium smegmatis*) and gram-negative (*Escherichia coli* BL21) strains have been tested in our device. The electric field thresholds for electroporation of *C. glutamicum* (5.20 ± 0.20 kV/cm), *M. smegmatis* (5.56 ± 0.08 kV/cm), and *E. coli* BL21 (3.65 ± 0.09 kV/cm) were determined using SYTOX® as the fluorescent marker.

We envision this microfluidic platform as a tool to rapidly optimize genetic transformation of intractable or previously challenging bacteria. This study focuses on prokaryotes but many of the processes developed could be applied to yeast or mammalian cells as well. This platform systematically samples a continuous spectrum of electric fields in a single experiment, differing from the traditional trial-and-error approach with discrete steps. Results of this study will broaden the scope of bacteria available for applications in synthetic biology and genetic engineering.

GPCR-Based Sensors to Accelerate the Engineering of Chemical-Producing Microbes

Peralta-Yahya, P.,
Originally presented on: 7/19/2016 10:30:00 - 11:00:00

Designer microbes able to convert inexpensive sugars into biofuels and value-added chemicals have the potential to provide a sustainable and cost-effective alternative to the synthesis of chemicals from petroleum. Today, the major challenge in the engineering of microbes for the production of non-colorimetric or non-fluorescent chemicals, such as biofuels, is to rapidly identify the highest chemical-producing microbe from a pool, akin to finding a needle in a haystack. Current methods of testing for chemicals synthesized by microbes are low-throughput (100 samples/day) as they rely on chromatography-based technologies. The Peralta-Yahya laboratory is pioneering the use of olfactory G-protein coupled receptors (GPCRs) in yeast as chemical sensors to enable the high-throughput screening ($>10^7$ microbes/day) of chemical producing microbes. Here, I will present recent work on the development of GPCR-based chemical sensors to detect advanced biofuels, discuss how we are using these sensors for the high- throughput screening of biofuel-producing microbes, and consider how this throughput now allow us to apply evolutionary approaches to the bioproduction of biofuels and other non- colorimetric chemicals.

Engineered Polyketide Synthases for Production of Commodity and Specialty Chemicals

Jay Keasling

Joint BioEnergy Institute
University of California, Berkeley
Lawrence Berkeley National Laboratory
Center for Biosustainability, DTU

Approximately 15-20% of every barrel of oil is used to produce chemicals

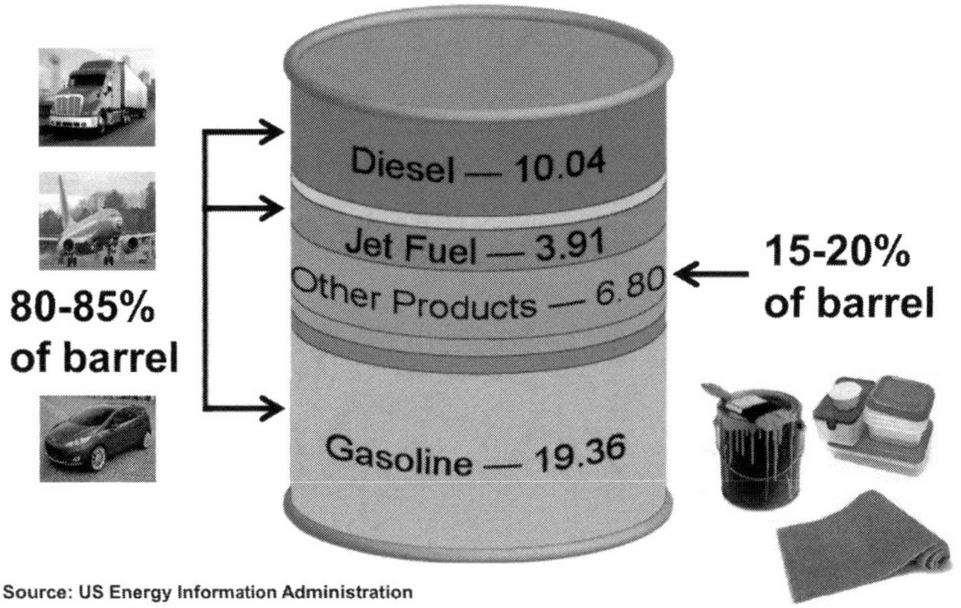

Source: US Energy Information Administration

Some products made from petroleum

Tents
Crayons
Parachutes
Telephones
Enamel
Pillows
Dishes
Cameras
Anesthetics
Artificial Turf
Artificial limbs
Bandages
Dentures
Model Cars
Folding Doors
Hair Curlers
Cold cream
Movie film
Soft Contact lenses
Drinking Cups
Fan Belts
Car Enamel
Shaving Cream
Ammonia
Refrigerators
Golf Balls
Toothpaste
Gasoline

Heart Valves
Candles
Trash Bags
House Paint
Water Pipes
Hand Lotion
Roller Skates
Surf Boards
Shampoo
Wheels
Paint Rollers
Shower Curtai
Guitar Strings
Luggage
Aspirin
Safety Glasses
Antifreeze
Football Helmets
Awnings
Eyeglasses
Clothes
Toothbrushes
Ice Chests
Footballs
Combs
CD's & DVD's
Paint Brushes
Detergents

Perfumes
TV Cabinets
Shag Rugs
Electrician's Tape
Tool Racks
Car Battery Cases
Epoxy
Paint
Mop

Lipstick
Denture Adhesive
Linoleum
Ice Cube Trays
Synthetic Rubber
Speakers
Plastic Wood
Electric Blankets
Glycerin

Cassettes
Dishwasher parts
Tool Boxes
Shoe Polish
Motorcycle Helmet
Caulking
Petroleum Jelly
Transparent Tape
CD Player
Faucet Washers
Antiseptics
Clothesline
Curtains
Food Preservatives
Basketballs
Soap
Vitamin Capsules
Antihistamines
Purses
Shoes
Dashboards
Cortisone
Deodorant
Footballs
Putty
Dyes
Panty Hose
Refrigerant

Dresses
Tires
Golf Bags
Percolators
Life Jackets
Rubbing Alcohol
Tennis Rackets
Rubber Cement
Fishing Boots
Vaporizers
Balloons
Sun Glasses
Solvents
Diesel fuel
Motor Oil
Bearing Grease
Ink
Floor Wax
Ballpoint Pens
Football Cleats
Upholstery
Sweaters
Boats
Insecticides
Bicycle Tires
Sports Car Bodies
Nail Polish
Fishing lures

How do we replace the 15-20% of every barrel of oil that is used to produce chemicals?

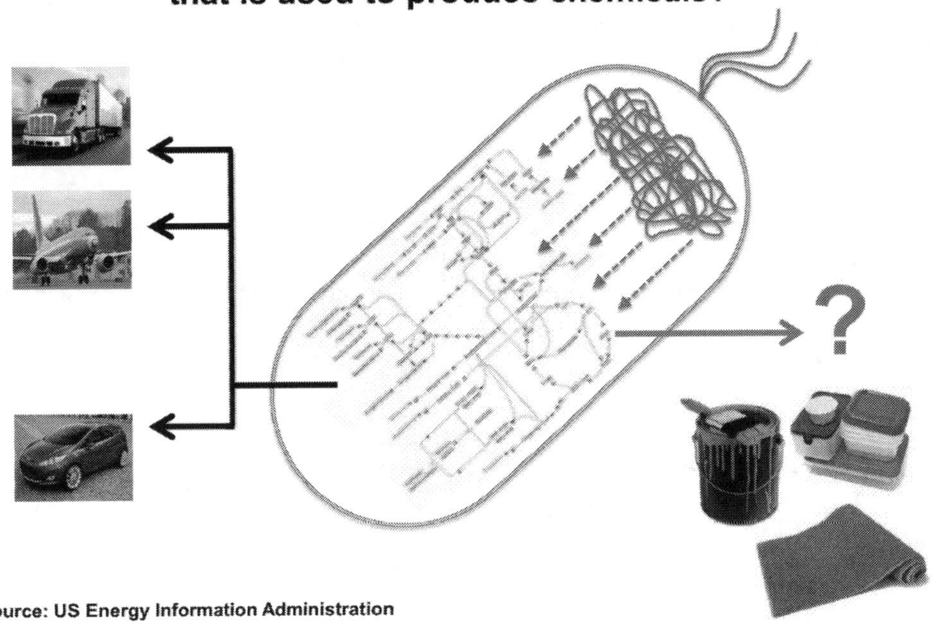

Source: US Energy Information Administration

The molecules needed to make fuels and materials are different

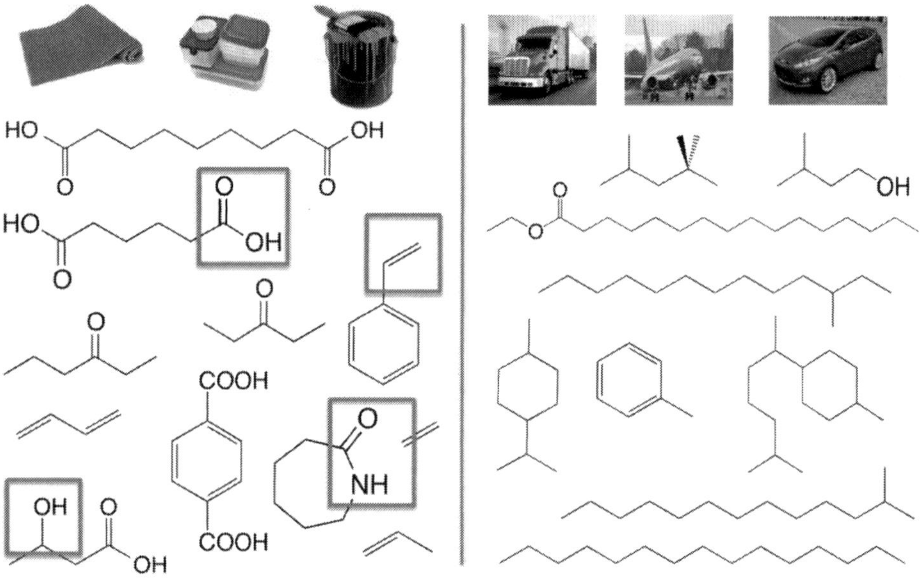

Polyketides have many of the functional groups we need for chemicals

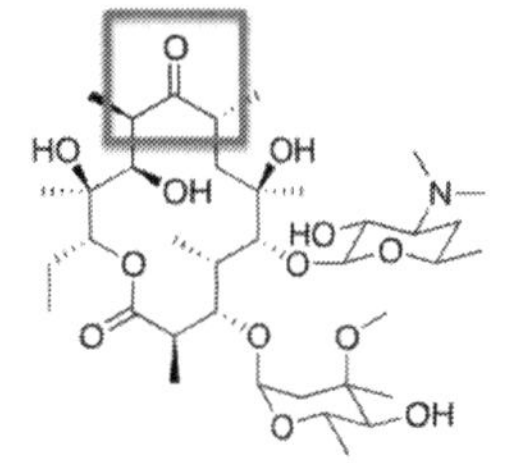

Erythromycin

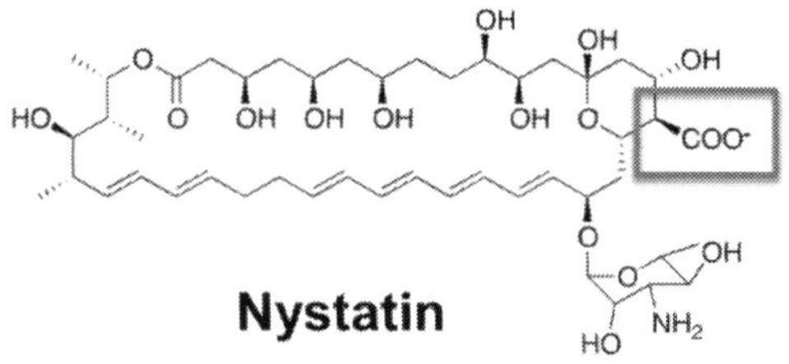

Nystatin

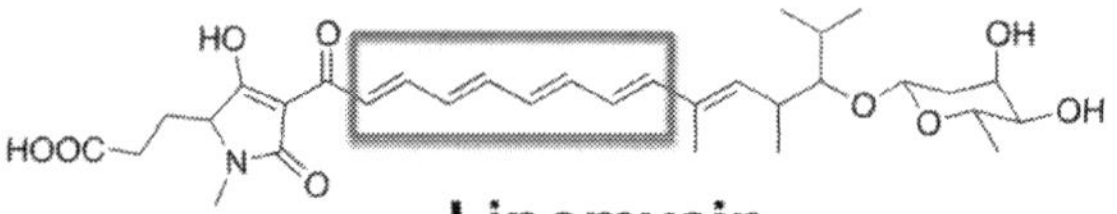

Lipomycin

Erythromycin synthase is a well-studied polyketide synthase (PKS)

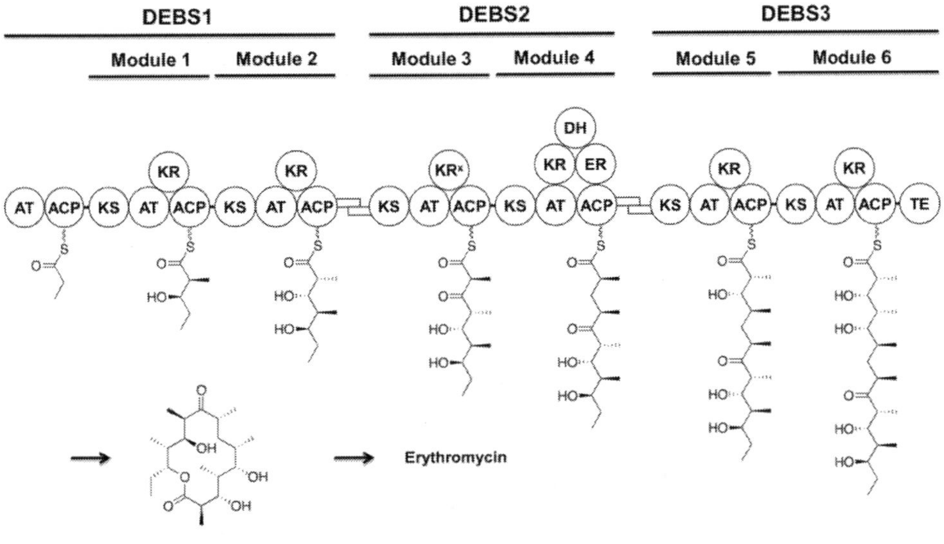

Donadio 1991 *Science*

Polyketide synthases (PKSs) are very large, multi-activity proteins

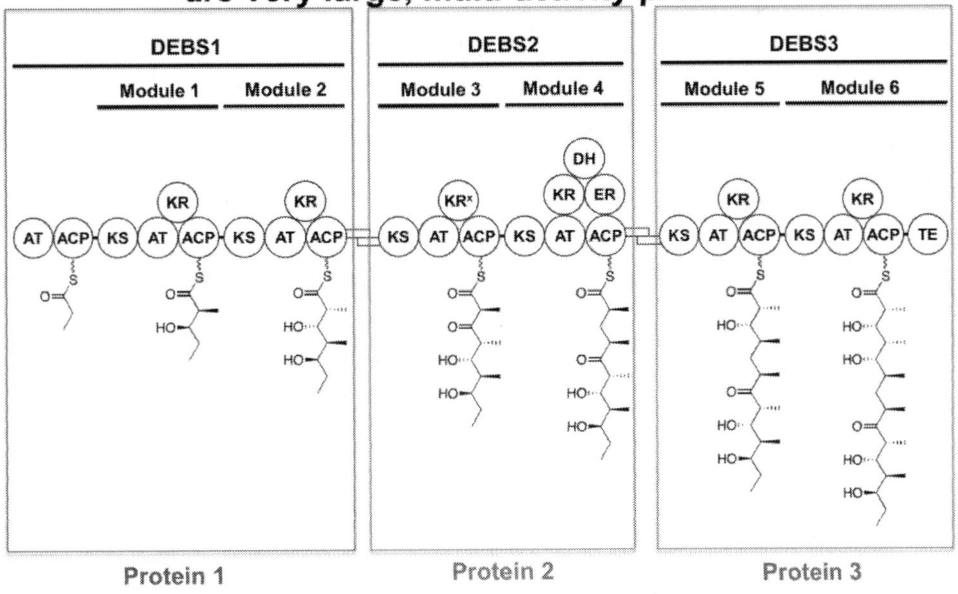

Polyketide synthases (PKSs) are very large, multi-activity proteins

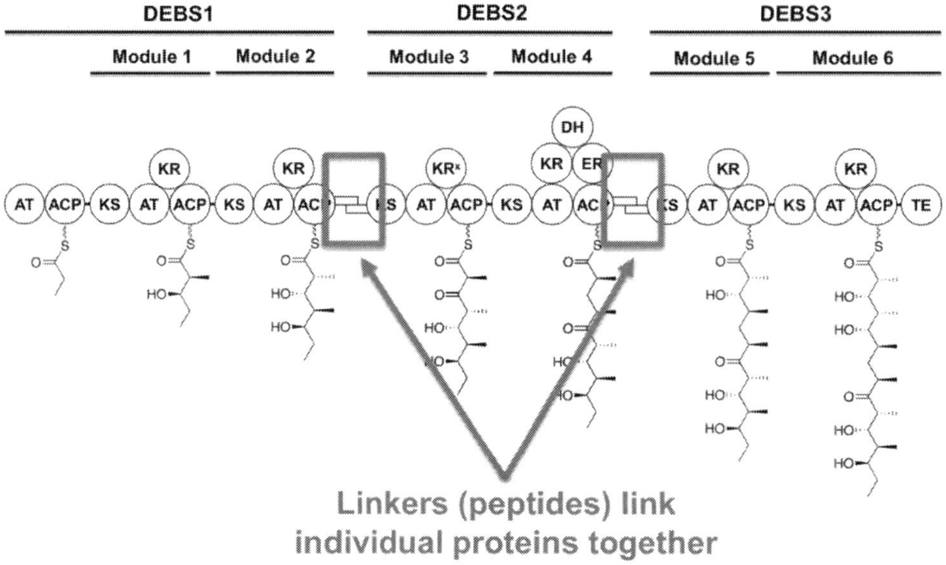

Each extension module adds 2 carbons (plus pendant groups) to the polyketide backbone

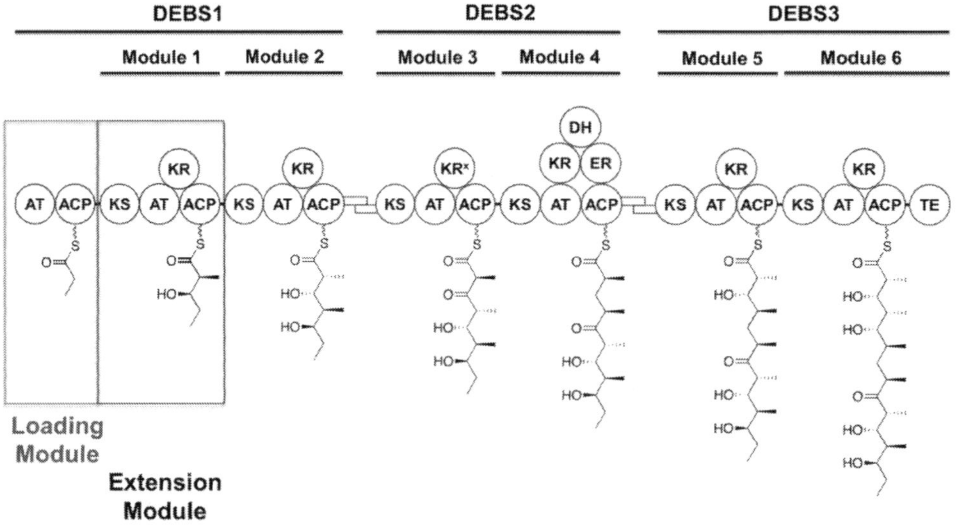

Polyketide synthases (PKSs) are very large, multi-activity proteins

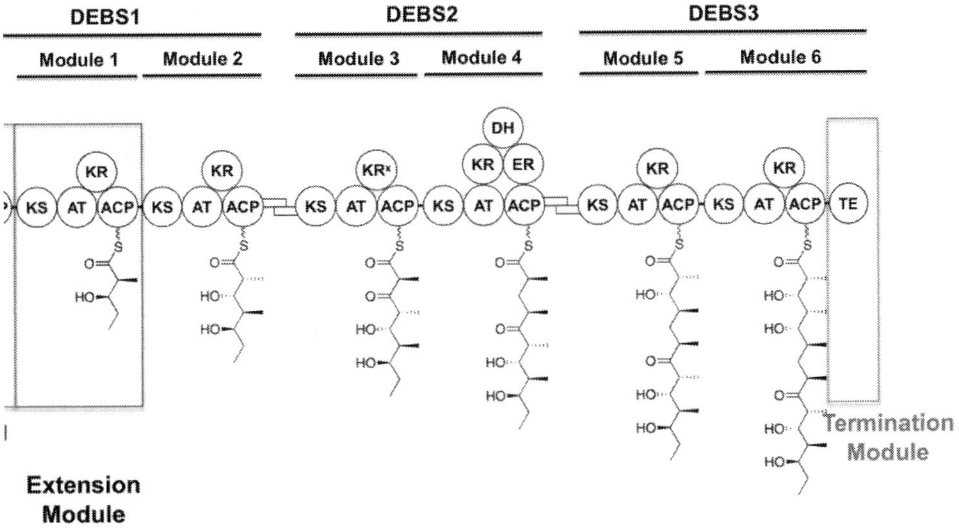

Each domain in a PKS has an individual enzyme activity

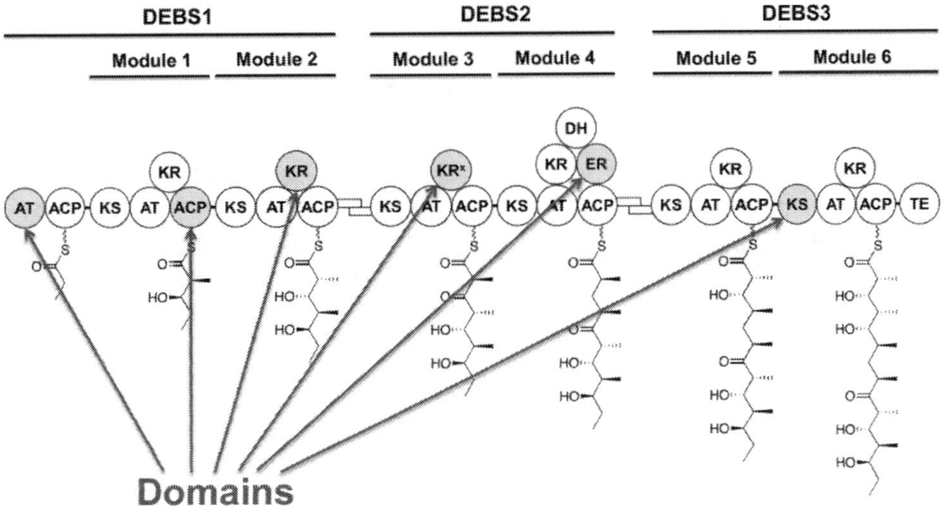

Basic PKS

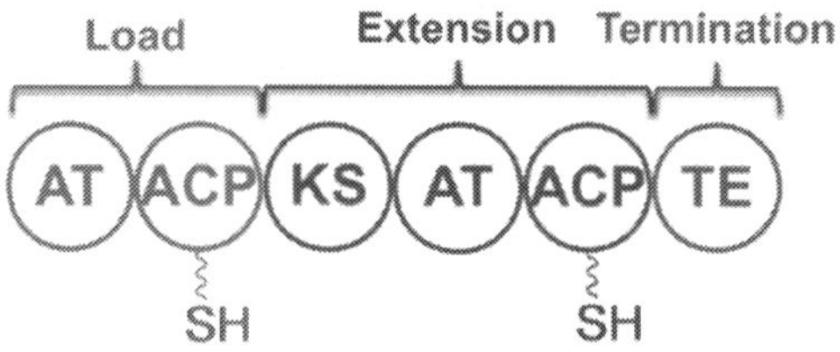

Basic PKS

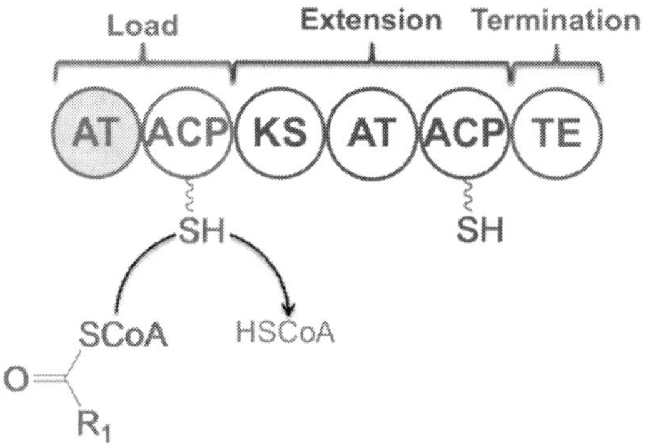

Basic PKS

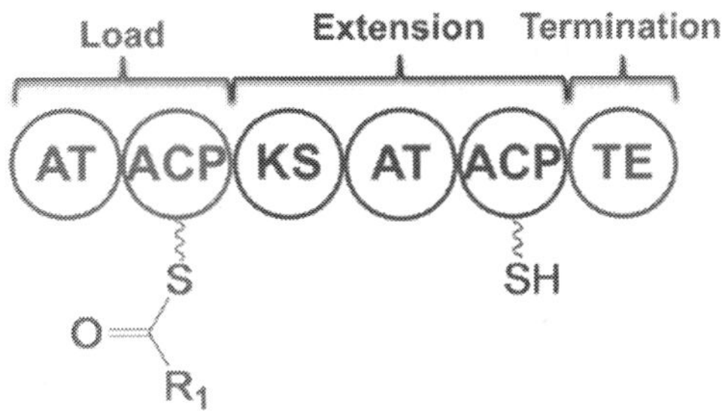

Basic PKS

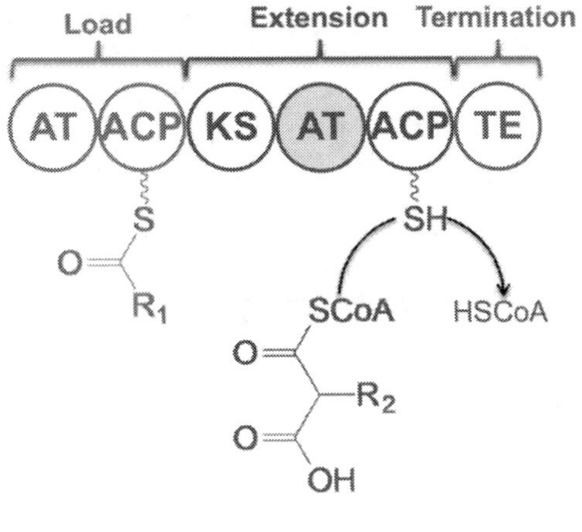

Basic PKS

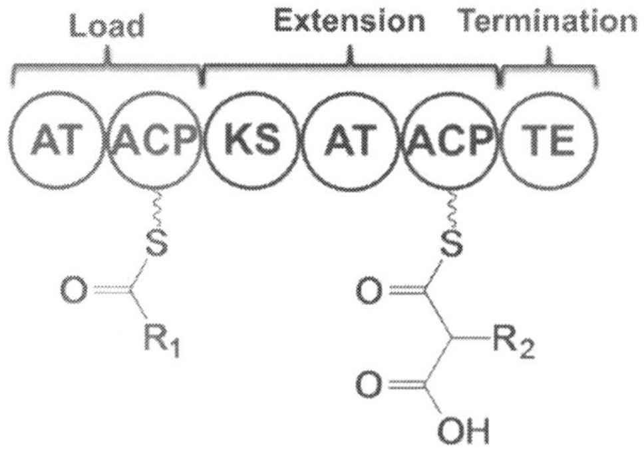

Basic PKS

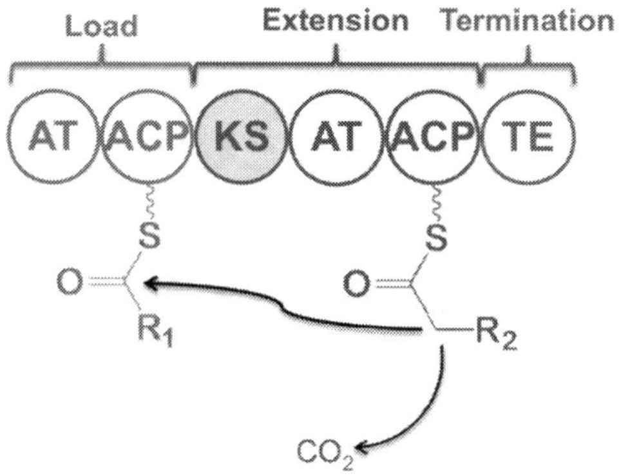

Basic PKS

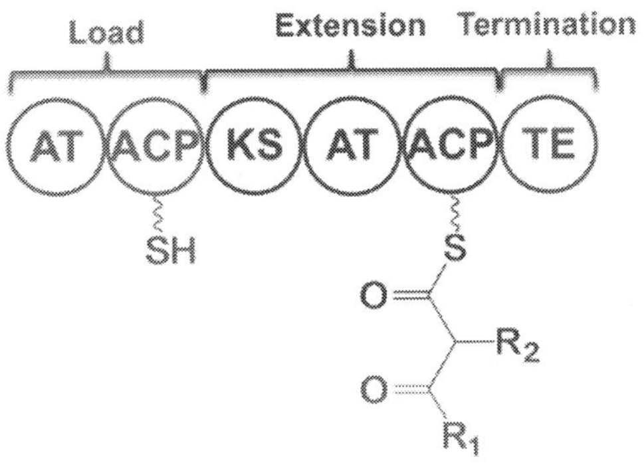

Basic PKS

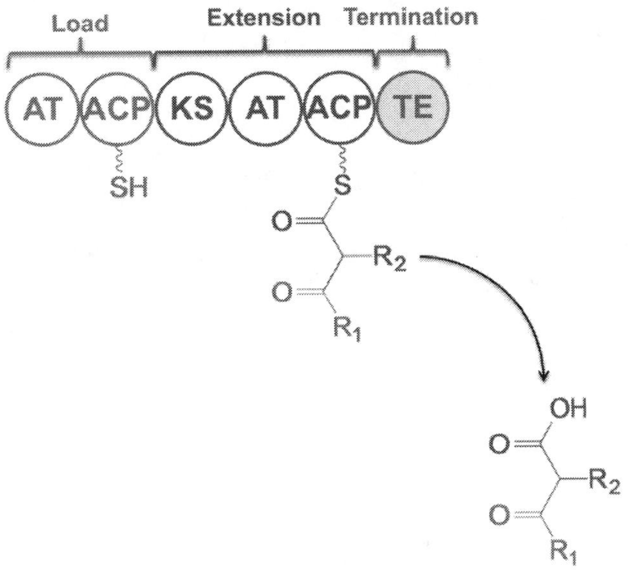

Basic PKS

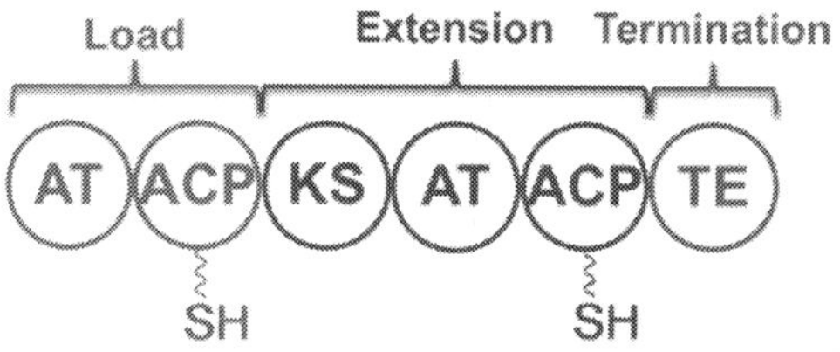

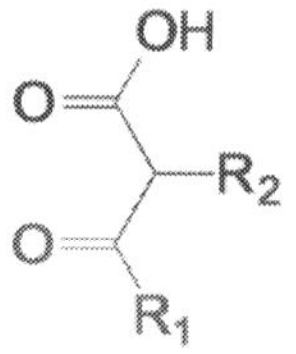

Basic PKS

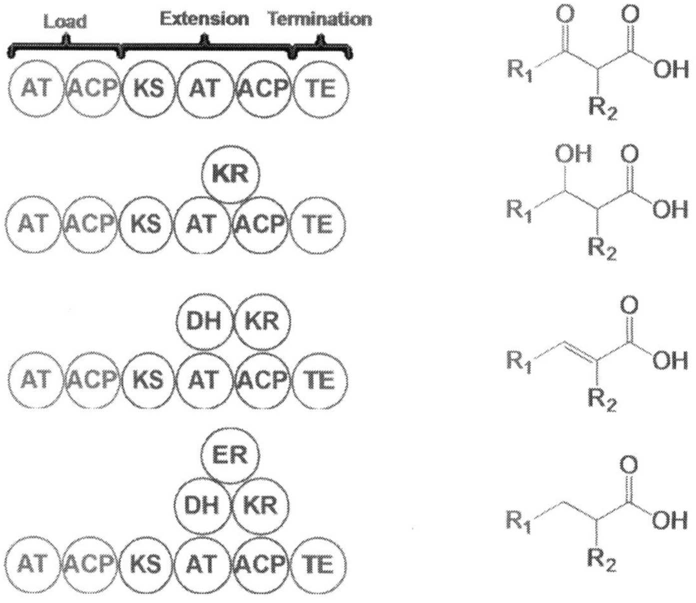

PKSs exist a dimers

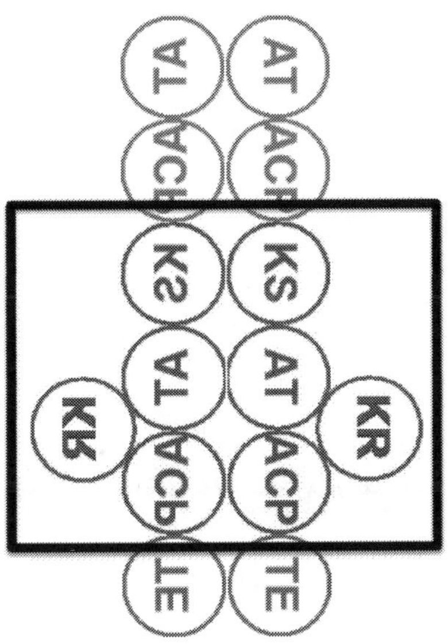

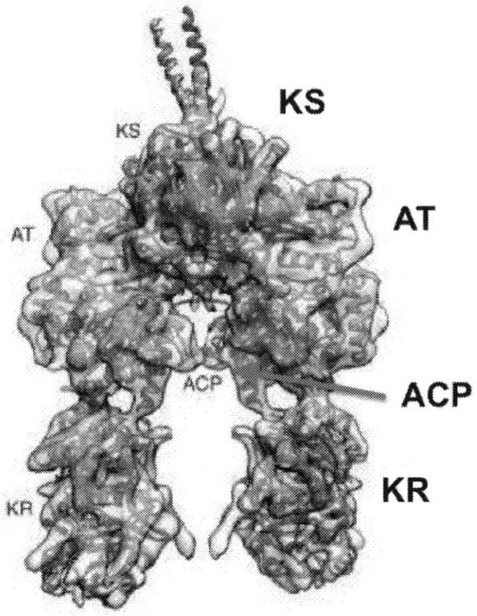

Whicher et al. 2014. *Nature*. 510(7506): 560–564

Lots of naturally occurring Load, Extension, and Termination modules

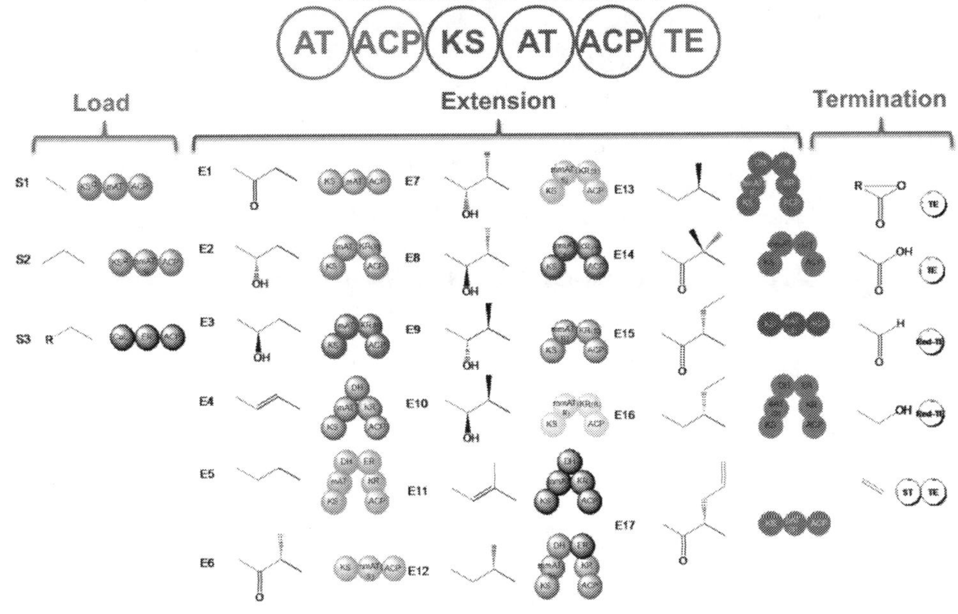

Polyketide synthases ...

- are unique enzymes whose sequence can be directly translated into chemical structures
- are capable producing diverse chemicals
- produce few off-target products due to the nature of the chemical reactions (decarboxylative condensation) and editing by catalytic domains in PKSs
- have specific synthesis rates in commercial antibiotic producers similar to fatty acid synthases

Goals of our work

- Determine unknown rules for PKS recombination to make target chemicals
- Build hybrid PKS-NRPS proteins that can produce valuable chemicals
- Construct one or more host capable of producing functional hybrid PKSs and produce the desired final product
  - Protein folding
  - Precursor production

Polyesters

- Used for
  - Cushioning, clothing, bottles, films, liquid crystal displays, holograms, filters, dielectric films, insulation for wires
  - Finishes on wood products, paints, etc.

Product Type	2002 (M tons/yr)	2008 (M tons/yr)
Textile-PET	20	39
Resin, bottle PET	9	16
Film-PET	1.2	1.5
Special polyester	1	2.5
Total	31.2	49

Constructing a polyketide synthase for production of polyesters

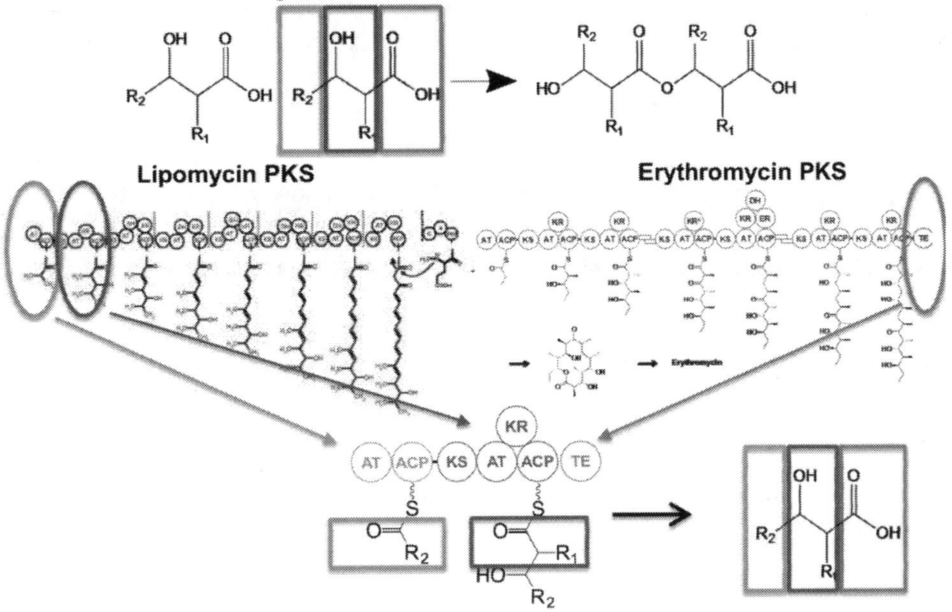

Lipomycin PKS

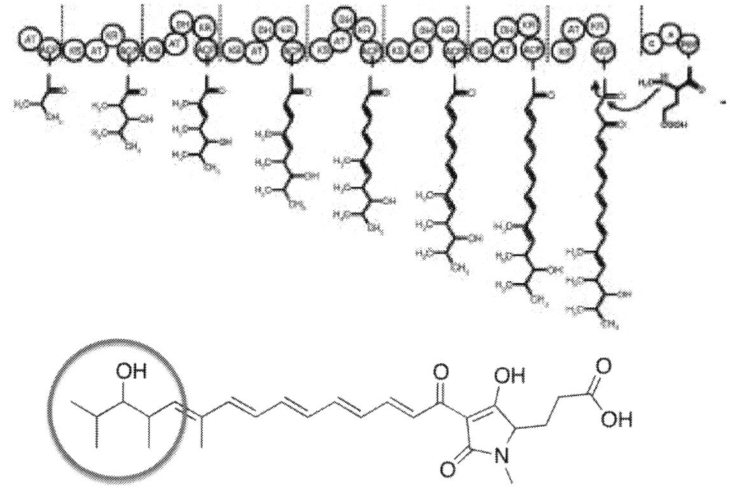

β-Lipomycin

Yuzawa et al. 2013. *Biochemistry* 52:3791-3793.

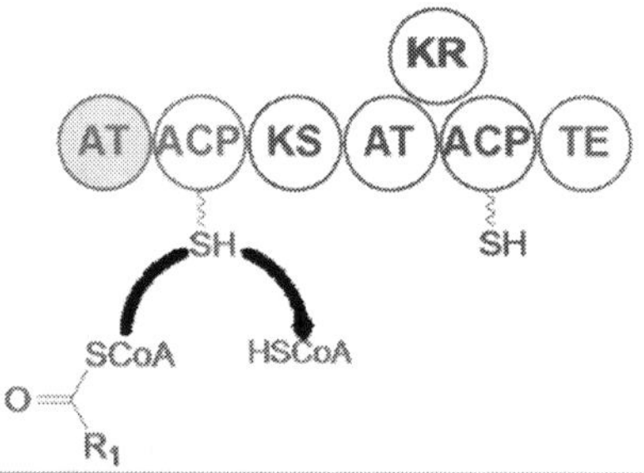

Lipomycin load has a broad substrate range

Substrate	kcat (min^{-1})	Km (uM)	kcat/Km (M^{-1}/s^{-1})
isobutyryl-CoA	0.053	2.9	304.1
propionyl-CoA	0.056	13.4	70.3
n-butyryl-CoA	0.036	26.4	22.7
2-methylbutyryl-CoA	0.126	8.8	237.0
isovaleryl-CoA	0.290	128.1	38.0
pivaloyl-CoA	0.002	8.8	4.1

Yuzawa et al. 2013. *Biochemistry* 52:3791-3793.

Production of 3-hydroxy-2-methyl acids with Lip PKS

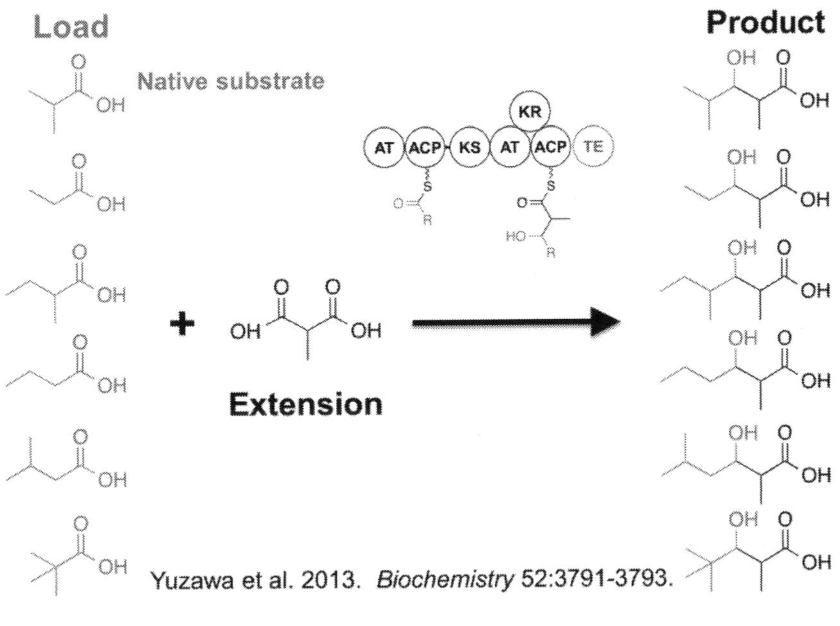

Yuzawa et al. 2013. *Biochemistry* 52:3791-3793.

The substituent in the 2-position can greatly impacts polymer properties

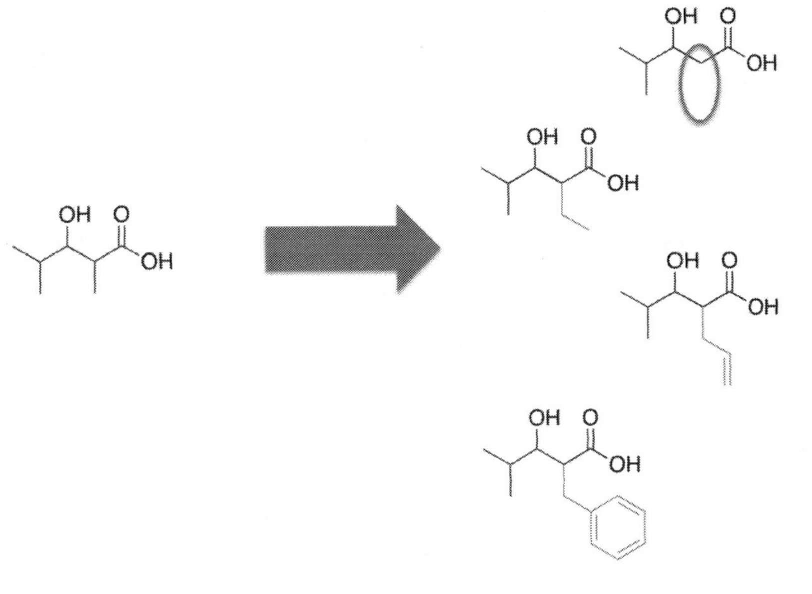

The acyltransferase (AT) determines the acyl-CoA loaded onto the ACP

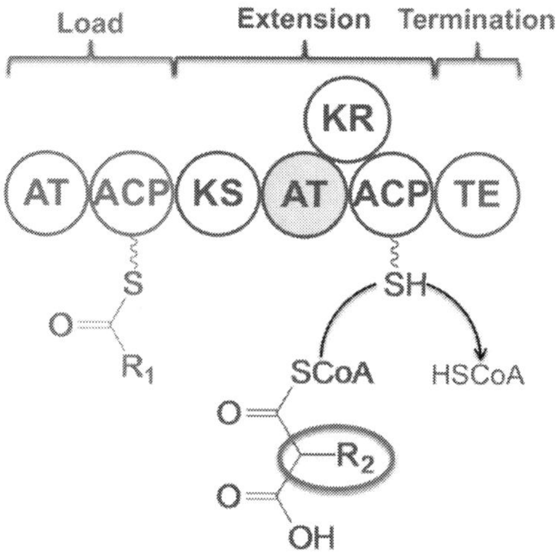

There are many AT's available

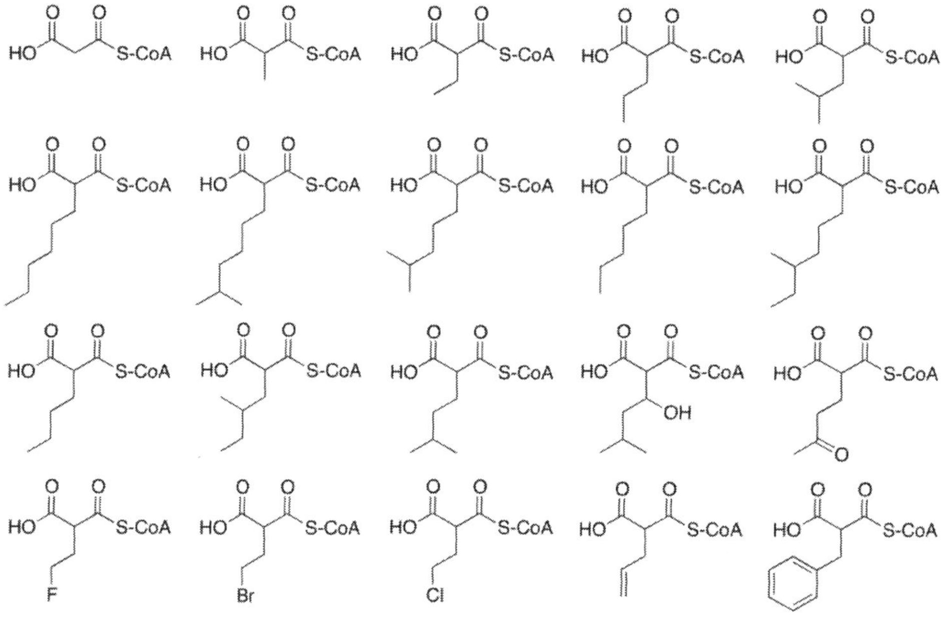

Acyl transferase (AT) specifies the acyl group that gets loaded onto the ACP

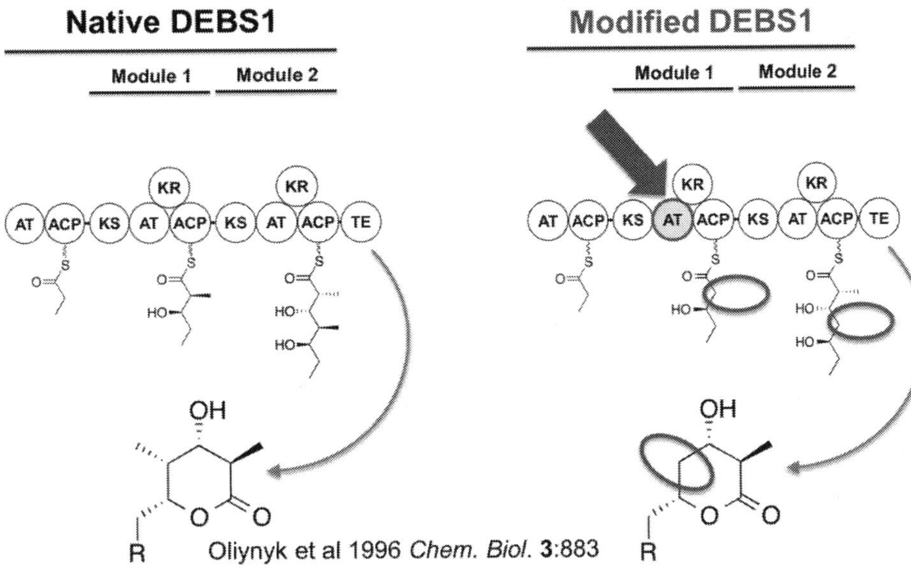

Oliynyk et al 1996 *Chem. Biol.* **3**:883

The acyltransferase (AT) determines the acyl-CoA loaded onto the ACP

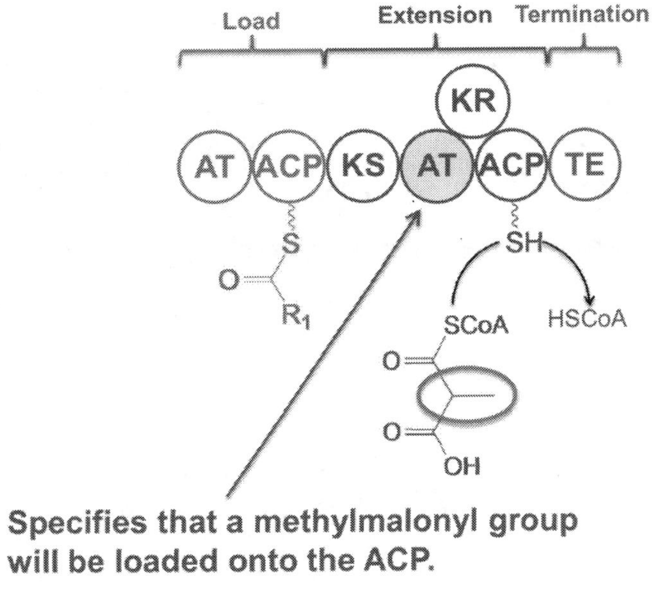

The acyltransferase (AT) determines the acyl-CoA loaded onto the ACP

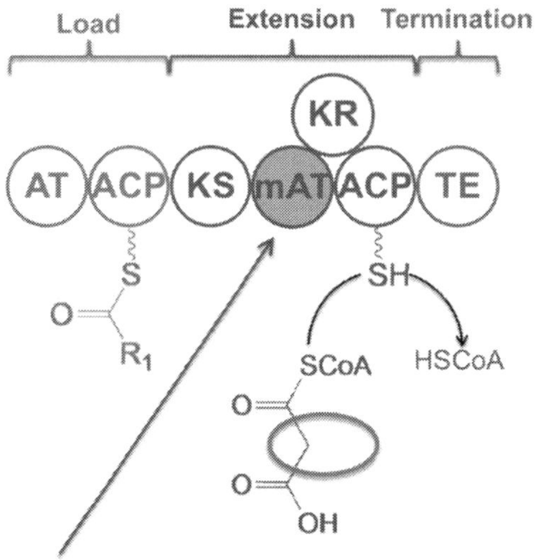

Specifies that a malonyl group will be loaded onto the ACP.

Choosing the right sites to swap in the new AT is essential to retain activity

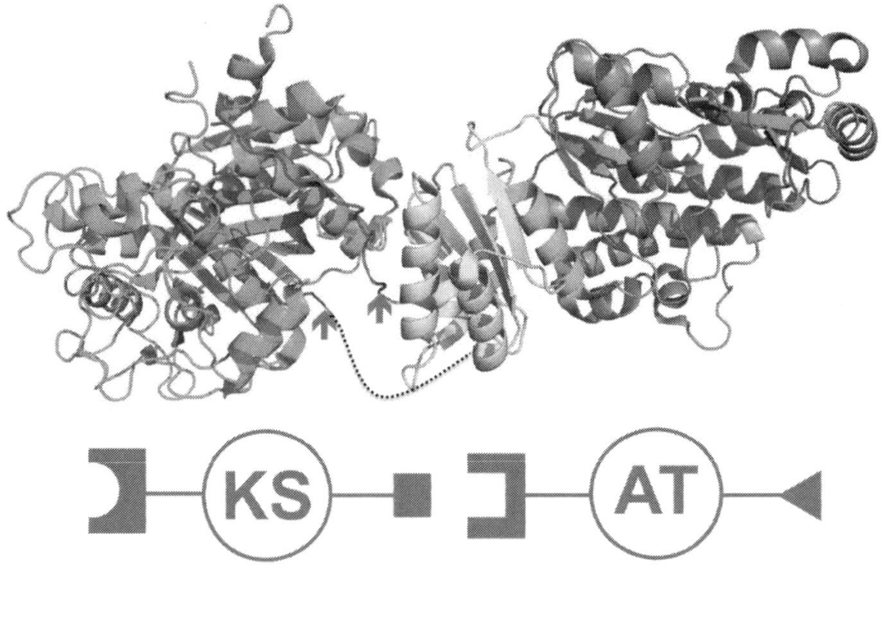

KS-AT structure

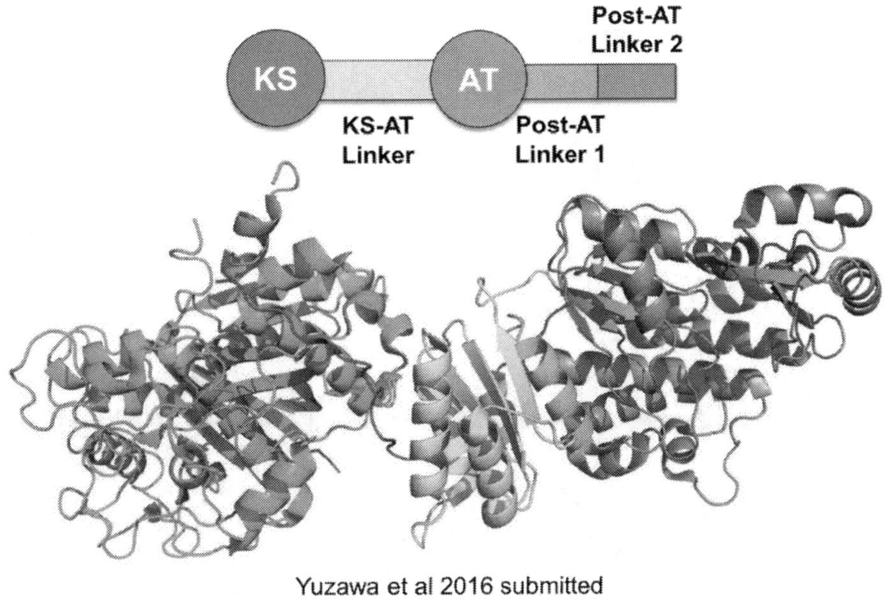

Yuzawa et al 2016 submitted

Substantial loss of activity results in transferring only the AT

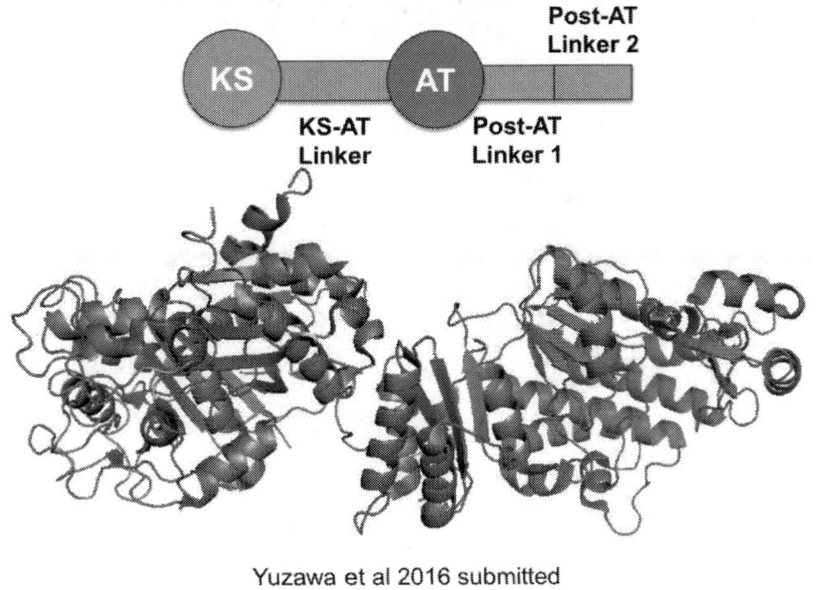

Yuzawa et al 2016 submitted

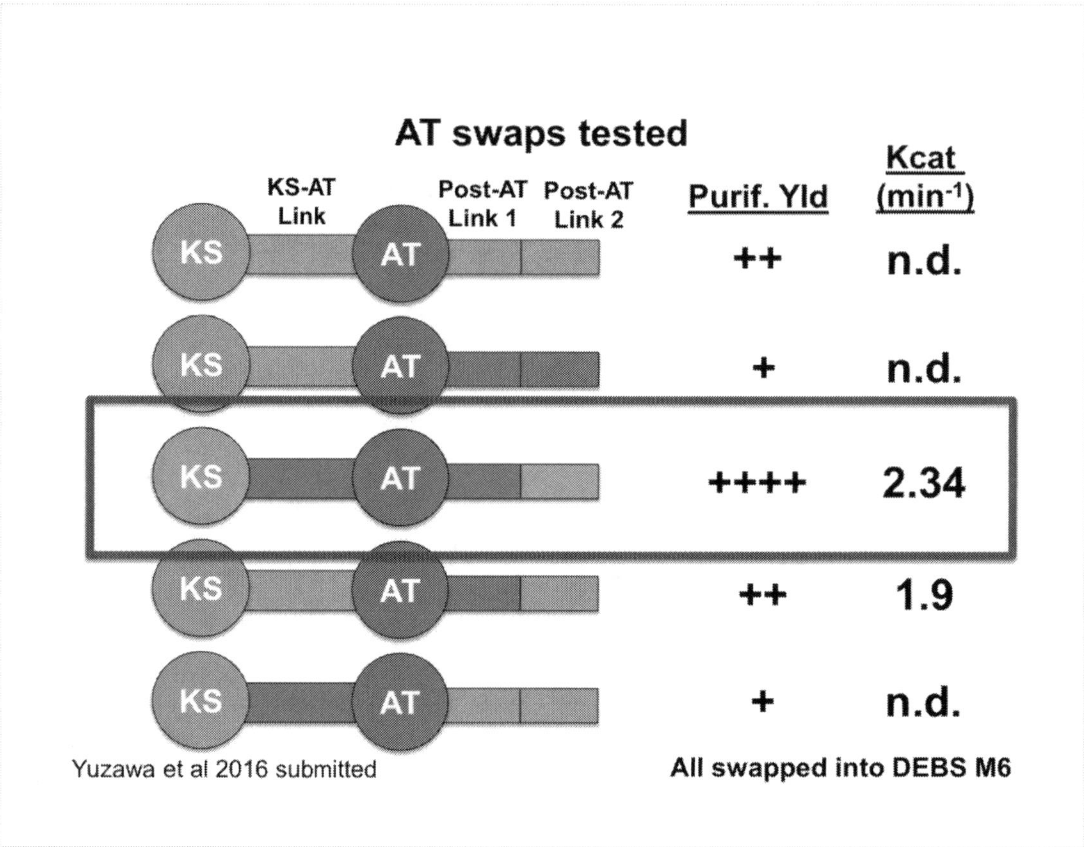
AT swaps tested
KS-AT Link
Post-AT Link 1
Post-AT Link 2
Purif. Yld
Kcat (min-1)
KS
AT
++
n.d.
+
n.d.
++++
2.34
++
1.9
+
n.d.
Yuzawa et al 2016 submitted
All swapped into DEBS M6

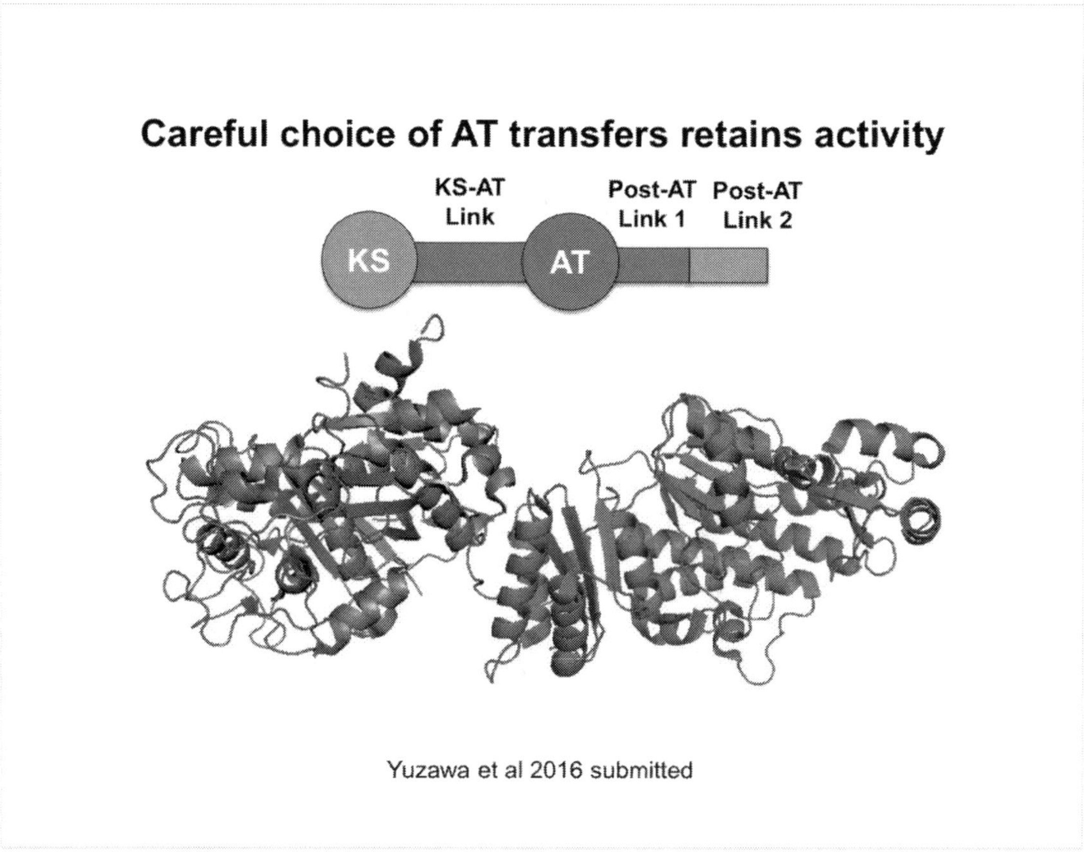
Careful choice of AT transfers retains activity
KS-AT Link
Post-AT Link 1
Post-AT Link 2
KS
AT
Yuzawa et al 2016 submitted

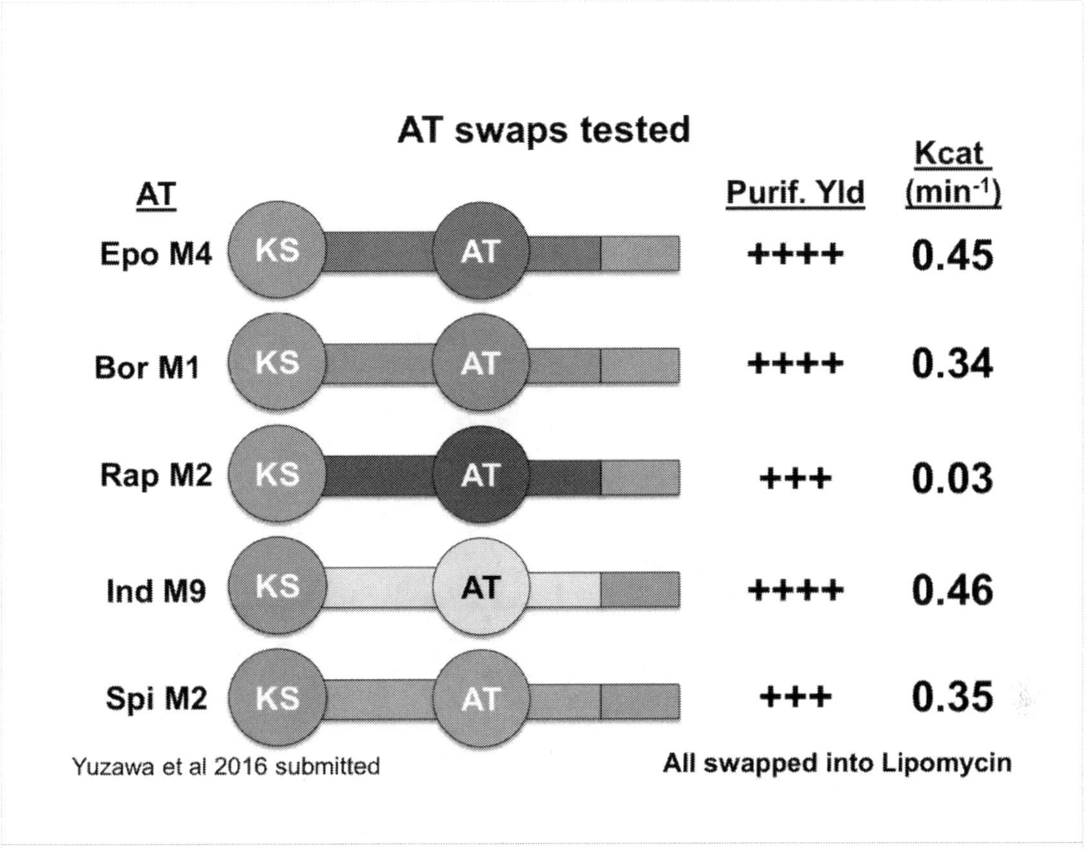
AT swaps tested
AT
Purif. Yld
Kcat (min-1)
Epo M4
KS
AT
++++
0.45
Bor M1
++++
0.34
Rap M2
+++
0.03
Ind M9
++++
0.46
Spi M2
+++
0.35
Yuzawa et al 2016 submitted
All swapped into Lipomycin

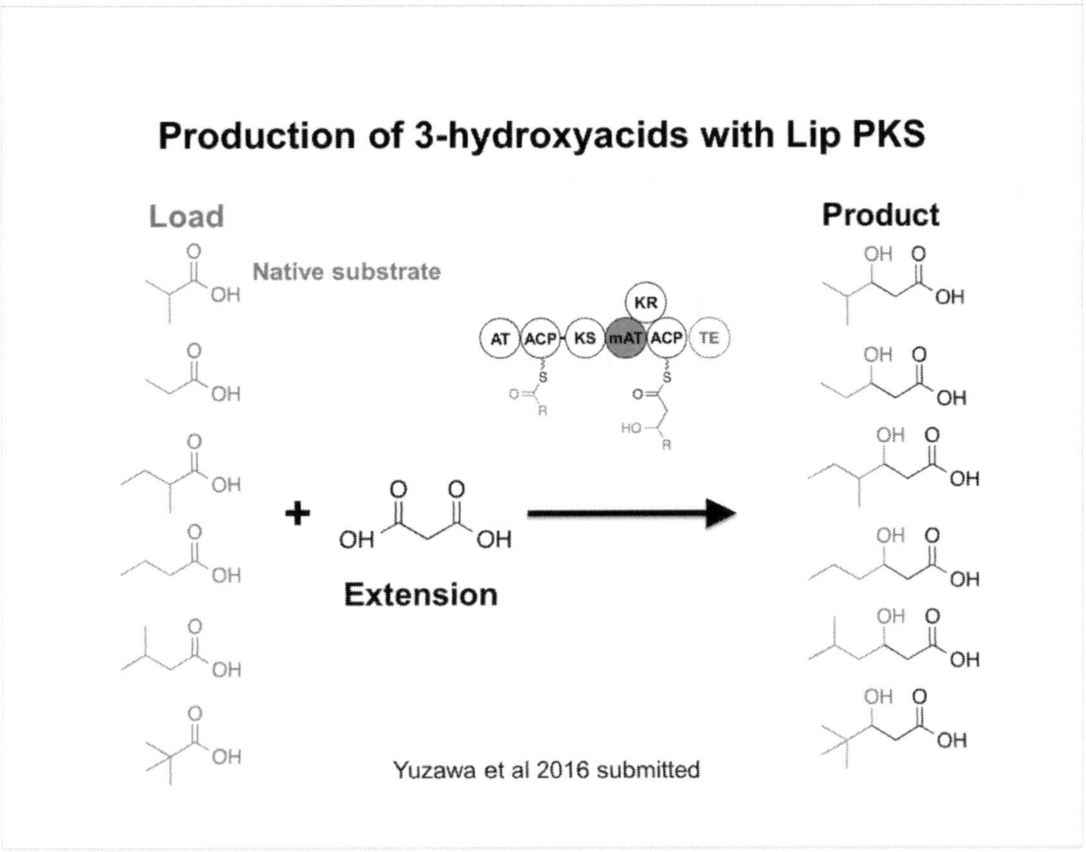
Production of 3-hydroxyacids with Lip PKS
Load
Native substrate
Product
KR
AT
ACP
KS
mAT
ACP
TE
+
Extension
Yuzawa et al 2016 submitted

Controlling stereochemistry

The AT will only load (*S*)-methylmalonyl group onto the ACP.

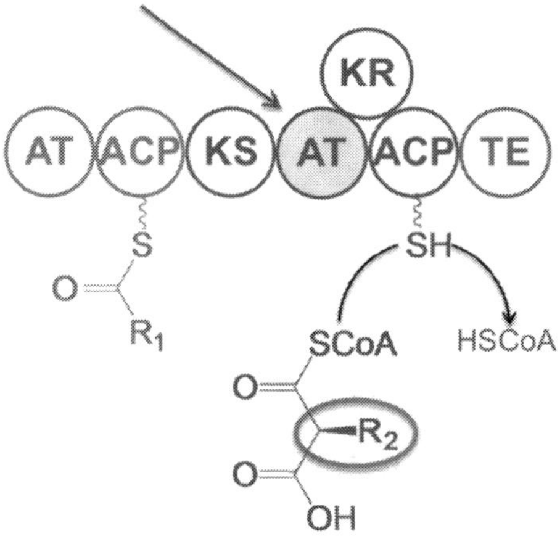

Controlling stereochemistry

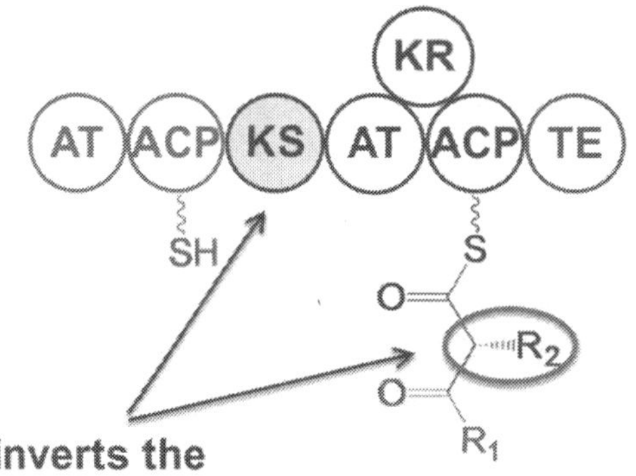

The KS inverts the stereochemistry during condensation.

Controlling stereochemistry

The KR controls the stereochemistry of the product.

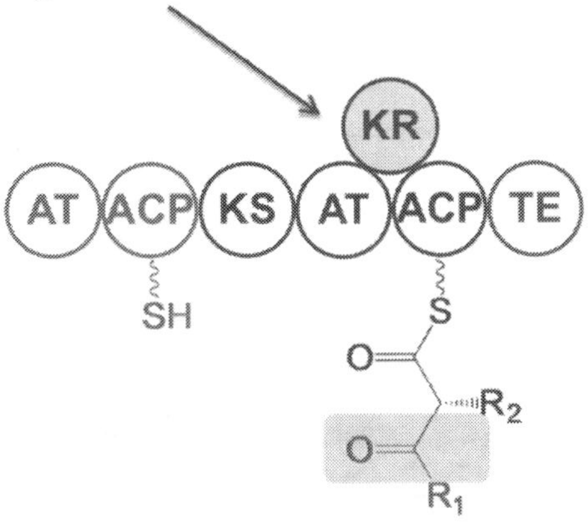

Controlling stereochemistry

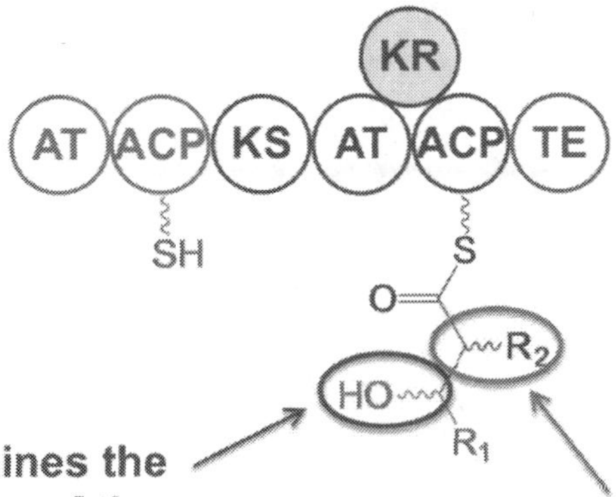

The KR determines the stereochemistry of the hydroxyl at the 3 position.

The KR can also epimerize the R group at the 2 position.

Stereochemistry of resulting 2- and 3-positions

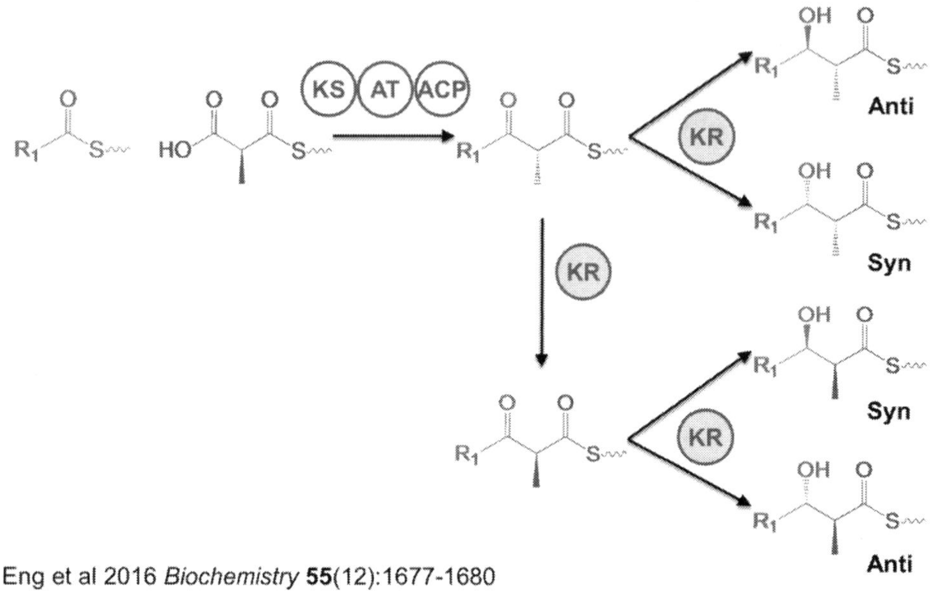

Eng et al 2016 *Biochemistry* **55**(12):1677-1680

Stereochemistry of resulting 2- and 3-positions

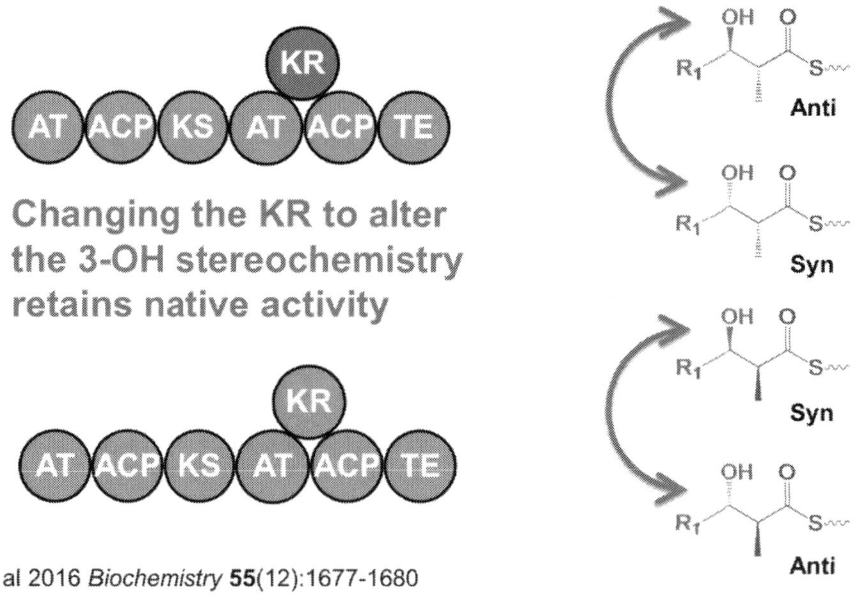

Eng et al 2016 *Biochemistry* **55**(12):1677-1680

Stereochemistry of resulting 2- and 3-positions

Great loss of activity when attempting to change stereochemistry of the 2-methyl

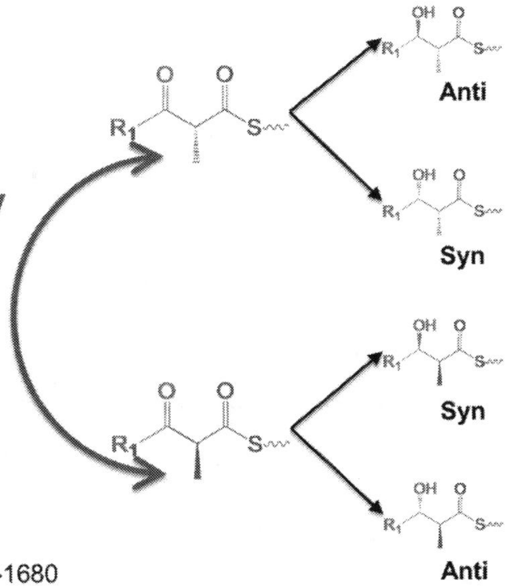

Eng et al 2016 *Biochemistry* **55**(12):1677-1680

The KR of lipomycin and some other PKSs has a dimerization element

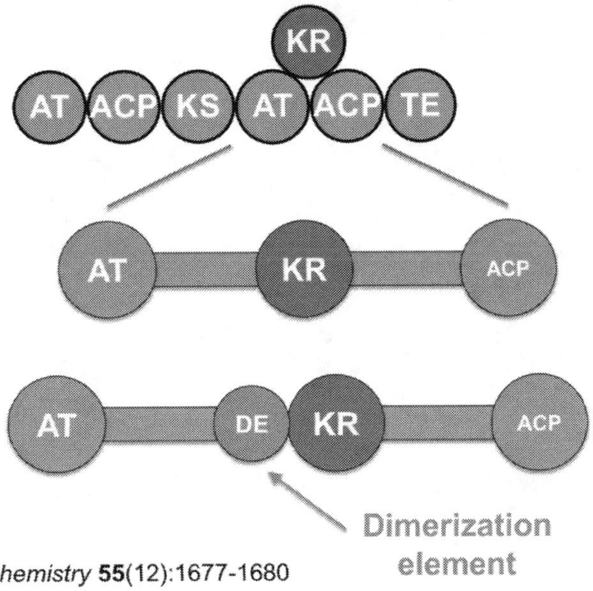

Eng et al 2016 *Biochemistry* **55**(12):1677-1680

DE element stabilizes the polyketide synthase structure

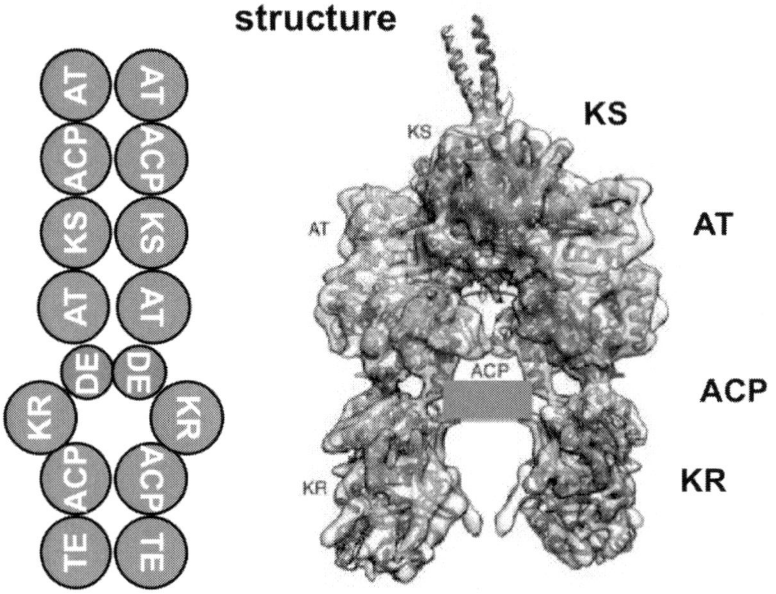

Eng et al 2016 *Biochemistry* **55**(12):1677-1680

Retaining the dimerization element (DE) in hybrid PKSs allows us to control stereochemistry *and* retain enzyme activity

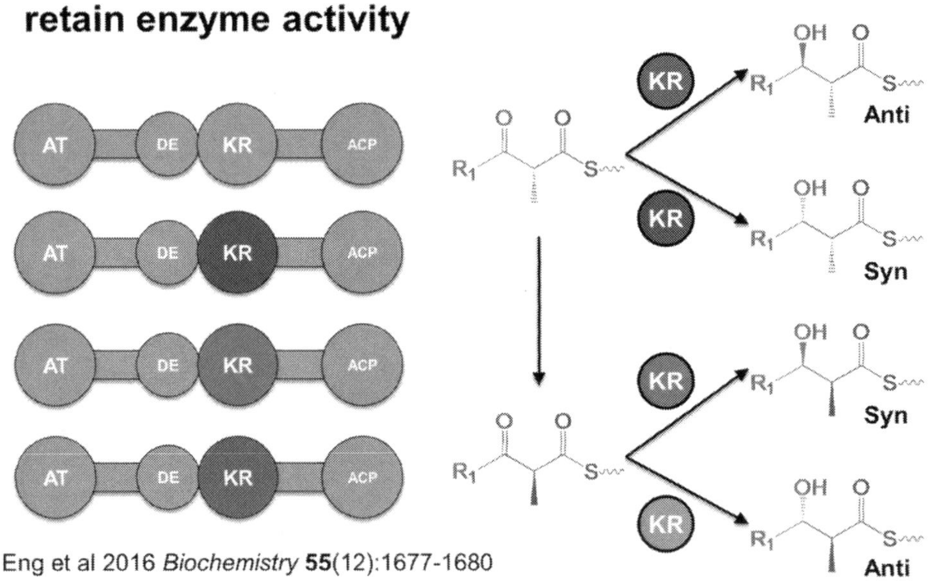

Eng et al 2016 *Biochemistry* **55**(12):1677-1680

Ketones

- Solvents
  - Methyl ethyl ketone (MEK)
  - Ethyl ethyl ketone (EEK)
- Flavors and fragrances
  - Long-chain alkyl groups
  - Stereochemistry matters
- Fuels
  - Diesels (C14-C16)

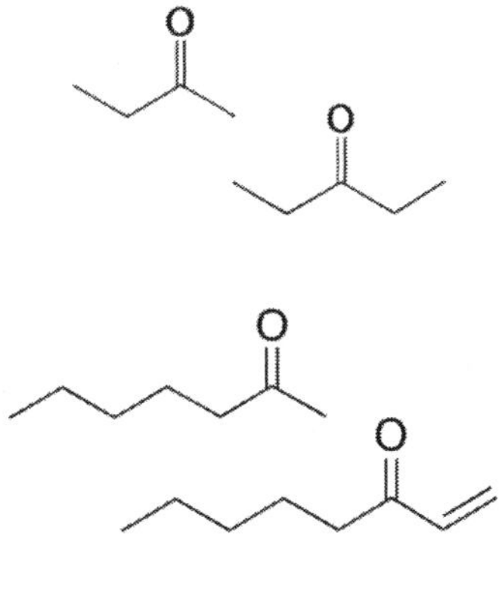

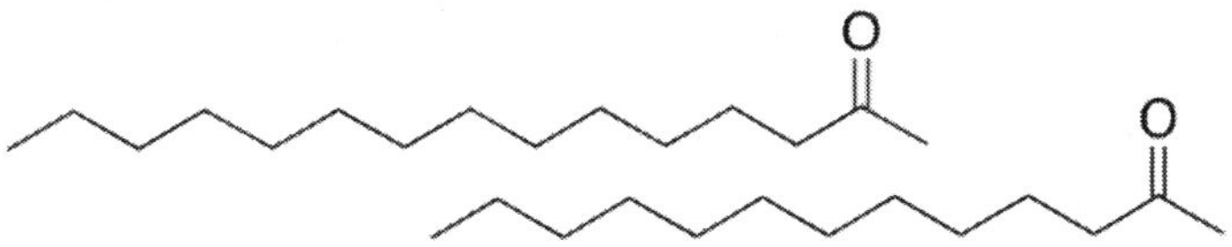

Knocking out the ketoreductase activity allows production of ethyl ketones

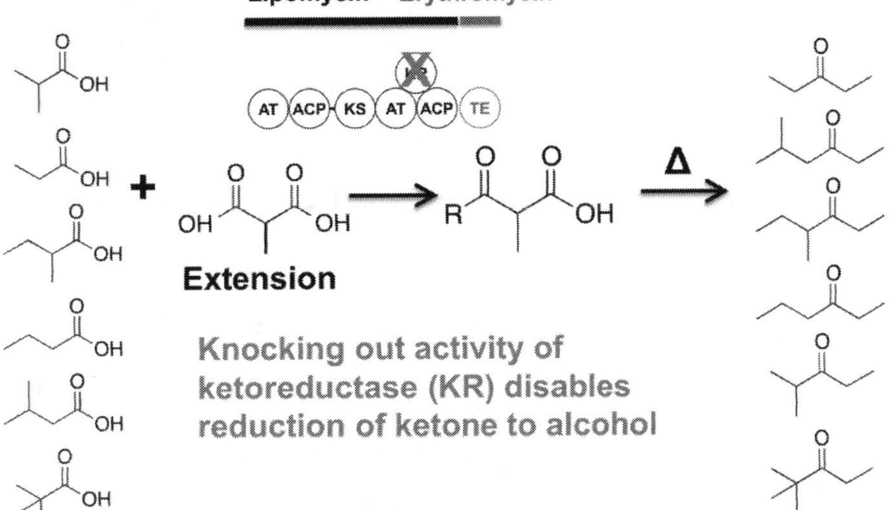

Yuzawa et al 2016 submitted

In vivo production of methyl ketones

Exchanging the native methylmalonyl-CoA—specific acyltransferase (AT) for a malonyl-CoA—specific AT enables production of methyl ketones

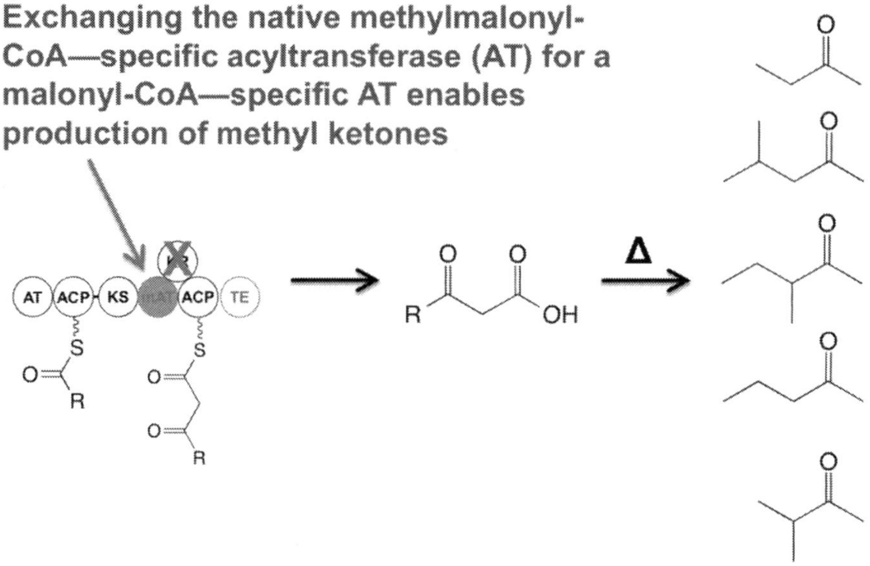

Yuzawa et al 2016 submitted

Adipic acid

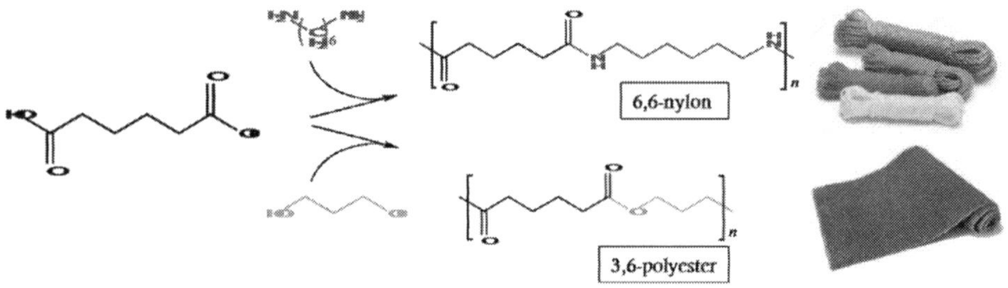

- Monomeric component of nylons, polyesters, and food ingredients
- 2.6 million tons produced worldwide in 2010
- Synthesis relies on petroleum precursors and generates copious NOx emissions…huge environmental impact!

Borrelidin PKS starts with an acid

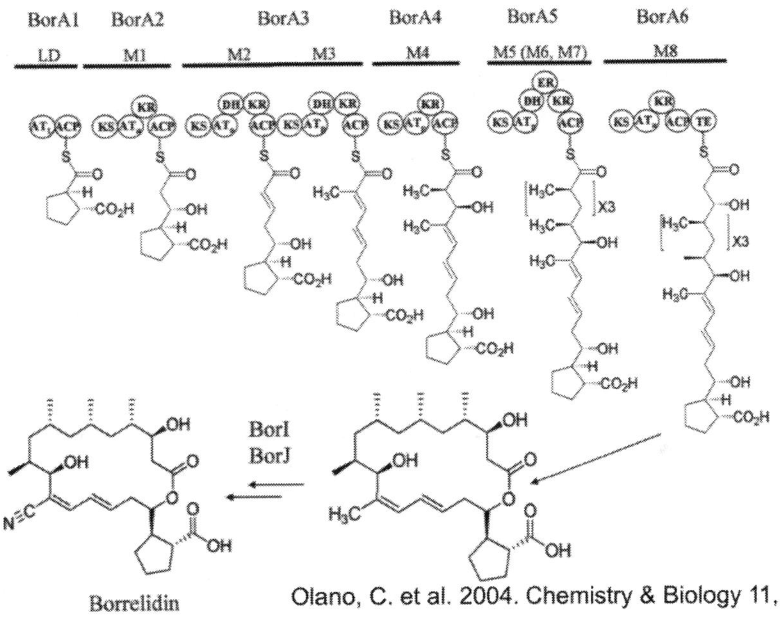

Olano, C. et al. 2004. Chemistry & Biology 11, 87–97.

Borrelidin PKS starts with an acid

- Predicted to load a carboxyacyl substrate, *t*1,2-CPDA-CoA
- Iterative use of module 5

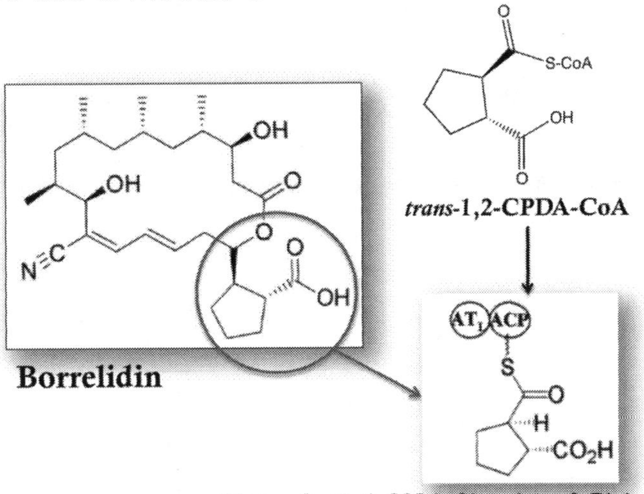

Olano, C. et al. 2004. Chemistry & Biology 11, 87–97.

First adipic acid PKS begins with BorA1 and BorA2

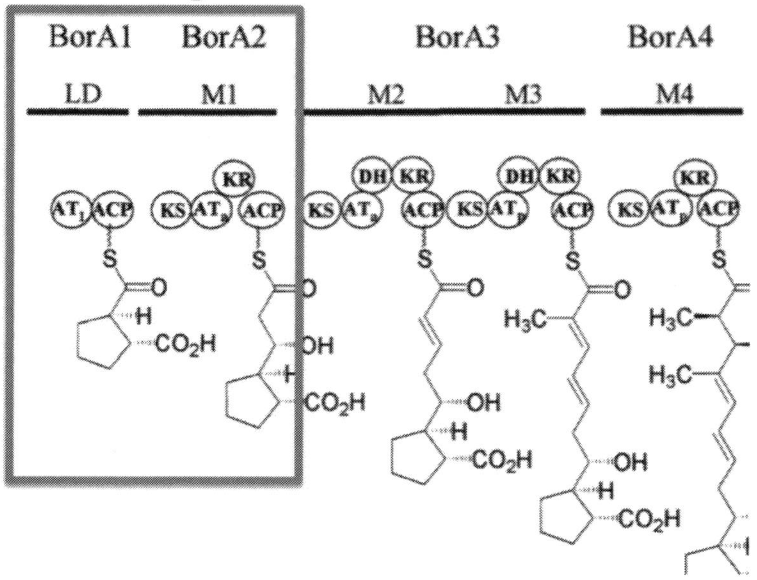

Hagen et al. 2014. *Biochemistry* **53**:5975-5977

Bor_{Mod1} shows substrate promiscuity

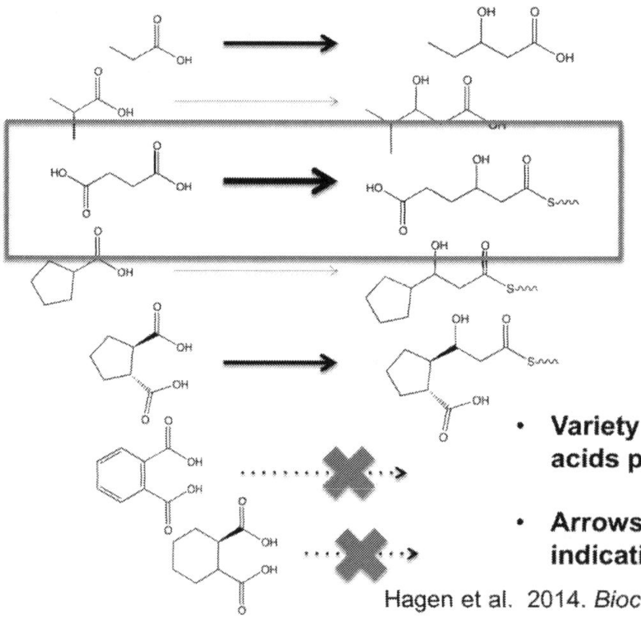

- **Variety of β-hydroxy acids produced**
- **Arrows are very rough indication of peak area***

Hagen et al. 2014. *Biochemistry* **53**:5975-5977

Reductive domain swap necessary for production of adipic acid

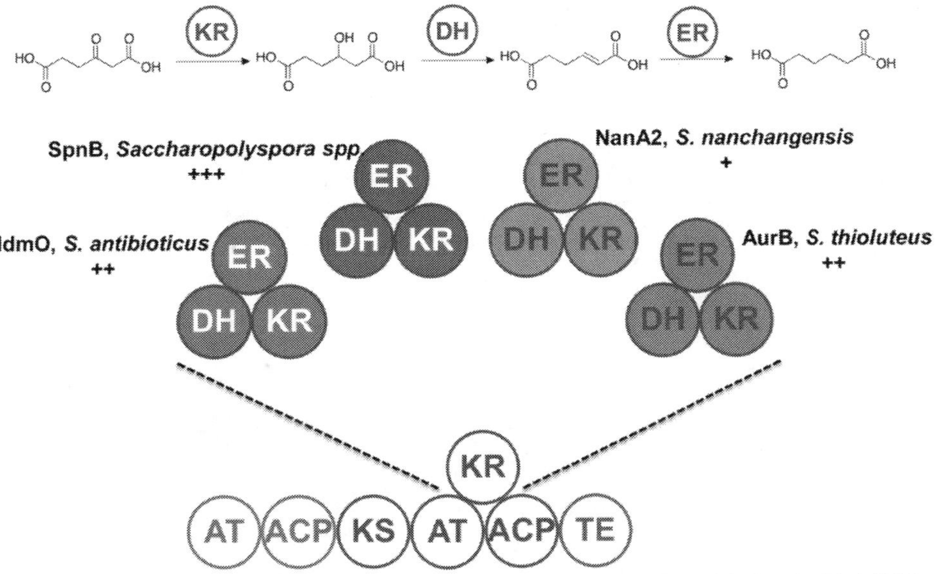

Hagen et al 2016 *ACS Synth. Biol.* **5**:21

Reductive loop replacement yields 3-hydroxyadipoyl-ACP

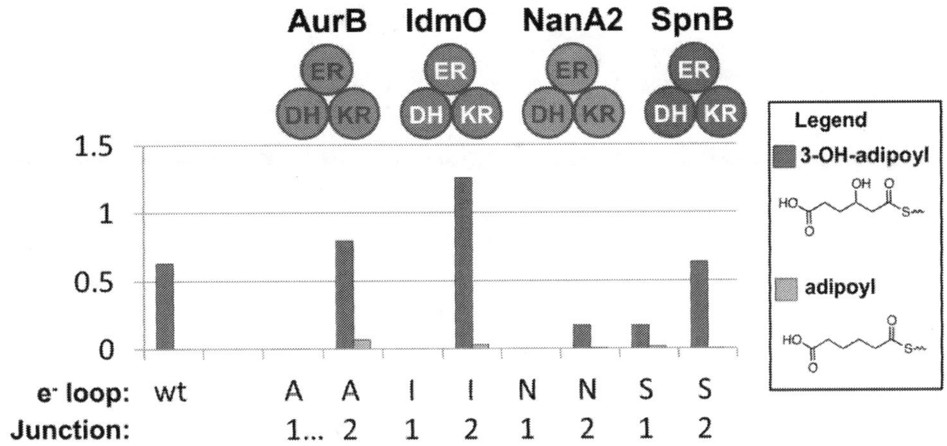

Hagen et al 2016 *ACS Synth. Biol.* **5**:21

Supplying the Bor2 dehydratase (BorDH2) in trans produces more adipoyl-ACP

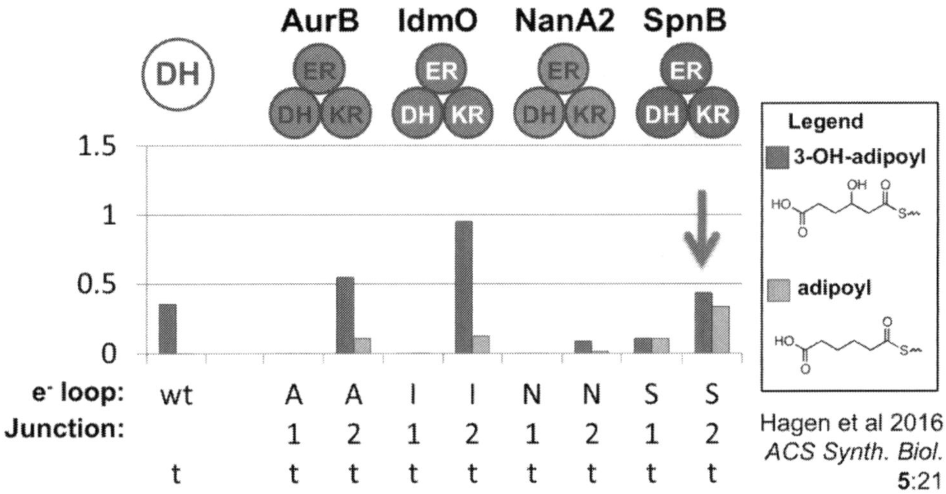

Hagen et al 2016
ACS Synth. Biol.
5:21

Replacing the SpnB DH with the BorDH2 produces even more adipoyl-ACP

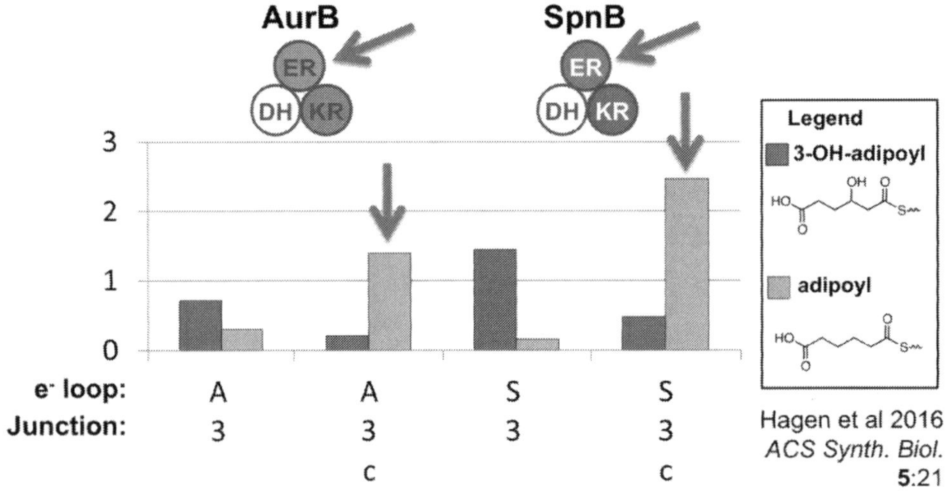

Hagen et al 2016
ACS Synth. Biol.
5:21

Adipic acid PKS made from parts of three PKSs

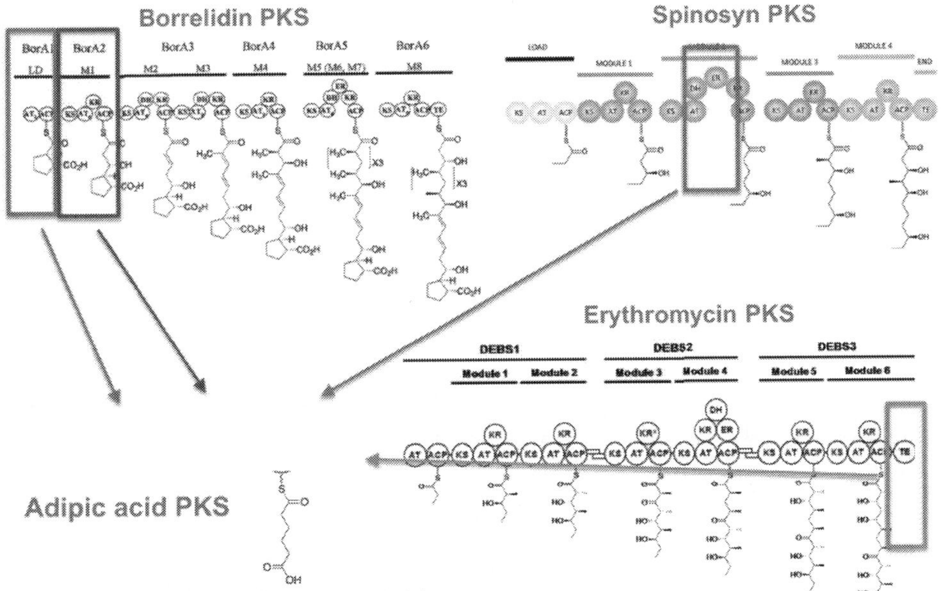

Hagen et al 2016 *ACS Synth. Biol.* **5**:21

New monomers can be generated by replacing the acyltransferase

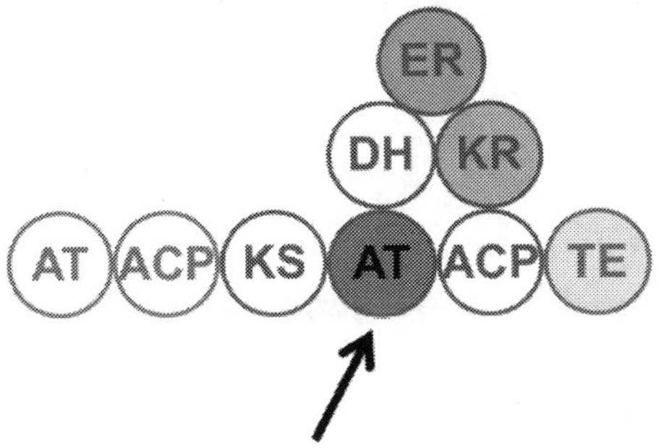

Replacing the acyl-transferase (AT) with others allows incorporation of side chains on adipic acid

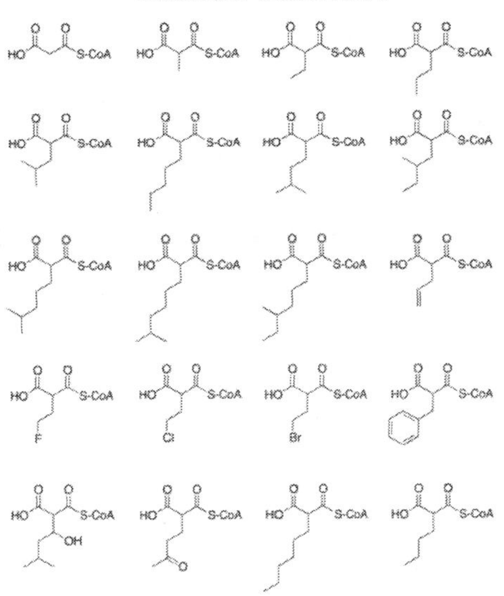

PKS engineering enables future materials that cannot be made from petroleum

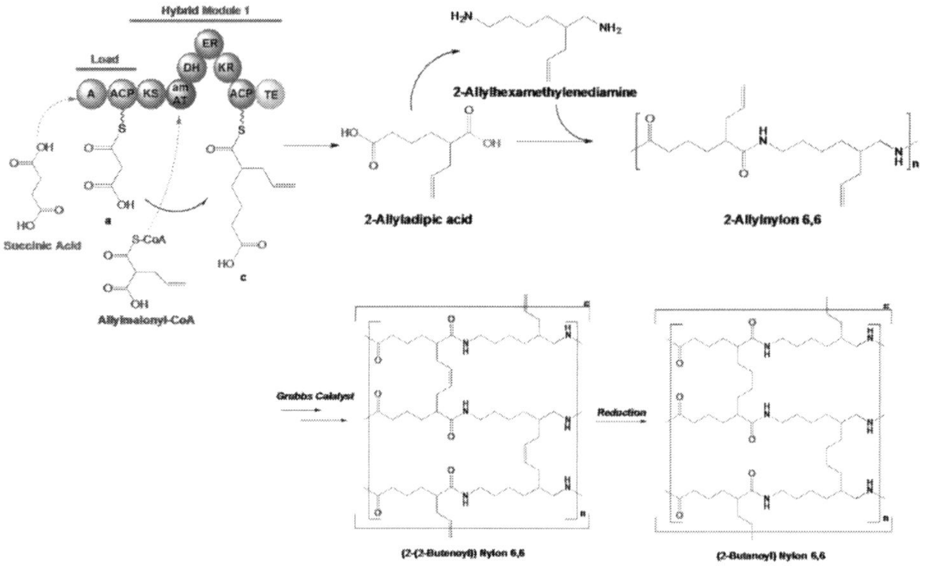

Lactams

- Butyrolactam (2-pyrrolidone)
  - Top value added chemical
  - Poly vinyl pyrrolidone
- Valerolactam
  - Pharmaceutical and pesticide intermediate
  - Nylon 5, nylon 6,5
- Caprolactam
  - Nylon 6
  - 7 million tons used annually

Several polyketides contain amine groups

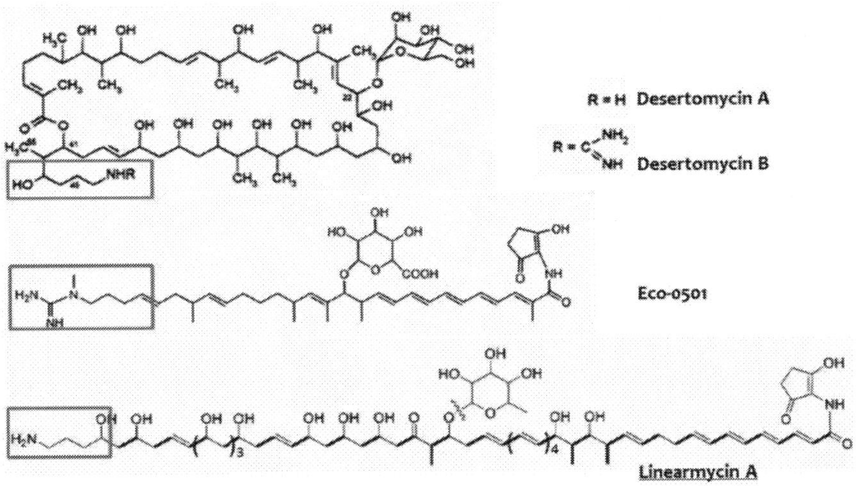

Retro-biosynthetic analysis of PKS products

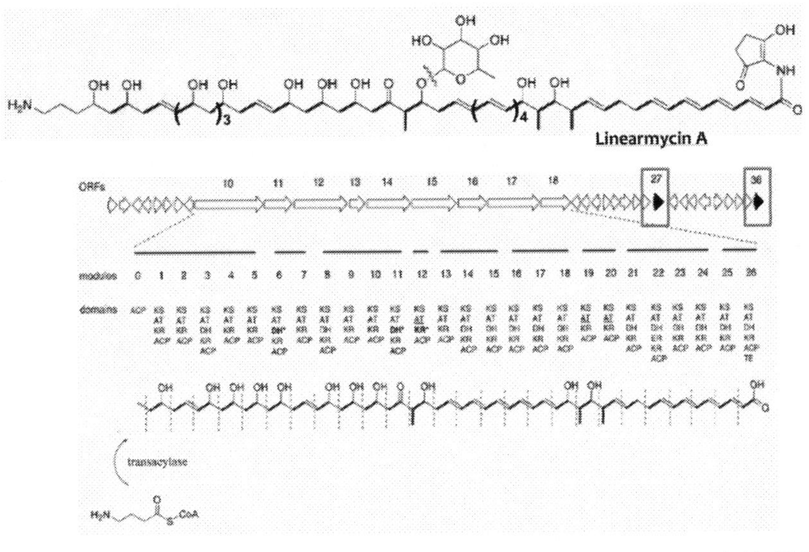

McAlpine et al. 2005 *J. Nat. Prod.* **68:**493-496

ORF27 Enzymology

- Enzymatic reaction

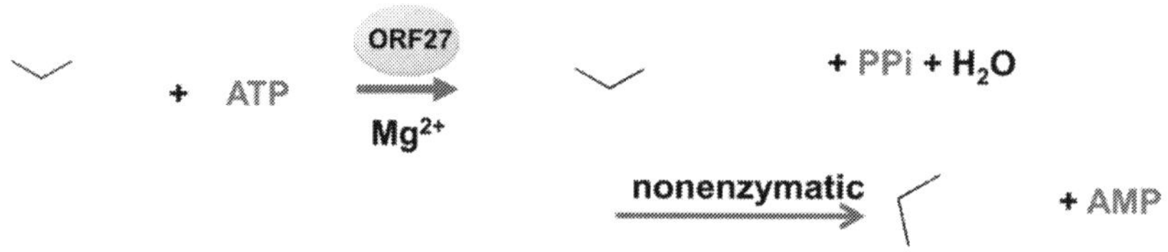

- Proposed total reaction:

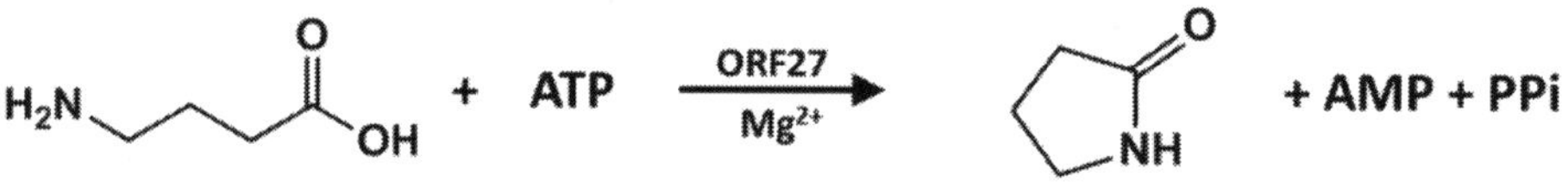

ORF27 is a general lactam synthase

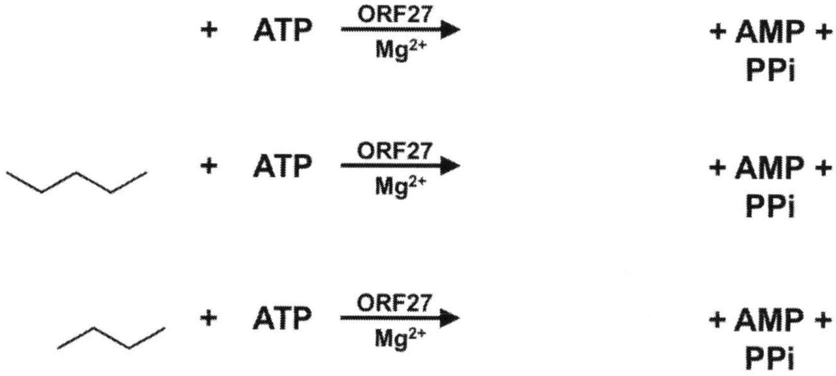

In vivo production of 2-pyrrolidone

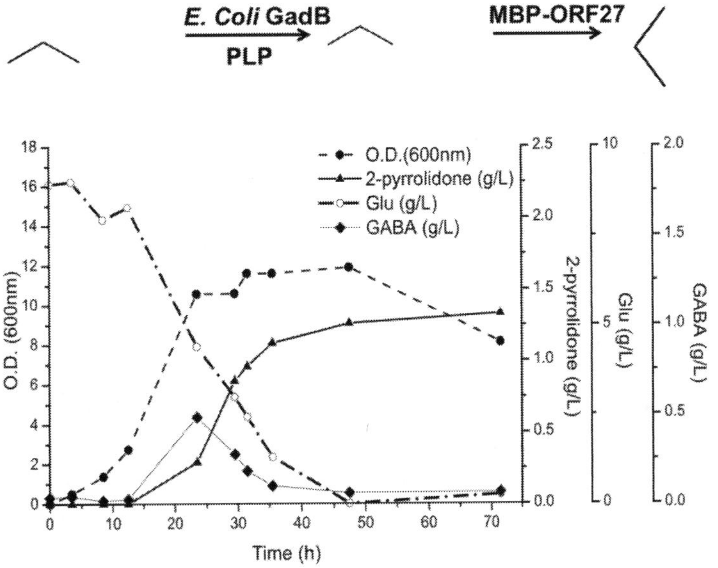

Hosts for production of polyketides

	Escherichia coli	*Streptomyces venezuelae*	*Saccharomyces cerevisiae*
Genetically tractable	✔	✔	✔
Well characerized biological tools	✔	?	✔
Well understood metabolism	✔	X	✔
Complex secondary metabolism			
PKS	X	✔	X
NRPS	X	✔	X
Isoprenoid	X	✔	✔
Shikimate	✔	✔	✔
Consolidated bioprocessing	X	✔	X
GC content	**50%**	**72%**	**38%**

Streptomyces venezuelae: orthogonal integration vector assembly

Phage-mediated Chromosomal Integration

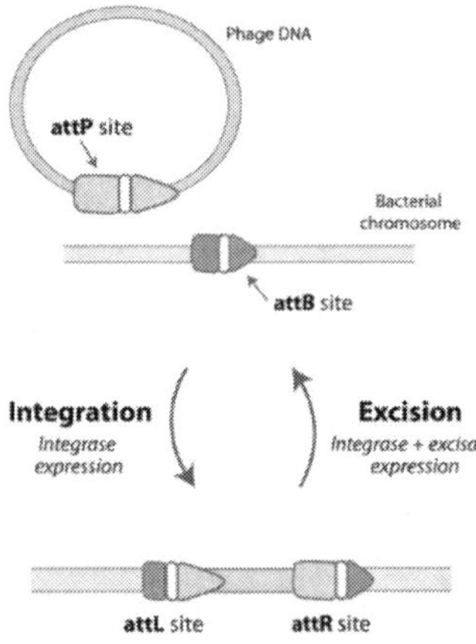

Library of Integration Vectors for *S. venezuelae* (42 total)

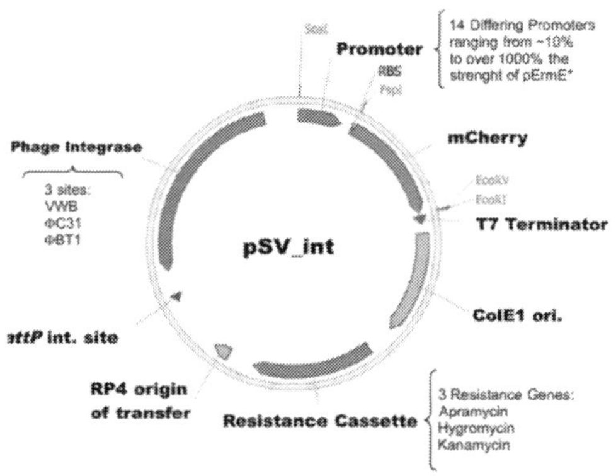

Expression from VWB and φC31 is equivalent

VWB Integration Site

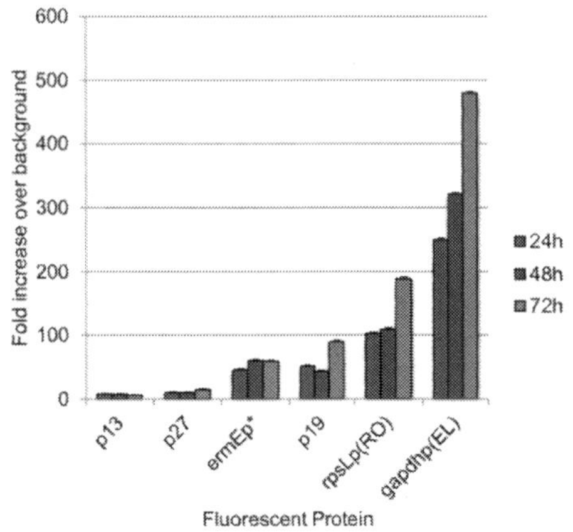

ΦC31 Integration Site

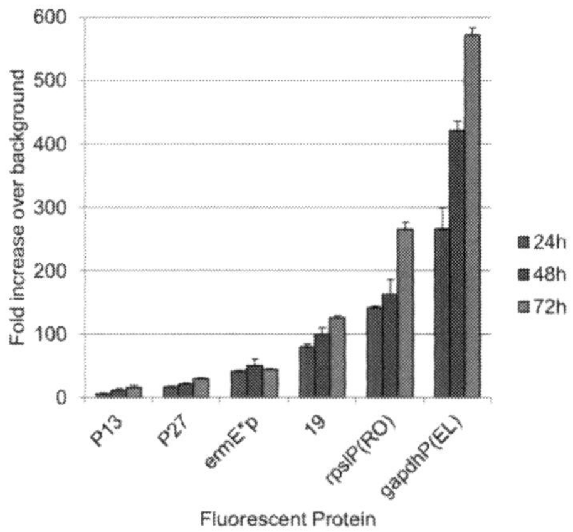

Integration into and expression from φBT1 is inferior to that from φC31 and VWB

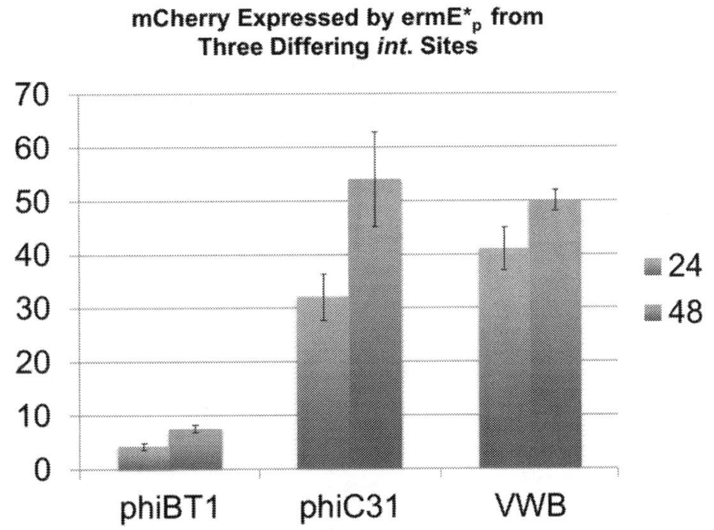

Integrate model PKS genes into *Streptomyces* genomes

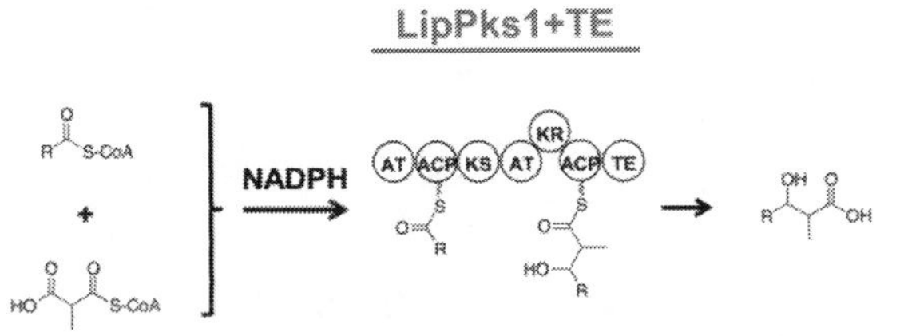

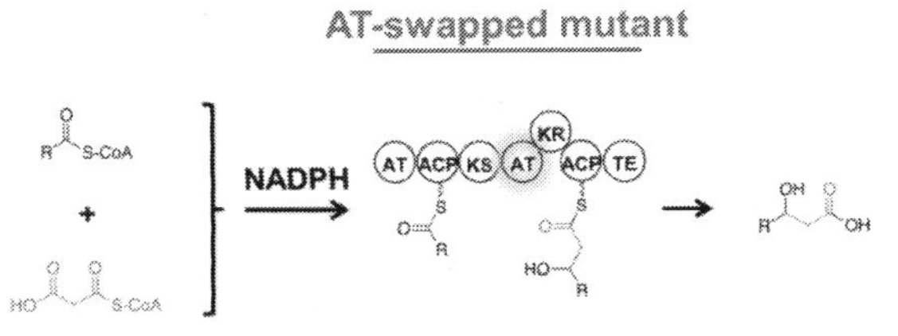

- Target *Streptomyces*:
  - ***S. venezuelae***
  - ***S. albus***
- Genome locations:
  - **Phi-BT1**
  - **Phi-C31**
  - **VWB**
- Promoters:
  - **gapdhp**
  - **kasOp***
  - **kasOp* variant-1**
  - **kasOp* varinat-2**
- Codon usage:
  - ***S. venezuelae***
  - ***E. coli***

Production of 3-hydroxy-2,4-dimethyl pentanoic acid in *S. venezuelae*

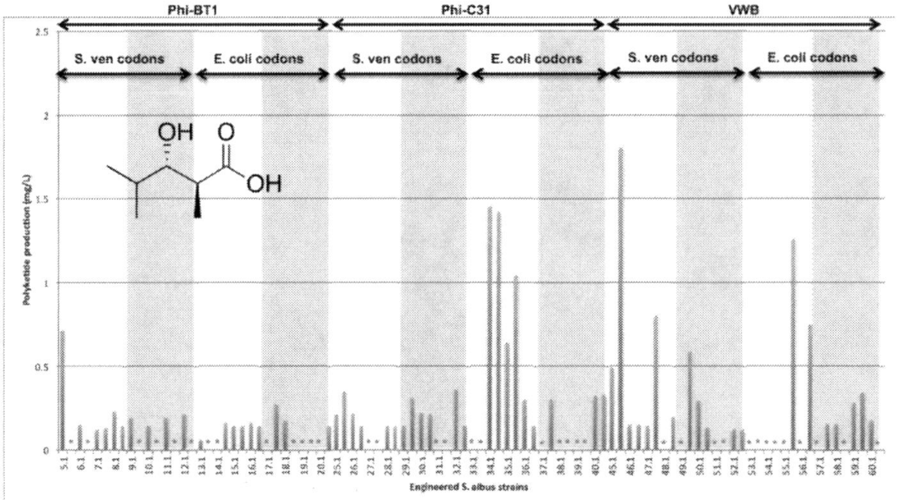

* = we have not obtained the data yet.
Red box = AT-swapped mutant

Production of 3-hydroxy-2,4-dimethyl pentanoic acid in S. *albus*

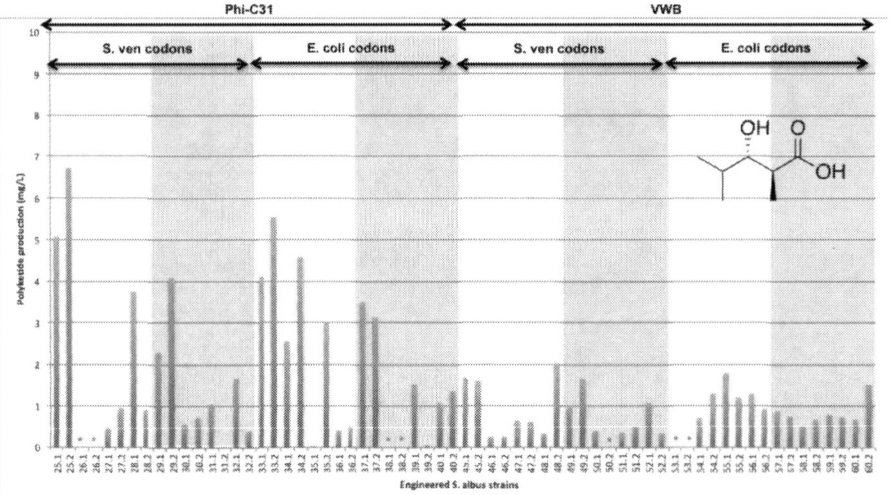

* = we have not obtained the data yet.
Red box = AT-swapped mutant

Conclusions

- We have constructed PKSs that produce several commodity and specialty chemicals
- We have determined some of the rules (and structures) for PKS recombination
- We have begun to develop several new hosts for PKS and PK production
- These hosts produce modest amounts of the desired final products without contamination of unwanted side products

Acknowledgements

Leonard Katz
Satoshi Yuzawa
Clara Eng
Ryan Phelan
Sean Poust
Andrew Hagen
Anna Lechner
Chris Petzold
Jesus Barajas
Tristan de Rond

jbei Joint BioEnergy Institute

crb The Novo Nordisk Foundation Center for Biosustainability

synberc

Funding: **Department of Energy**
National Science Foundation
Novo Nordisk Foundation

[#]Jay Keasling has a financial interest in Amyris and Lygos.

ENGINEERING BIOLOGY RESEARCH CONSORTIUM

Douglas Friedman, Executive Director

ebrc

www.ebrc.org

DNA Repair Profiling Reveals Nonrandom Outcomes at Cas9-Mediated Breaks

Haurwitz, R. E.,
van Overbeek, M.,
Capurso, D.,
Carter, M. M.,
Thompson, M. S.,
Frias, E.,
Russ, C.,
Reece-Hoyes, J. S.,
Nye, C., Caribou Biosciences, Inc.
Gradia, S.,
Vidal, B.,
Zheng, J., Caribou Biosciences, Inc.
Hoffman, G. R.,
Fuller, C. K.,
May, A. P.,
Originally presented on: 7/19/2016 14:00:00 - 14:30:00

TheÂ repairÂ outcomesÂ at site-specificÂ DNAÂ double-strandÂ breaksÂ (DSBs) generated by the RNA-guidedÂ DNAÂ endonucleaseÂ Cas9Â determine how gene function is altered. Despite the widespread adoption of CRISPR-Cas9Â technology to induce DSBs for genome engineering, the resultingÂ repairÂ products have not been examined in depth and have been thought to result in randomÂ outcomes. In this study, theÂ DNAÂ repairÂ profiles of over 200 sites in the human genome are presented and demonstrate that the pattern ofÂ DNAÂ repairÂ followingÂ Cas9Â cutting at each site is, in fact,Â nonrandomÂ and consistent across experimental replicates, cell lines, and reagent delivery methods. Furthermore, genomic context does not dictateÂ repairÂ outcomes; outcomesÂ are determined by the targeted DNA sequence,Â elucidating a strategy to anticipate the functional outcome of gene editing experiments and to use â??error-proneâ?□Â DNA-repairÂ machinery to generate precise edits.

Reading, Writing and Editing a Genome: An Emerging Frontier in Synthetic Biology

Zhao, H., University of Illinois at Urbana-Champaign
Originally presented on: 7/19/2016 14:30:00 - 15:00:00

Advances in reading, writing and editing genomes have greatly expanded our ability to reprogram biological systems at the resolution of a single nucleotide and on the scale of a whole genome. Such capacity has drastically accelerated the cycle of design, build and test in synthetic biology to construct organisms for both basic and applied biological research and medicine. In this presentation, I will highlight our recent work in the development of new strategies and tools for genome editing and our attempt in automating the design, build and test cycle. Examples include but are not limited to: (a) automated design and synthesis of TALENs and their applications in treatment of sickle-cell disease and mitochondrial diseases (1-4); (b) CRISPR based multiplexed genome engineering in *Saccharomyces cerevisiae* and Streptomycetes (5-6); and (c) rapid cellular prototyping of *S. cerevisiae* via genome-scale engineering based on either RNAi or CRISPR technologies (7-9).

References:

1. N. Sun, J. Liang, Z. Abil, and H. Zhao. "Optimized TAL Effector Nucleases (TALENs) for Use in Treatment of Sickle Cell Disease." *Molecular Biosystems*, 8, 1255 – 1263 (2012).
2. J. Liang, R. Chao, Z. Bao, N. Sun, Z. Abil, and H. Zhao."FairyTALE: A High-throughput TAL Effector Synthesis Platform." *ACS Synthetic Biology*, 3, 67-73 (2014).
3. L. Cuculis, Z. Abil, H. Zhao, and C. Schroeder. "Direct Observation of TALE Protein Dynamics Reveals a Two-state Search Mechanism." *Nature Communications*, 6: 7277 (2015).
4. P. Reddy, A. Ocampo, K. Suzuki, J. Luo, S. R. Bacman, S. L. Williams, A. Sugawara, D. Okamura, Y. Tsunekawa, D. Lam, X. Xiong, N. Monsterrat, C. Rodriguez, I. Sancho-Martinez, D. Manau, S. Civico, M. del Mar O'callaghan, J. Campistol, H. Zhao, J. M. Campistol, C. T. Moraes, J. C. Izpisua-Belmonte. "Selective Elimination of Mitochondrial Mutations in the Germline by Genome Editing." *Cell*, 161, 459–469 (2015).
5. Z. Bao, H. Xiao, J. Liang, X. Xiong, N. Sun, T. Si, and H. Zhao. "Homology Integrated CRISPR-Cas (HI-CRISPR) System for One-step Multi-gene Disruptions in *Saccharomyces cerevisiae*." *ACS Synthetic Biology*, 4, 585–594 (2015).
6. R.E. Cobb, Y. Wang, and H. Zhao. "High-efficiency Multiplex Genome Editing of Streptomyces Species using an Engineered CRISPR/Cas System." *ACS Synthetic Biology*, 4, 723–728 (2015).
7. T. Si, H. Xiao, and H. Zhao. "Rapid Prototyping of Microbial Cell Factories via Genome-scale Engineering." *Biotechnology Advances*, 33, 1420–1432 (2015).
8. T. Si, Y. Luo, Z. Bao, and H. Zhao. "RNAi-Assisted Genome Evolution in *Saccharomyces cerevisiae* for Complex Phenotype Engineering." *ACS Synthetic*

Biology, 4, 283–291 (2015).
9. Z. Bao, R.E. Cobb, and H. Zhao. "Accelerated Genome Engineering through Multiplexing." *Wiley Interdisciplinary Reviews: Systems Biology and Medicine*, 8, 5–21 (2016).

Crispr Logic Circuits to Enable Next Generation Crispr-Based Gene Therapies

Kiani, S., M.I.T.
Chavez, A.,
Ebrahimkhani, M. R., M.I.T.
Beal, J., Raytheon BBN Technologies
Hall, R. N., M.I.T.
Church, G.,
Weiss, R., M.I.T.
Originally presented on: 7/19/2016 15:00:00 - 15:30:00

Cas9 is an RNA guided DNA endonuclease of the bacterial Clustered Regularly Interspaced Short Palindromic Repeat (CRISPR) system that has been adopted for programmable genome editing and transcriptional regulation. Since the complexity of genetic interrogations with the CRISPR system is rapidly increasing, there is a need for novel genetic tools to gain control over Cas9 activity or expression or regulate guide RNA (gRNA) function. To this end, we apply the design principles of synthetic biology to develop genetic logic gates, layered genetic circuits and kill switches in mammalian cells. RNA Pol II driven gRNAs enable cell and context specific expression of gRNA. This strategy combined with synthetic promoter engineering allow us to develop genetic circuits to control the timing and sequence of function of catalytically active or inactive Cas9 protein or the gRNA. In parallel, we have devised a strategy to readily switch between catalytically active and null Cas9 protein, allowing knocking out and activating target genes in the same cells using distinct gRNAs. By truncating the gRNA from 5' end, we have shown that gRNAs with 14nt guide sequences ablate the nuclease activity of Cas9, while leaving its DNA targeting capacity intact. By fusing a Cas9 nuclease protein to a potent activation domain, VPR or application of gRNAs with MS2 binding loops, we can use a same Cas9 protein for transcriptional activation (14nt gRNAs) or gene disruption (20nt gRNAs). By incorporating this capacity within our layered circuits, we have developed and tested several synthetic gene circuits that can incorporate and modulate the desired functionality of this Cas9 complex, including DNA cleavage or transcriptional modulation or both. Considering the growing clinical application of DNA-based gene therapies and vaccines, the developed CRISPR logic circuits pave the way towards developing safer and more specific gene therapies and allow us to perform more sophisticated biological discoveries.

jbei
Joint BioEnergy Institute

Transcript switches for fine-tuning of transgene expression

Yan Liang 2016 SEED

U.S. DEPARTMENT OF ENERGY Office of Science

jbei
Joint BioEnergy Institute

Why studying plants ?

Food

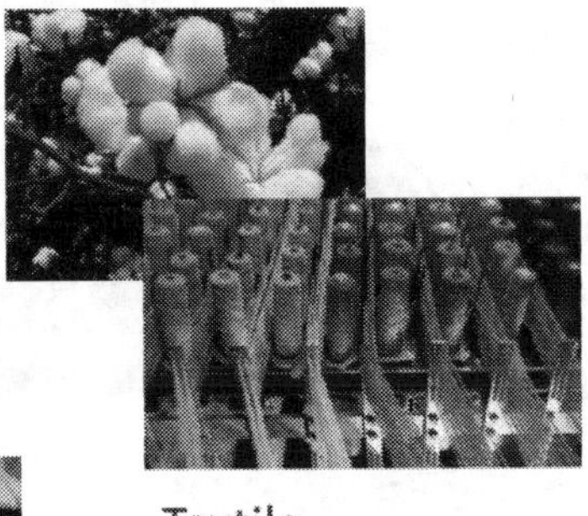

Textile

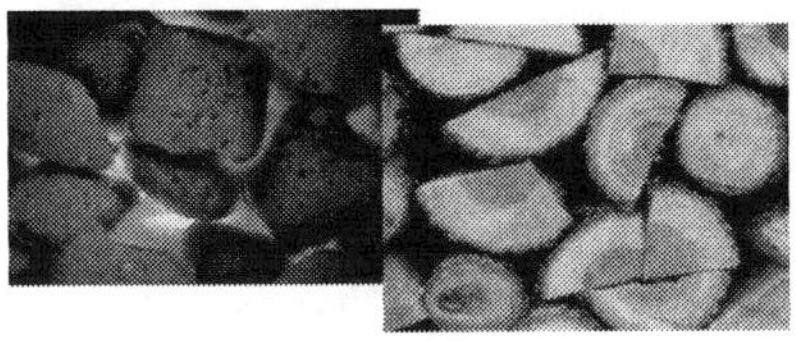

Wood

8/18/2016 2

U.S. DEPARTMENT OF ENERGY Office of Science

jbei
Joint BioEnergy Institute

Plants for biofuel production

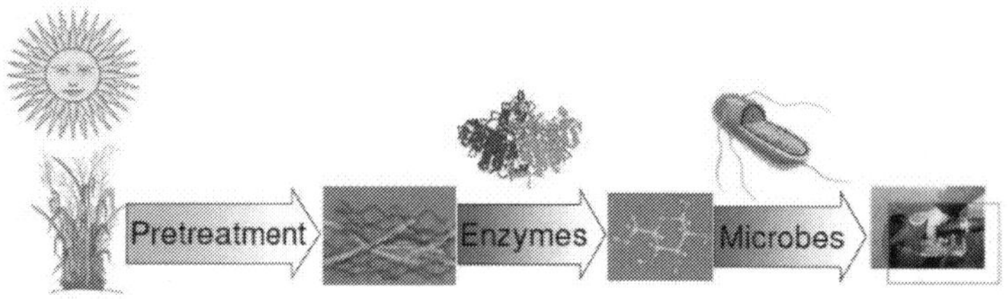

Lignocellulose

jbei
Joint BioEnergy Institute

Transgene expression needs to be fine tuned

Growth defect **Undesirable Traits**

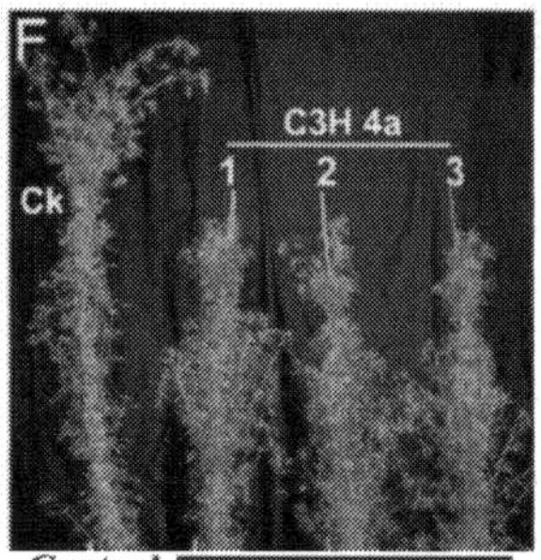

Control

Ralph et al. 2006 JBC

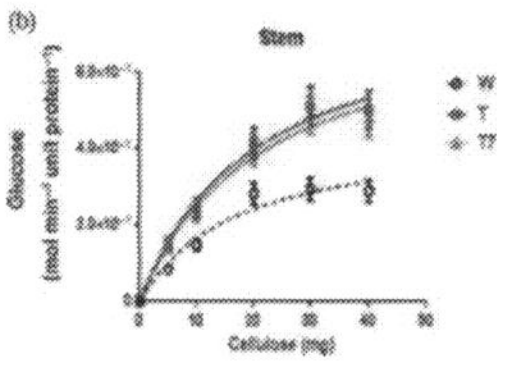

Harris et al. PNAS 2012; Sahoo et al. Plant Biotech J. 2012

jbei Transcript switches for transgene regulation

Joint BioEnergy Institute

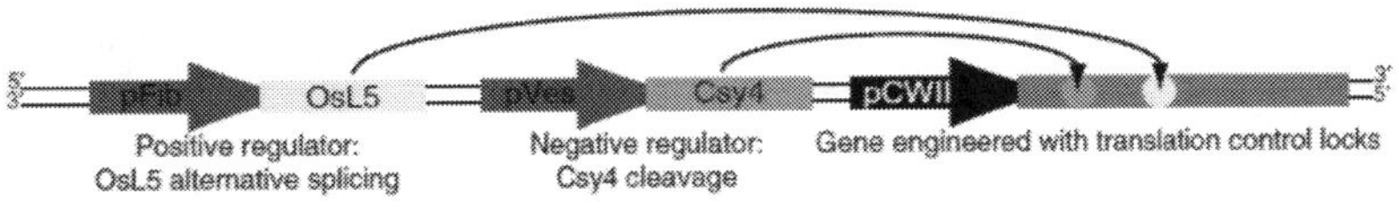

Eude et al. 2014 Curr Opin Biotech

- Two-component regulation system
- Functions at the post-transcriptional level
- Transcript **activation**: OsL5 and HyP5SM
- Transcript **repression**: CSY4 and its cognition sequence

jbei Transgene activation by alternative splicing of a suicide exon

Joint BioEnergy Institute

HyP5SM: Plant 5s rRNA mimic, a structured *cis*-regulatory RNA element in TFIIIA
OsL5: ribosomal protein L5 from rice, a splicing factor

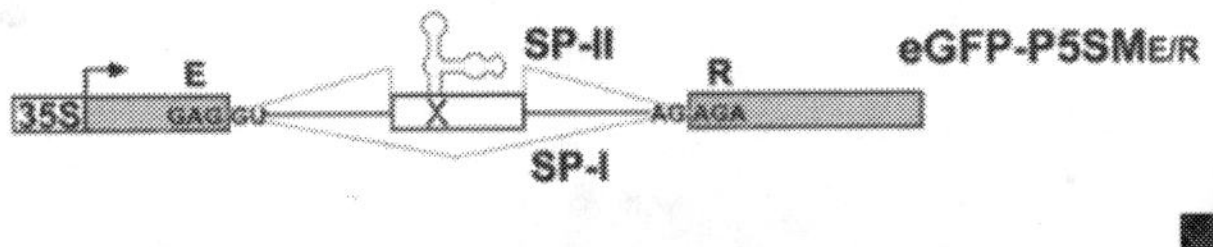

eGFP-HyP5SM

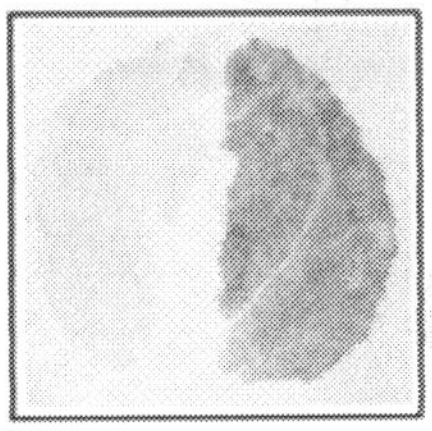

-OsL5 +OsL5

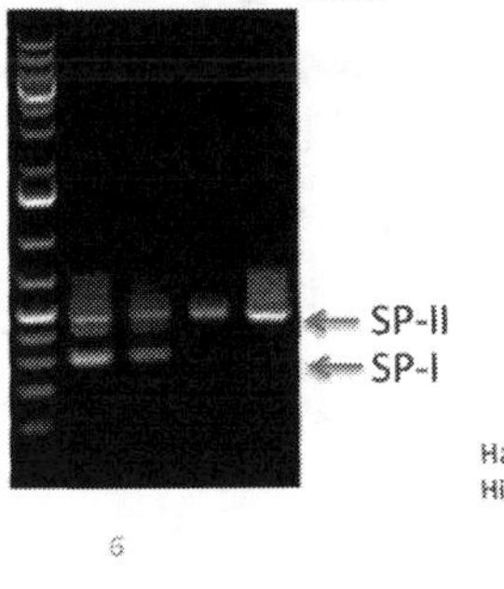

Hammond et al. 2009 Nat Struct Mol Biol
Hickey et al. 2012 Nucleic Acids Res

jbei Tight control by the two-input regulatory system

Joint BioEnergy Institute

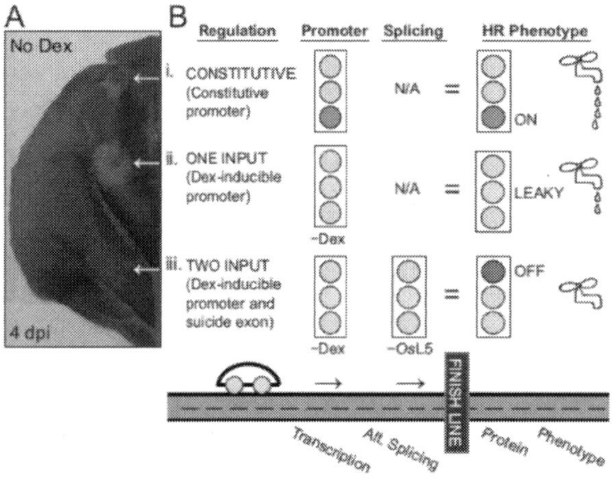

Gonzalez et al. 2015 Nucleic Acids Res

U.S. DEPARTMENT OF ENERGY Office of Science

jbei Transgenic plants containing both the resistance gene and the dual regulated effector gene

Joint BioEnergy Institute

Gonzalez et al. 2015 Nucleic Acids Res

U.S. DEPARTMENT OF ENERGY Office of Science

jbei Joint BioEnergy Institute — Current systems for transgene repression

- Lac or Tet repressor-operator systems
  Needs to modify the transgene promoter with operators
  Up to 10 fold repression
- Cre-lox system
  Nonreversible
- RNAi
  Specific design for each transgene
  Lack of tissue specificity
- dCAS9
  Specific design for each transgene

8/18/2016 9

U.S. DEPARTMENT OF ENERGY Office of Science

jbei Joint BioEnergy Institute — CSY4 recognizes and cuts specific RNA sequence

- CSY4 is identified from CRISPR/CAS system in *Pseudomonas aeruginosa*
- CSY4 is an endoribonuclease for sequence- and structure-specific RNA processing
- CSY4 is composed of 200aa. The cognition sequence of CSY4 is 28bp

A

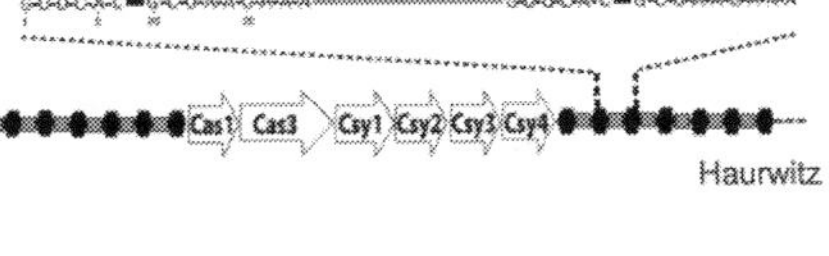

Haurwitz et al. 2010 Science

U.S. DEPARTMENT OF ENERGY Office of Science

jbei Working model

Joint BioEnergy Institute

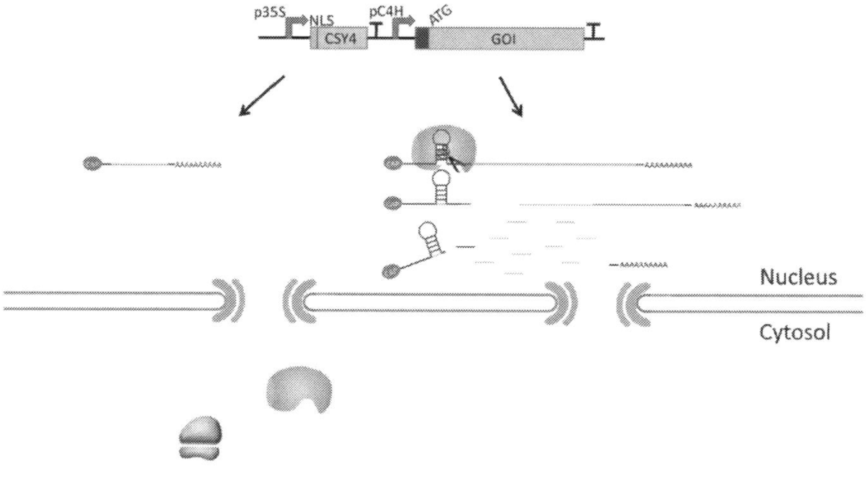

8/18/2016 11

U.S. DEPARTMENT OF ENERGY Office of Science

jbei CSY4 represses cogGFP or cogRFP expression in tobacco leaves

Joint BioEnergy Institute

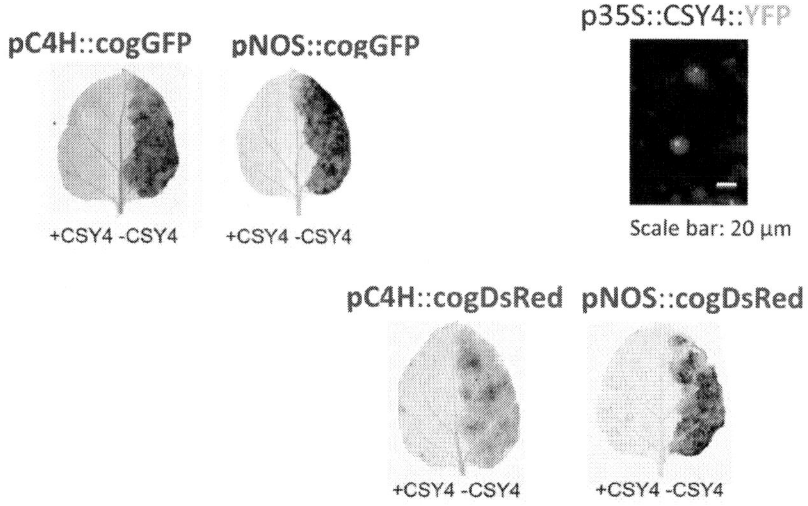

U.S. DEPARTMENT OF ENERGY Office of Science

jbei Transcript analysis of cogGFP and cogDsRed

Joint BioEnergy Institute

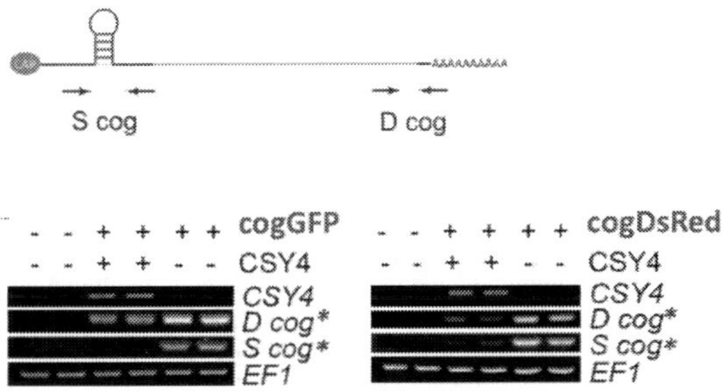

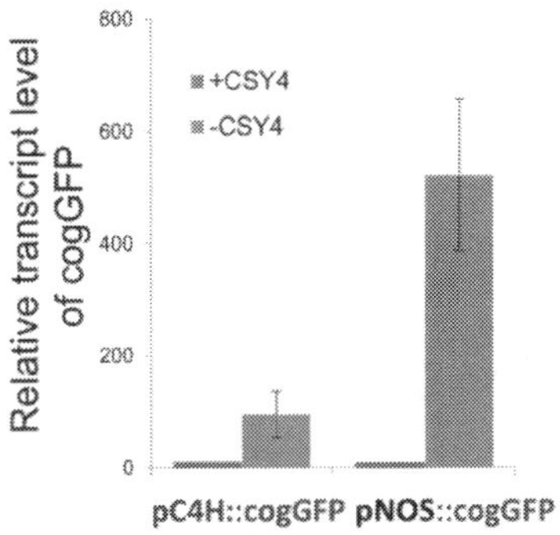

**S cog* and *D cog* designate the regions spanning (S) and downstream (D) of the CSY4 cognition site, respectively.

8/18/2016 13

U.S. DEPARTMENT OF ENERGY Office of Science

jbei Simultaneous repression of cogGFP and cogDsRed by CSY4

Joint BioEnergy Institute

pAct1::cogDsRed-p35S::cogGFP

GFP Scan

DsRed Scan

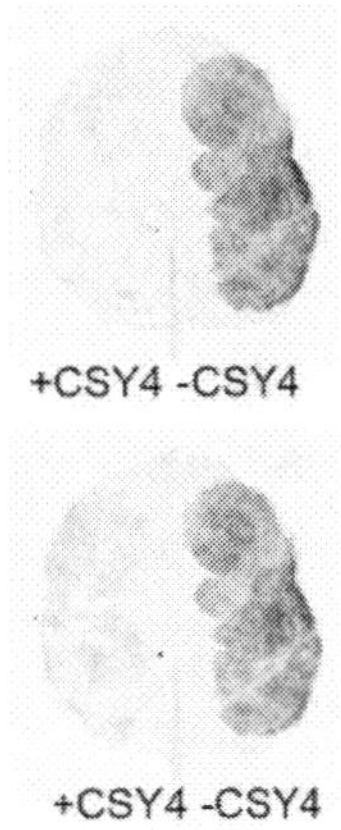

U.S. DEPARTMENT OF ENERGY Office of Science

jbei Joint BioEnergy Institute

Inducible repression of cogFLuc by CSY4

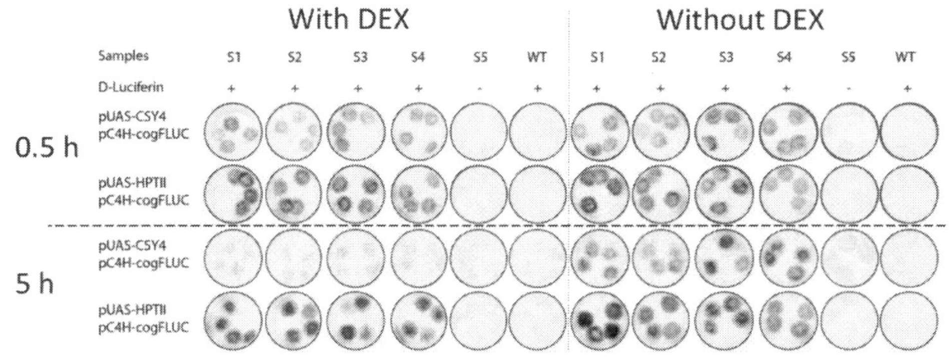

8/18/2016 15

U.S. DEPARTMENT OF ENERGY Office of Science

jbei Joint BioEnergy Institute

Cell type specific repression of cogGFP by CSY4

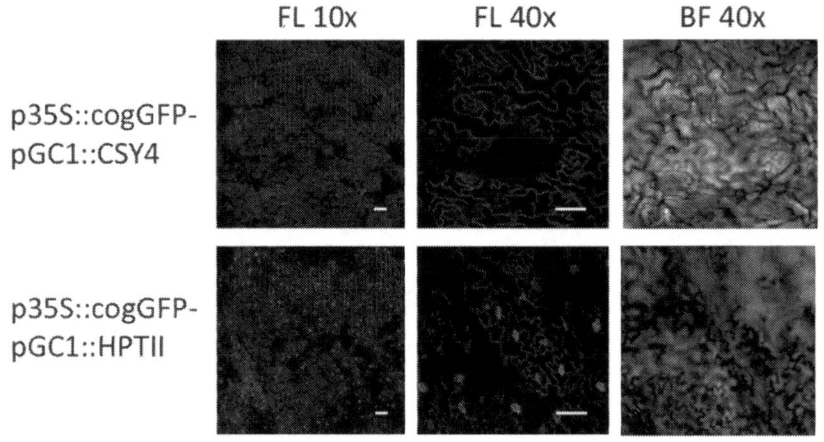

Scale bar: 50 μm

8/18/2016 16

U.S. DEPARTMENT OF ENERGY Office of Science

jbei Joint BioEnergy Institute

CSY4 represses cogGFP expression in rice protoplast

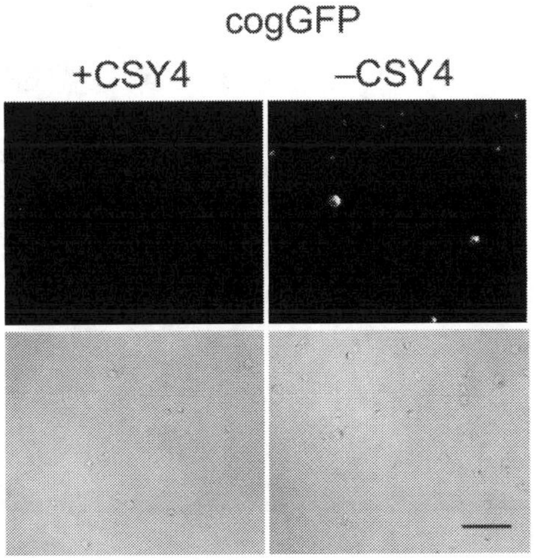

Scale bar: 50 μm

8/18/2016 17

U.S. DEPARTMENT OF ENERGY Office of Science

jbei Joint BioEnergy Institute

p35S:CSY4 transformed Arabidopsis did not show defects in growth and development

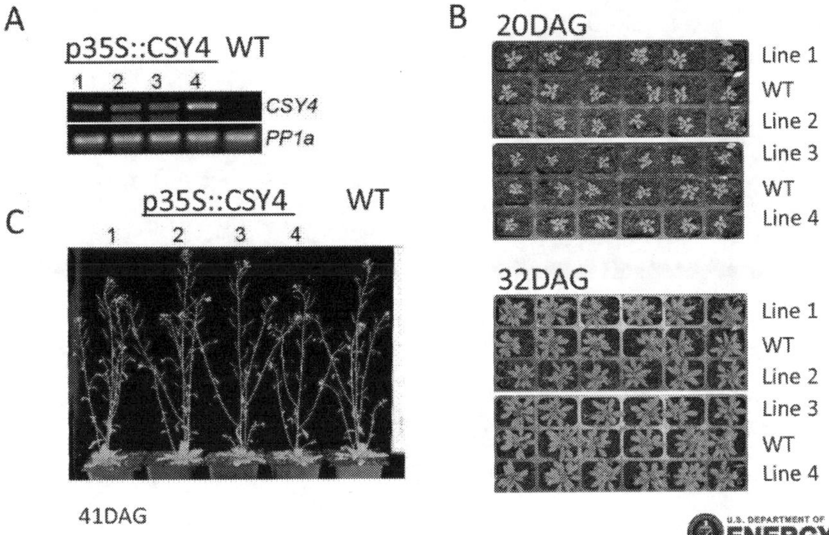

U.S. DEPARTMENT OF ENERGY Office of Science

jbei Joint BioEnergy Institute

Identification of CSY4 orthologs

U.S. DEPARTMENT OF ENERGY Office of Science

jbei Joint BioEnergy Institute

CSY4 orthologs showing repression activity in tobacco leaves

pC4H::cogGFP

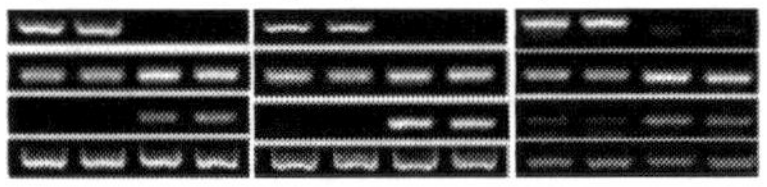

U.S. DEPARTMENT OF ENERGY Office of Science

jbei Joint BioEnergy Institute

CSY4 orthologs showing no cleavage activity in tobacco leaves

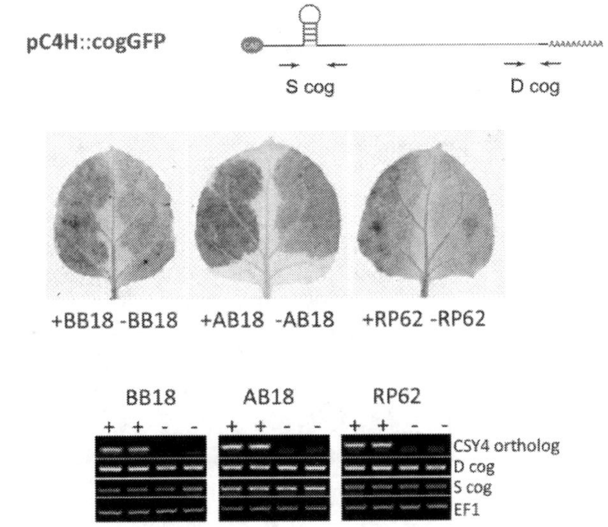

jbei Joint BioEnergy Institute

Prospective application of transgene switches

- Increased spatial and temporal control
- Inducible and reversible regulation
- Versatile tool for synthetic biology

jbei Joint BioEnergy Institute

Acknowledgement

Ming Hammond
Tania L. Gonzalez

Jay Keasling
Sarah Richardson

NIH DIRECTOR'S NEW INNOVATOR AWARD

BURROUGHS WELLCOME FUND

Regents' *and* Chancellor's Scholarship Program Berkeley

CWE group at JBEI

U.S. DEPARTMENT OF ENERGY Office of Science **Early Career Research Program**

8/18/2016 23

U.S. DEPARTMENT OF ENERGY Office of Science

Mechanisms and applications of membrane-breaking protein needles

Jessica Polka, @jessicapolka
Postdoc, lab of Pam Silver
Department of Systems Biology, Harvard Medical School

1

Can we control biological motion with literal protein machines?

2

R bodies are simple protein actuators

High pH (>5.7)

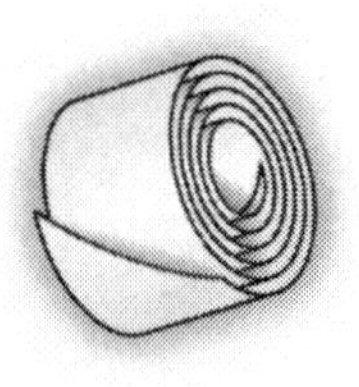

3

R bodies are simple protein actuators

Low pH (<5.7)

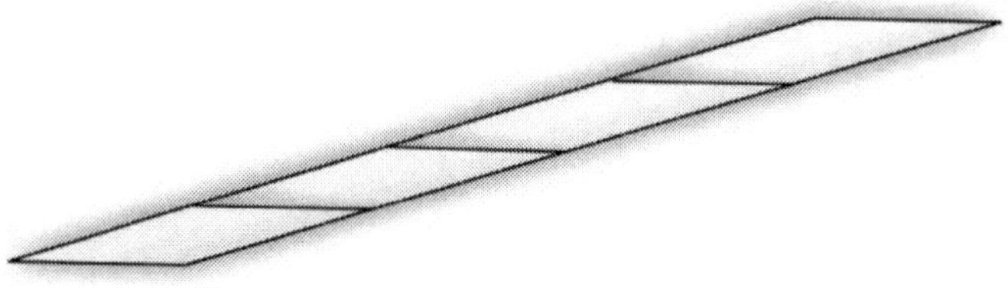

4

R bodies are natural toxin delivery devices

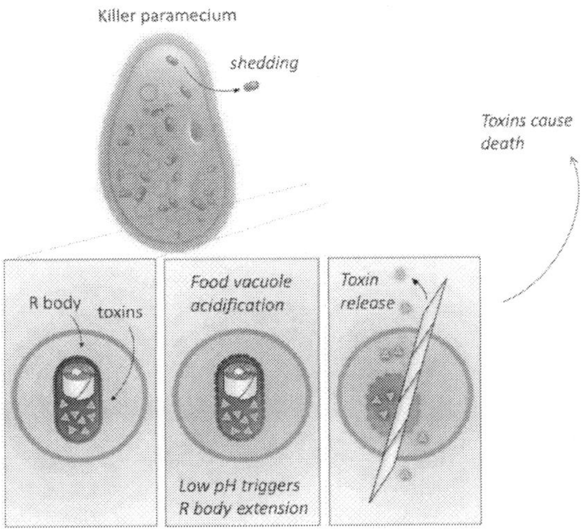

Sonneborn et al 1938
Dippell 1958
Mueller 1965

Purified R bodies can undergo a pH transition

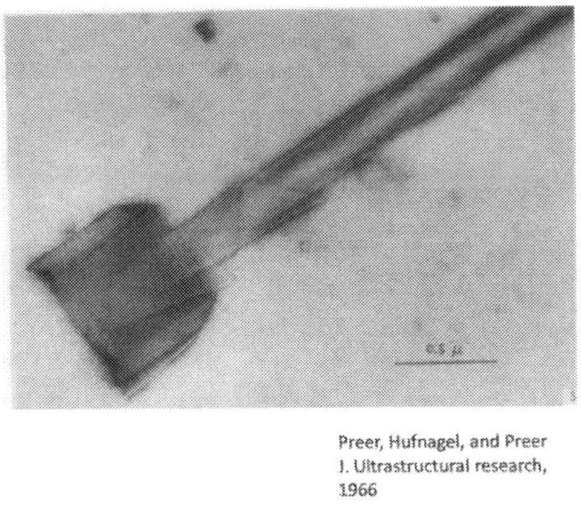

Preer, Hufnagel, and Preer J. Ultrastructural research, 1966

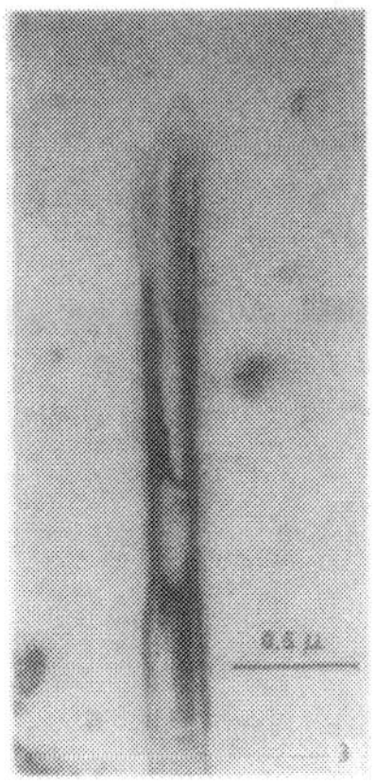

A giant structure from 2 tiny proteins

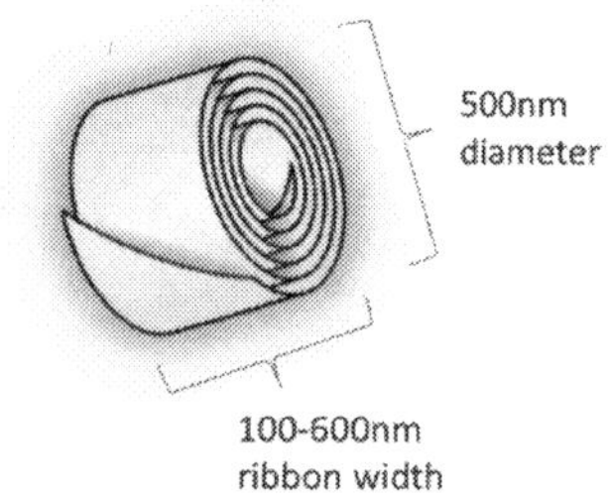

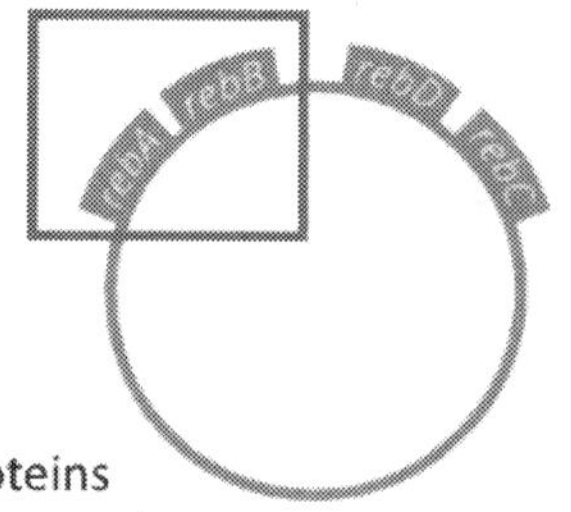

4 proteins
~100aa each
Tens to hundreds of thousand of copies

Kanabrocki et al 1986
Heruth et al 1994

7

Outline

- How R bodies are built
- How R bodies extend
- Practical uses for R bodies

8

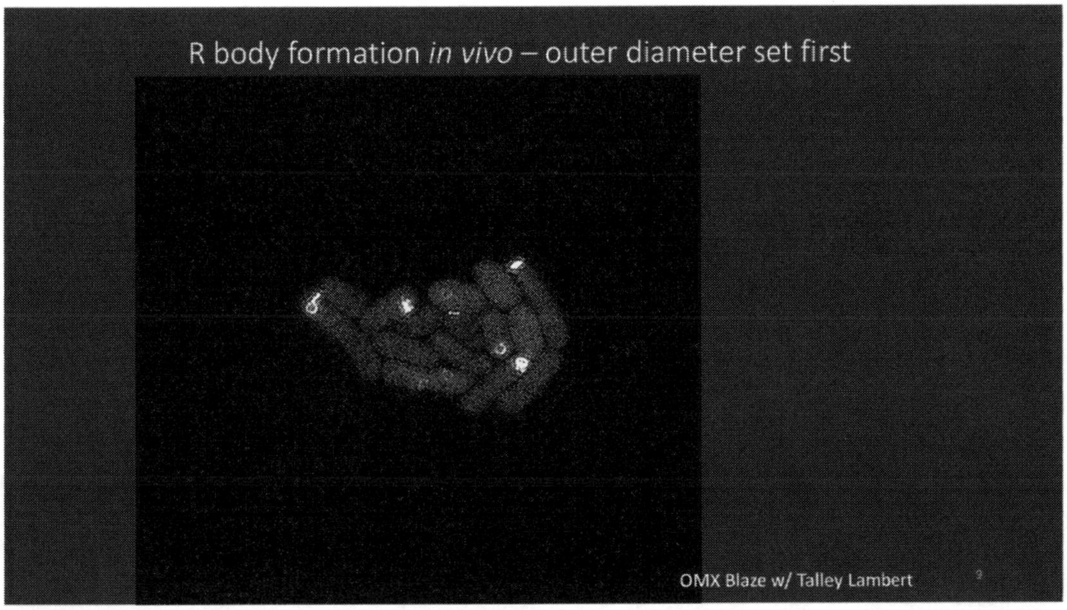

Is temporal control of patterning possible?

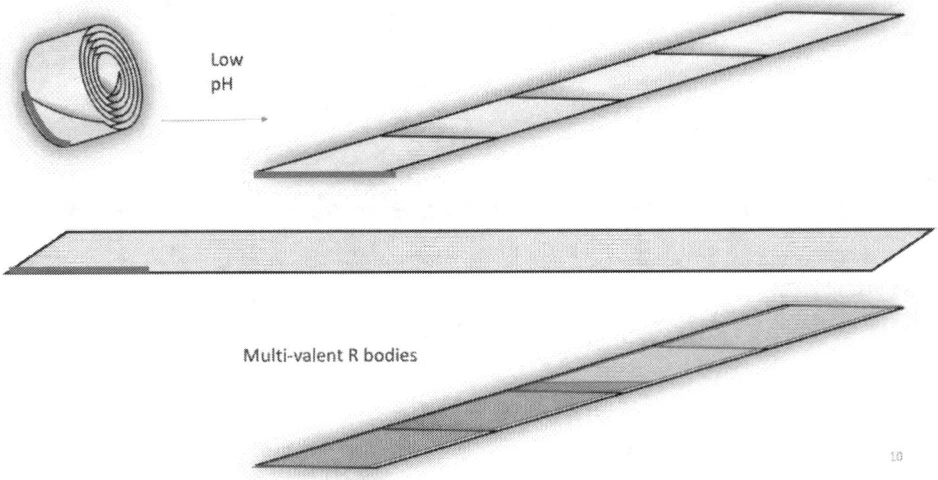

10

R body proteins assemble covalently

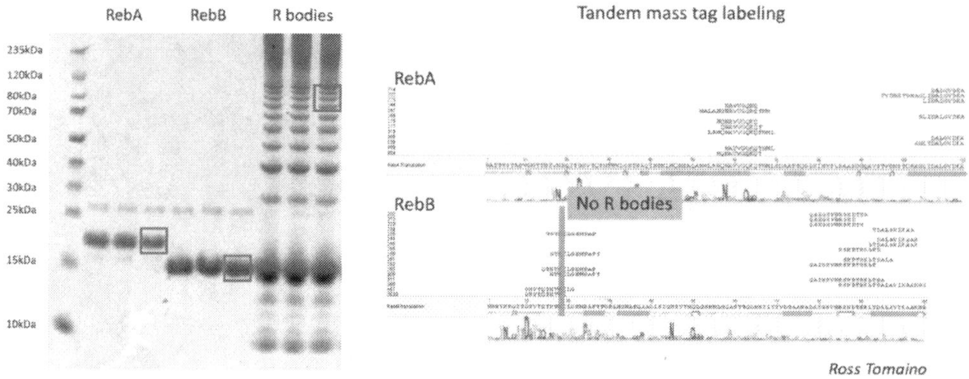

Ross Tomaino
Taplin MS facility

11

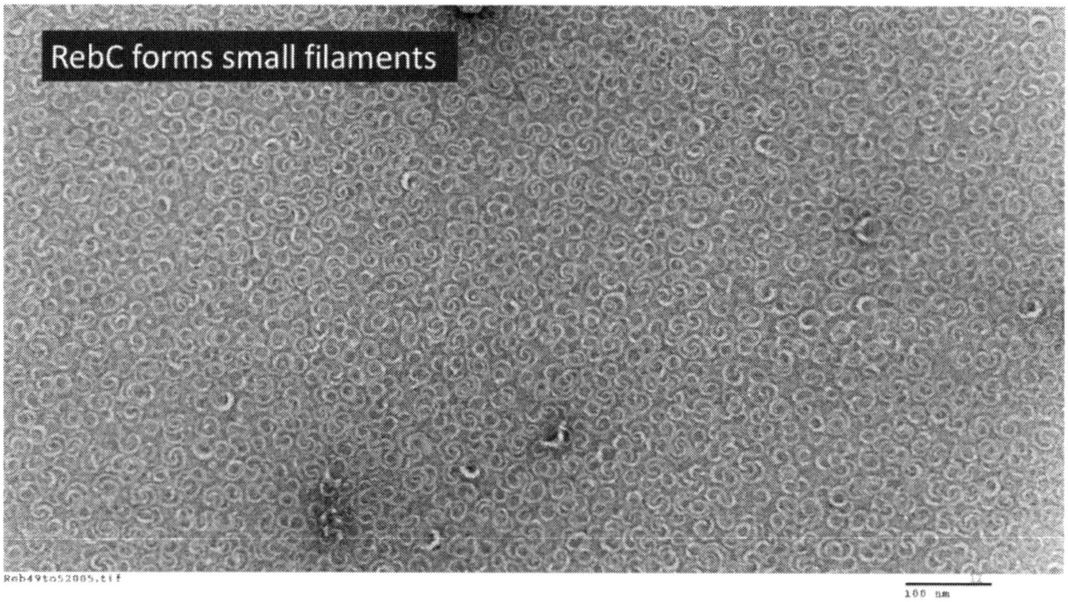

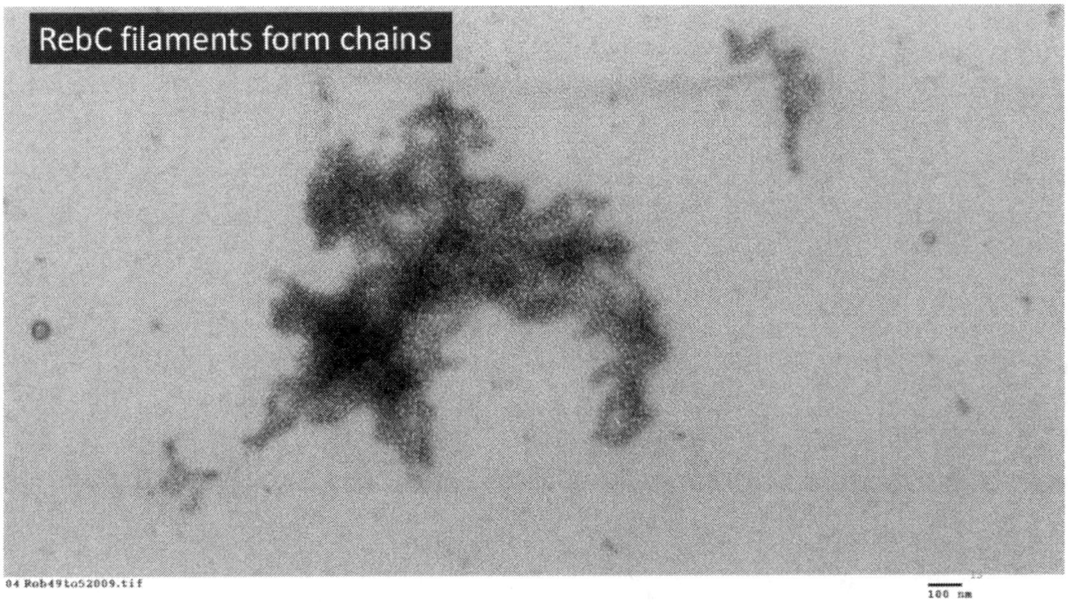

Outline

- How R bodies are built
- How R bodies extend
- Practical uses for R bodies

A Tunable Protein Piston That Breaks Membranes to Release Encapsulated Cargo
Jessica K. Polka[†‡] and Pamela A. Silver[*†‡]
ACS Synth. Biol., Article ASAP
DOI: 10.1021/acssynbio.5b00237
Publication Date (Web): January 26, 2016

14

R bodies are fully reversible

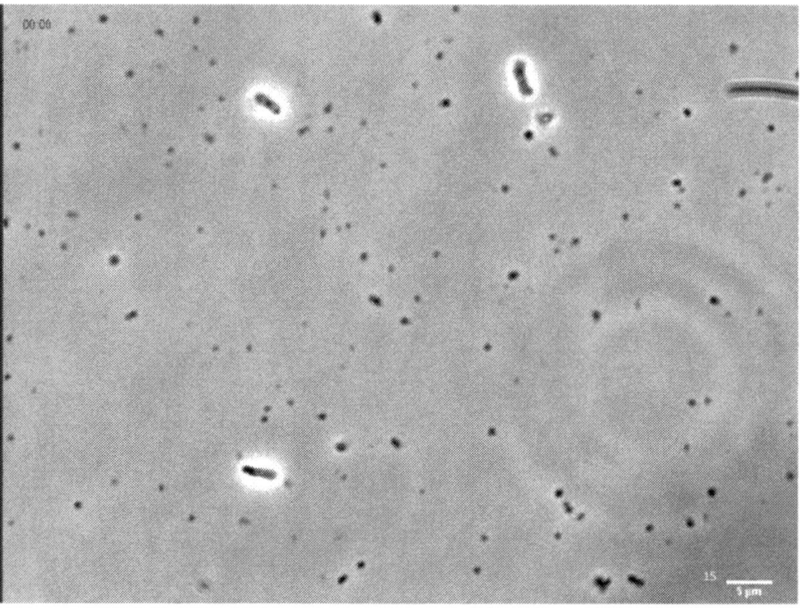

15

What's the molecular mechanism?

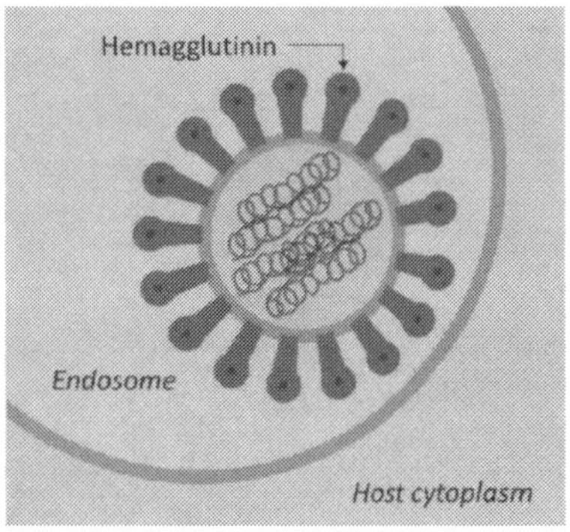

Endosomal membrane

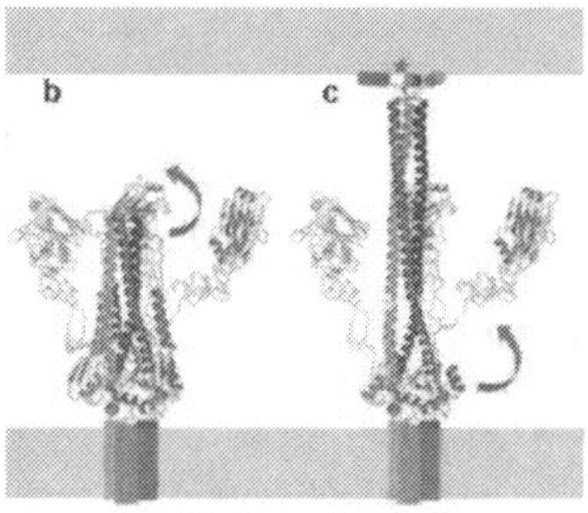

Viral membrane

Viral membrane fusion
Stephen C Harrison (2008)

16

Circular dichroism: More alpha helical at low pH

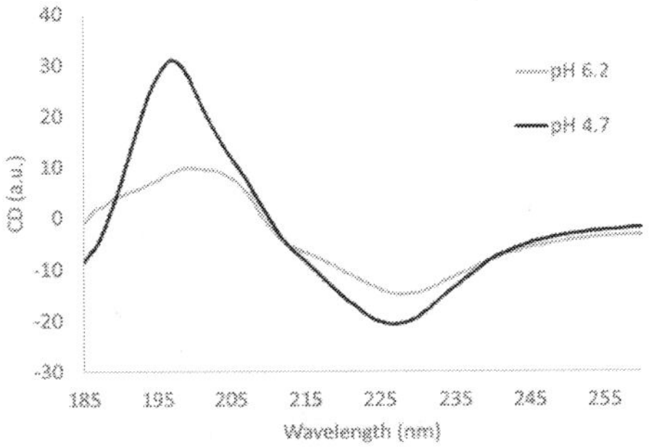

	Helix1	Helix2	Strand1	Strand2	Turns	Unordered	Total
pH 6.2 (NRMSD: 0.001)	0.43	0.12	0.11	0.11	0.06	0.18	1.01
pH 4.7 (NRMSD: 0.000)	0.43	0.14	0.10	0.10	0.06	0.18	1.01

17

Is there evidence for a helical conversion in R bodies?

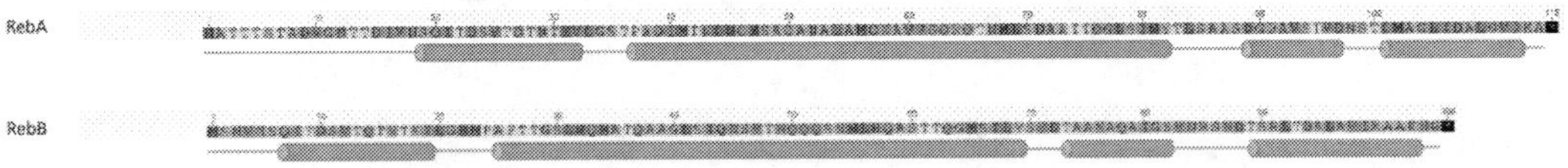

Need a screen – and a high-throughput assay

18

pH differences are manifested macroscopically

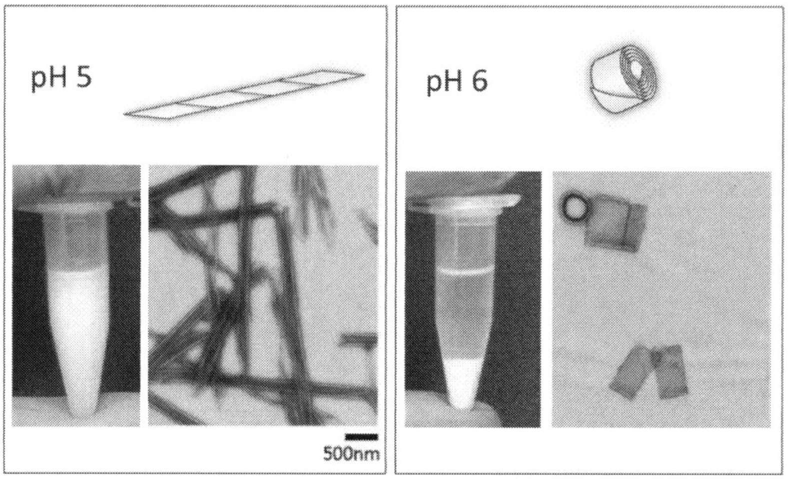

19

Visually identifying hits

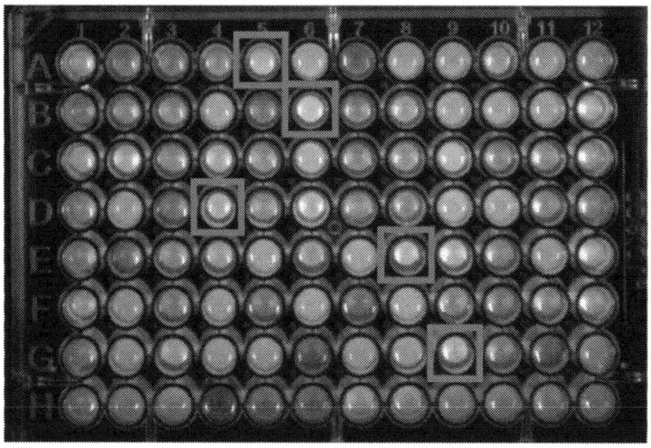

20

Validation of hits (subset)

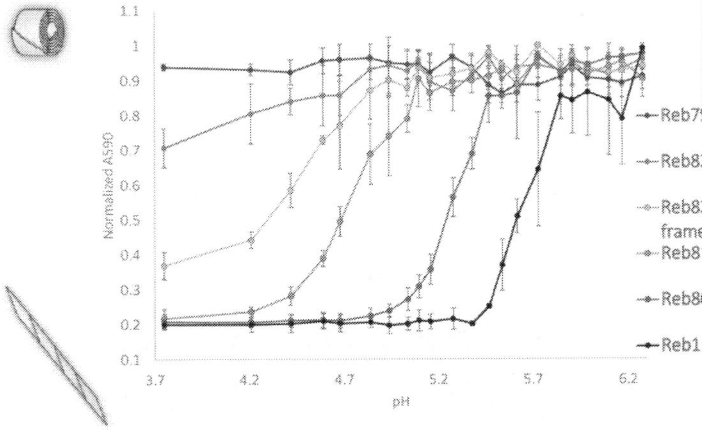

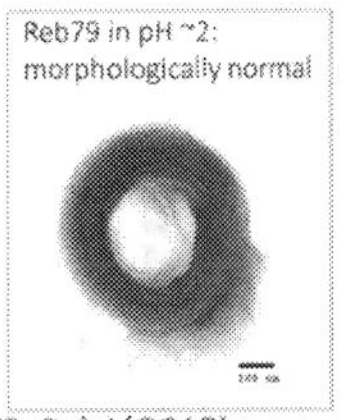

21

Most severe mutants (pH conversion <3)

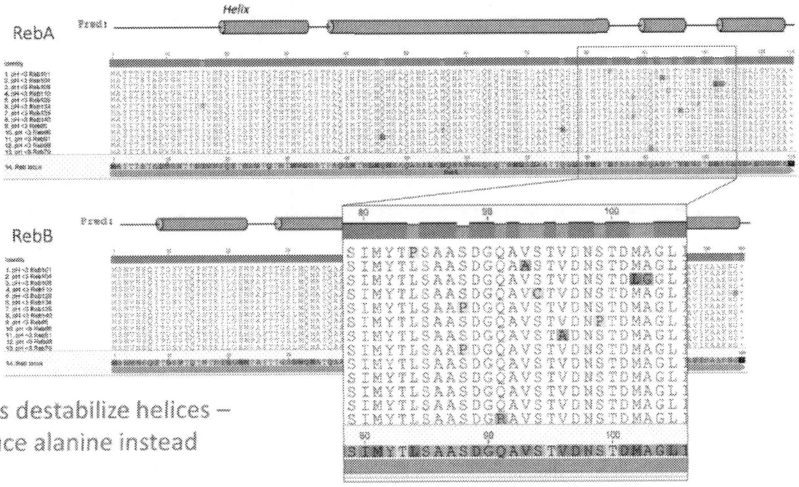

Prolines destabilize helices – introduce alanine instead

22

High pH mutant validation

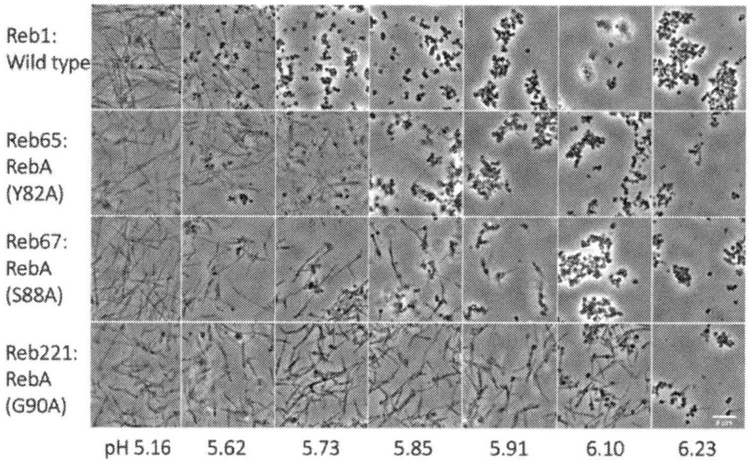

23

Hypothesis

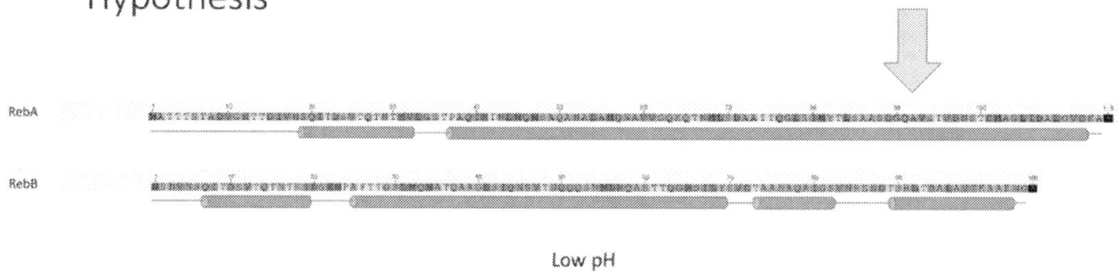

24

Outline

- How R bodies are built
- How R bodies extend
- **Practical uses for R bodies**

- Stable structures
- Functionalize with cargo
- Tune the pH sensitivity up and down

25

Goal: enhance endosomal escape with R bodies

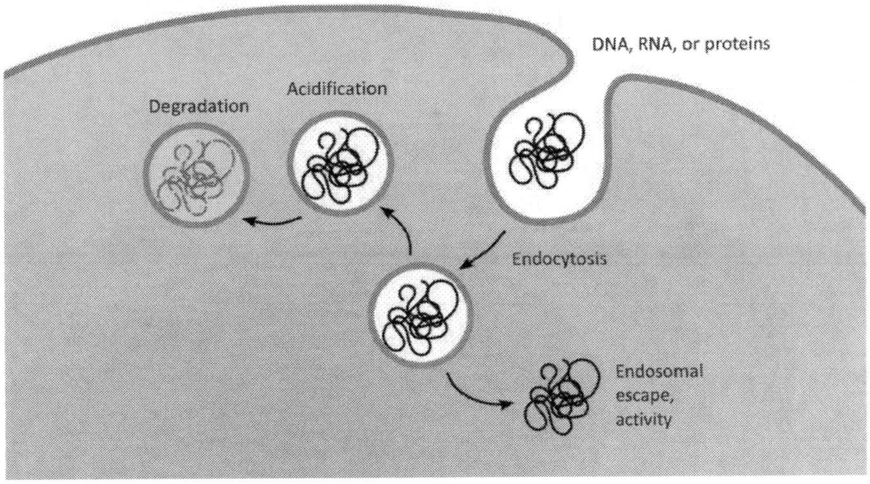

26

Goal: enhance endosomal escape with R bodies

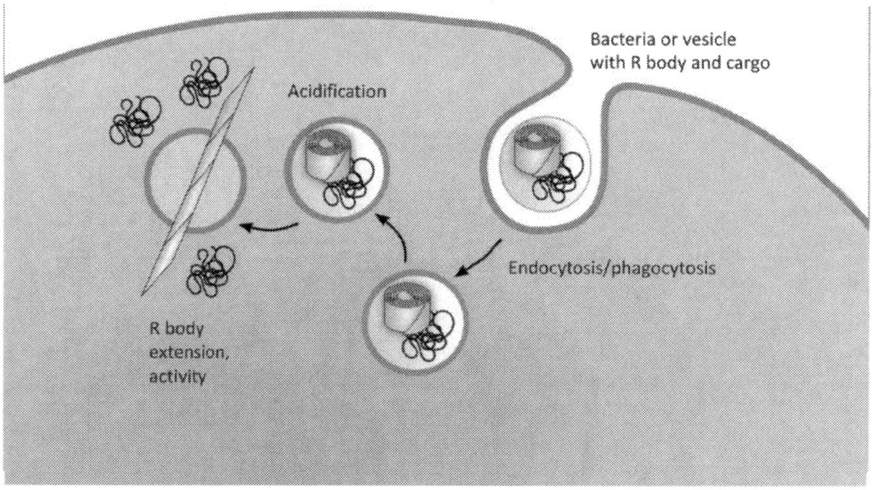

27

Can R bodies rupture foreign membranes?

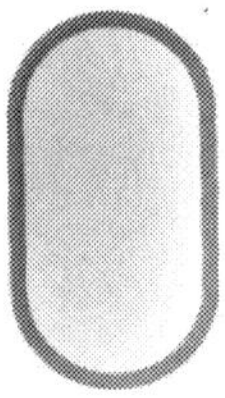

28

R bodies can lyse spheroplasts

Non-functional R bodies

Functional R bodies

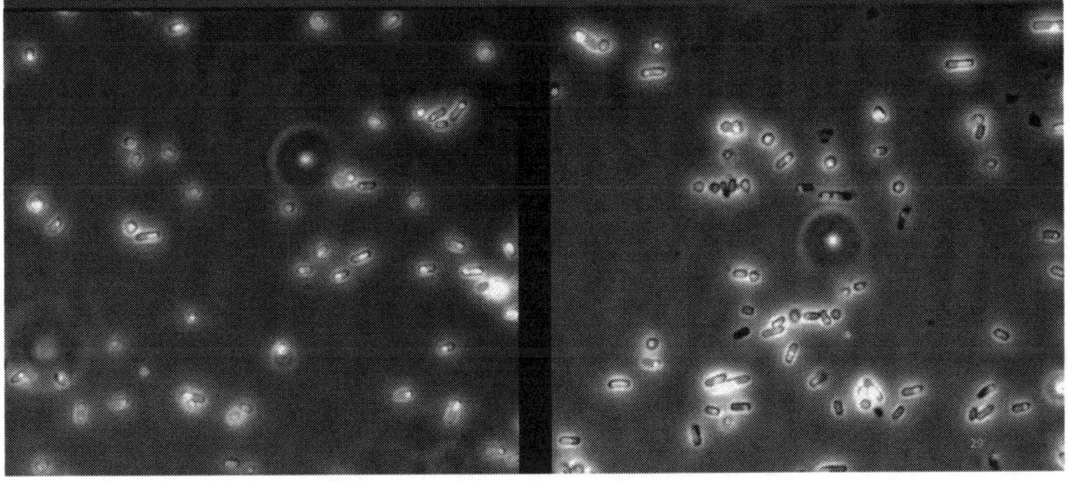

Extension in HEK293 cells

Transmitted | R bodies | Lysotracker | Merge

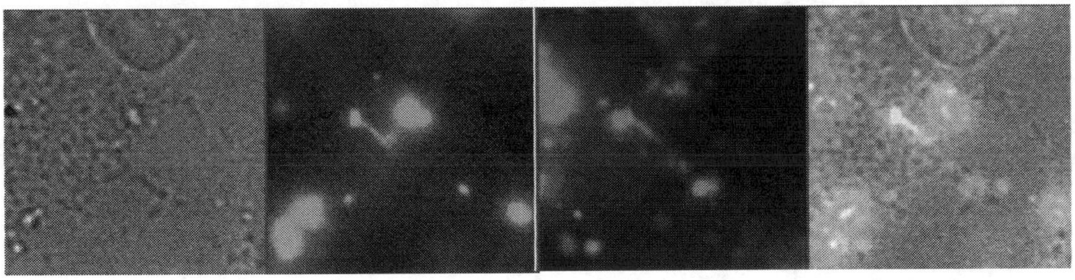

With Mike Vahey at MBL

Delivery next?

30

Let's talk about preprints

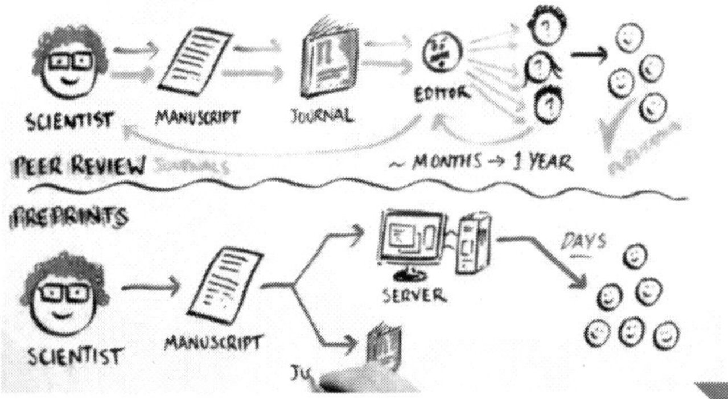

- Questions?
- Ideas?
- Concerns?

ASAPbio.org

31

Thank you!

HMS
Pam Silver
Jeff Way
Tim Mitchison
Genevieve Dobihal
Nathan Rollins
Bryan Hsu
Ethan Garner
Michael Baym
Tyler Ford
L. Mahadevan
Talley Lambert (NIC)

SGI

Addgene

Woods Hole
Dan Fletcher
Michael Vahey
Peter Foster
Mathijs Vleugel
Franziska Decker
Rob Phillips

Justin Kollman (UW)
Julien Bergeron (UW)

Jane Coffin Childs Memorial Fund for MEDICAL RESEARCH

32

Stoichiometry of covalent assembly

One protein His tagged

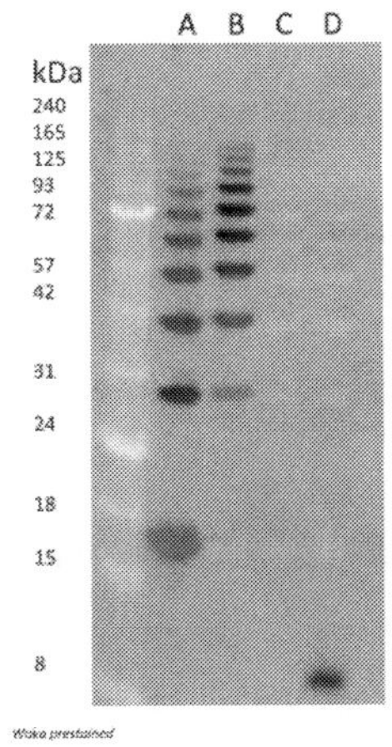

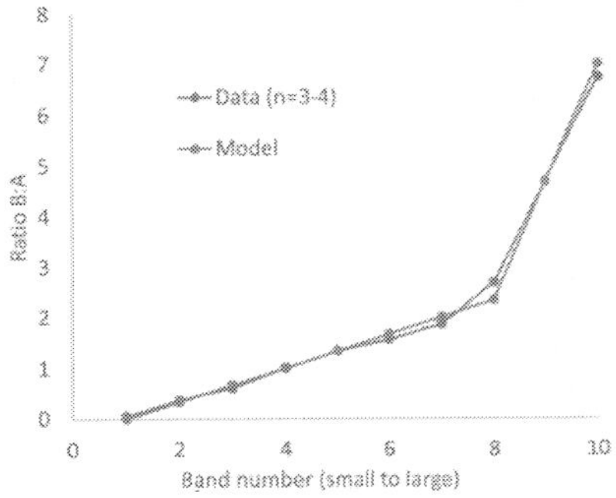

Stoichiometry of covalent assembly

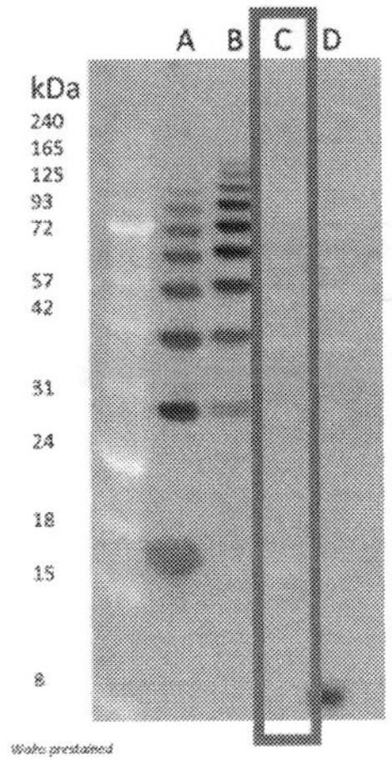

Is there a ruler?

Stabilizing Genetic Programs Against Changes in Context

Segall-Shapiro, T., Massachusetts Institute of Technology
Sontag, E., Rutgers-The State University of New Jersey
Voigt, C. A., University of California, San Francisco
Originally presented on: 7/19/2016 16:30:00 - 16:45:00

Synthetic genetic programs often require fine tuning of expression levels to achieve optimal function. They can therefore be quite sensitive to changes in context, as perturbations of gene expression can break this tuning and lead to system failure. To address such fragility, we are creating stabilized promoters that incorporate simple regulatory systems to buffer gene expression against changes in genetic and cellular context. Using TALE proteins as transcriptional repressors in E.coli, we have built a promoter regulated by an incoherent feedforward loop that is nearly insensitive to its copy number (a greater than 20-fold change in gene expression due to copy number is reduced to a less than 50% change in expression). This stabilized promoter allows movement of genetic systems onto most commonly used plasmid backbones with little change in expression. Additionally, it facilitates more predictable gene expression from libraries of random genomic insertions.

The regulatory motifs used in our stabilized promoters are also expected to buffer gene expression against other perturbations, such as a change in the number of free ribosomes or RNAPs within a cell. We are exploring how stabilized promoters buffer gene expression against transient changes in cellular context, including shifts in growth rate due to a media change, or local ribosome depletion caused by protein overexpression. While less striking than the effect on genetic context, the stabilized promoter largely eliminates small perturbations caused by these factors, suggesting that it may yield more predictable gene expression in varying cellular environments. Due to their ability to buffer gene expression against context changes, we hope that stabilized promoters will prove valuable as a foundation for building more robust synthetic genetic programs.

Targeted DNA Methylation in Human Cells Using Engineered Cas9-Methyltransferases

Glenna Meister

2016 SEED Conference

7-19-2016

DANA-FARBER CANCER INSTITUTE | Department of Cancer Immunology and Virology

Epigenome

DNA methylation

- Mammals methylate cytosine in CpG sequences
- Methylation represses gene expression

Histone modification

- Signals chromatin structure

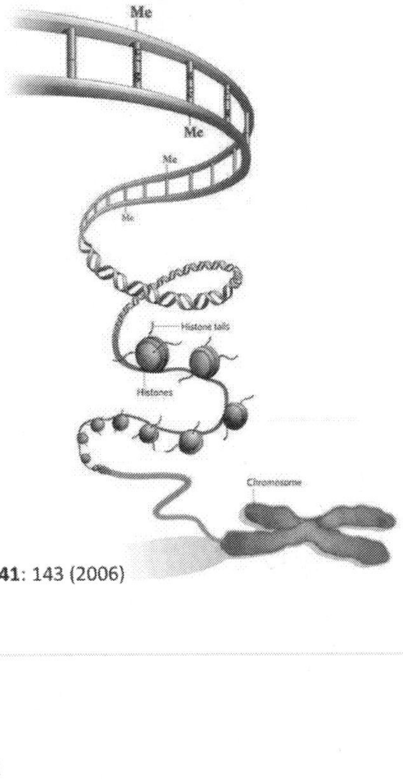

Nature **441**: 143 (2006)

Epimutations: DNA methylation patterns change in cancer cells

Example is SALL2 silencing by methylation

- *SALL2* is silenced by methylation in ~90% of high-grade serous carcinomas
- *SALL2* appears to be protected from tumor genetic alterations

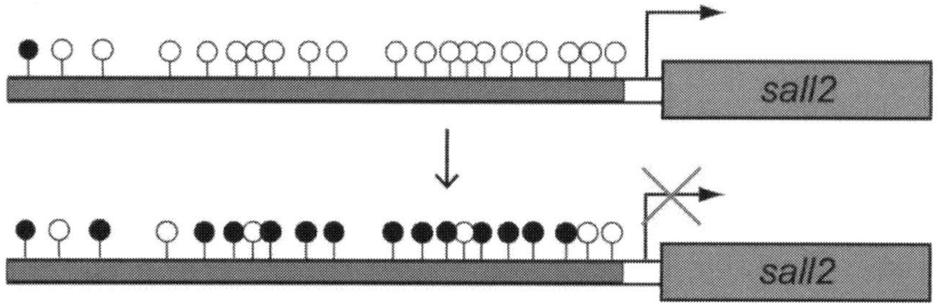

Is abnormal methylation a driver of cancer or is it a byproduct?

Previous DNA methyltransferae (MTase) fusions resulted in off-target methylation

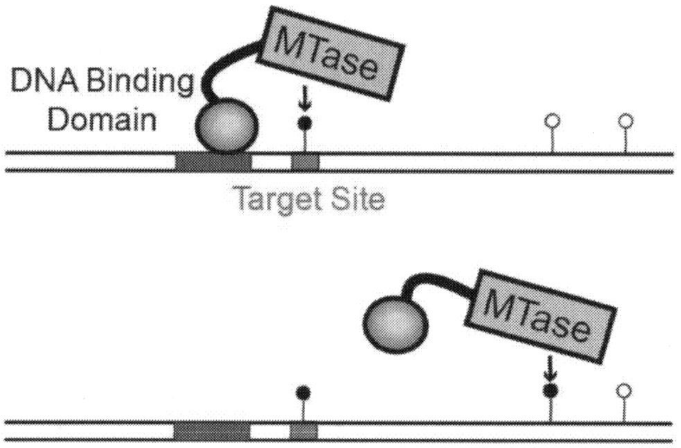

Bestor Lab, Ford Lab, Jeltsch Lab, Kadde Lab, Rots Lab, Zoldos Lab

Split MTase (sMTase) system significantly reduces off-target effects

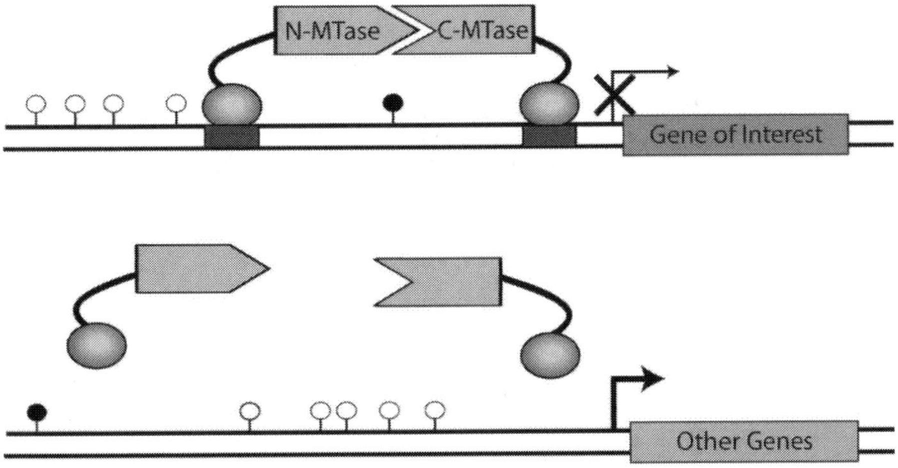

Ostermeier Lab: Wonchae Choe, Dave Paschon, Glenna Meister and Brian Chaikind

Proof-of-concept in human cells: Zinc finger-CpG MTase split fusions for site-specific methylation on genomic DNA

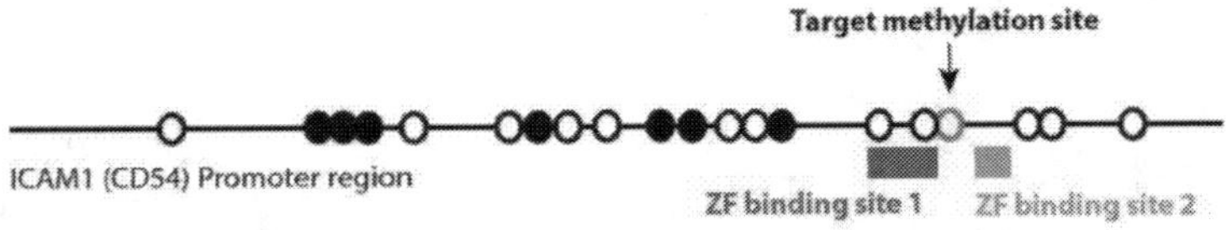

	Target Methylation	Total Promoter Methylation
Mock Control	13%	31.7%
Mutant Split M.SssI CpG Methyltransferase	67%	30.4%

Targeting of multiple sites on a CpG island requires multiple DNA binding domains

Zinc finger or TALE DNA binding domains

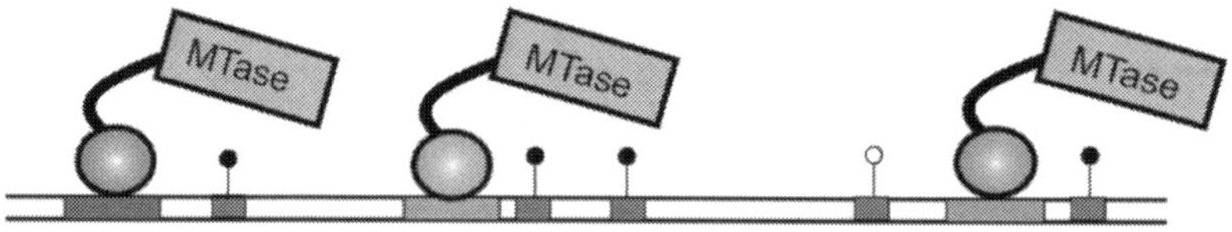

Crispr/Cas9 is more user friendly RNA guided DNA binding domain

Zinc finger or TALE DNA binding domains

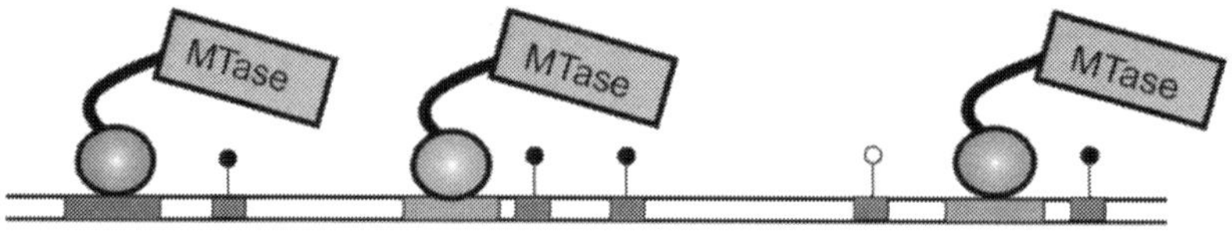

Cas9 RNA guided DNA binding domains

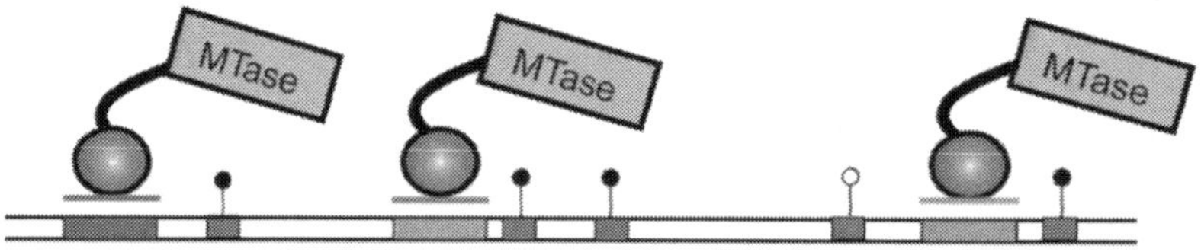

Creating new dCas9-sMTase in E. coli

Tina Xiong

Poster # 208
Wednesday, July 20th

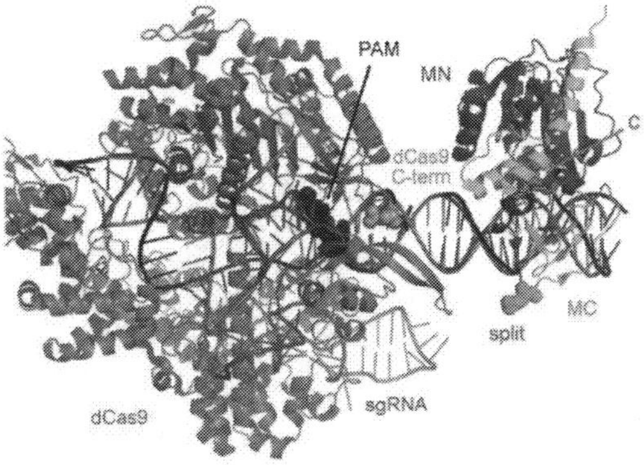

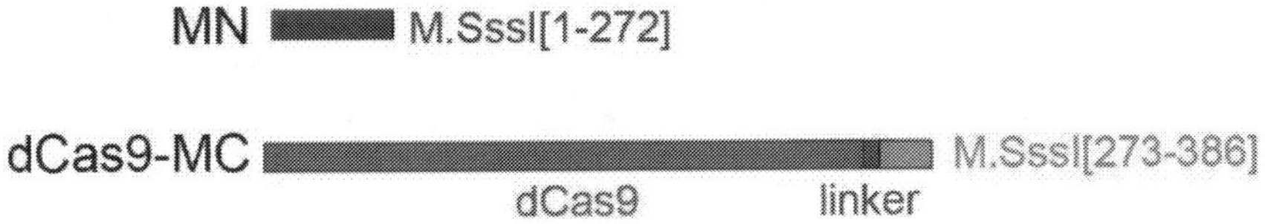

The dCas9-sMTAse system was optimized for human cells

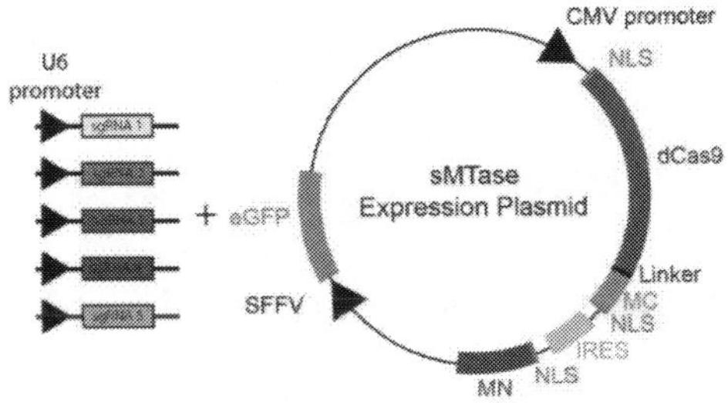

Pyrosequencing to test methylation activity of the sMTase system in human cell lines

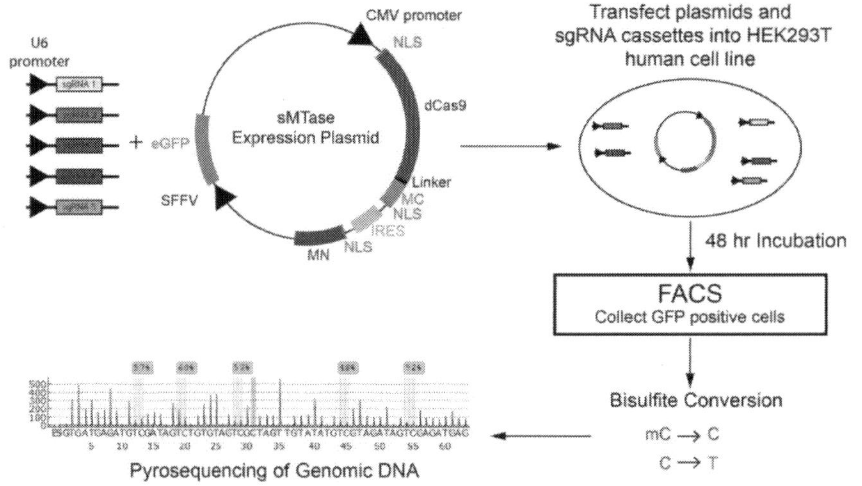

Proof of concept testing on the SALL2 promoter with dCas9-sMTase system

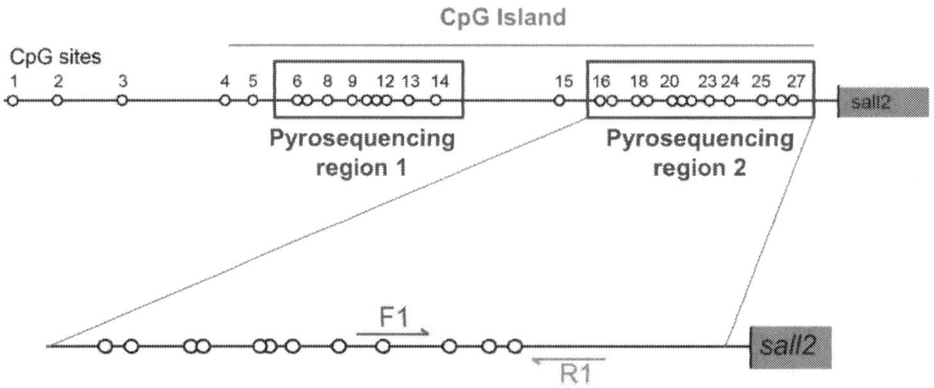

The SALL2 promoter is normally hypomethylated in HEK293T cells

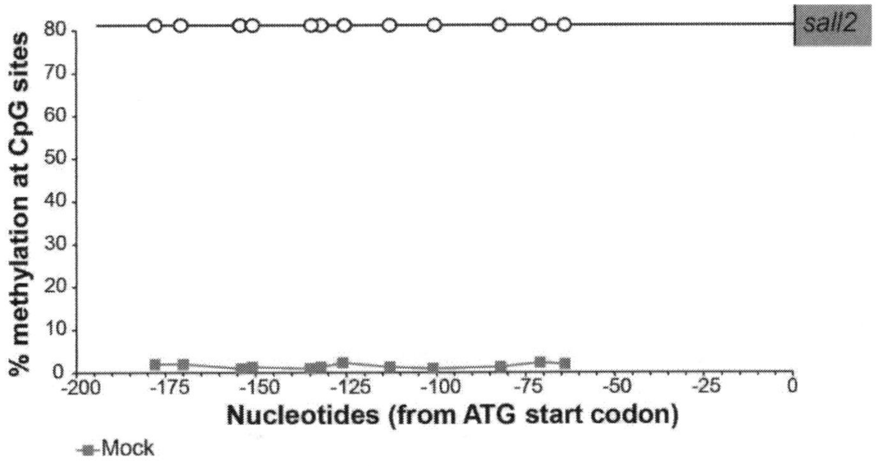

A dCas9-MC only expression vector does not have any methylation activity

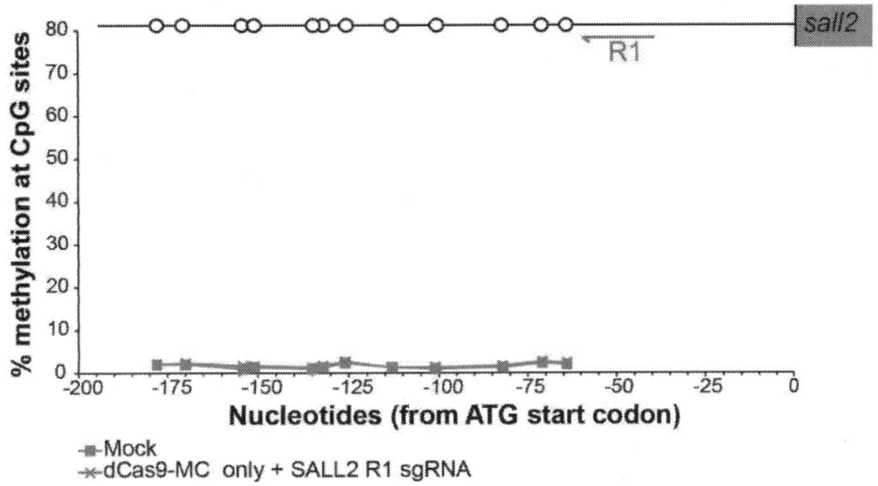

The sMTase expression vector has minimal activity with a scambled control sgRNA

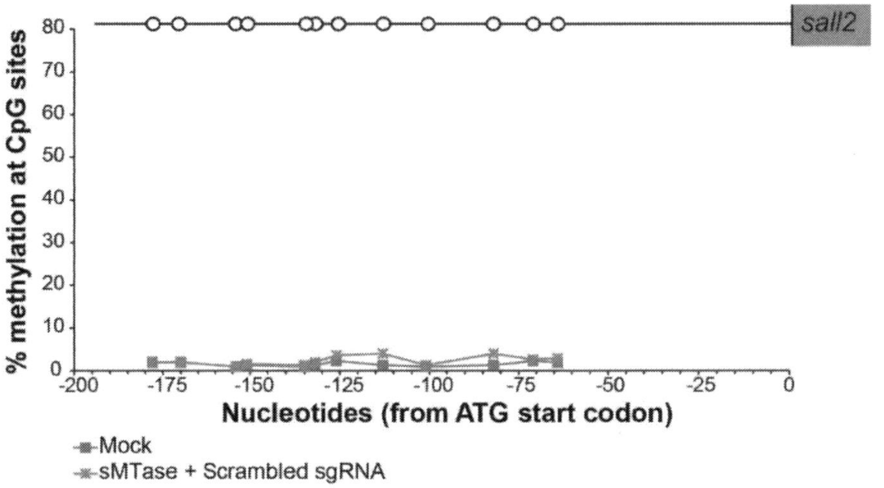

Targeted methylation 22 nt away from PAM binding site with the F1 guide strand

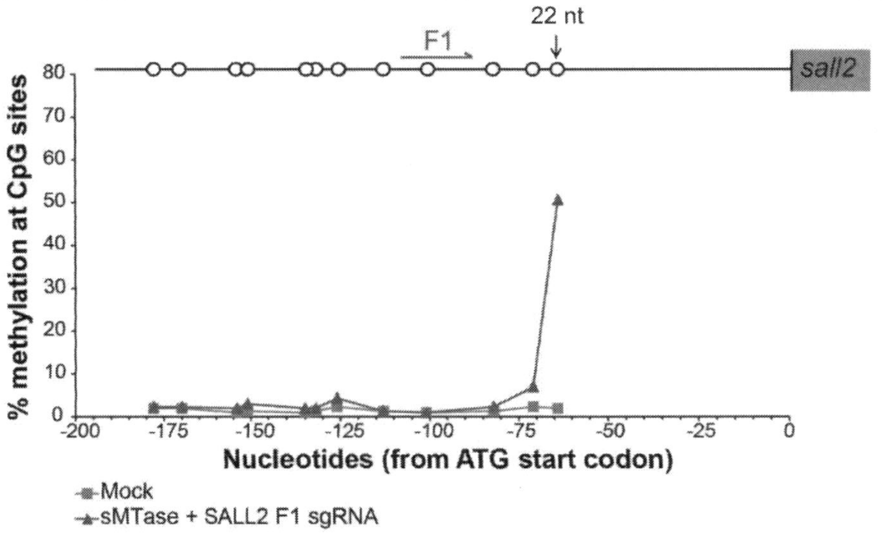

Targeted methylation can be changed by using different guide strand

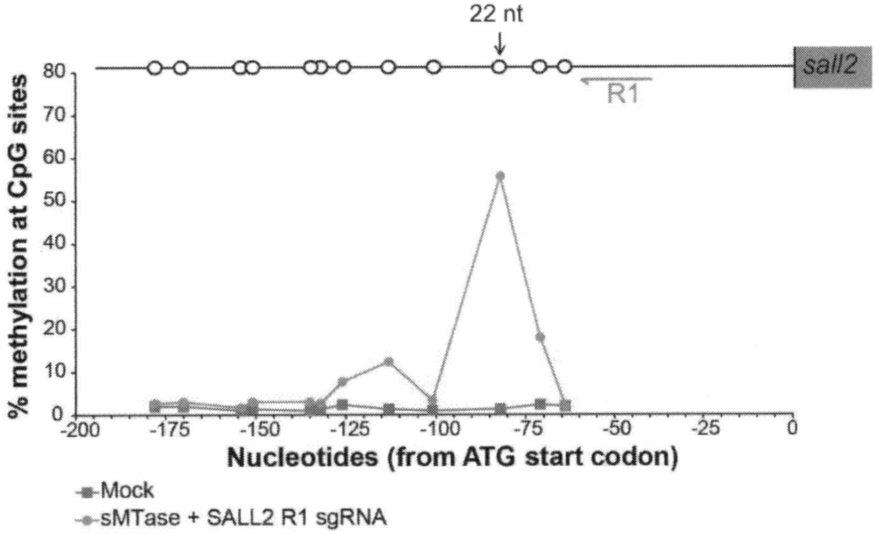

However combining guide strands can lead to interference decreasing methylation

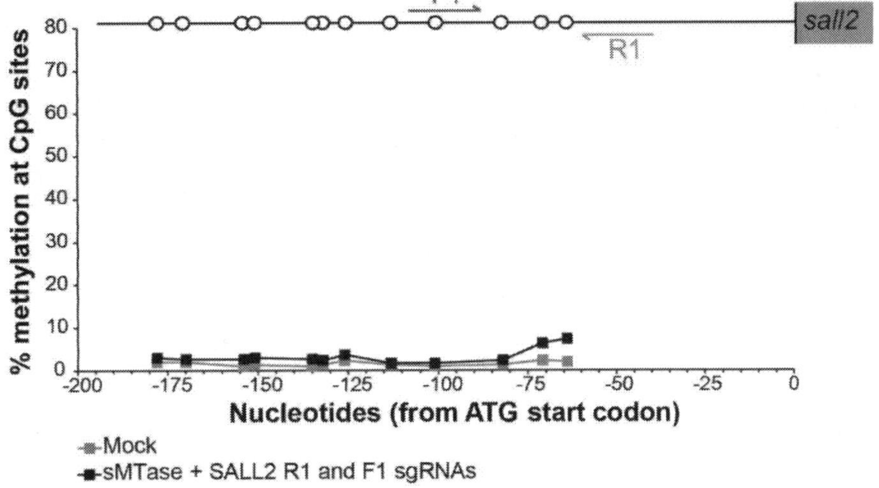

Proper spacing and orientation are required to reduce dCas9 interference

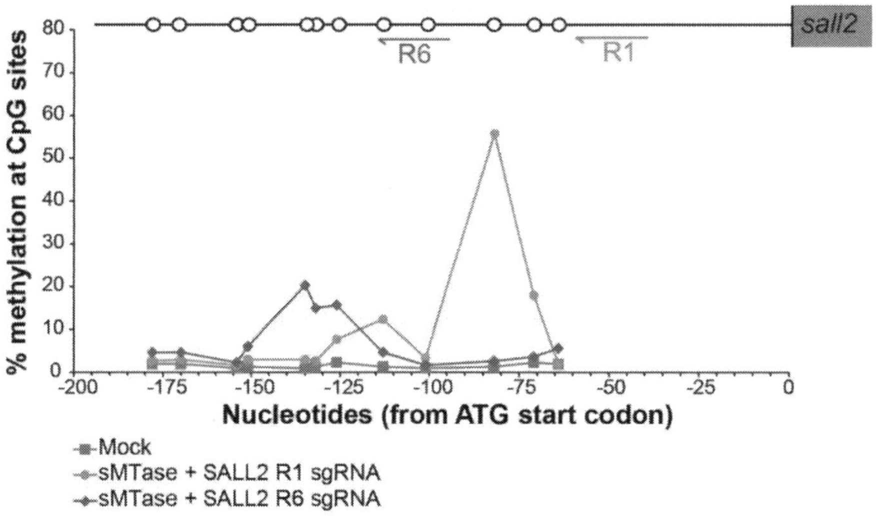

Proper spacing and orientation are required to reduce Cas9 interference

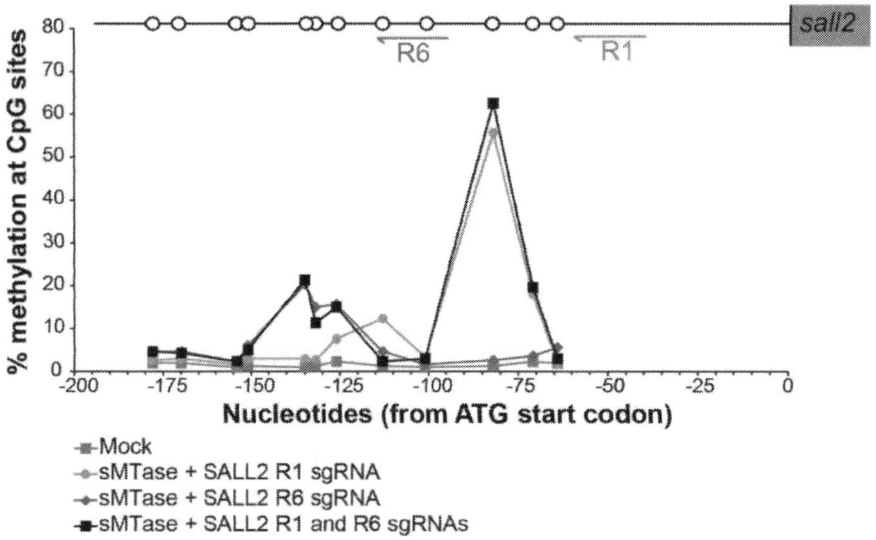

sgRNAs can be multiplexed to target multiple sites on the SALL2 promoter

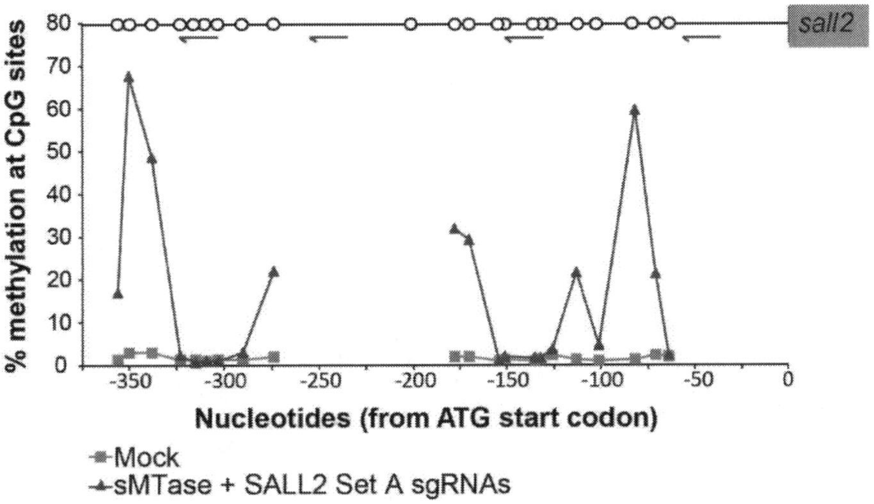

Targeted methylation sites can be changed with new sgRNA sets

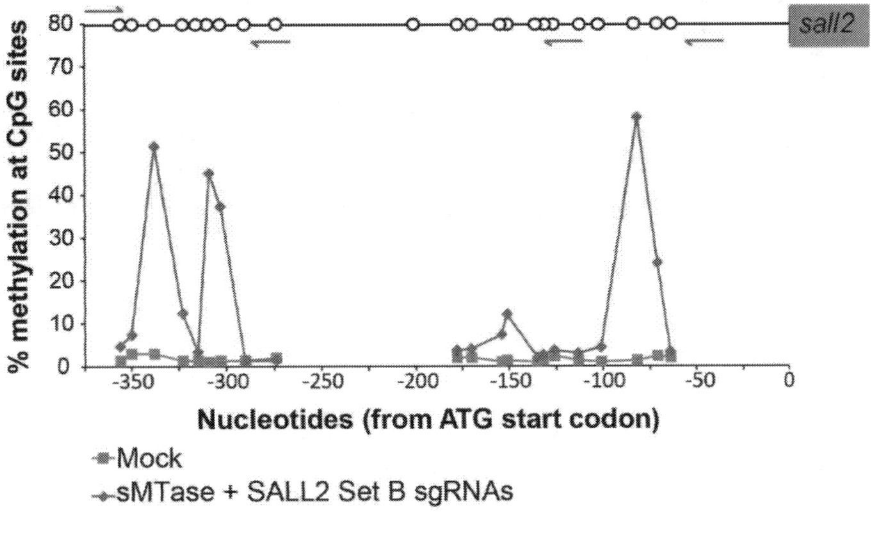

Targeted methylation sites can be changed with new sgRNA sets

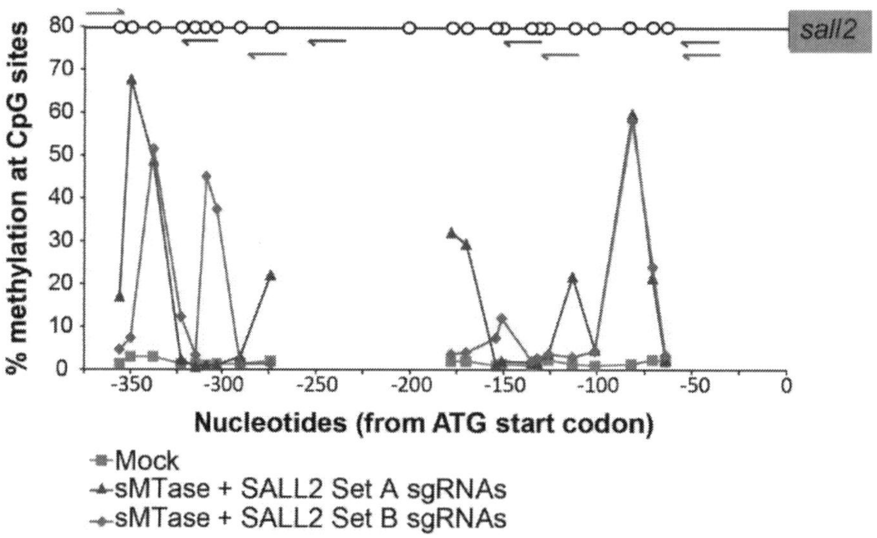

dCas9-sMTase system targets methylation in reproducible patterns on genomic DNA

- dCas9-sMTase constructs are able to target methylation on genomic DNA
  - Different sgRNA's target methylation to different sites with varying efficiency based on based on distance from cas9 binding site
  - Able to multiplex and target different sites on promoter (or separate promoters)

- Overlapping or closely spaced sgRNA's interfere with methylation activity

Acknowledgments

DFCI/Harvard Medical School

- Carl Novina
- Ronny Drapkin (University of Pennsylvania)
- Ursula Matulonis
- Tom Benjamin

Johns Hopkins University

- Marc Ostermeier
- **Tina Xiong**
- Brian Chaikind
- Dave Paschon
- Wonchae Choe
- Srinivasan Chandrasegaran

Novina Lab

- **Mike Spellberg**
- **Chadene Tremaglio**
- Meghan Bushway
- Alberto Nobili
- Johanna Carroll
- Cailin Joyce
- Karyn Schmidt
- Dolly Thomas
- Akinori Yoda
- Elaine Yee
- **Fulya Turker**

Funding

NSF

NIH National Institutes of Health

Mix and Match Sweets: Modular, Coordinated Cell-Free Transcription, Translation, and Glycosylation of Proteins Using Selectively Enriched Escherichia coli Cell Lysates

Stark, J. C., Northwestern University
Jaroentomeechai, T., Cornell University
DeLisa, M. P., Cornell University
Jewett, M. C., Northwestern University
Originally presented on: 7/19/2016 17:15:00 - 17:30:00

Glycosylation, or the attachment of glycans (sugars) to proteins, is one of the most abundant post-translational modifications in nature and plays a pivotal role in protein folding, sorting, and activity. In molecular medicine, the compositions and structures of glycans on recombinant therapeutic glycoproteins are known to impact pharmacokinetics and drug activity. However, unlike DNA, RNA, and protein synthesis, protein glycosylation is not a template-driven process. Instead, glycans are assembled by coordinated expression of glycosyltransferase enzymes and attached to proteins by oligosaccharyltransferase (OST) enzymes. The complexity of this biosynthetic process results in inherent heterogeneity of glycans produced in living organisms. The inability to precisely control protein glycosylation in eukaryotic protein expression systems (*e.g.*, yeast and Chinese hamster ovary cells) represents a key challenge in the fields of glycoprotein synthesis and glycoprotein therapeutics. Bacterial glycoengineering, or the expression of orthogonal glycosylation machinery in *Escherichia coli*, is an alternative strategy for production of proteins with controllable glycosylation. However, *in vivo* bacterial glycoengineering is limited by i) the need to maintain cell viability during glycoprotein synthesis while simultaneously overexpressing multiple non-native, membrane-bound glycosylation enzymes and ii) lengthy design-build-test cycles for novel glycosylation pathway engineering. To address these limitations, we have developed a bacterial cell-free glycoprotein synthesis (CFGpS) system with the potential to enable controllable, user-specified protein glycosylation. Uniquely, in our CFGpS system, i) all the biosynthetic machinery for protein synthesis and glycosylation is supplied by *E. coli* lysates and ii) transcription, translation, and glycosylation occur in an all-in-one *in vitro* reaction. We engineered glycosylation chassis strains that are optimized for glycosylation and produce ~1.5 g/L protein in cell-free protein synthesis, which represents a 50% increase in potential glycoprotein yields compared to state-of-the-art *E. coli* glycosylation chassis and recombinant protein expression strains. We then demonstrated that *E. coli* cell lysates can be selectively enriched with active glycan precursors and/or the OST from *Campylobacter jejuni via* overexpression of orthogonal genes in the chassis strain. Next, we used these lysates individually and as a mixture to carry out one-pot synthesis of glycoproteins. By mixing and matching lysates, we also used the CFGpS system to characterize the *in vitro* activity of four additional OSTs with varying sequence homology to the archetypal *C. jejuni* OST. The CFGpS platform is modular, flexible, and has promising applications as a high-throughput prototyping platform for glycosylation pathway design and synthesis of glycoproteins of biotechnological interest. This system promises to advance our understanding of bacterial glycosylation and aid development of new glycoengineering tools.

Expanding the Synthetic Capabilities of Yeast

Cornish, V. W., Columbia University

Originally presented on: 7/20/2016 8:30:00 - 8:55:00

*In vitro*Â directed evolution allows biomolecules with new and useful properties to be engineeredâ??mimicking natural evolution on an experimentally accessible time scale by creating large libraries of DNA mutants using PCR and then carrying out a high-throughput assay for variants with improved function. To provide a breakthrough in the complexity of libraries that can be readily searched experimentally and to allow systems to be directly engineered in the cell, my laboratory is engineeringÂ *S. cerevisiae*Â so that both the mutagenesis and selection steps of directed evolution can be carried out entirelyÂ *in vivo*, under conditions of sexual reproduction. We have built a modular chemical complementation assay, which provides a selection for diverse chemistry beyond that natural to the cell using themes and variations on the yeast two-hybrid assay. In addition, we devised a heritable recombination system, for simultaneous mutagenesis and selectionÂ *in vivo*Â under conditions of sexual reproduction. Finally, we have begun to utilize these mutagenesis and selection technologies to engineer yeast to carry out new functions themselves ranging from being a biosensor to a self-organizing community.

Molecular Systems Engineering with DNA

Dietz, H.,

Originally presented on: 7/20/2016 10:00:00 - 10:30:00

It is notoriously difficult to observe, let alone control, the position and orientation of molecules because of their small size and the constant thermal fluctuations that they experience in solution. Molecular self-assembly with DNA provides a route for placing molecules and constraining their fluctuations in user-defined ways and with up to Angstroem-scale precision. These positioning options open attractive and unprecedented avenues for scientific and technological exploration, in particular with respect to the creation of artificial molecular machines. In my talk I will introduce some of the key concepts and methods, and highlight a number of recent developments.

Next generation cell-free TXTL for synthetic biology

Vincent Noireaux
University of Minnesota

SEED, Chicago, July 2016

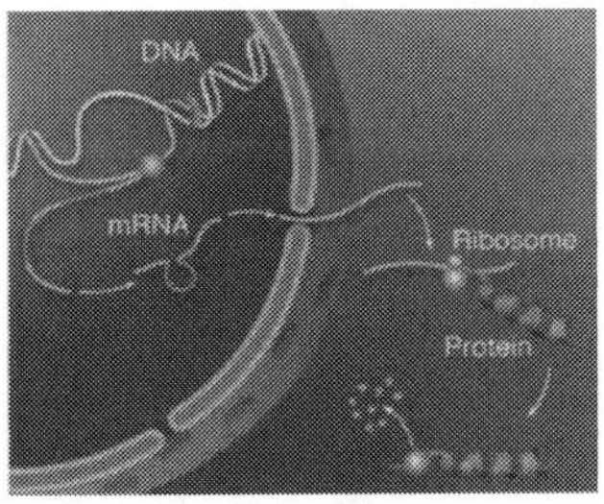

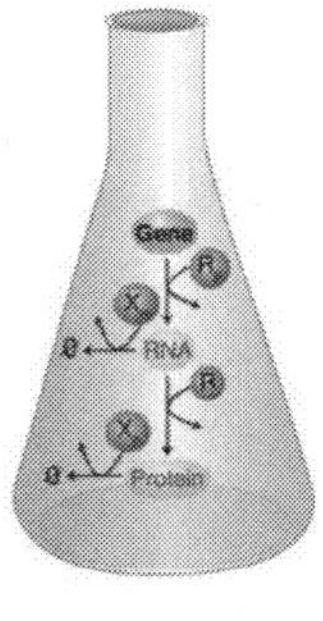

Outline:

- Introduction - Motivations.
- Cell-free TX-TL.
- Synthetic gene circuits and phage synthesis.
- Minimal cell.

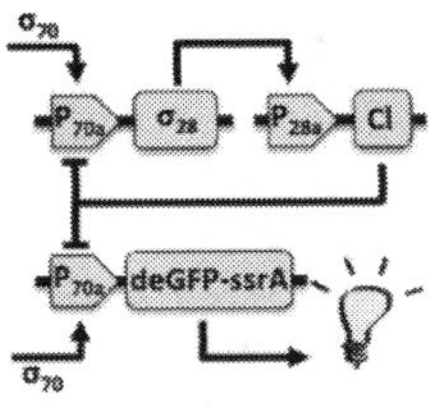

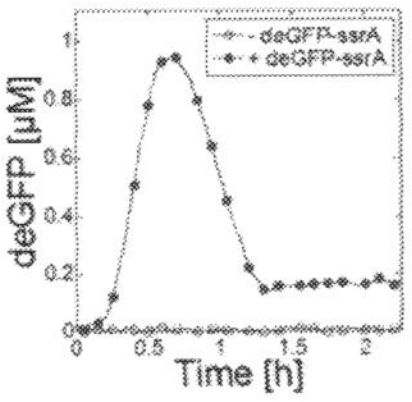

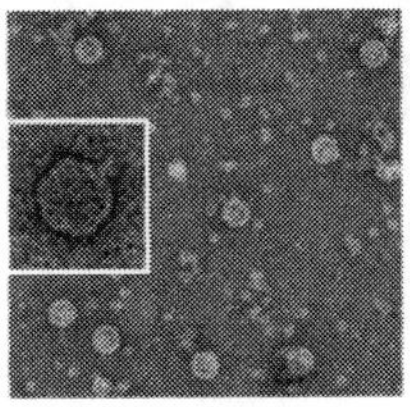

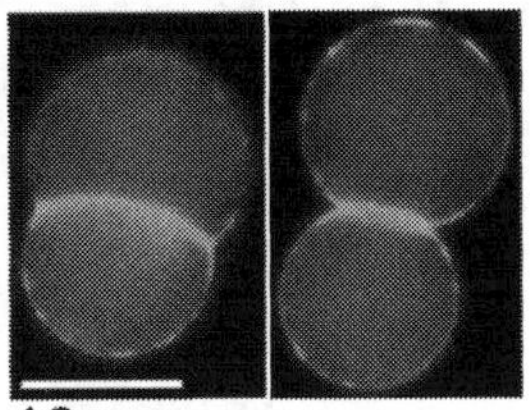

1
Introduction – Motivations

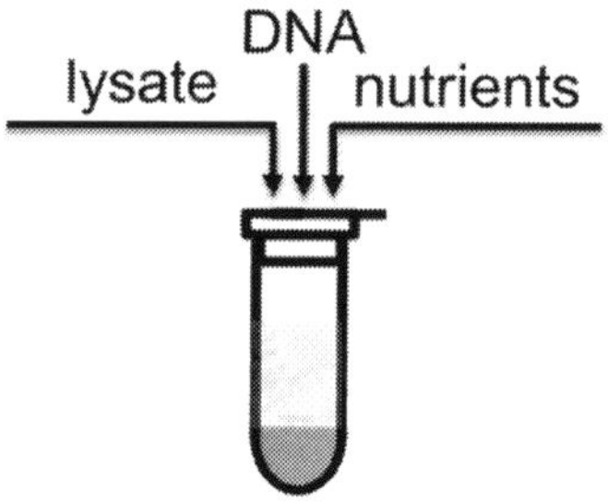

Synthetic biology

It's programming with DNA to direct and control biological systems:

- to understand gene regulation and make simple computations
- to use them either as molecular-scale factories
- to create new hybrid materials

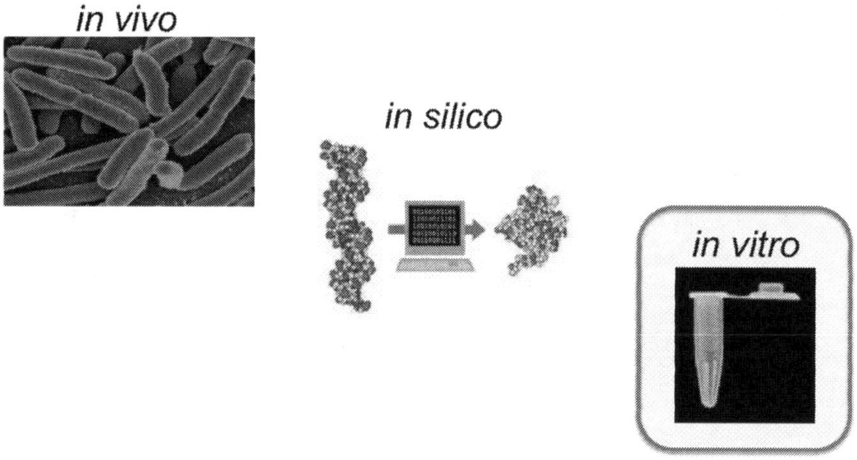

Synthetic biology in a test tube
(cell-free synthetic biology)

Construct biochemical systems in a test tube from DNA programs.

- bottom-up, reductionist and constructive approach.
- no endogenous information.
- no interference and response from an organism.
- more freedom of control and design compared to in vivo.

Living cell to cell-free TX-TL
(bacteria E. coli)

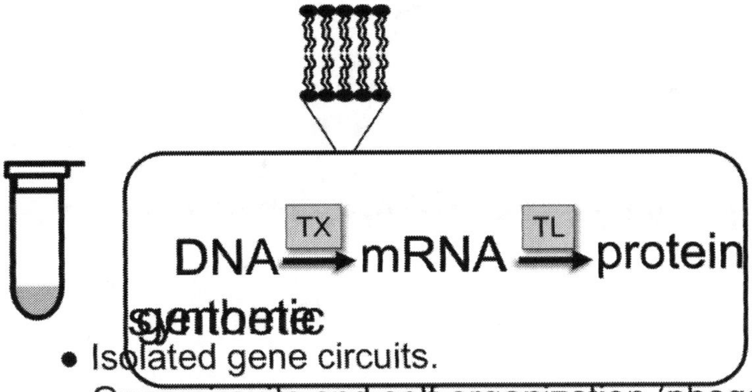

- Isolated gene circuits.
- Gene circuits and self-organization (phage).
- Bottom-up minimal cell.

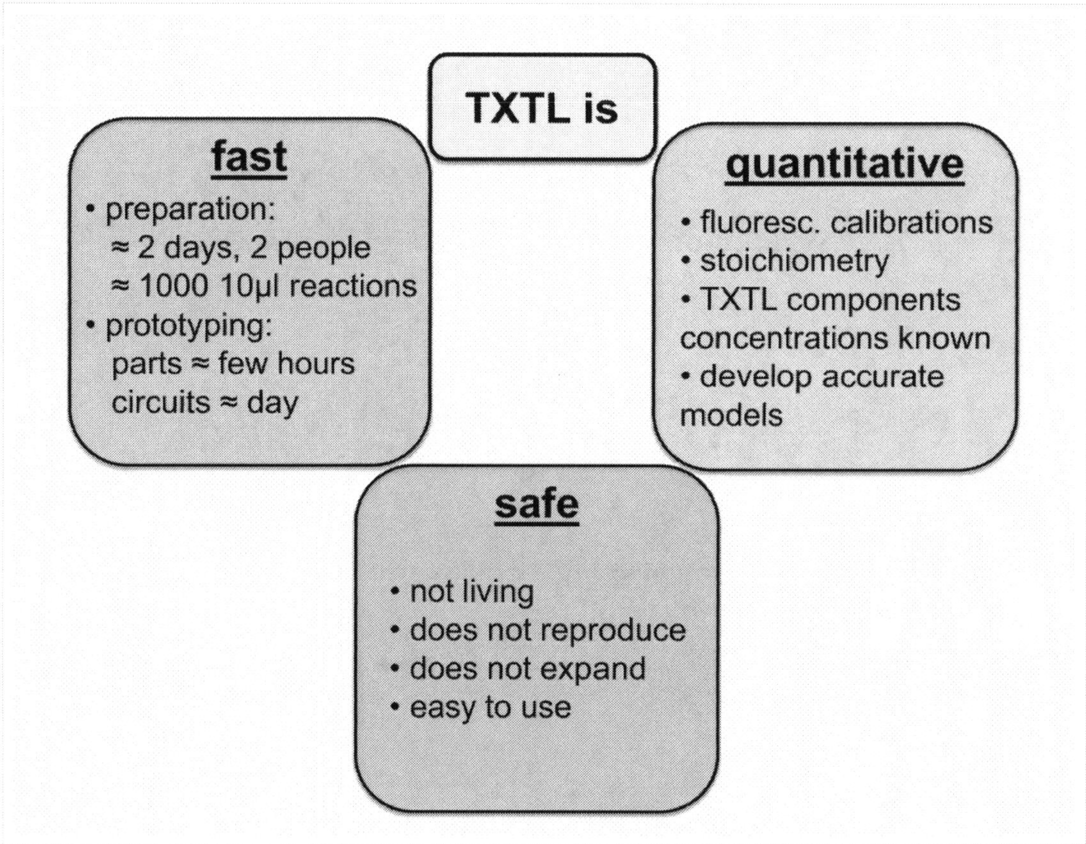
TXTL is
fast
• preparation:
≈ 2 days, 2 people
≈ 1000 10μl reactions
• prototyping:
parts ≈ few hours
circuits ≈ day
quantitative
• fluoresc. calibrations
• stoichiometry
• TXTL components concentrations known
• develop accurate models
safe
• not living
• does not reproduce
• does not expand
• easy to use

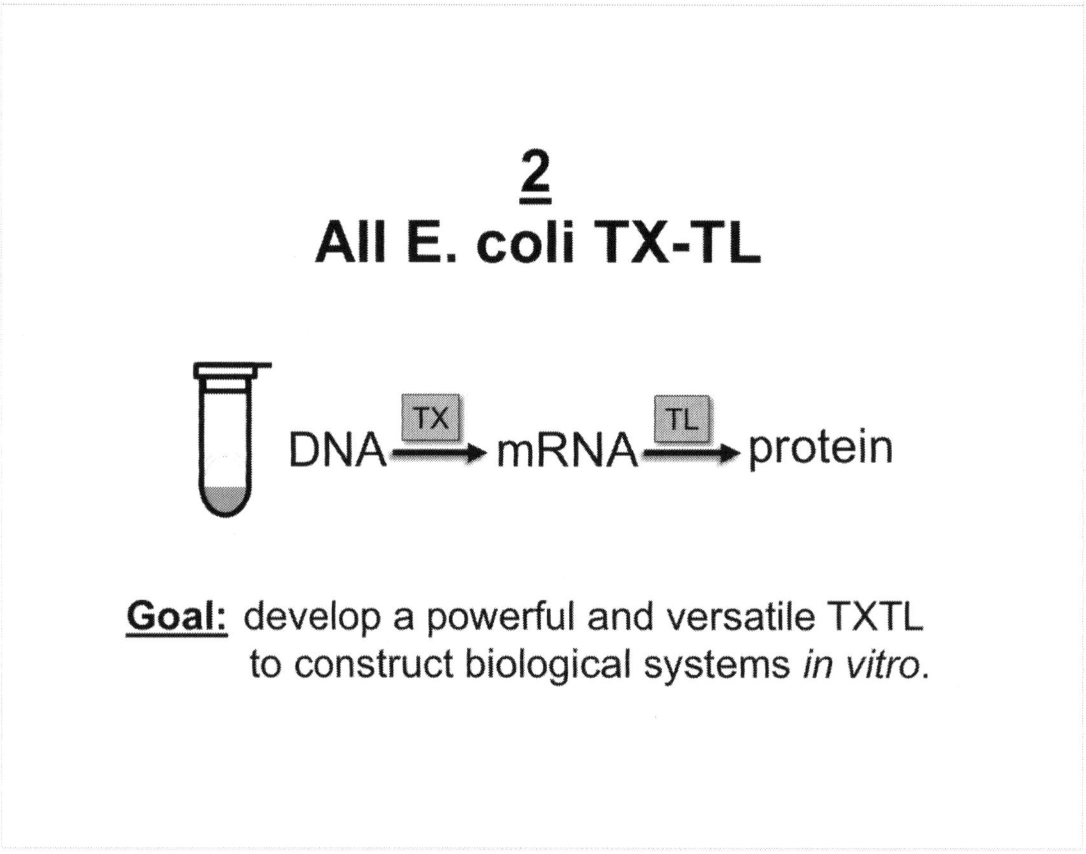
2
All E. coli TX-TL
DNA
TX
mRNA
TL
protein
Goal: develop a powerful and versatile TXTL to construct biological systems in vitro.

Powerful-versatile cell-free TX-TL

TX:
expanded to 100s of E. Coli regulatory parts

Protein yield:
5 nM plasmid → ≈ 80 μM eGFP
Amplification: ≈ 10 000

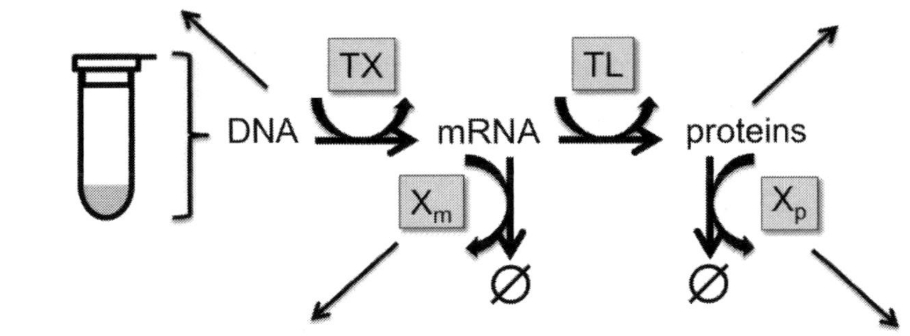

mRNA degradation (mazF):
1st order kinetics
Tuned from 25 to 0 min (eGFP)

Protein degradation (ClpXP):
0th order kinetics
Tuned from ∞ to 200 nM/min (eGFP)

Shin and Noireaux, 2010, 2012 (JBE, ACS SB), Caschera and Noireaux 2013 (Biochimie), Garamella, Marshall, Rustad, Noireaux (ACS SB 2016)

TX repertoire: 100s of parts

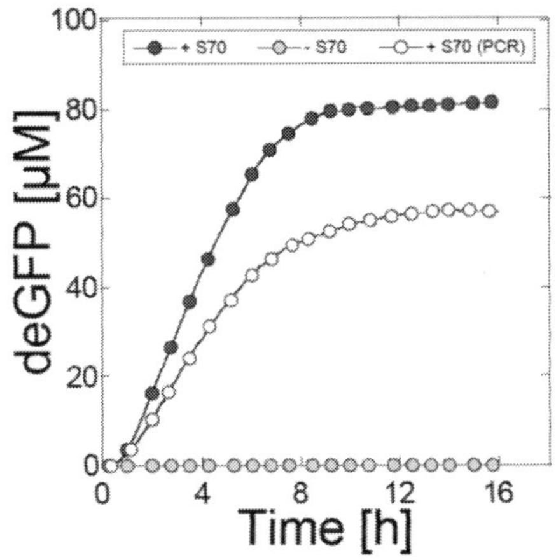

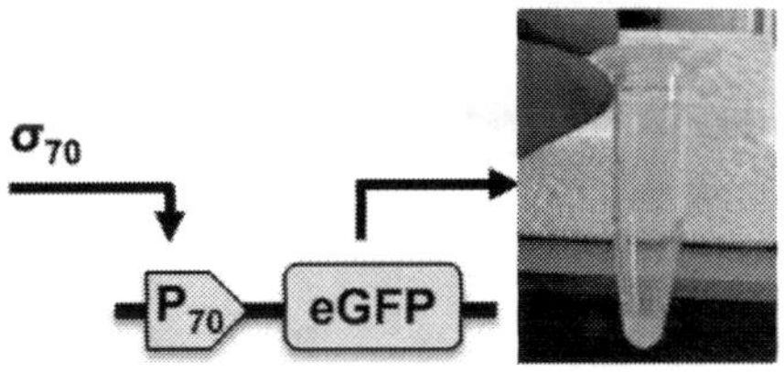

- 6-8 hours of expression.
- 2-2.5mg/ml of protein synthesized (80-100μM eGFP).

Garamella et al - 2016

TX repertoire: 100s of parts

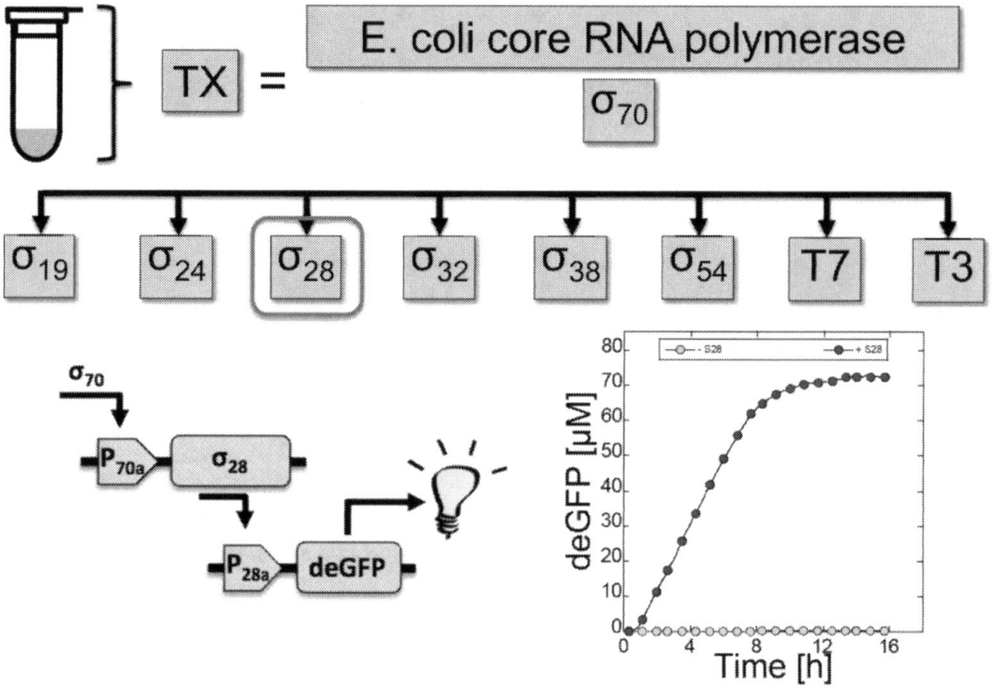

TXTL is easy to use

Typical TXTL system: mix component A with component B with component C, ….

Easy TXTL: all components pre-packaged

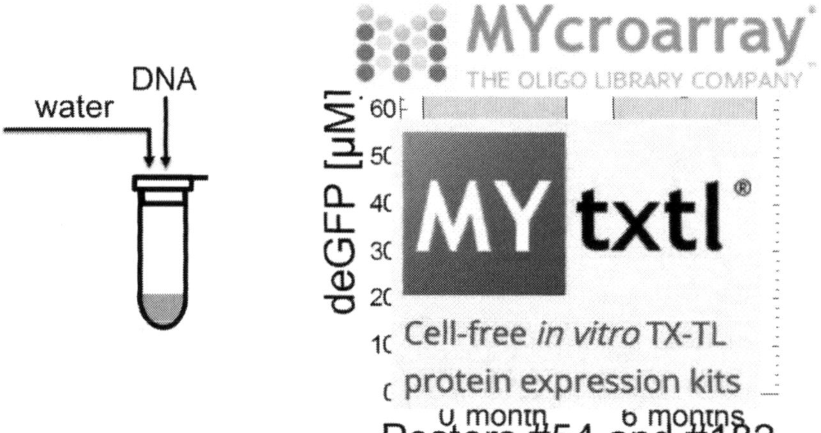

Posters #54 and #182

3

Prototyping

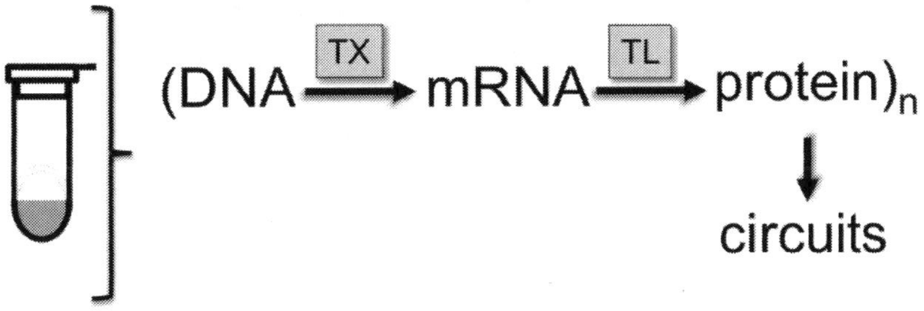

Goal: accelerate the design-build-test-cycle of gene circuits

'In vivo' riboregulator testing

(Julius Lucks and James Chappell, Northwestern)

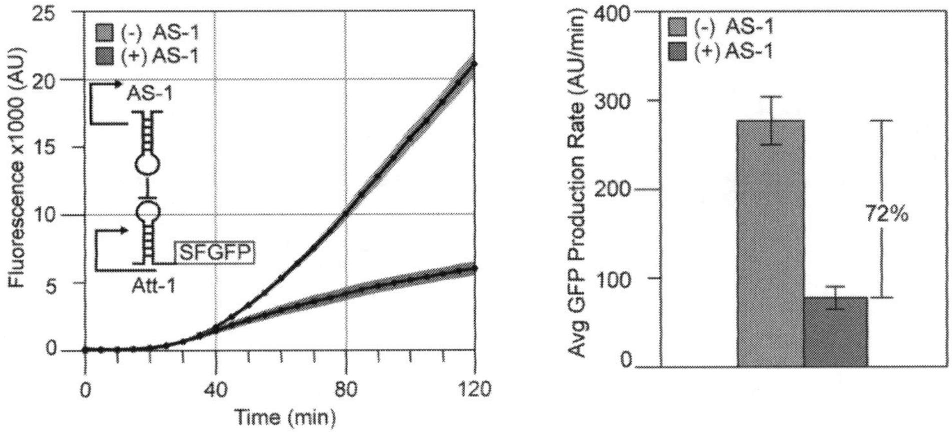

Takahashi et al. 2014 (ACS SB)

Examples of cell-free circuits

(Richard Murray and Zach Sun, Caltech)

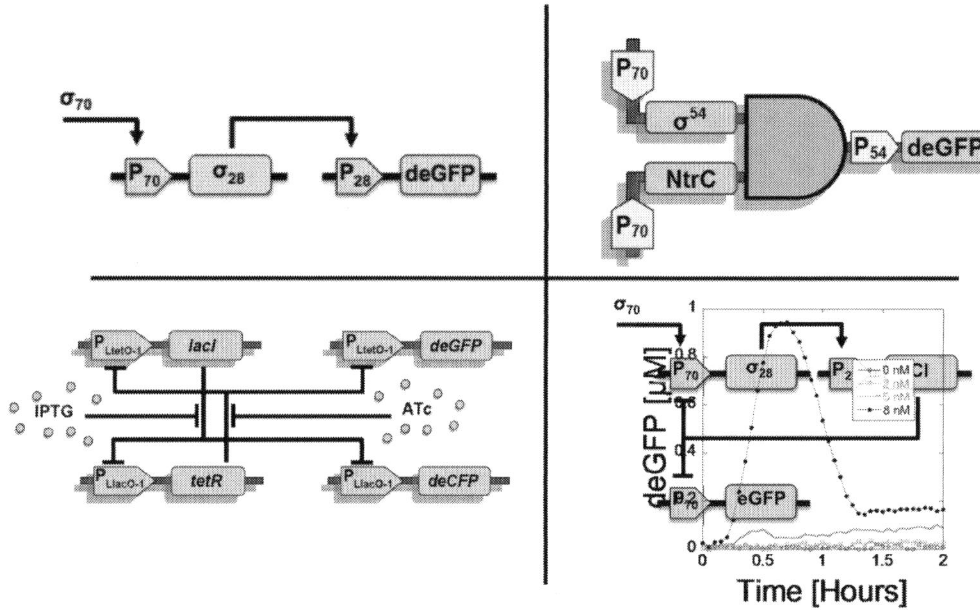

'In vivo' circuit testing

(Chris Voigt, MIT)

– IPTG	1
+ IPTG	0

Chunbo et al - 2012

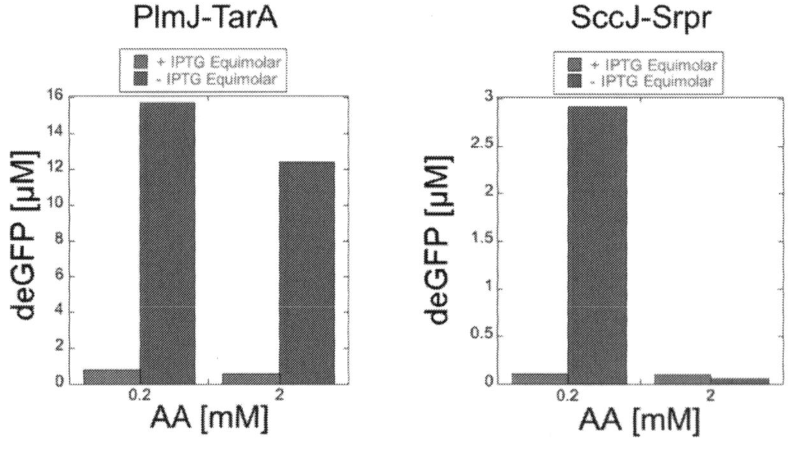

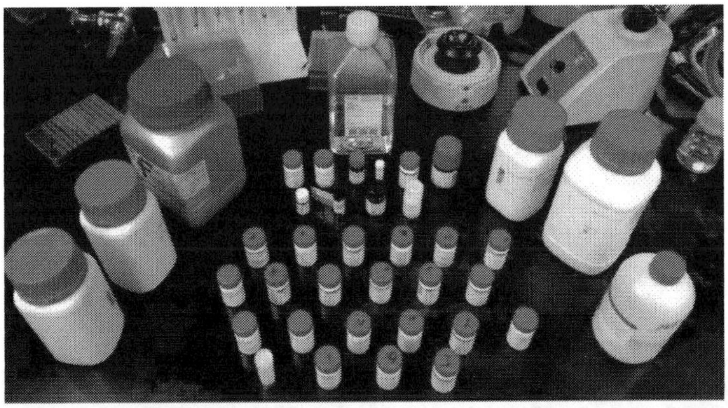

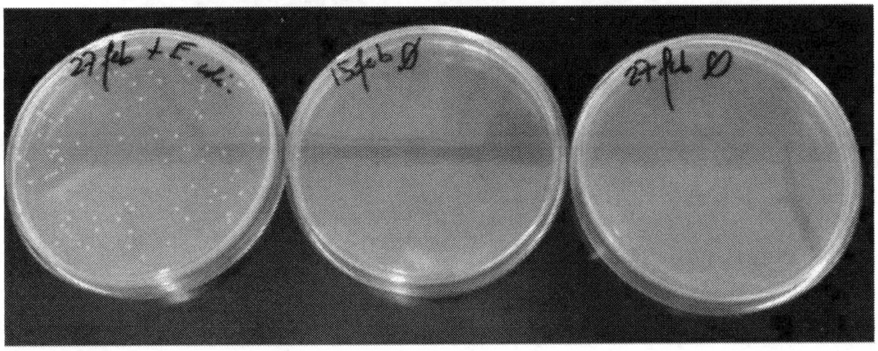

Cell-free biosynthesis of chemicals

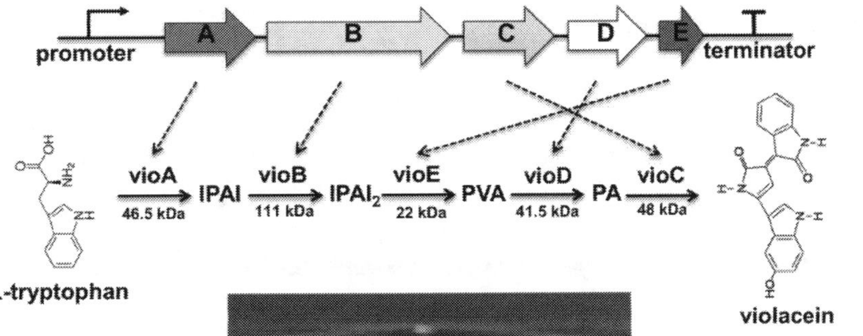

TXTL in microfluidic chip

(Roy Bar-Ziv, Eyal Karzbrun, Alex Tayar, Weizmann)

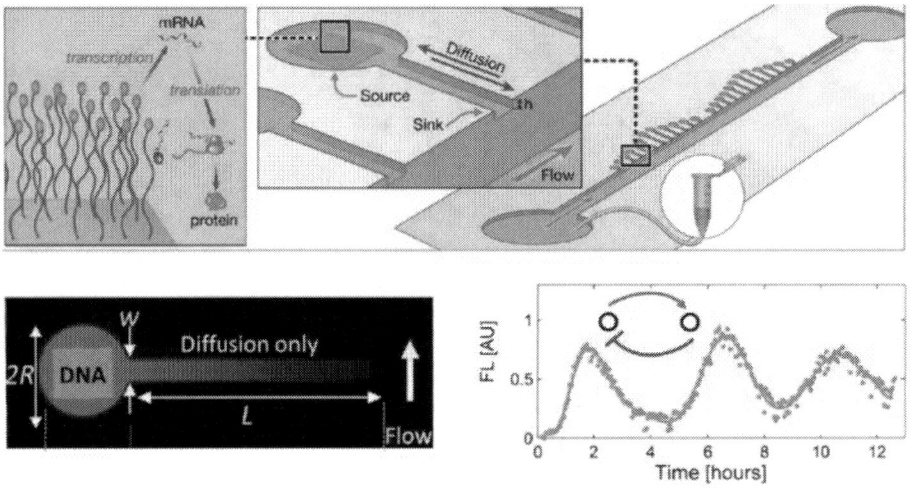

Karzbrun, Tayar, Noireaux, Bar-Ziv - 2014 (Science)
Tayar, Karzbrun, Noireaux, Bar-Ziv - 2015 (Nature Physics)

CRISPR dCas9 TXTL

(Chase Beisel and Michelle Luo, NCSU)

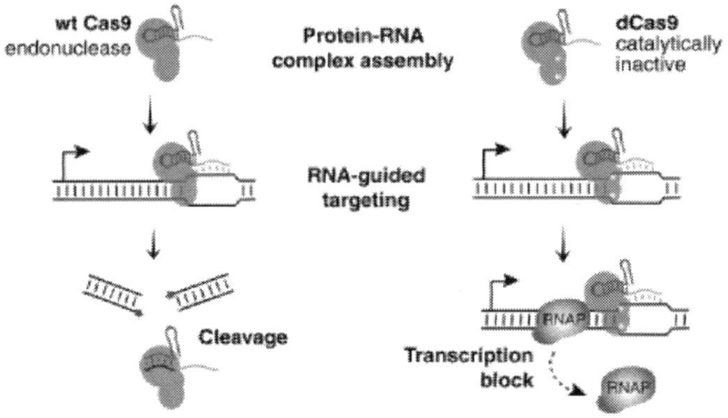

- ☑ Preparation of a dCas9 loaded TXTL
- ☑ Quantification of dCas9 in TXTL (50-100 nM)
- ☑ E. Coli promoter repression ≈ 500-1000 fold

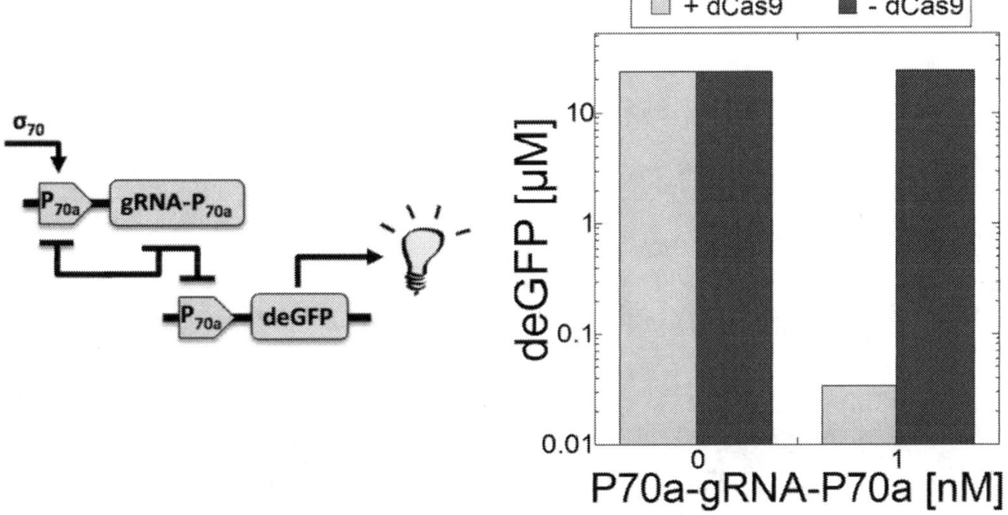

4
Genome scale TXTL

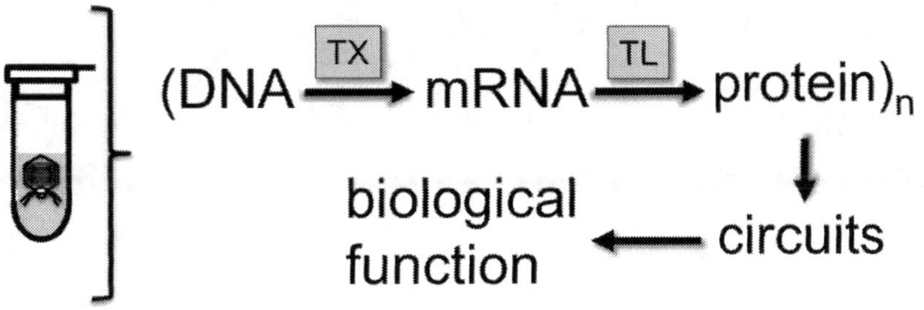

Goal: extend TXTL to biophysics
develop phage-related biotechnologies

Genome scale circuits
(information and self-organization)

• What is the real capacity of the system to construct circuits and living systems?

CFR batch mode: [Protein] = 80μM
E. coli: $[Protein]_{ave}$ = 500nM
} ≈ 100-200 genes

• Test the system with genome-sized information.

• Bacteriophages:
- search for genomes composed of 4-200 genes.
- condition/bottleneck: complexity of the interaction with the host beyond TX-TL.

Total synthesis of
(Paul Jardine, UMN)

T7

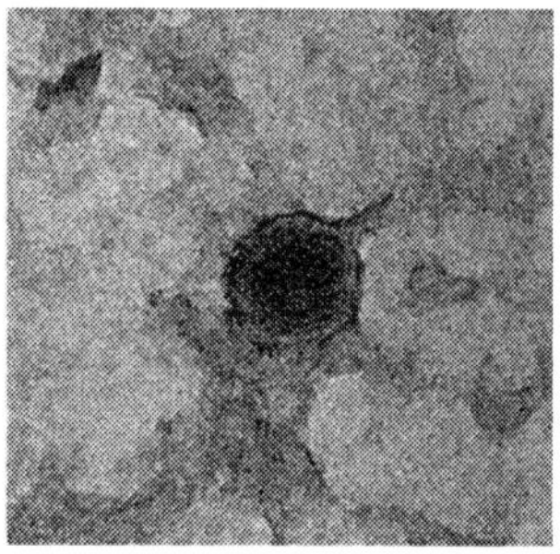

phix174

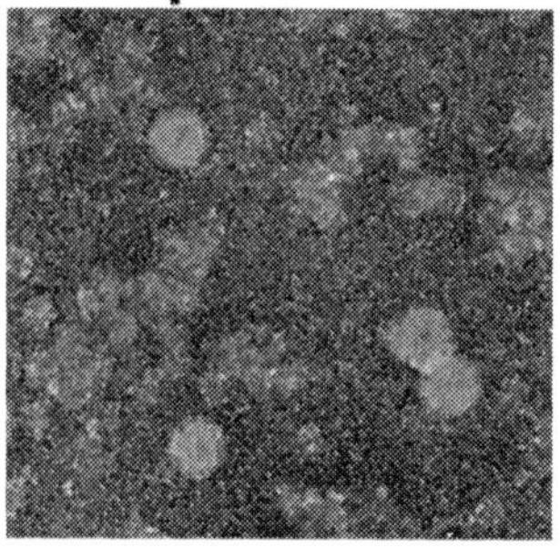

MS2

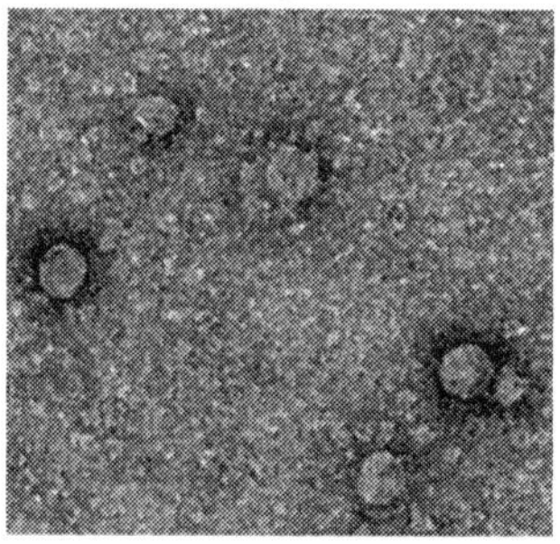

	Phage characteristics		
	T7	ΦX174	MS2
Genome type	linear dsDNA	ssDNA (Group II)	ssRNA
Genome size (bp or b)	39937	5386	3569
# genes	57	11	4
Viral type	lytic	lytic	lytic
Physical size	60 nm	32 nm	27 nm
	Cell-free synthesis		
# Phage/ml	$3.35\ 10^{11}$	$1.9\ 10^{12}$	$4.23\ 10^{12}$
# Phage/genome	2.2	0.6	0.047
Genome replication	Yes	No	undetermined

Garamella *et al* – ACS SB 2016

Cell-free synthesis of phage T7

(Shin et al 2012, Garamella et al 2016)

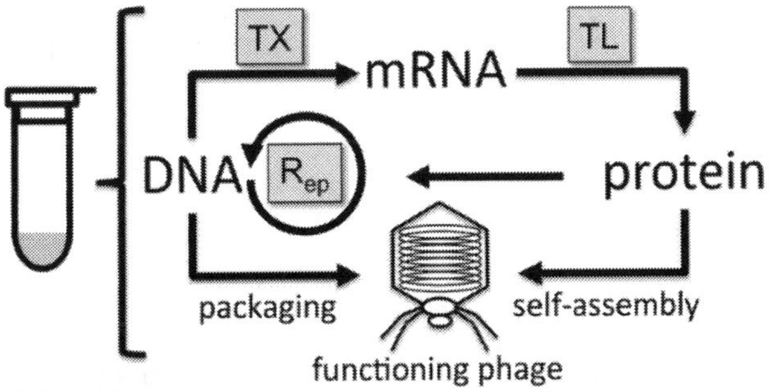

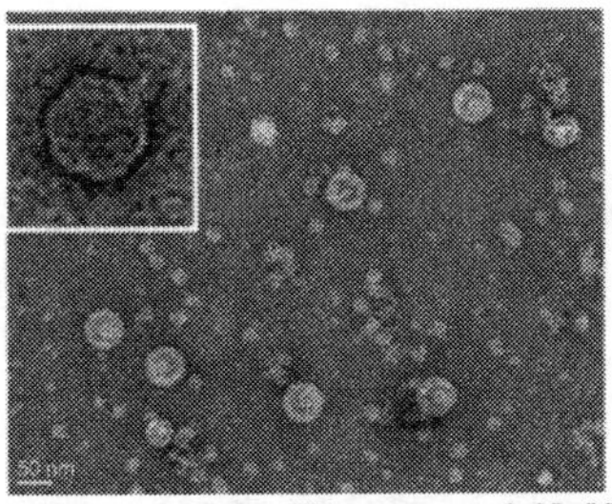

EM by Wei Zhang, UMN.

10^{11} phages/ml
(≈ 2 phage/genome).

A model of information to self-assembly process

T7 - E. coli Infection test

No difference observed between phages synthesized *in vivo* and *in vitro*.

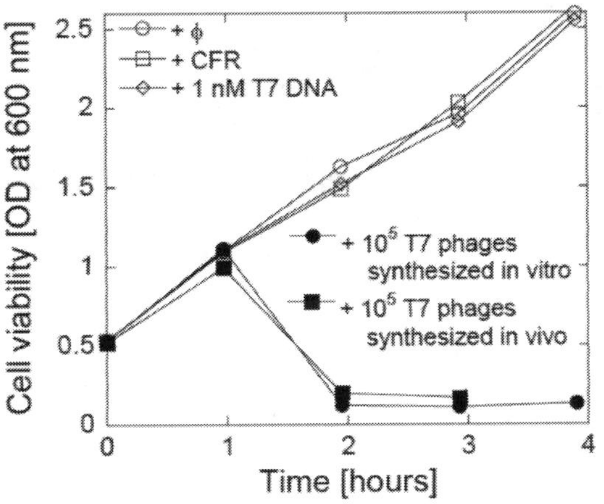

- phages per cell ≈ 100.
- phage cycle ≈ 25 min.
- E. coli division ≈ 30 min.

Molecular Crowding and phage T7

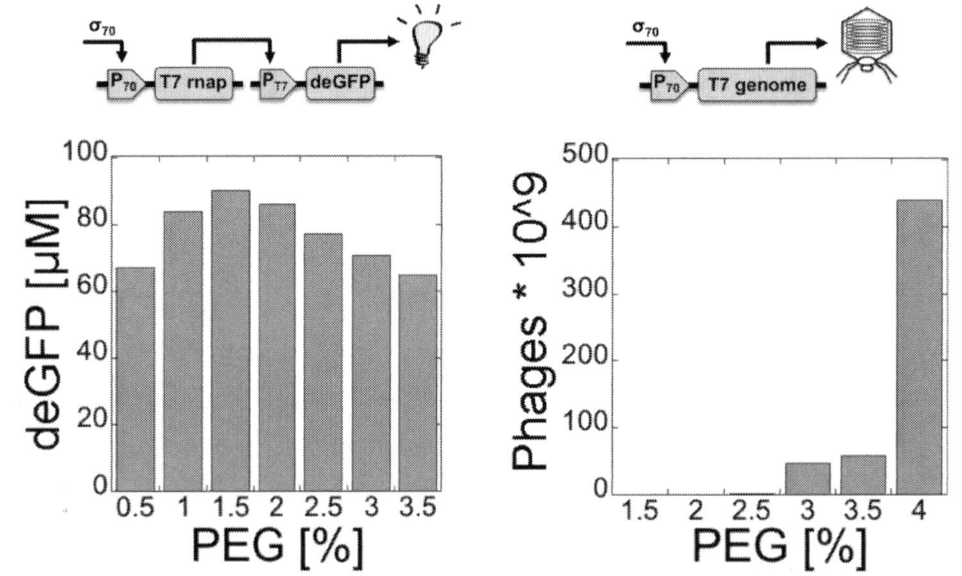

(1% PEG8000 = 1.25 mM, 250 mg/ml protein = 4 mM)

5
Synthetic cell

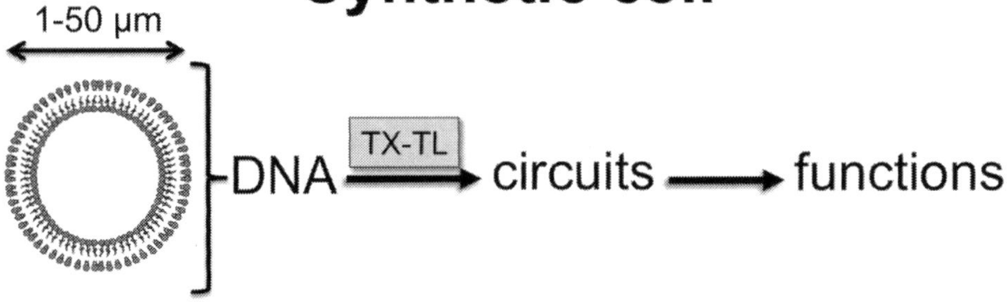

Goal: construct and characterize biological functions using TXTL in a cell-sized compartment

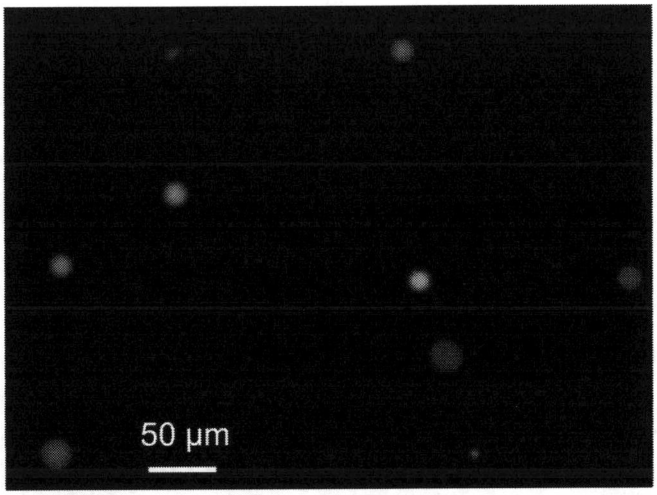

1 – 50 µm vesicles

Membrane channels (feeding)

Alpha Hemolysin

- toxin Staph. Aureus
- soluble monomer
- membrane heptamer
- channel of 1.4nm: 2-3kD

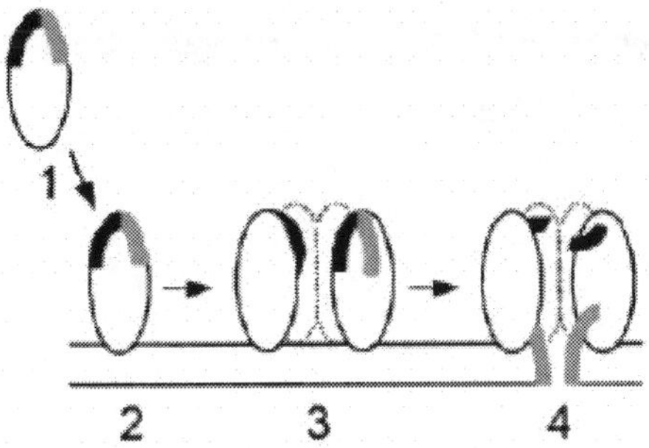

Song et al - 1997

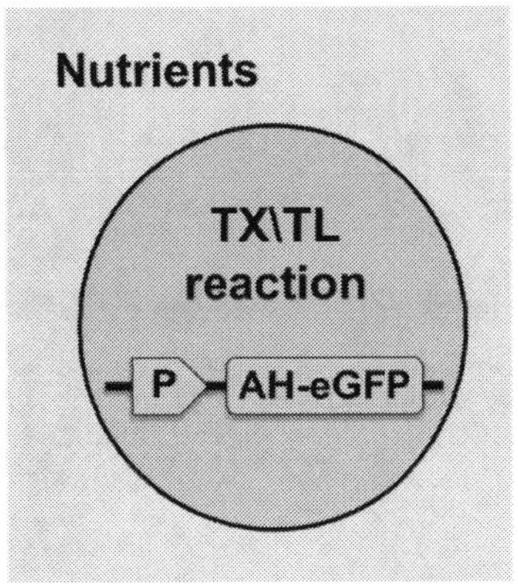

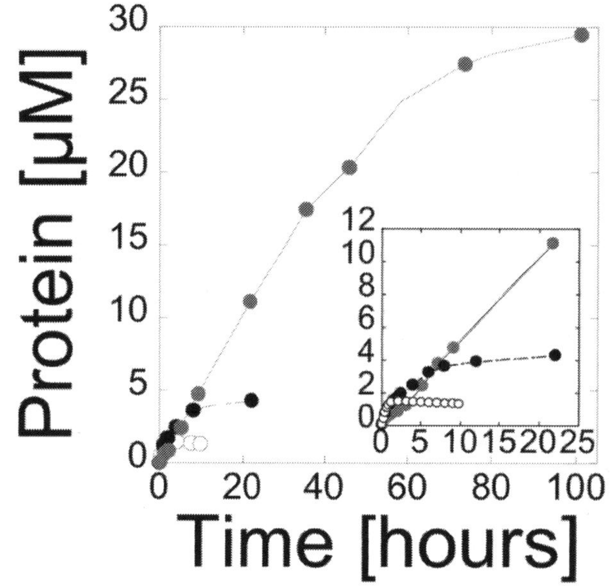

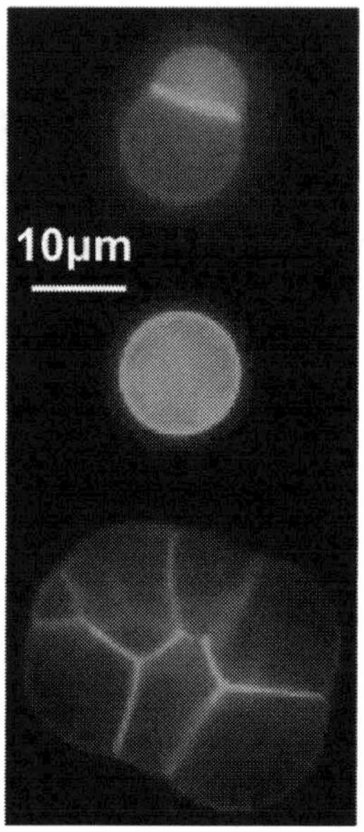

Noireaux and Libchaber 2004, Garamella et al - 2016

MreB Cytoskeleton (cell rigidity)

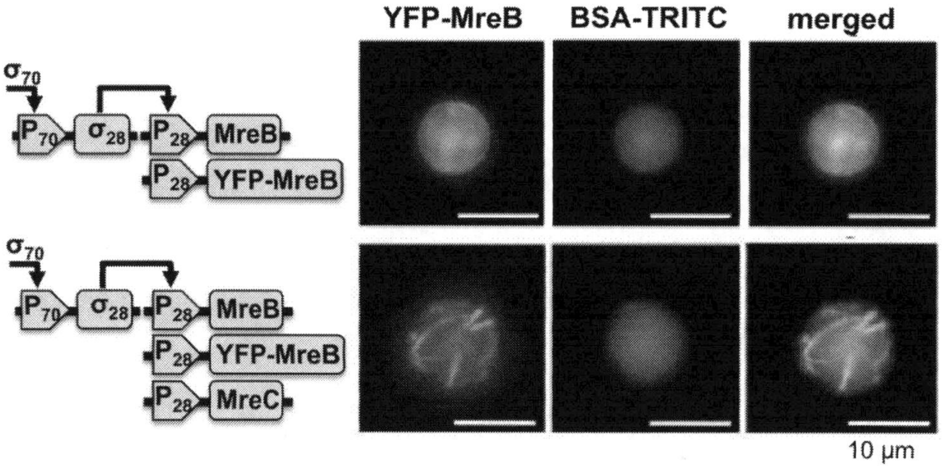

Maeda et al – 2012 (ACS SB)

FtsZ Cytoskeleton (cell division)

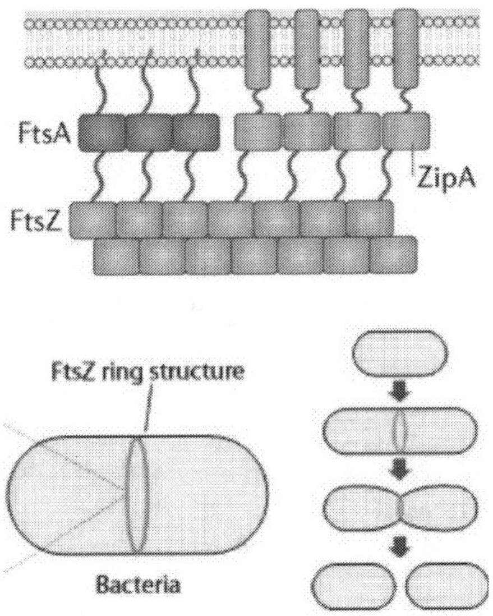

Conclusion

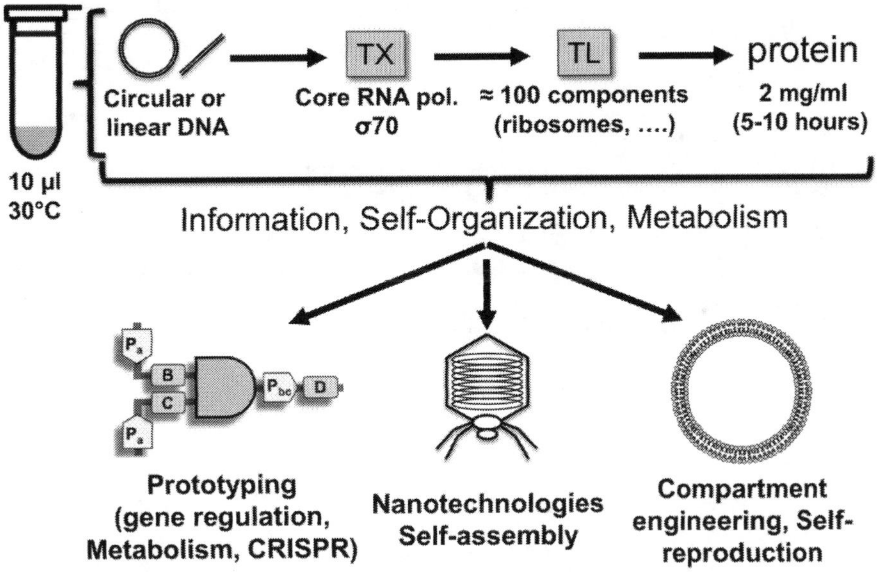

CSHL Summer Course in Synthetic Biology

Cold Spring Harbor Laboratory

Chase Biesel (NC State U), Vincent Noireaux (U Minnesota)
Mary Dunlop (U Vermont), Harris Wang (Columbia), Mo Khalil (Boston U), Michael Smanski (UMN)

Confirmed speakers: Megan McClean (Wisconsin), Doug Densmore (Boston U), Reshma Shetty (Ginkgo Bioworks), Samira Kiani (Arizona), Matt DeLisa (Cornell), Lingchong Yu (Duke), Chang Liu (UC Irvine),Caroline Ajo-Franklin (LBNL)

Topics: SynBio Cloning Strategies, Gene Expression Modeling, Genetic Circuits in TX-TL, RNA Engineering, Chromatin Engineering

August 2-14, 2016

Work done by:

Jon Garamella
Ryan Marshall
Mark Rustad
Kate Ciampa
Jonghyeon Shin (2008-2012)
Filippo Caschera (2012-2014)

postdoc positions opened

noireaux@umn.edu

Work sponsored by:

BSF United States - Israel Binational Science Foundation

NSF

HFSP

DARPA

ONR

Collaborations

Parts and Circuits
- Richard Murray (Caltech)
- Julius Lucks (Cornell)
- Christopher Voigt (MIT)
- Eduardo Sontag (Rutgers)
- Chase Beisel (NCSU)
- Elisa Franco (UCR)
- Mary Dunlop (U Vermont)
- Vishwesh Kulkarni (U Warwick)

Microfluidics/biophysics:
- Roy-Bar-Ziv (Weizmann).
- Doug Densmore (BU).
- Yusuke Maeda (Kyushu).
- Alexis Prevost (Paris).

Metabolism – Biosynthesis:
- Lital Alfonta (Ben Gourion)
- Eriko Takano (U Manchester)
- Albrecht Ott (Saarbrucken).
- Andreas Mershin (MIT).

Synthetic cell:
- Allen Liu (Michigan)
- Albert Libchaber (Rockefeller)
- Daniel Ladant (Pasteur)
- Giovanni Zocchi (UCLA)

Companies:
- Mycroarray (Michigan)
- Synthelis (France)
- Nexdot (France)

Past:

Jonghyeon Shin
PhD 2007-2012

Filippo Caschera
Postdoc 2012-2014

Present:

Mark Rustad

Ryan Marshall

Jon Garamella

Discovery and Engineering of Metabolic Pathways in Plants

Sattely, E., Stanford University

Originally presented on: 7/20/2016 11:00:00 - 11:30:00

Plants produce an impressive array of molecules important for both plant and human health. The discovery of biosynthetic pathways for plant natural products has classically been a slow process; as a consequence, few complete pathways are known and even fewer have been engineered. New plant genome sequences and synthetic biology tools have opened the door to three transformative research areas under investigation in my lab: 1) Identifying the minimum set of enzymes required to make known plant-derived molecules and non-natural derivatives through metabolic engineering, 2) discovering new molecules from plants, and 3) developing new strategies to sustainably enhance plant fitness. In this talk, I will describe some of our recent efforts to accelerate the discovery and engineering of complete plant pathways for known and novel molecules, not only in the model plant Arabidopsis but also in non-model plants.

Plants as Self-Sustaining Platforms for Synthetic Biology

Medford, J. I., Colorado State University

Originally presented on: 7/20/2016 11:30:00 - 12:00:00

Plant synthetic biology promises immense technological benefits, including the potential development of a sustainable bio-based economy. This goal will require the ability to design and produce synthetic gene circuits with predictable outputs. Such circuits are built from quantitatively characterized genetic parts, but can be slow in plants due to the time required for stable transformation. We have developed a method for rapid characterization of parts using transient expression in protoplasts and dual luciferase outputs. Using this method, we characterized hundreds of quantitatively characterized synthetic parts. We also developed descriptive models and normalization methods to account for stochastic effects. Our approach will speed the development of well-characterized genetic parts that can then be modeled, assembled and transferred to plants to produce predictable functions and controls such as toggle switches and positive feedback circuits. Another type of circuit is an environmental biosensor that enables a plant to detect a small molecule such as digoxin and initiate a transcriptional response. A transcriptional activator was fused to a computationally designed ligand binding domain; the fusion is unstable and inactive unless in the presence of digoxin. Upon binding the ligand, the activator activates transcription of a reporter or gene of interest. Such circuits can be precisely tuned through mathematical modeling and incorporating well-characterized and well-balanced parts, and can be used to control expression of desirable traits or the production of materials in plants.

Hidden interactions arising from resource sharing in genetic circuits

Domitilla Del Vecchio
Mechanical Engineering
MIT
SBC@MIT
Synthetic Biology Center @ MIT

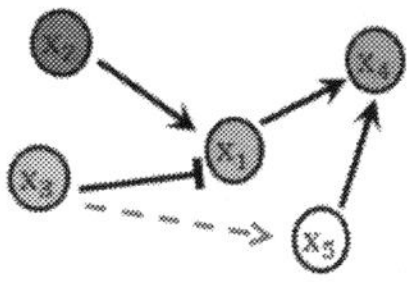

SEED, July 2016

Biological circuits are designed bottom-up

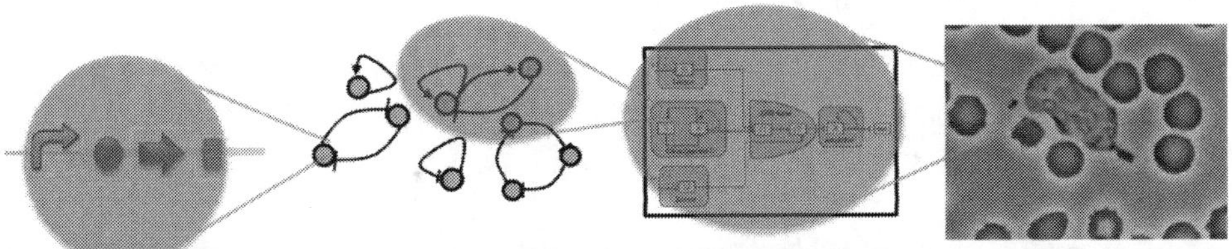

Library of parts
Promoters
RBS
coding sequence
terminators
...

Library of functional modules and logic gates
Oscillators
toggle switches
AND, OR, NOR
...

Multi-module systems
Molecular signature classifier
...

Programmed cells
For medical, energy, environment applications

2

Functional modularity is crucial for bottom-up design...but it is not a natural property

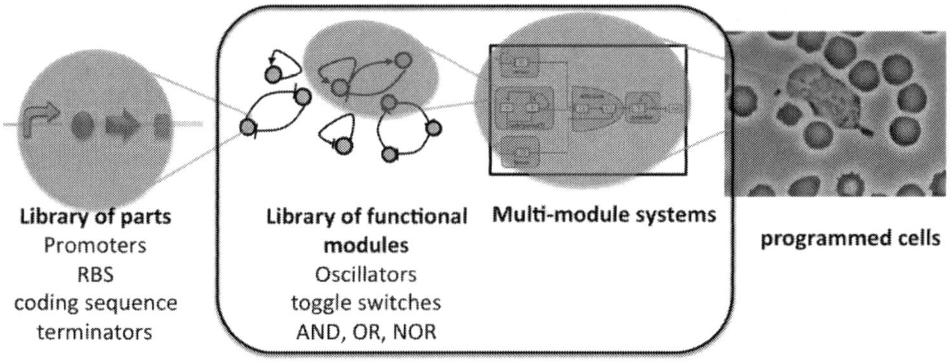

The technology to build arbitrarily complex circuits is there, but these work unpredictably as the complexity grows

D. Del Vecchio. *Trends in Biotech.* 2015

3

Functional modules are context-dependent

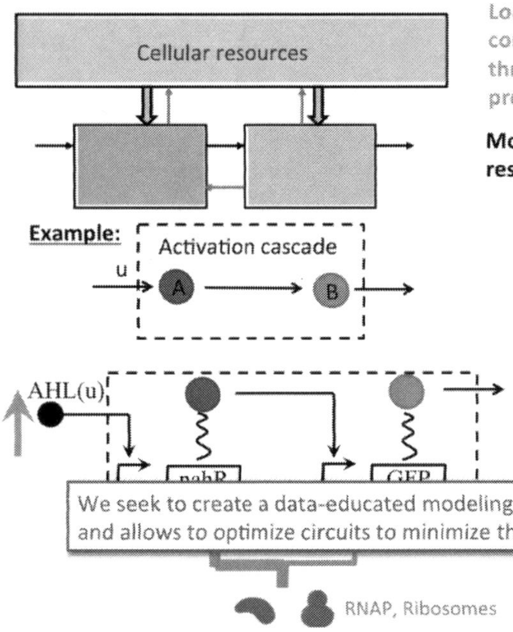

Load between modules is one source of context dependence (formally modeled through retroactivity) – tackled in our previous work

Modules also apply a load to shared cellular resources: creates subtle couplings

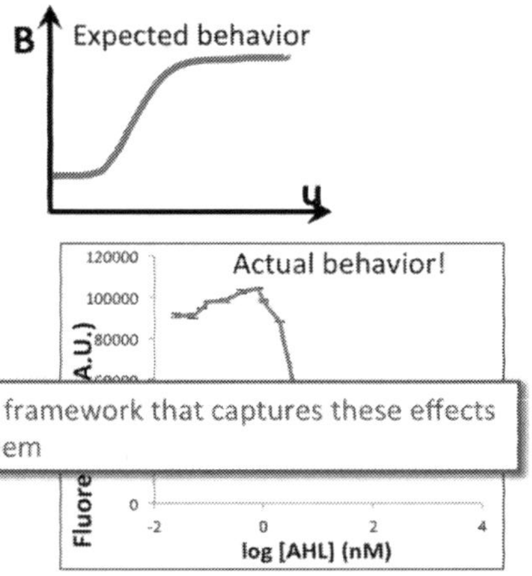

We seek to create a data-educated modeling framework that captures these effects and allows to optimize circuits to minimize them

Outline

Effects of sharing limited resources at the node level

Effects of sharing limited resources at the network level

5

Model of resource sharing among unconnected genes

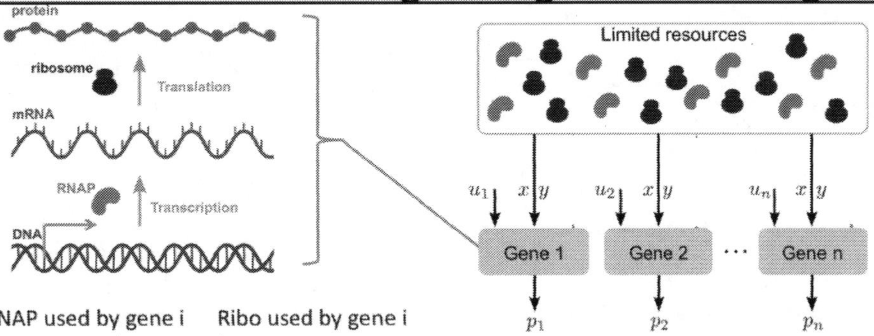

RNAP used by gene i — Ribo used by gene i

$$p_i = u_i q_i \frac{x}{x+\kappa_i}\frac{y}{y+k_i}$$

Prop to DNA copy number — Promoter dissociation constant — RBS dissociation constant

Conservation of total resources

$$x_T = x + \sum x_i, \quad y_T = y + \sum y_i$$

Steady state behavior (simulation)

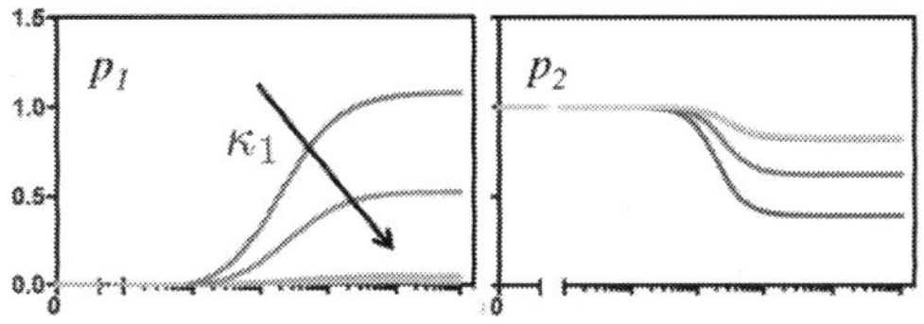

p_2 decreases by up to 70%:
Gene expression becomes coupled 6

Data confirms gene expression becomes coupled

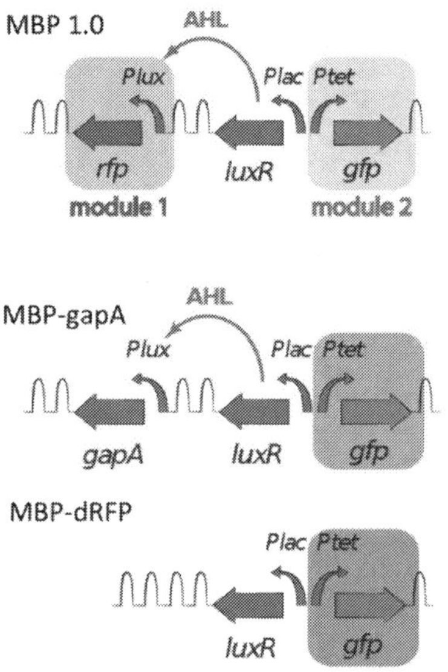

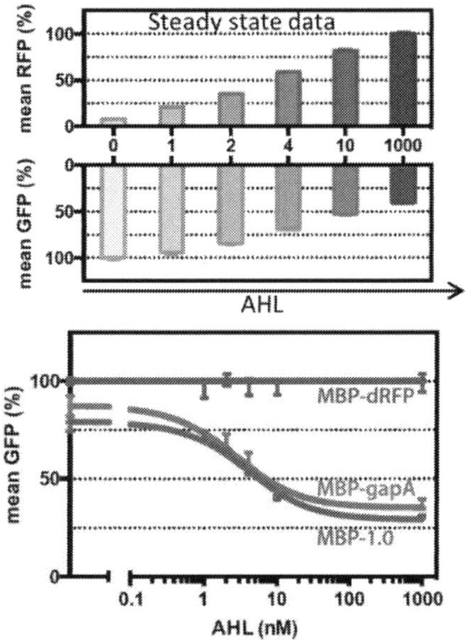

Gyorgy et al. *Biophysical J.* 2015

The expression of a constitutive gene reduces up to 70% when a different gene is induced because of RNAP/ribosome sharing. Both are on medium plasmid copy number (about 60 copies)

7

Model: The coupling is captured by *isocost* lines

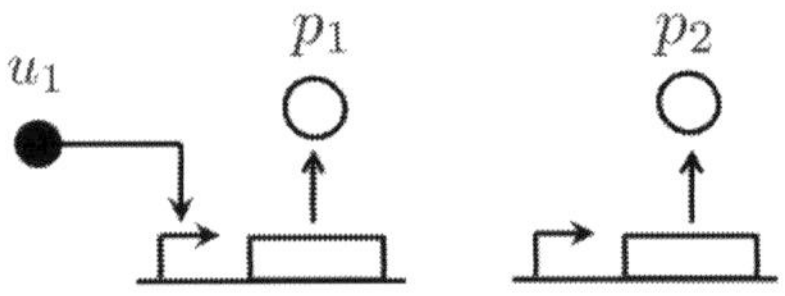

$x \ll \kappa_i, \quad y \ll k_i$ linear binding assumption

$$\alpha p_1 + \beta p_2 = y_T$$

allowed ribosome budget

price of p_1 price of p_2

price of p_i increases as its RBS strength decreases

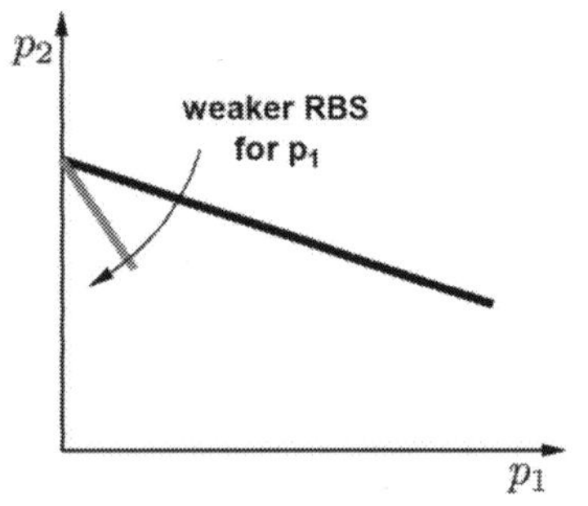

price of p_i increases as copy number decreases

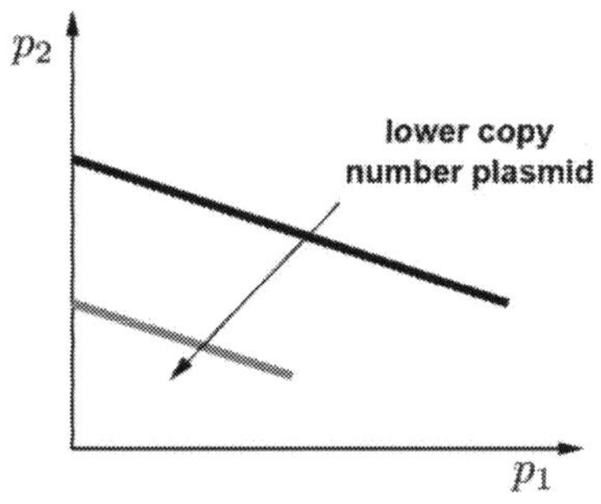

Gyorgy et al. *Biophysical J.* 2015

Experiments confirm the existence of isocost lines

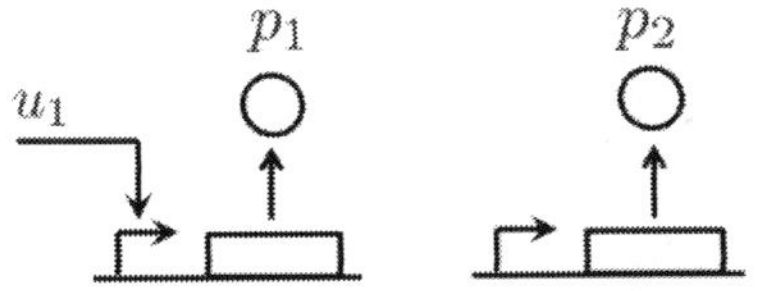

price of p_i increases as its RBS strength decreases

$$\alpha p_1 + \beta p_2 = y_T$$

price of p_1 — price of p_2 — allowed ribosome budget

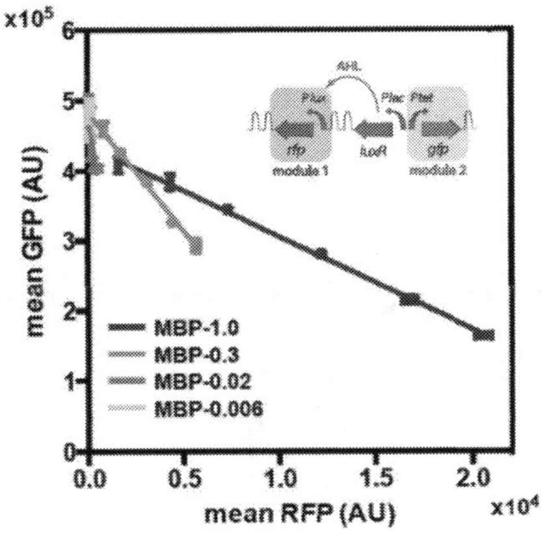

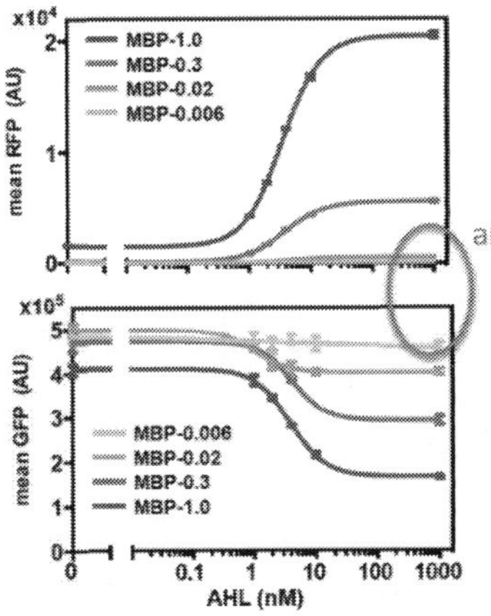

ribosomes are responsible for resource limitations (not RNAP)

Gyorgy et al. *Biophysical J.* 2015

9

Experiments confirm the existence of isocost lines

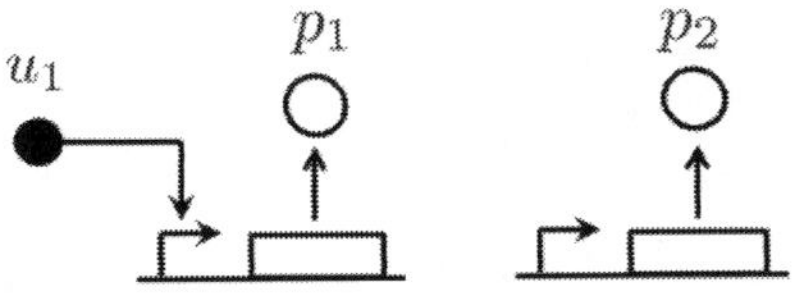

price of p_i increases as copy number decreases

$$\alpha p_1 + \beta p_2 = y_T$$

price of p_1 — price of p_2 — allowed ribosome budget

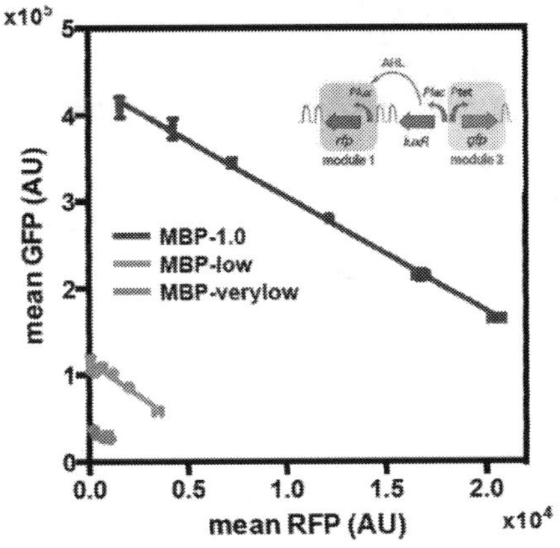

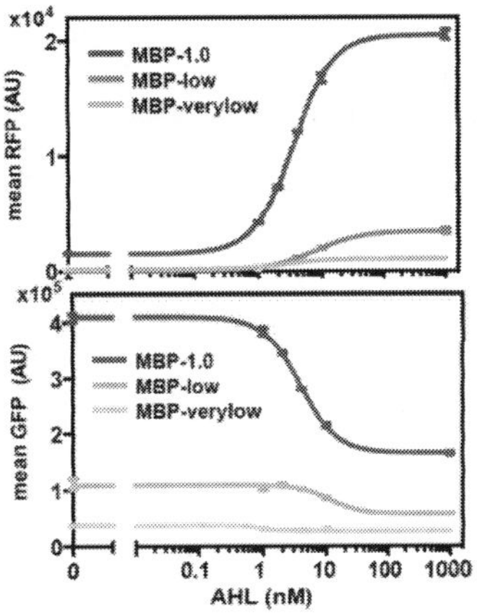

Gyorgy et al. *Biophysical J.* 2015

Chromosomal integration suggests resources are *somewhat* partitioned

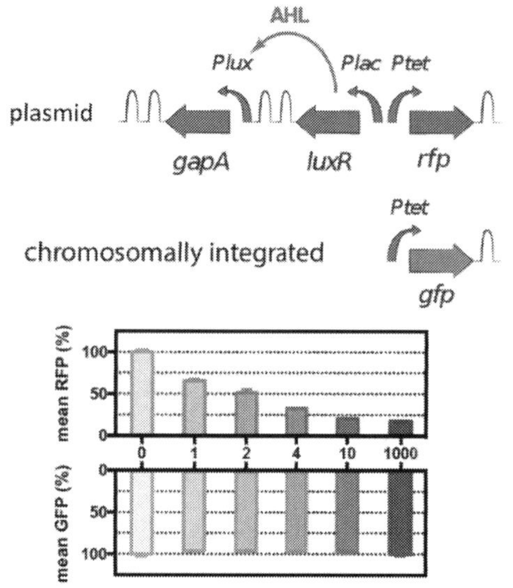

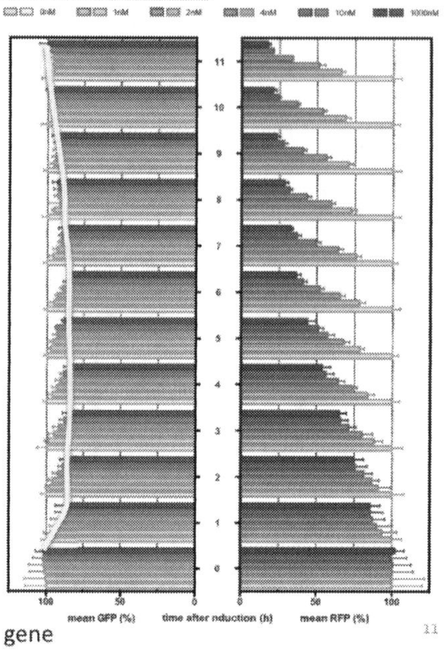

A constitutive chromosomal gene is only
Transiently affected by expression of a plasmid gene

Gyorgy et al. *Biophysical J.* 2015

11

Outline

Effects of sharing limited resources at the node level

Effects of sharing limited resources at the network level

12

Network-level effects of limited resources

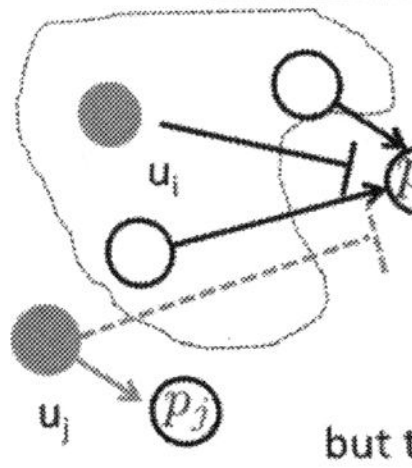

$$\dot{p}_i = \underbrace{\alpha_i F_i(u_i)}_{H_i(u_i)} - \delta_i p_i \qquad \alpha_i \propto \frac{x}{\kappa_i}\frac{y}{k_i}$$

$H_i(u_i)$ Hill function

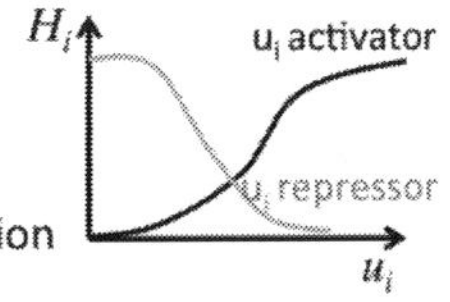

standard models assume x *(RNAP)* and y *(Ribo)* are constant parameters

but they are not: $x_T = x + \sum x_i, \quad y_T = y + \sum y_i$

New model

$$\dot{p}_i = \underbrace{\bar{\alpha}_i \frac{F_i(u_i)}{1 + J_i F_i(u_i) + \sum_j J_j F_j(u_j)}}_{G_i(u)} - \delta_i p_i$$

new interactions emerge (not due to regulatory links)

$$J_k \propto \frac{d_k}{k_k}\frac{x_T}{\kappa_k}$$ quantifies the demand for resources of node k

effective interaction graph

$$\frac{\partial G_i(u)}{\partial u} = \frac{\bar{\alpha}_i \partial F_i/\partial u}{(1 + \sum_j J_j F_j)^2} - \frac{\bar{\alpha}_i F_i \sum_j J_j \partial F_j/\partial u}{(1 + \sum_j J_j F_j)^2}$$

re-scaling of existing regulatory links (they are weakened) — first term

new interactions — second term

13

Network-level effects of limited resources

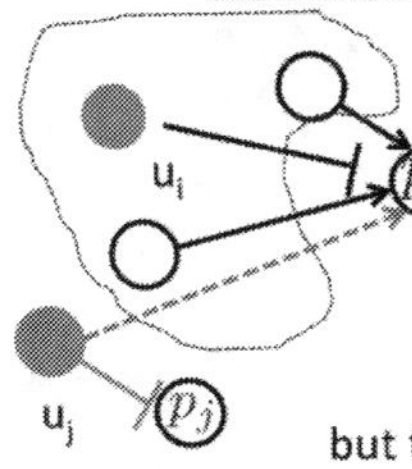

$$\dot{p}_i = \underbrace{\alpha_i F_i(u_i)}_{H_i(u_i)} - \delta_i p_i \qquad \alpha_i \propto \frac{x}{\kappa_i}\frac{y}{k_i}$$

$H_i(u_i)$ Hill function

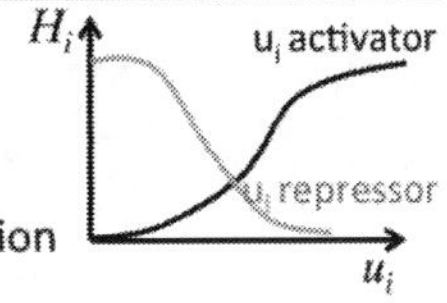

standard models assume x *(RNAP)* and y *(Ribo)* are constant parameters

but they are not: $x_T = x + \sum x_i, \quad y_T = y + \sum y_i$

New model

$$\dot{p}_i = \underbrace{\bar{\alpha}_i \frac{F_i(u_i)}{1 + J_i F_i(u_i) + \sum_j J_j F_j(u_j)}}_{G_i(u)} - \delta_i p_i$$

new interactions emerge (not due to regulatory links)

$$J_k \propto \frac{d_k}{k_k}\frac{x_T}{\kappa_k}$$ quantifies the demand for resources of node k

effective interaction graph

$$\frac{\partial G_i(u)}{\partial u} = \frac{\bar{\alpha}_i \partial F_i/\partial u}{(1 + \sum_j J_j F_j)^2} - \frac{\bar{\alpha}_i F_i \sum_j J_j \partial F_j/\partial u}{(1 + \sum_j J_j F_j)^2}$$

re-scaling of existing regulatory links (they are weakened) — first term

new interactions — second term

14

Effective interactions arise

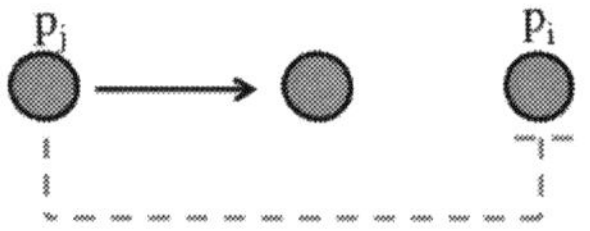

If p_j is an activator, it is an effective repressor for any node not regulated by it

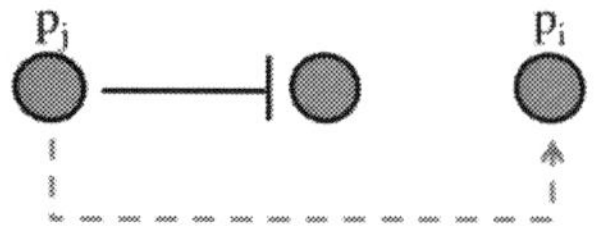

If p_j is a repressor, it is an effective activator for any node not regulated by it

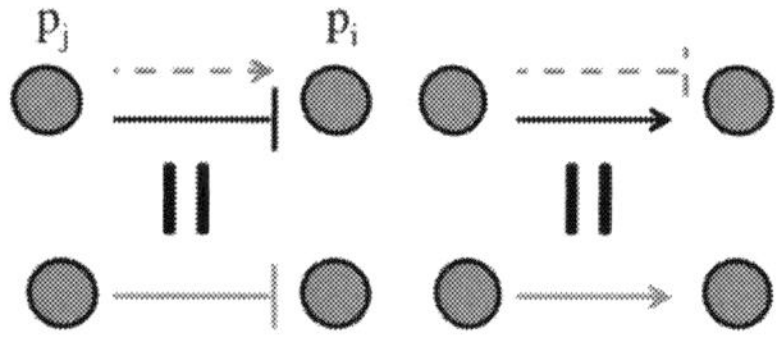

The overall effect of p_j on a node p_i regulated by it is not changed if p_i is the only node regulated by p_j (but it is weaker)

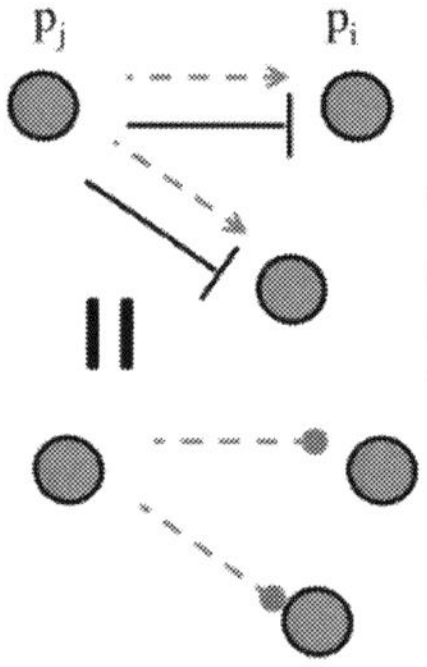

The overall effect of p_j on a node p_i regulated by it is undetermined if p_i Is not the only node regulated by p_j

Qian and Del Vecchio, *Proc. American Control Conference* 2015 15

Summary

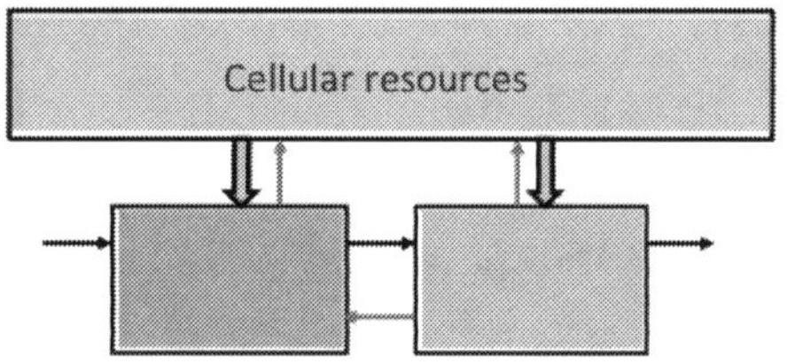

Modules apply a load to shared cellular resources: creates subtle couplings

We have uncovered and modeled the node and network-level effects of sharing TX/TL resources in gene circuits

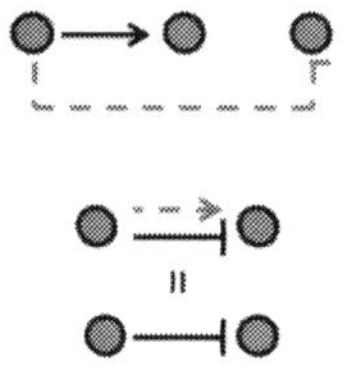

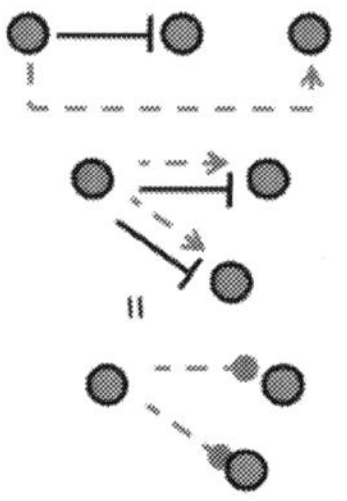

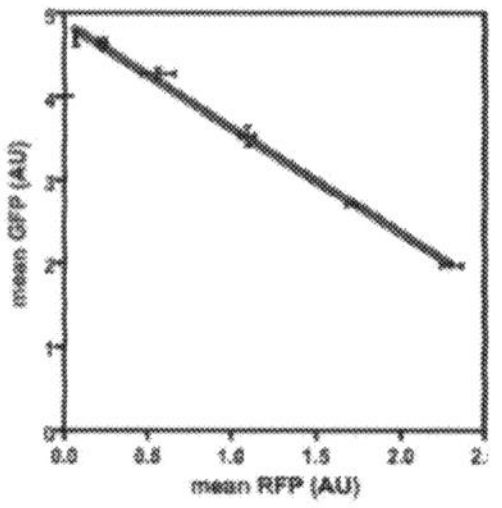

Currently: How to fix this?

distributed feedback controllers to make each node expression adapt to changes in free ribosomes (quasi-integral feedback)

17

Text Book

Biomolecular Feedback Systems

***Princeton* (2014)**

by

Domitilla Del Vecchio

and Richard Murray

https://www.cds.caltech.edu/~murray/amwiki/BFS

18

Constructing Programmable Genetic Circuits through Generalizable RNA Tools

Lee, Y. J., Washington University in St. Louis
Hoynes-O'Connor, A., Washington University in St. Louis
Leong, M., Washington University in St. Louis
Moon, T. S., University of California, San Francisco
Originally presented on: 7/20/2016 14:00:00 - 14:30:00

The past decade has witnessed the tremendous power of RNA regulators for precise and orthogonal gene expression control. Although gene regulation research has focused more on protein regulators than RNA regulators, RNA regulation has many advantages over protein regulation. RNA regulators are easier to design than protein regulators, due to the simplicity of base-pairing rules and our increased understanding of RNA folding and functions. In addition, RNA regulation is faster and has a lower metabolic burden than protein regulation because regulatory RNAs are shorter than regulatory proteins, and RNA regulators do not undergo the cellular resource-intensive process of translation. Still, synthetic RNA regulators for precise and tunable gene expression control are in great demand, and we aim to provide generalizable RNA regulators that allow for rational reprogramming of cellular functions. To this end, we have employed the CRISPR interference system (CRISPRi) from *Streptococcus pyogenes* and synthetic antisense RNAs (asRNAs) to repress or derepress target genes in a programmable manner. Specifically, we demonstrated that the repressed target output by the CRISPRi system can be derepressed by activating transcription of asRNAs that sequester their target small guide RNAs (sgRNAs). Our RNA regulators are unique in that they consist of both the nuclease-deficient Cas9 (dCas9)-customized sgRNA system and asRNAs, which can control the dynamics of gene expression. Furthermore, these RNA-based genetic circuits have been optimized by varying design parameters (e.g., the hybridization free energy of the sgRNA-asRNA complex), and tunable levels of derepression have been achieved (up to 95%). These RNA regulators can also target multiple genes orthogonally, allowing for multiplex gene regulation. To our knowledge, this is the first demonstration of an RNA circuit that can integrate multiple types of RNA regulators in bacteria. This new system will allow for rational reprogramming of cellular processes, facilitating diverse synthetic biology applications.

Rewiring the DNA Binding Domains of Bacterial Two-Component System Response Regulators

Tabor, J. J., Rice University

Originally presented on: 7/20/2016 14:30:00 - 15:00:00

Two-component systems (TCSs) are the most ubiquitous sensing modality in biology, and the primary means by which bacteria sense environmental signals. However, most TCSs have unknown functions, and virtually no progress has been made in repurposing them as engineered sensors for medical, industrial, and scientific applications. The major limitations are that the bacteria containing them cannot be grown or genetically manipulated in the laboratory, TCSs are subject to complex and poorly understood genetic and environmental regulation, and there are no methods to adjust TCS sensitivity to desired levels.

My group has recently developed technologies to overcome these limitations. In this talk, I will describe the development of technologies for 'mining' TCSs from nature, identifying the inputs they sense, and recapitulating their natural function in non-native host bacteria. I will also describe new technologies to increase their often poor dynamic range hundreds of fold – to levels sufficient for robust *in vivo* sensing, and to rationally tune their sensitivities from the sub-micromolar to millimolar range, overcoming a long-standing challenge in biological sensor design. Finally, I will discuss early progress in using our engineered bacteria to diagnose and study inflammation in the colon, which is linked to pathogen infection, obesity, cancer, and other diseases.

Reverse-Engineering Human Gut Microbiome Synthetic Ecologies

Venturelli, O.,
Hsu, R.,
Fisher, G.,
Carr, A.,
Ng, A.,
Justice, N.,
Jenkins, S.,
Northen, T., Joint BioEnergy Institute
Arkin, A. P., University of California at Berkeley
Originally presented on: 7/20/2016 15:00:00 - 15:30:00

The human gut microbiome is a diverse and interconnected network that has a significant impact on health, nutrition and development. Developing the capability to design functional, stable, energy efficient and resilient multi-species communities to mediate human health is an obtainable grand challenge. Central to this problem is systematically mapping unknown inter-species interactions that generate emergent community-level properties and the development of predictive computational models of ecosystem function and dynamics. We integrated experiment and computational modeling to decipher inter-species interactions among members of a human gut microbiome synthetic ecology by leveraging a bottom-up and top-down combinatorial community assembly approach. These communities demonstrated a myriad of population dynamic responses and patterns of co-existence. Three hub organisms were identified in the network that we hypothesize play a central role in shaping community function and dynamics. Measurement of metabolites that were produced or consumed by single species as a function of time provided key insights into the mechanisms driving inter-species interactions. 50% of interactions could be explained by resource competition or metabolite cross-feeding among species based on known metabolites. We systematically characterized the response of a multi-species community to temporal perturbations based on the hypothesis that diverse species growth parameters and inter-species interactions could generate differential sensitivity to periodic environmental shifts. The abundance of specific species varied significantly as a function of transfer period, demonstrating that time-dependent inputs can be used to modulate community diversity and stability. In sum, these methods will be used to extract generalizable design principles for ecosystem engineering including stability, resilience to perturbations and resistance to invasion.

YEAST-BASED BIOSENSORS AS A LOW-COST HEALTHCARE DIAGNOSTIC

Keith EJ Tyo
Assistant Professor
Department of Chemical & Biological Engineering
Northwestern | McCORMICK SCHOOL OF ENGINEERING

Point-of-care Testing

Clinical outcomes

Outcome
Faster decision making
Starting treatment earlier
Improved adherence to treatment
Reduced incidence to complications
Quicker optimization of treatment

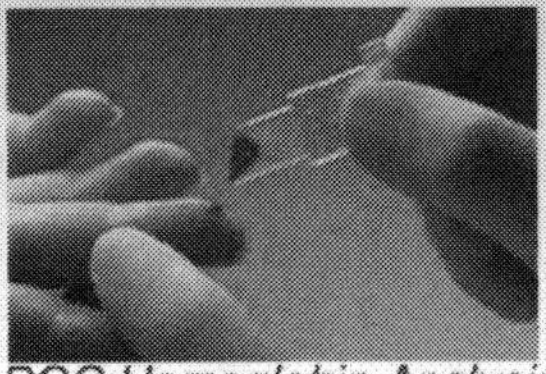

POC Hemoglobin Analysis

Economics outcomes

Outcome
Reduced # of clinical visits
Reduced length in hospital
Fewer unnecessary interventions
Better optimized drug treatment
Better resource allocation

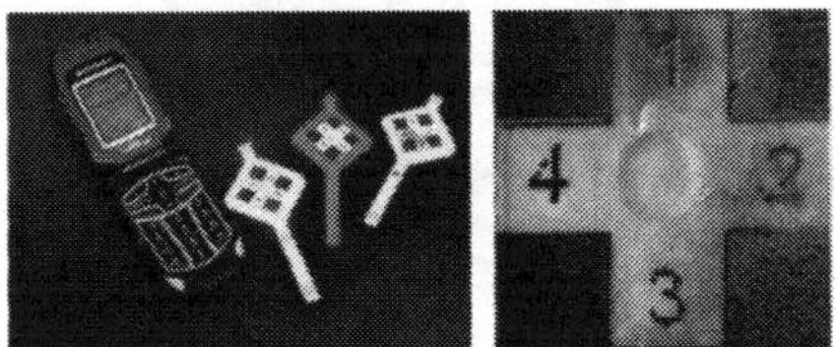

At home TB drug compliance test

BMJ **322**, 1285–1288 (2001)

Testing in Resource-poor Locations

Infant HIV testing

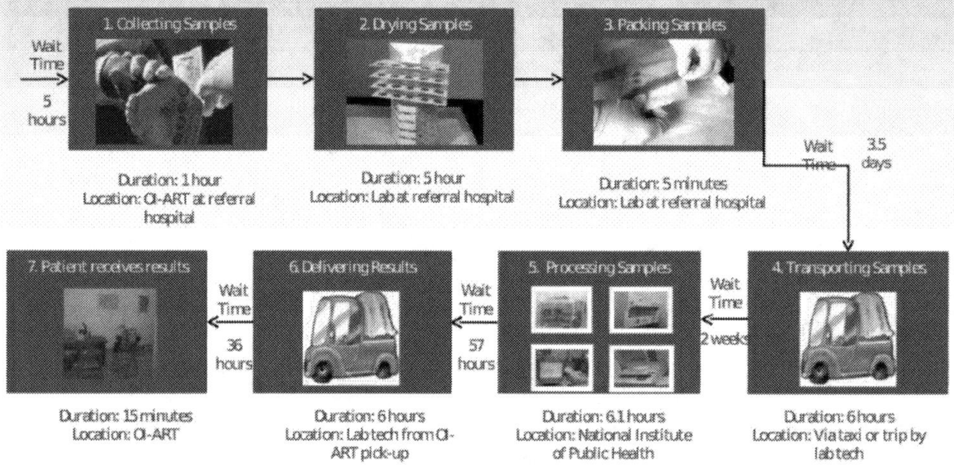

Courtesy: Prof. Kara Pallamountain

Renal Failure & Chronic Kidney Disease

Worldwide: 5-7% has Stage 3-5

Stage 5 is End-stage

US Chronic Kidney Disease Population

9.9 million

End-stage Renal Failure is Expensive

2-3 % of US health budget

(Dialysis and kidney transplants)

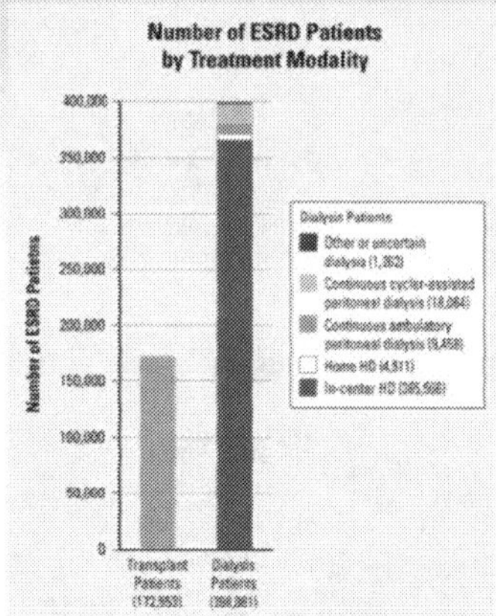

NIDDK Statistics (NIH)
Clin Kidney J (2015) 8: 54–60

Diagnosing CDK

Serum Creatinine

Filtered by kidney

Accumulates in renal disease

24 h Urine Creatinine

Amount filtered

Reduced clearance in renal disease

Urine Albumin/Creatinine Ratio (ACR)

Albumin retained by kidney

Albumin↑ Creatinine↓ in renal disease

24% of C/ACR (+) results are false positive
2.4 million people receiving unnecessary treatment

16% of C/ACR (-) results are false negatives
1 of 100 mortalities in >60 yr related to undiagnosed CDK

Northwestern | ENGINEERING 5 JAMA, April 20, 2011—Vol 305, No. 15 1545
Yeast based biosensors

Cystatin C

Creatinine

Breakdown product of creatine phosphate in muscle

Filtered from blood by kidneys

Affected by

- Kidney filtration rate
- Muscle mass
- Age
- Sex
- Race

Cystatin C

Protease inhibitor produced by all cells in the body

Filtered from blood by kidneys

Affected by

- Kidney filtration rate
- % of lean body mass

<u>No POC diagnostic available</u>

Northwestern | ENGINEERING 6 Am J Kidney Dis 48: 712–719 (2006)
Yeast based biosensors

Cystatin Diagnostics

Cystatin = 120 amino acids, 13.3 kDA
Kidney disease = μM urine levels

Nephelometry

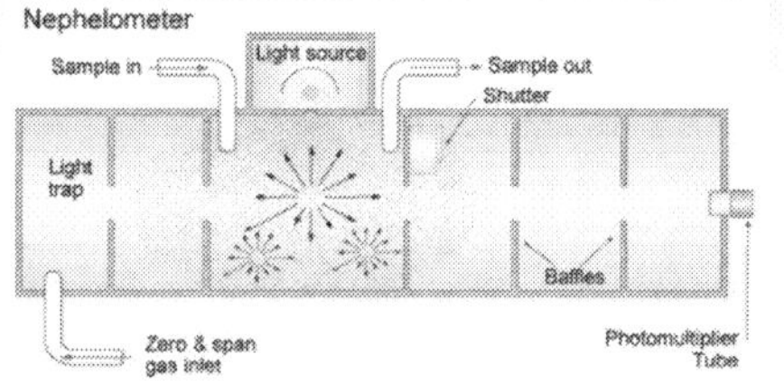

Northwestern | ENGINEERING   Qld.gov/au

http://en.wikipedia.org/wiki/Nephelometer

Yeast-based Biosensors (YBBs)

- Long shelf life at room temperature
- Easily transportable
- Cheap to manufacture
- ***Programmable (we can readily modify DNA)***

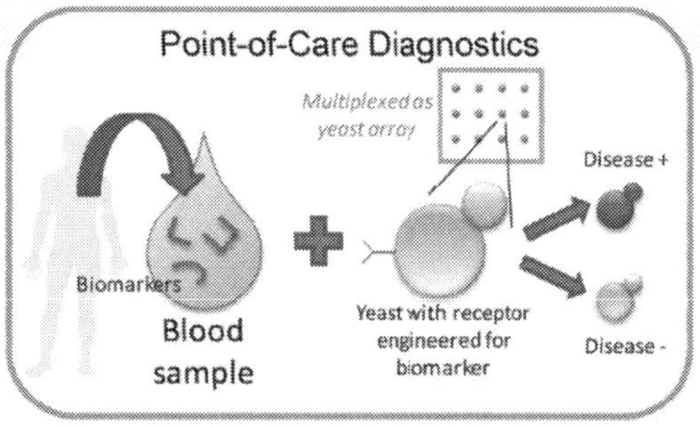

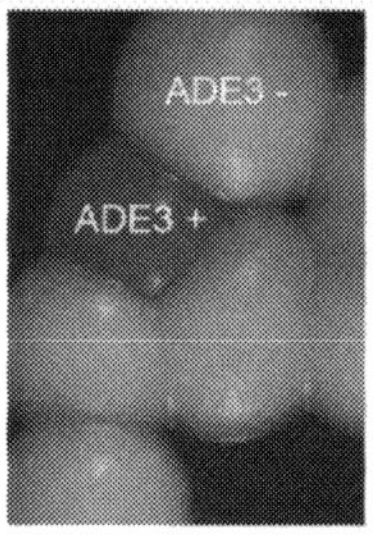

Northwestern | ENGINEERING   8

BMC Genetics 2002, **3**:4

Other Organism-based Biosensors

Carbon monoxide sensor

Opioid assay

G-Protein Coupled Receptors

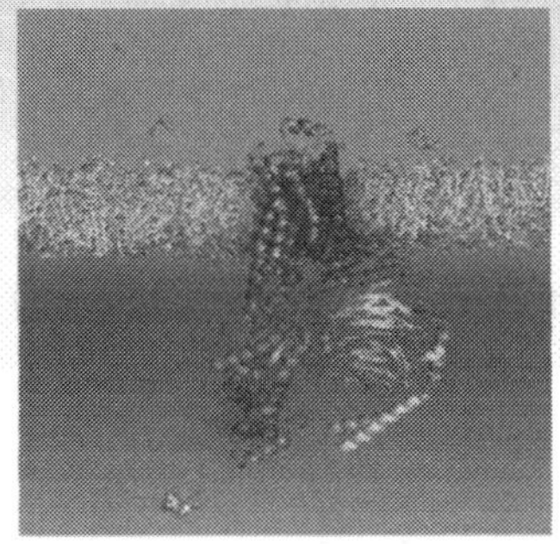

Nature's biosensor

- Small molecules
- Protein/peptides
- Hormones
- Olfactory
- Light

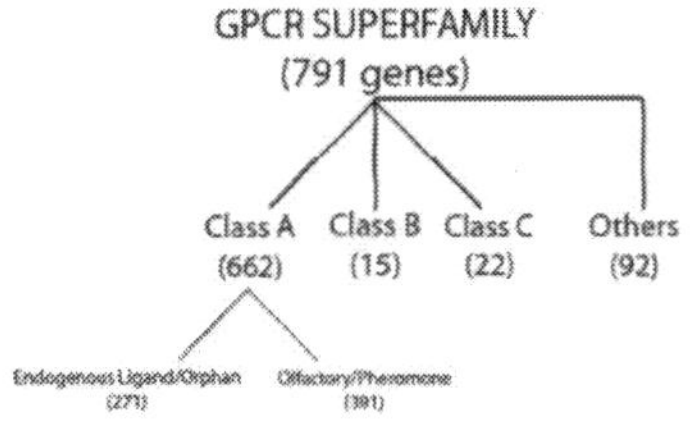

7 transmembrane structure is highly evolvable

Genomics 88 (3): 263–73 (2006)

The Yeast Pheromone GPCR System

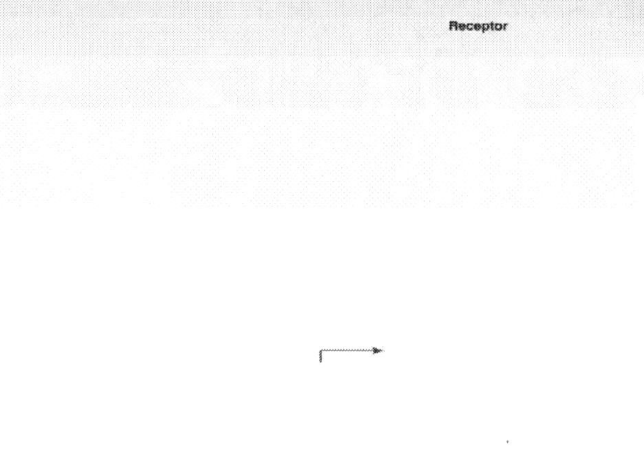

α-factor
13 amino acid peptide

EC_{50} = 18 nM

Trends Biotechnol, 2005. **23**(7): p. 367-73.
Yeast based biosensors

Diagnostic Process

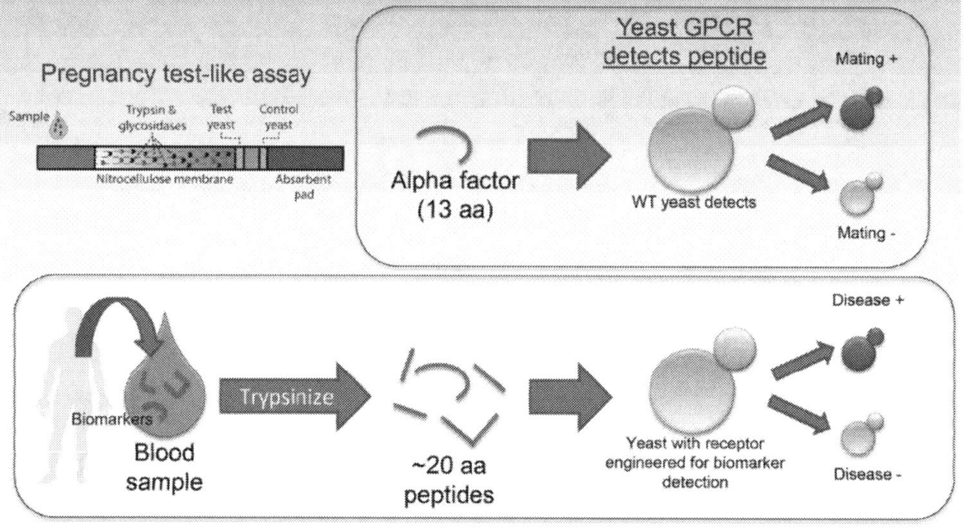

Technical Challenges of Engineering a GPCR

Mutations to alter GPCR's specificity are not known

No crystal structures for Ste2

Some sequence information about GPCR residues that interact with native ligand

Northwestern | ENGINEERING 13

Source: OldEurope
Yeast based biosensors

Directed Evolution

(1) Error-prone PCR to generate mutations in DNA
(2) Sort cells with activated receptor by expression of fluorescent protein
(3) Sequence and generate dose-response curves

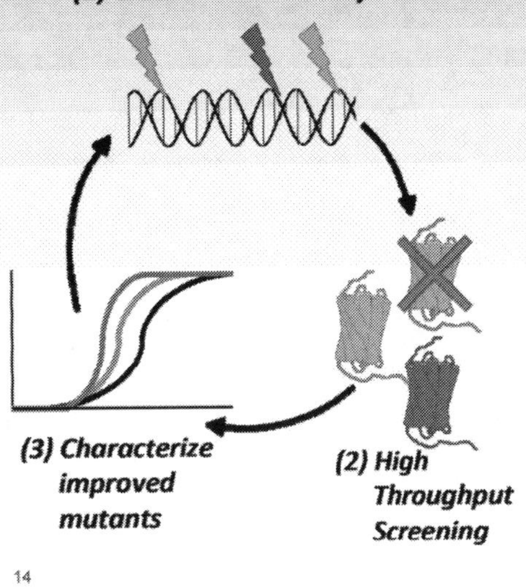

Northwestern | ENGINEERING 14

Evolving a Cystatin YBB

α-factor	WHWLQLKPGQPMY
Cystatin	--ALDFAVGEYNK

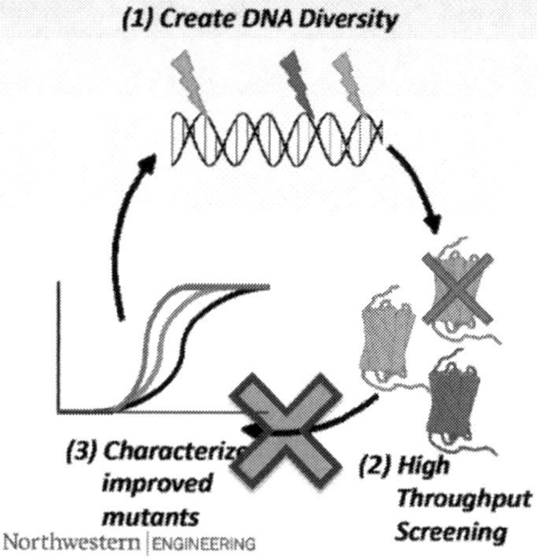

Unsuccessful at isolating a receptor for the cystatin peptide

Is the sequence of the target too far from α-factor?

Evolutionary Stepsize

Library Size ~ 10^6 mutations
STE2 ~ 400 aa (1200 bp)

All single mutations – 400 x 20 = ~10^4
All double mutations = ~10^8

Evolution likes baby steps

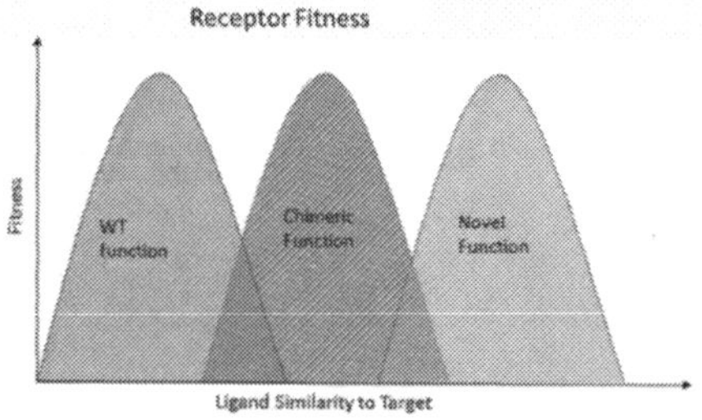

(in revision)

How Large of a Step Can We Take?

Substrate walking

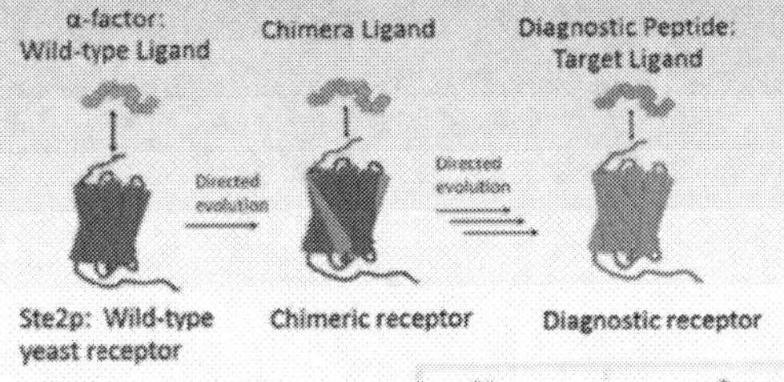

Design several peptides of varying length

Test if we can evolve a receptor for the peptide in one step

Name	Sequence	Step Size
α-factor	WHWLQLKPGQPMY	0.00
Cys1	-HALALKPGEPMY	3.46
Cys2	--ALALKPGEPMY	4.92
Var 44	WAWLQLKPGQPMY	0.77
Var 56	WHYLQLKPGQPMY	0.69
TBBI	WHWLQLKPGEPLYGR	2.15
TBSI	LHLLALKPGQPMY	2.46
TBE	LHLLALKPGQPLYGR	4.38
TBM	LHLLAGQPGESLYGR	6.15

(in revision)
Yeast based biosensors

Heuristic Step-size Determination

Step sizes beyond ~5 are too large

(Cystatin peptide was ~8.3)

Design several steps that are less than 5

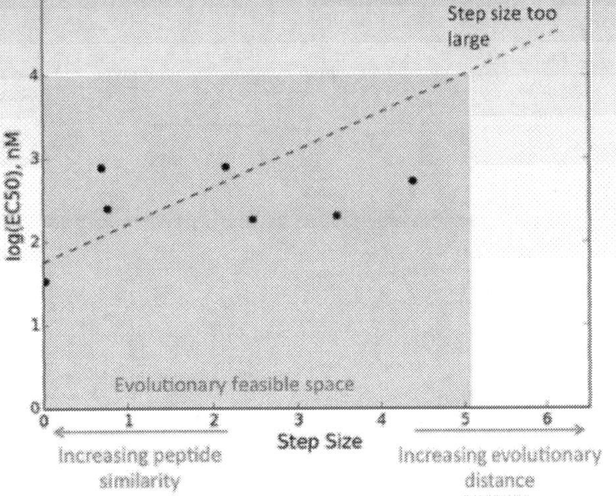

(in revision)
Yeast based biosensors

A conservative substrate walk

Name	Sequence	From previous step	From □ factor
α-factor	WHWLQLKPGQPMY	0.0	0.0
Cys1	-HALALKPGEPMY	3.5	3.5
Cys2	--ALALKPGEPMY	1.6	4.9
Cys3	--ALDFKPGEPMY	0.9	5.2
Cys4	--ALDFKPGEPMY	1.9	6.0
Cystatin peptide	--ALDFAVGEYNK	2.7	8.3

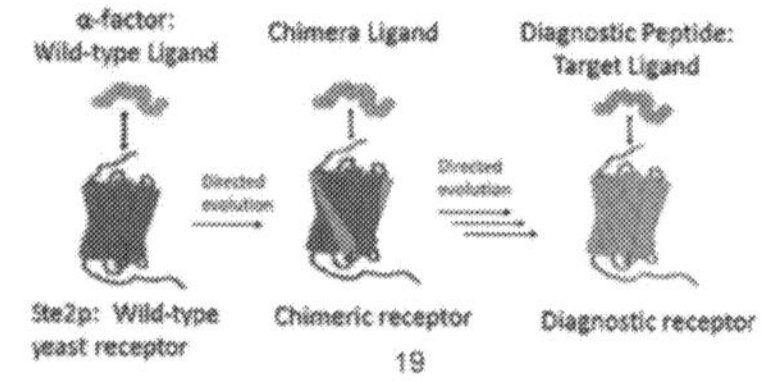

Northwestern | ENGINEERING 19 *(in revision)* Yeast based biosensors

A cystatin receptor

~10 μM sensitivity

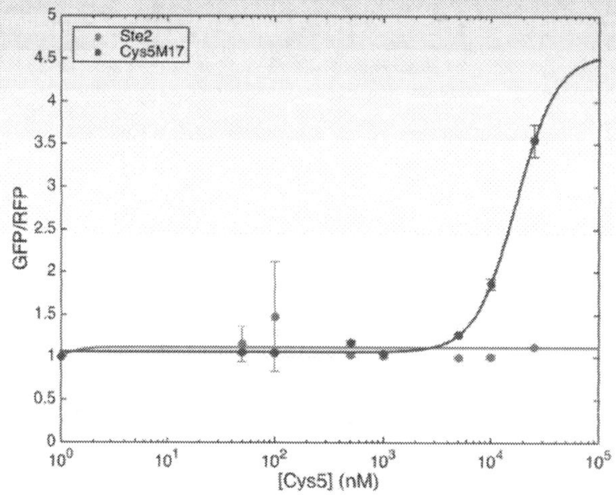

Northwestern | ENGINEERING 20 *(in revision)* Yeast based biosensors

Evolving Along the Substrate Walk

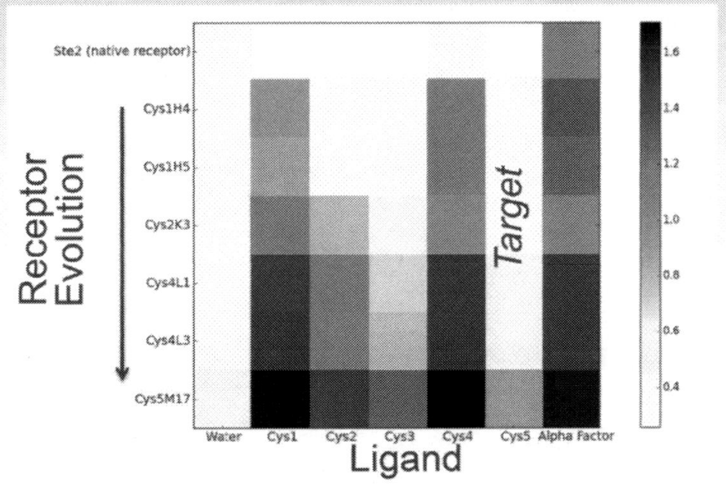

Northwestern | ENGINEERING   21   *(in revision)* Yeast based biosensors

Cystatin specificity in urine

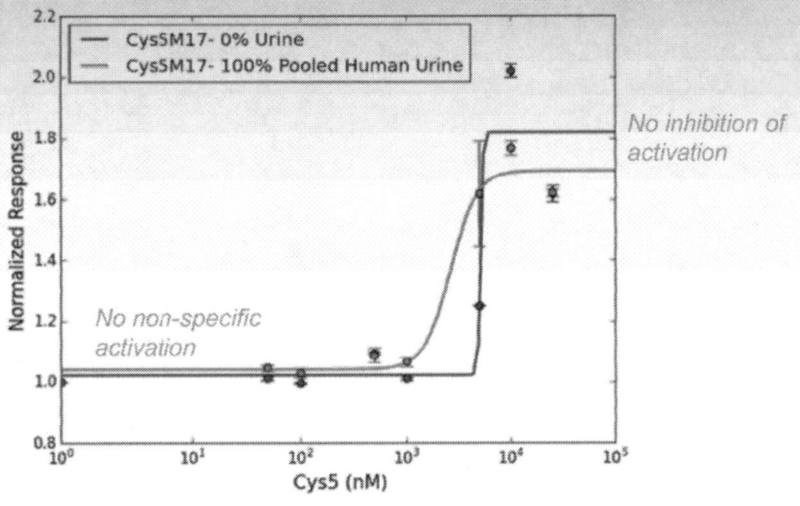

Northwestern | ENGINEERING   22   *(in revision)* Yeast based biosensors

Cystatin specificity in serum

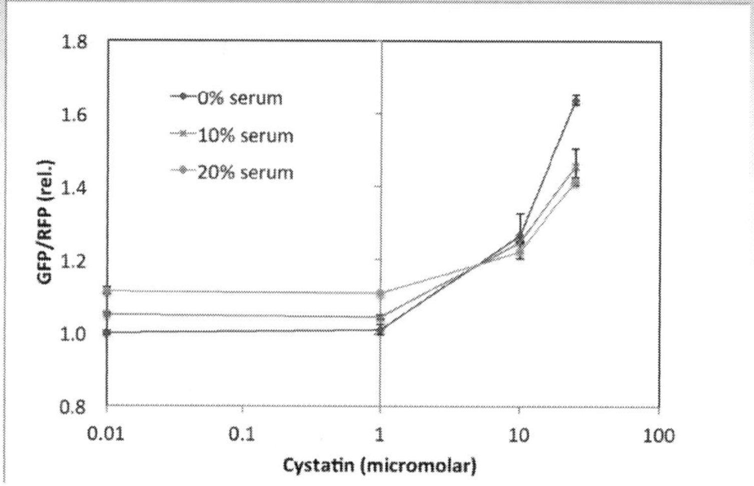

Specificity considerations

For directed evolution we want…
low specificity

For diagnostic, we want…
high specificity

Our current strategies evolve sensitivity, not specificity.

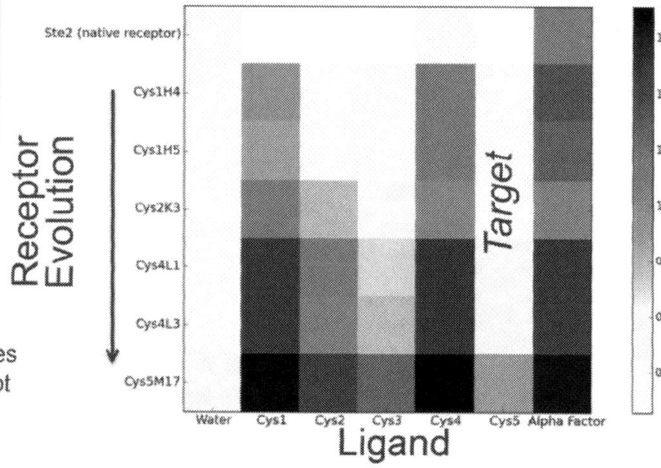

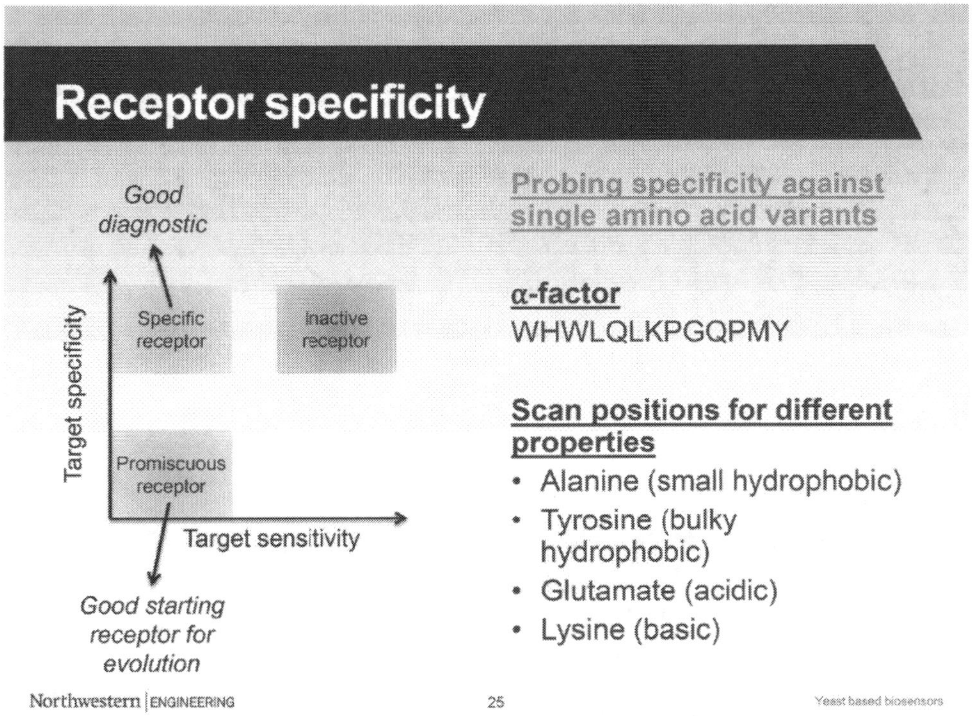
Receptor specificity
Good diagnostic
Specific receptor
Inactive receptor
Promiscuous receptor
Target specificity
Target sensitivity
Good starting receptor for evolution
Probing specificity against single amino acid variants
α-factor
WHWLQLKPGQPMY
Scan positions for different properties
• Alanine (small hydrophobic)
• Tyrosine (bulky hydrophobic)
• Glutamate (acidic)
• Lysine (basic)
Northwestern | ENGINEERING
25
Yeast based biosensors

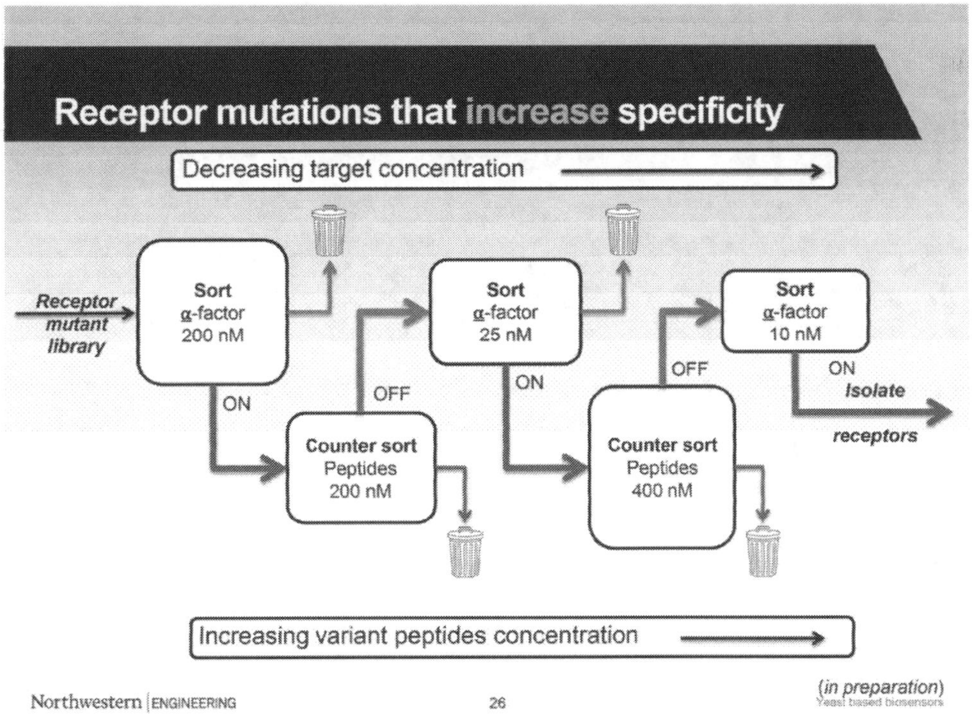
Receptor mutations that increase specificity
Decreasing target concentration
Receptor mutant library
Sort α-factor 200 nM
ON
Counter sort Peptides 200 nM
OFF
Sort α-factor 25 nM
ON
Counter sort Peptides 400 nM
OFF
Sort α-factor 10 nM
ON
Isolate receptors
Increasing variant peptides concentration
(in preparation)
Northwestern | ENGINEERING
26
Yeast based biosensors

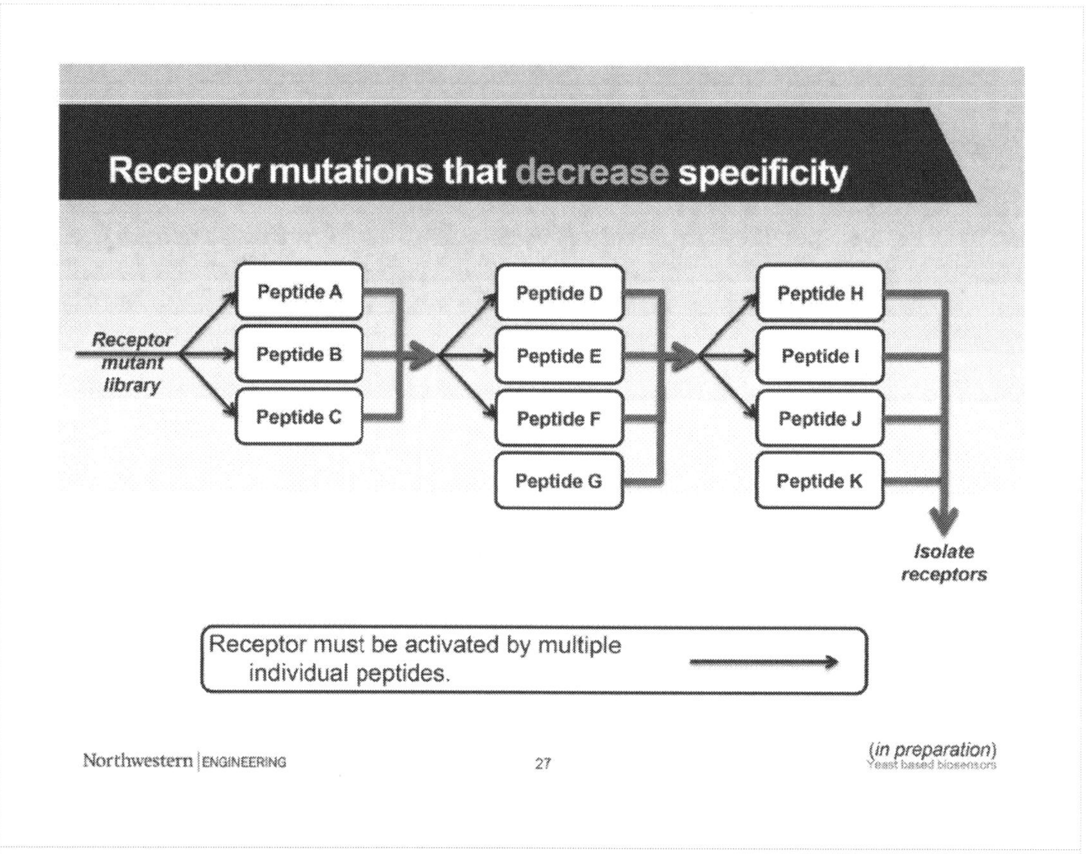
Receptor mutations that decrease specificity
Receptor mutant library
Peptide A
Peptide B
Peptide C
Peptide D
Peptide E
Peptide F
Peptide G
Peptide H
Peptide I
Peptide J
Peptide K
Isolate receptors
Receptor must be activated by multiple individual peptides.
Northwestern | ENGINEERING
27
(in preparation)
Yeast based biosensors

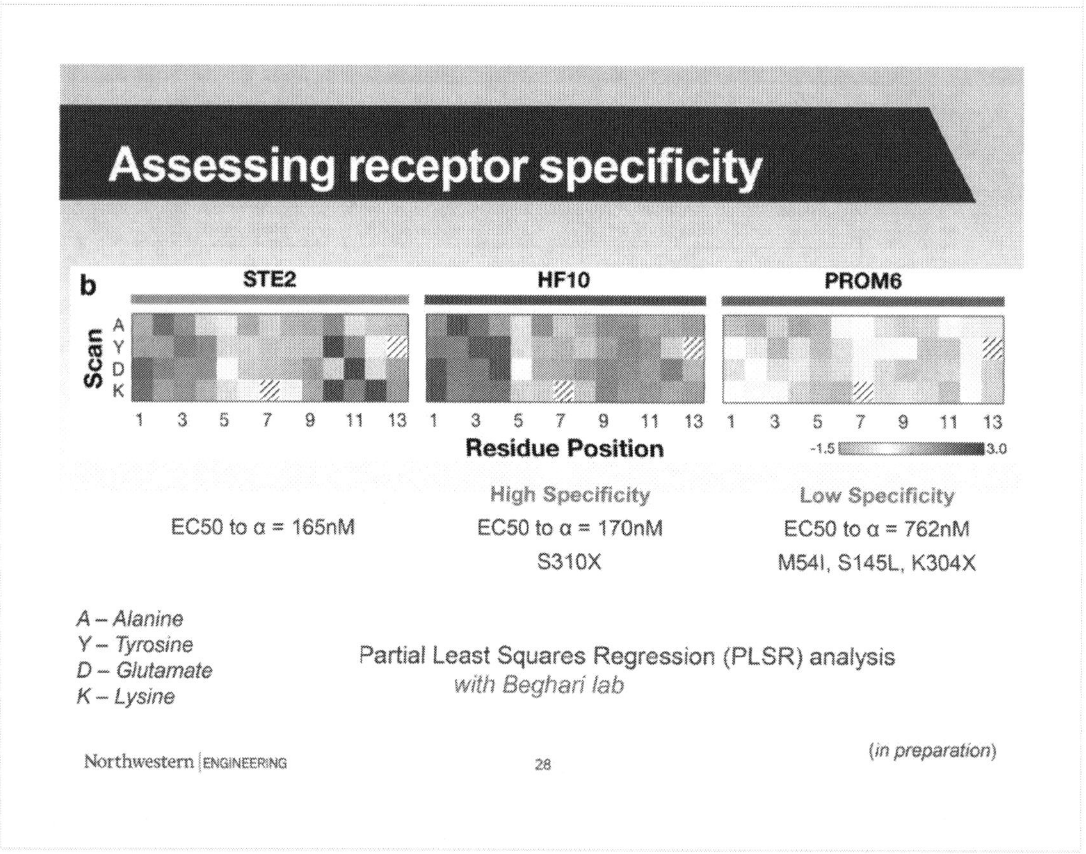
Assessing receptor specificity
b
STE2
HF10
PROM6
Scan
A
Y
D
K
1 3 5 7 9 11 13
Residue Position
-1.5
3.0
High Specificity
Low Specificity
EC50 to α = 165nM
EC50 to α = 170nM
S310X
EC50 to α = 762nM
M54I, S145L, K304X
A – Alanine
Y – Tyrosine
D – Glutamate
K – Lysine
Partial Least Squares Regression (PLSR) analysis
with Beghari lab
Northwestern | ENGINEERING
28
(in preparation)

What properties and residues are most predictive?

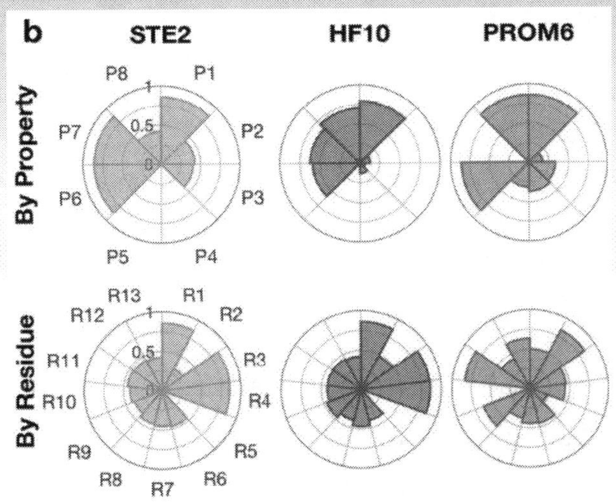

P1-2 – Hydrophobic
P3-4 – Steric
P5-8 – Electronic

R1 – N-terminus
R13 – C-terminus

(*in preparation*)

Conclusions

- Yeast-based biosensors have a large opportunity to impact diagnostics
- Ste2 can be mutated to detect cystatin, a clinically relevant peptide.
- Small evolutionary leaps are highly feasible
  - Large leaps may be more difficult.
- Receptor specificity can be altered by directed evolution

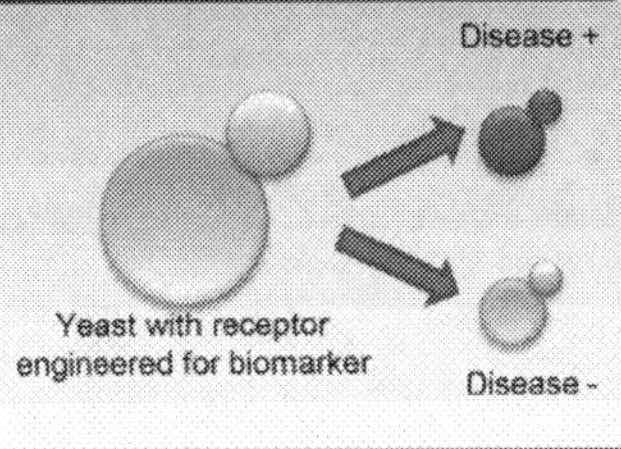

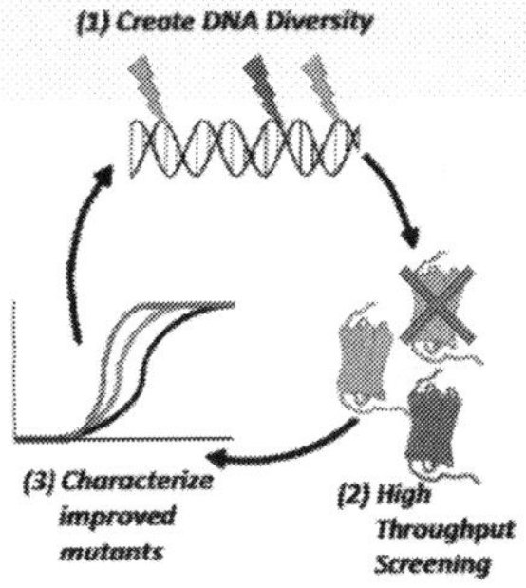

Acknowledgements

Tyo Lab @ NU

Using synthetic biology to improve resource-poor healthcare and sustainability

Dante Pertusi
Jennifer Greene
Adebola Adenirin
Will Bothfield
James Jeffryes
Alexandra de Paz
Sarah Stainbrook
Bradley Biggs
Peter Su

Dr. Namita Bhan
Dr. Jesse Collins

Jessica Yu
Justin Newton

Aaron Anderson
Tom Aunins

Collaborators
Linda Broadbelt
Josh Leonard
Neda Bagheri
Chris Henry (Argonne)
George Church (Harvard)
David Weitz (Harvard)
Ed Boyden (MIT)
Konrad Kording (RIC)
Ajikumar Parayil (Manus Bio)

Funding
Gates Foundation (OPP1044008)
NIH BTP (T32)
NSF (CAREER 1452549)
Argonne LDRD
TR01 (NIMH R01 MH103910)
Chicago Biomed. Cons.

Domain-Insertion Profiling Reveals Allosteric Hotspots for Protein Engineering

Dave Savage
Savage Lab
Departments of MCB and

Berkeley
University of California

A grand challenge of synbio is engineering complex genotypes

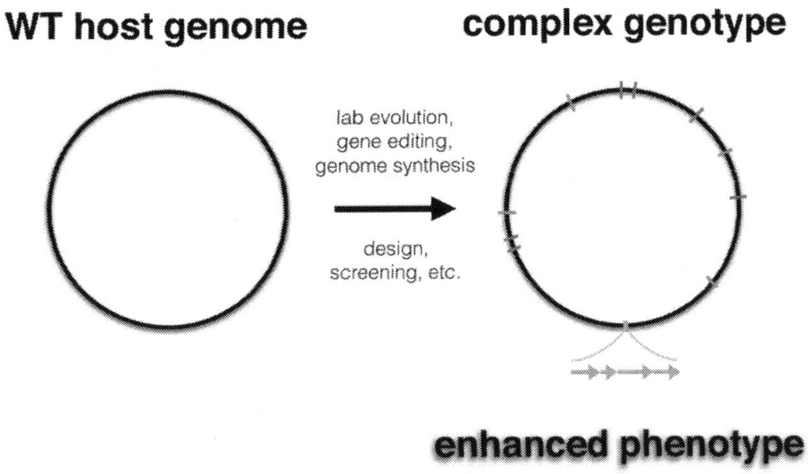

E.g. Wang et al. 2009 Nature, Warner et al. 2011 Nat. Biotech.

Goal of the Savage Lab: understanding and improving CO_2 assimilation

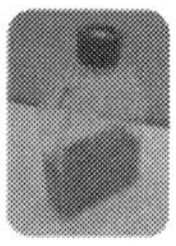

Cyanobacterial CO_2 assimilation is much more efficient than higher plants

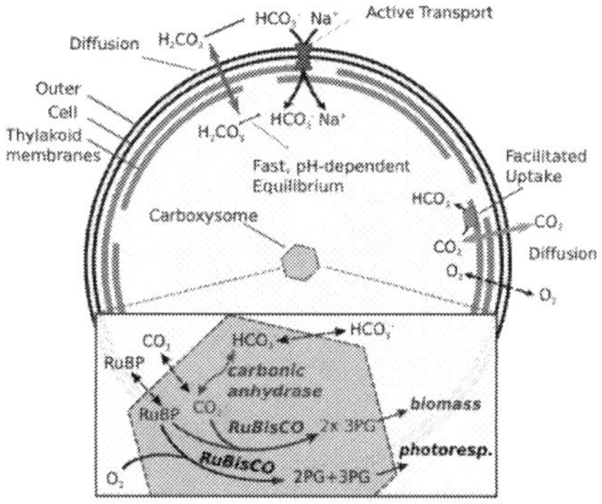

port machinery of CO_2 assimilation (e.g. 20+ genes)

Develop enhanced organisms with superior CO_2 assimilation phentopyes

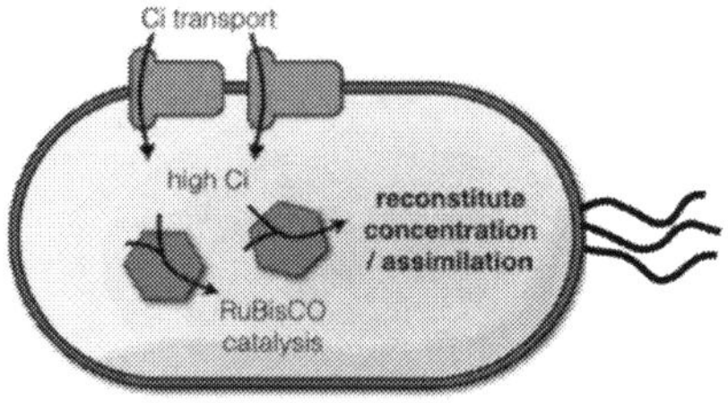

Bonacci et al. 2012 PNAS, Chaijarasphong et al. 2016 JMB, Hood et al. 2016 PNAS In press, Mangan et al. 2016 PNAS In press.

We need enabling technologies for measuring and perturbing physiology

Reading, writing... ...and understanding genomes

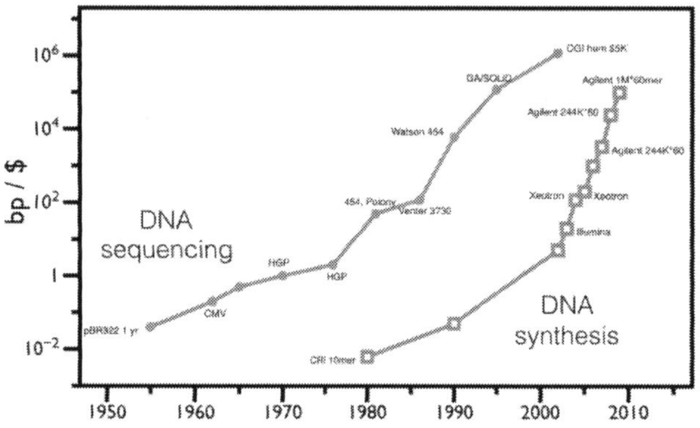

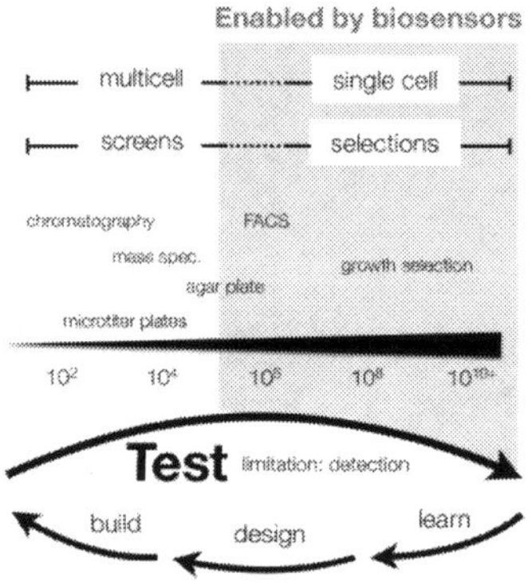

We are harnessing **allostery** to construct i) small moleucule **fluorescent biosensors** and ii) finely tuned **genome editing tools** using protein engineering

Carr and Church 2009

Morgan et al. 2016 Cur. Opinion Chem. Bio. In press.

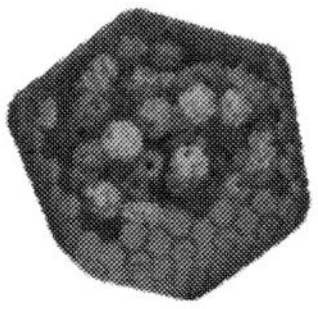

Roadmap

General methods for probing allostery

Rapid construction of metabolite biosensors

New tools for perturbing the genome

Engineering allostery is a common problem in synthetic biology projects

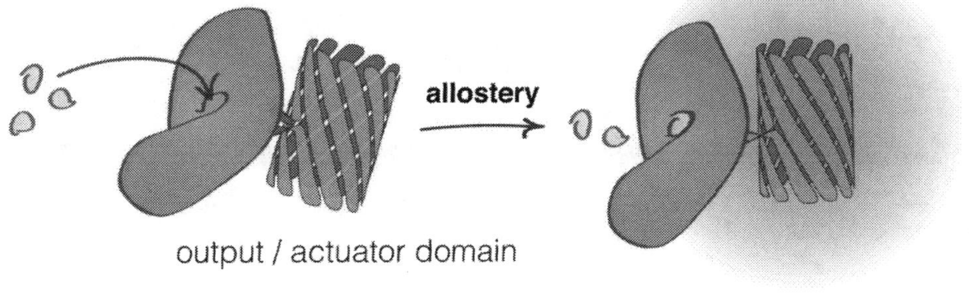

- Rational design of allostery is difficult -> 1. directed evolution
- Needs to function well *in vivo* -> 2. screen under conditions we want them to work
- But, proteins evolve by domain insertion -> **3. let's just speed up the process**

Lim, Dueber, Ranganathan, Ostermeier, Baker, Kortemme, Raman Groups et alia.

We set out to create an informative approach based on recombination

Human protein complexity is based on domain recombination

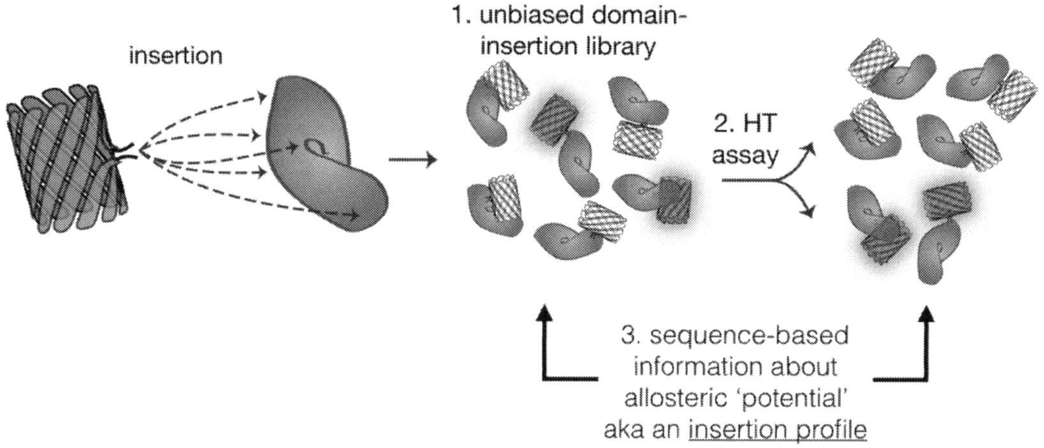

Lander et al. 2001 Nature.

Transposons can create random DNA insertions

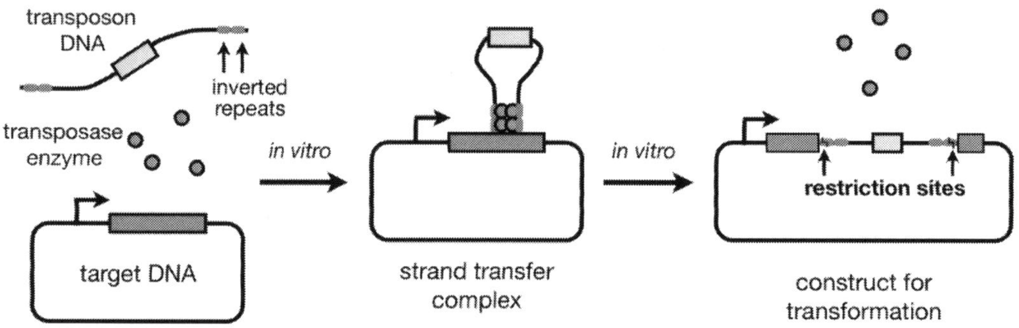

An engineered transposon sequence with designed restriction sites facilitates randomized domain insertion

Ostermeier, Jones, and Silberg Groups

Modified transposons create domain-insertion libraries

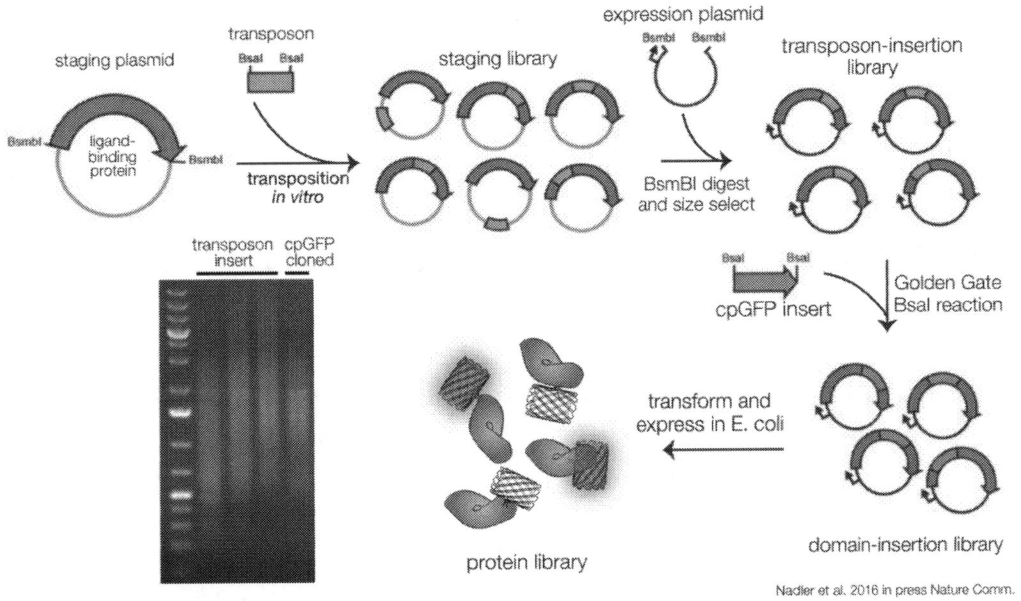

Nadler et al. 2016 in press Nature Comm.

These can be quantitatively assessed via HT assays and DNA sequencing

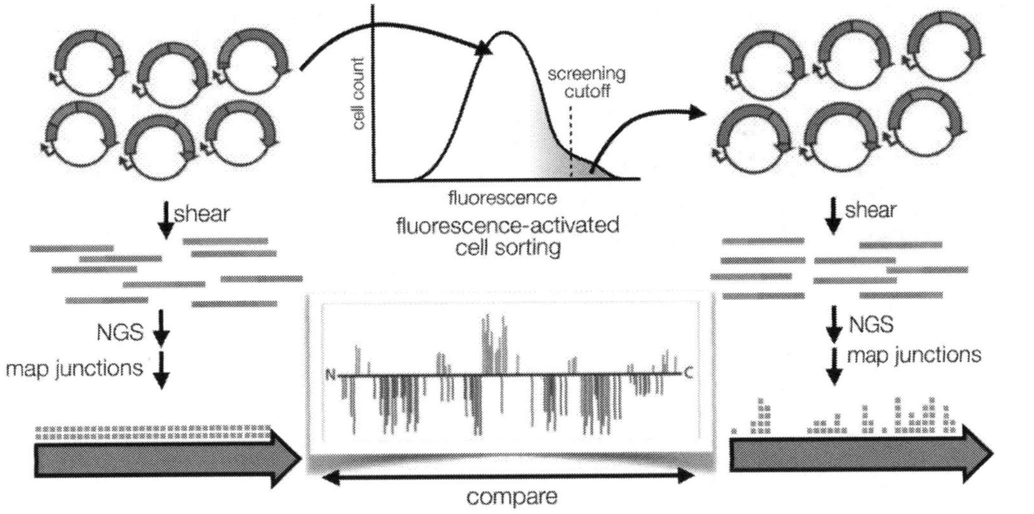

We call this process Domain-Insertion Profiling with sequencing or **DIP-seq**

Nadler et al. 2016 in press Nature Comm.

Roadmap

General methods for probing allostery

Rapid construction of metabolite biosensors

New tools for perturbing the genome

Measuring metabolism is a bottleneck in synthetic biology

Write and Read DNA at the scale of 10^8-10^9 base pairs

≠

100 assays per day

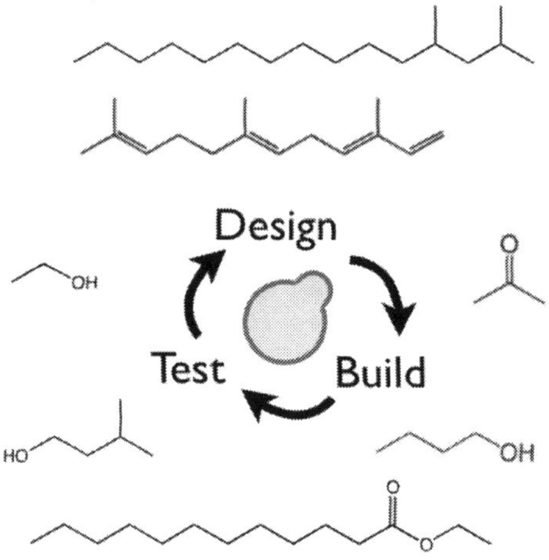

Thus, we need faster and non-invasive metabolite probes

GFP can be made into a ligand-binding biosensor

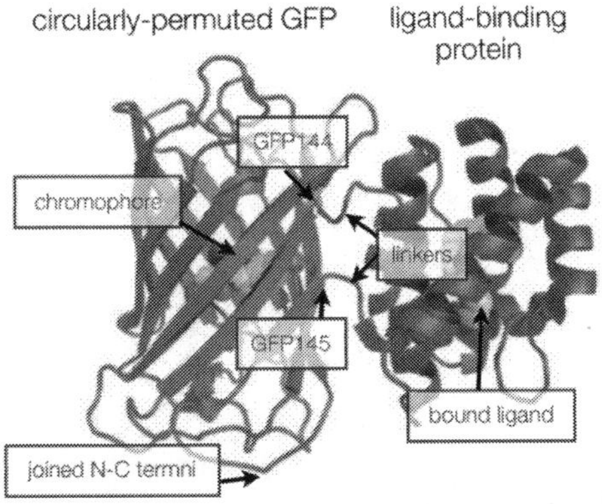

G. S. Baird, D. A. Zacharias, R. Y. Tsien, *Proc. Natl. Acad. Sci. U.S.A.* **96**, 11241–11246 (1999).

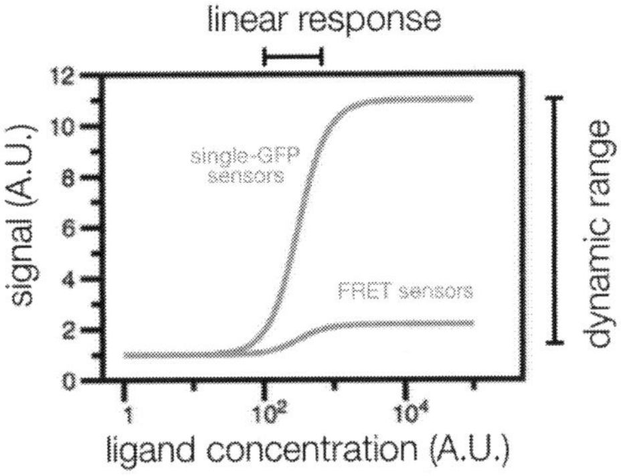

Advantages: high dynamic range, portable, real-time, imaging

Problem: Hard to predict domain insertion capable of allostery

Trehalose is an important and convenient proof of concept

- disaccharide containing two glucose units joined by a α,α-1,1-glycosidic linkage
- regulatory sugar
- signaling molecule that directs/controls certain metabolic pathways
- can protect proteins and membranes from inactivation and denaturation caused by stress (desiccation, dehydration, heat, cold and oxidation)
- *Thermococcus litoralis* TMBP

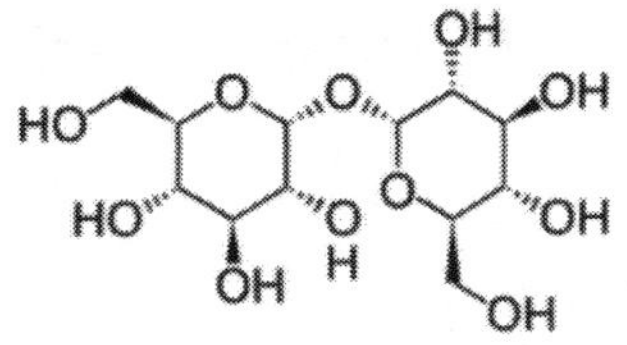

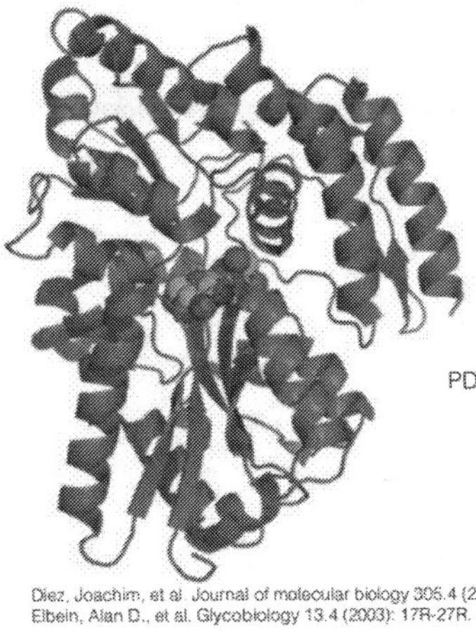

PDB ID: 1EU8

Diez, Joachim, et al. Journal of molecular biology 305.4 (2001): 905-915.
Elbein, Alan D., et al. Glycobiology 13.4 (2003): 17R-27R.

TMBP was subjected to DIP-seq using FACS

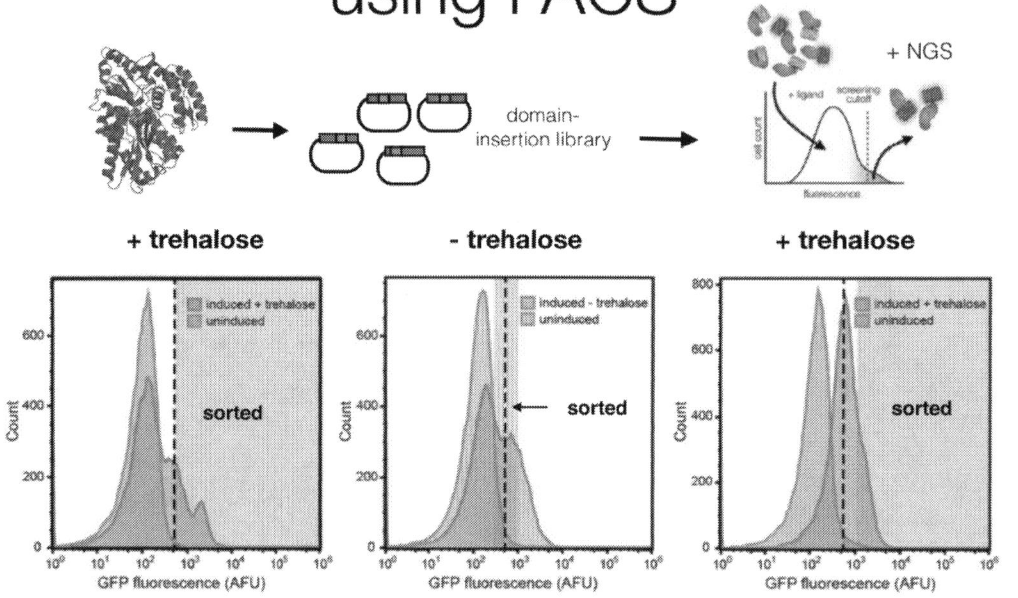

Nadler et al. 2016 in press Nature Comm.

DIP-seq reveals hotspots of enrichment within TMBP

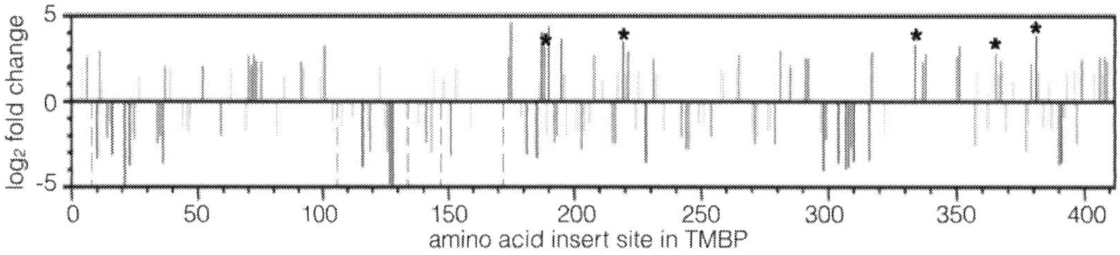

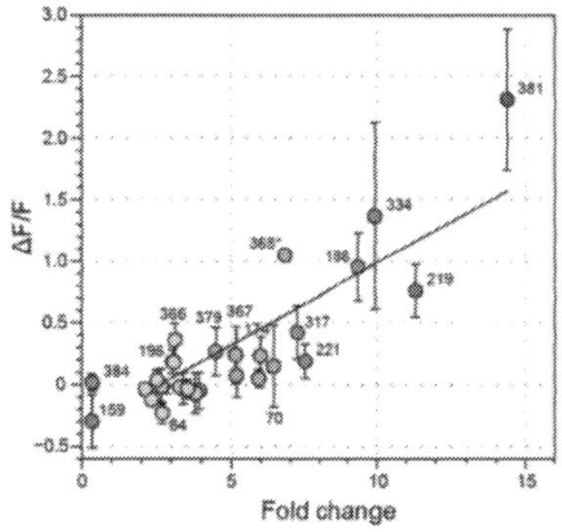

These hotspots are predictive of biosensor function

Enriched and functional sites cluster in two regions of TMBP structure

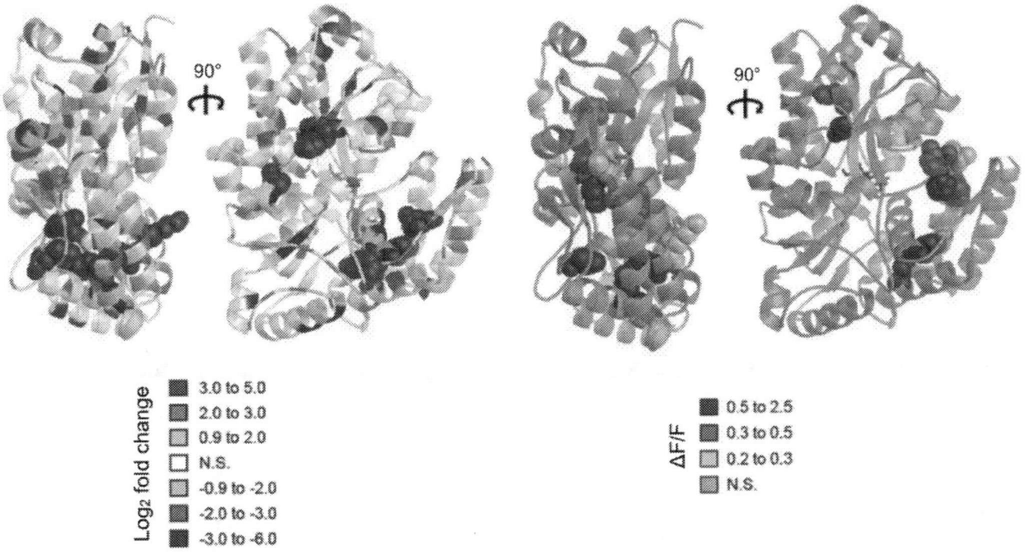

Linker-sequence optimization improves biosensor dynamic range

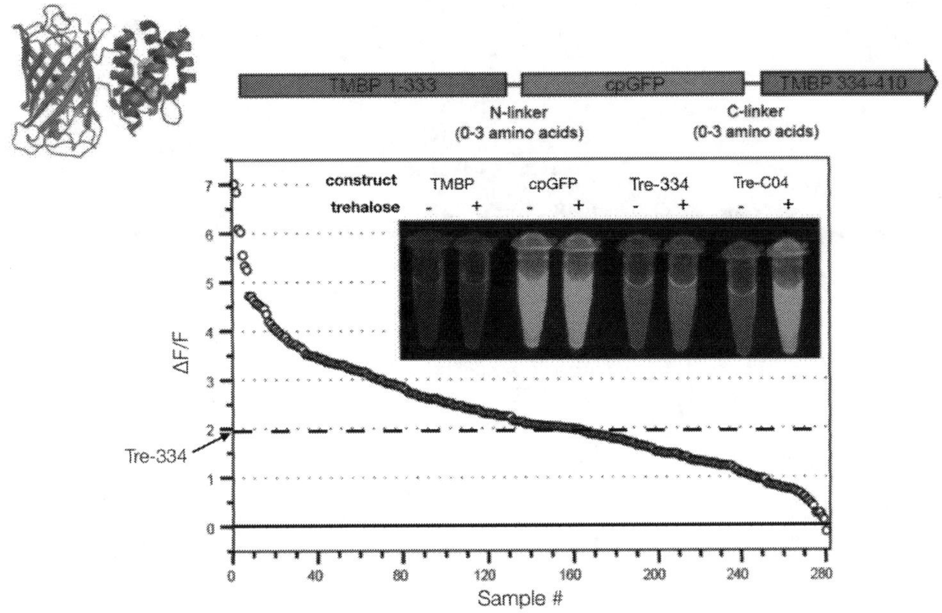

Tre-C04 reports dynamic changes in trehalose concentration *in vivo*

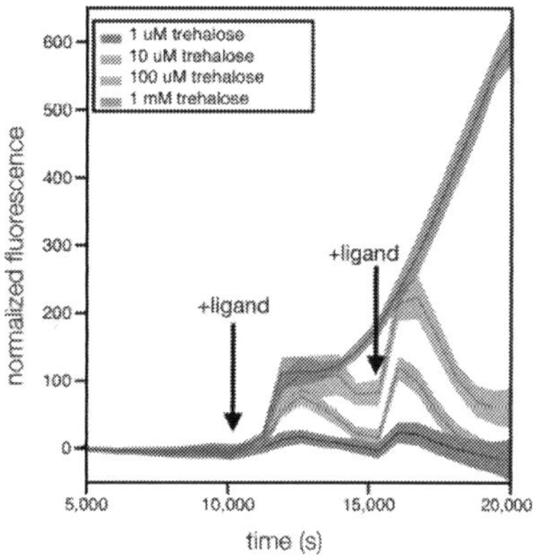

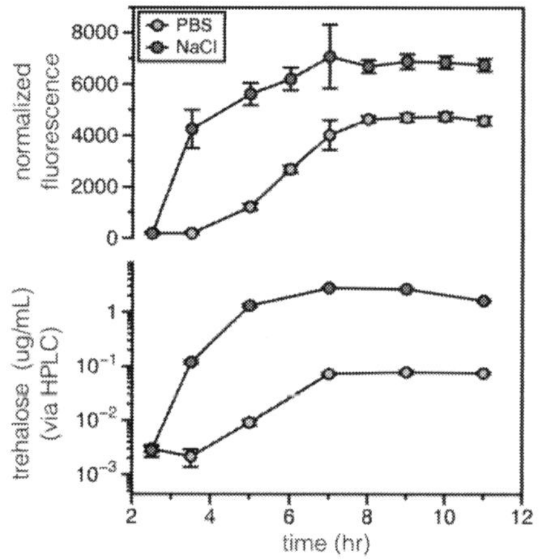

Thus, DIP-seq rapidly facilitates the process of constructing metabolite biosensors

Interrogating chemistry and disease

<u>Biosensor Construction</u>

Cancer

Gene	Binding Target
mglB	glucose
glnBP	glutamine
pdhR	pyruvate
lldR	lactate

Chemicals

Gene	Binding Target
fadR	fatty acyl-CoA
psrA	free fatty acids
alkS	alkanes (C6-C12)
bmoR	alcohols (C2-C8)

Metabolic Interests

Gene	Binding Target
malE	maltose
tmbp	trehalose
lbp	phenylalanine
citA	glutarate

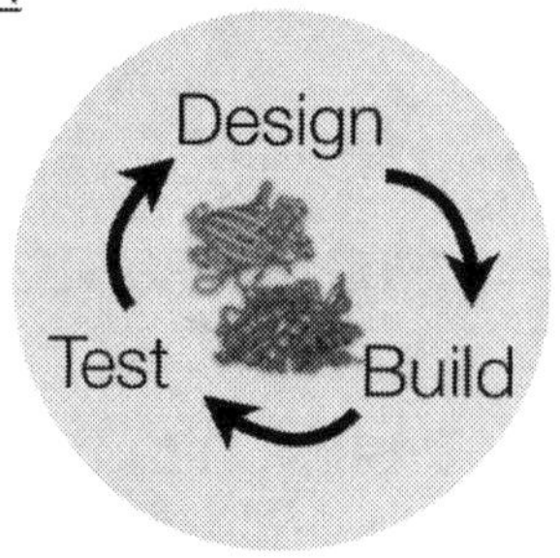

<u>Methods development</u>

Yeast display
Genome editing

<u>Biological questions</u>

high-throughput single cell quantification

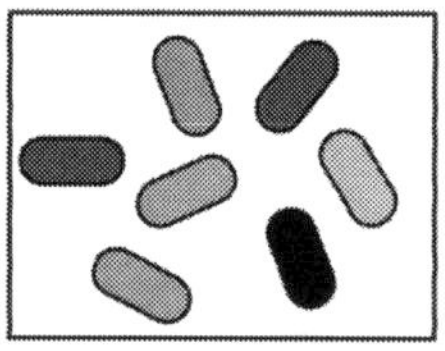

Whole animal atlas of metabolism

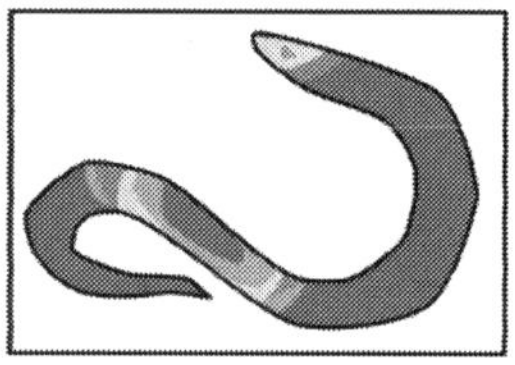

Roadmap

General methods for probing allostery

Rapid construction of metabolite biosensors

New tools for perturbing the genome

DIP-seq is a useful tool for identifying hotspots in any protein

S. pyogenes Cas9

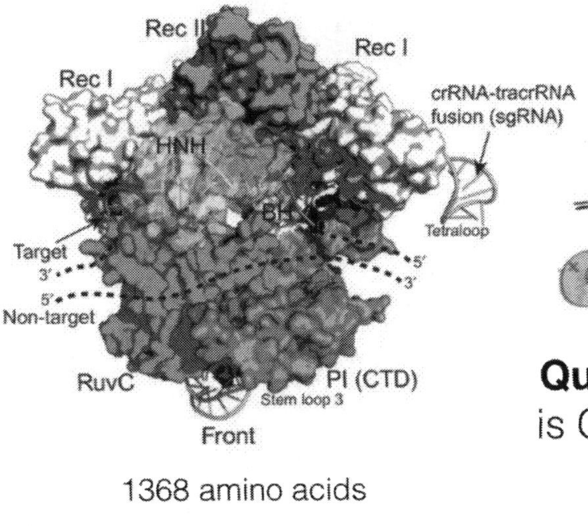

1368 amino acids

Cas9 function can be selected for in a variety of ways

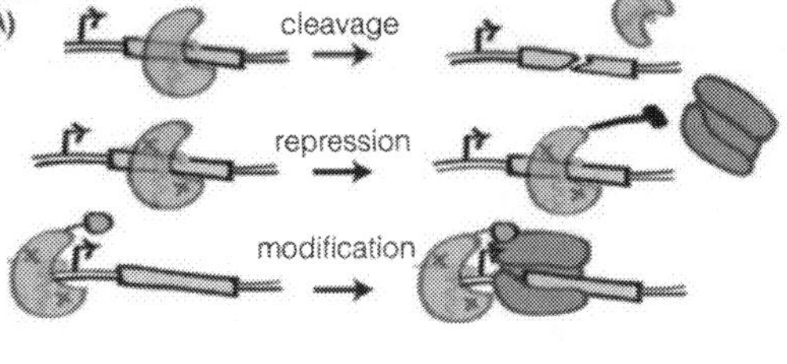

Question: how malleable is Cas9 to genetic fusion?

Oakes, B.L., et al. and Savage, D.F., 2014. Methods in Enz.

The structure of Cas9 restricts potential engineering

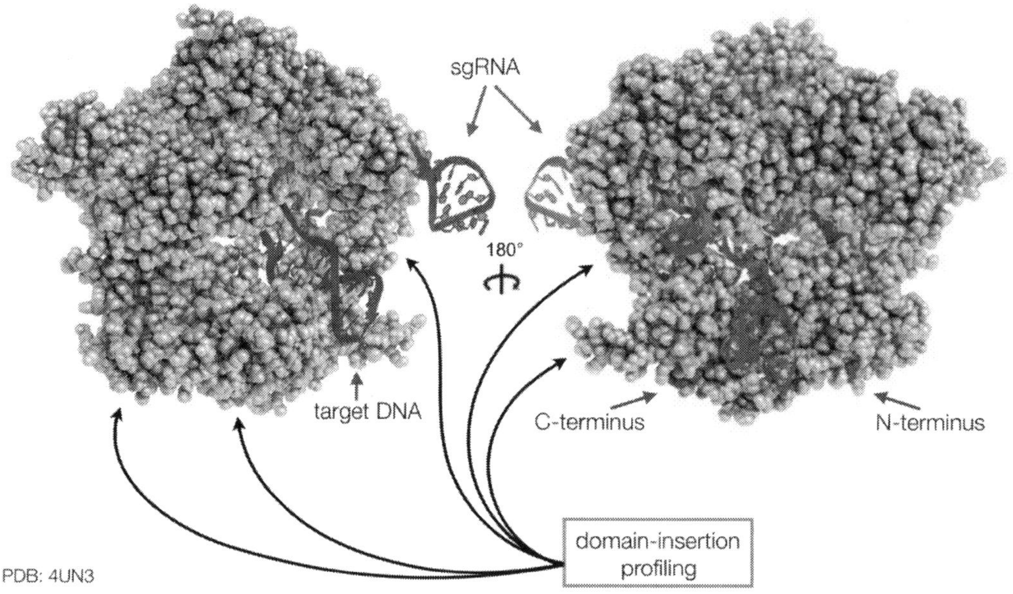

PDB: 4UN3

CRISPRi provides a simple screen for functional Cas9 in *E. coli*

background: *E. coli* w/chromosomal GFP/RFP

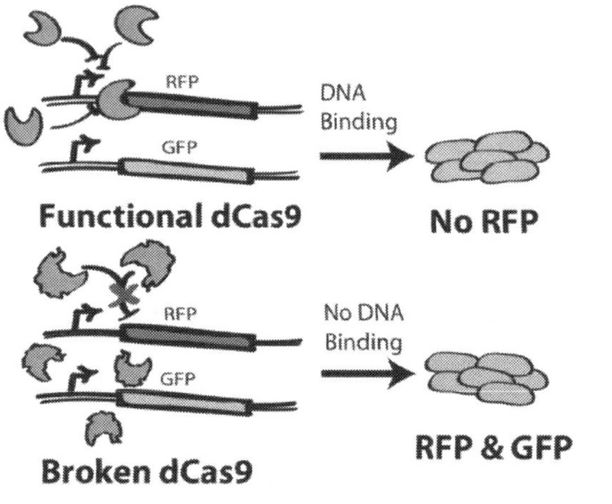

DIP-seq of Cas9

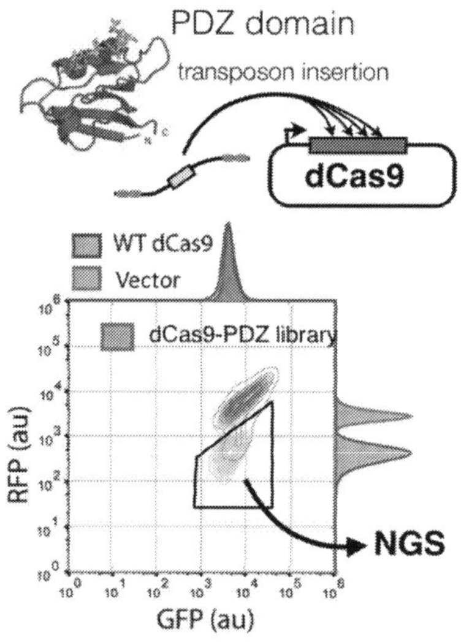

Qi, L.S., et al. 2013. Cell 152, 1173–1183.

DIP-seq with PDZ domain reveals hotspots in the Cas9 structure

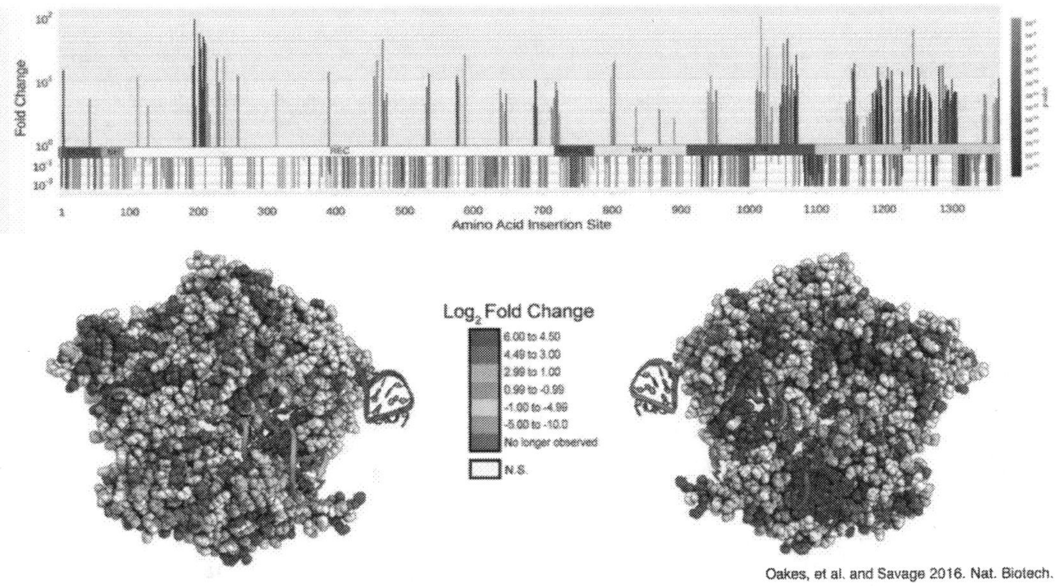

Oakes, et al. and Savage 2016. Nat. Biotech.

Analysis reveals surprising and prescient insertion points

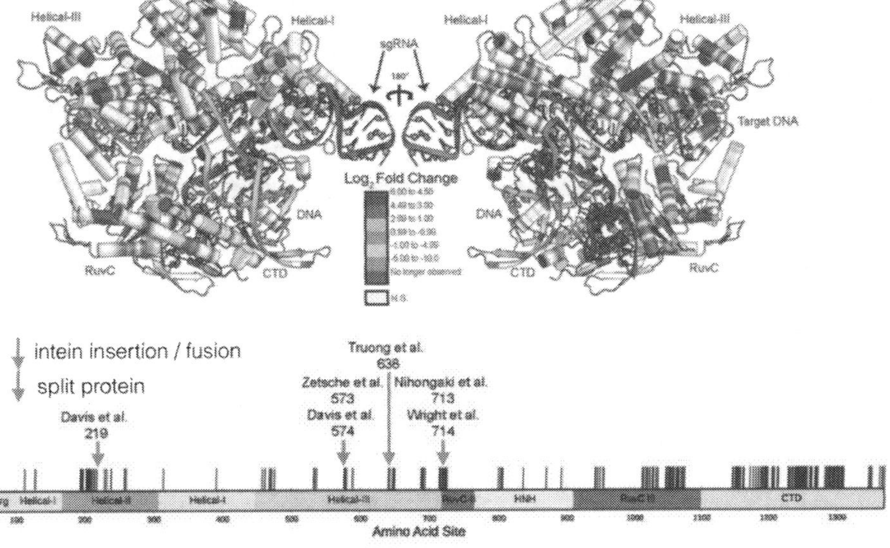

Activity is correlated with fold-enrichment

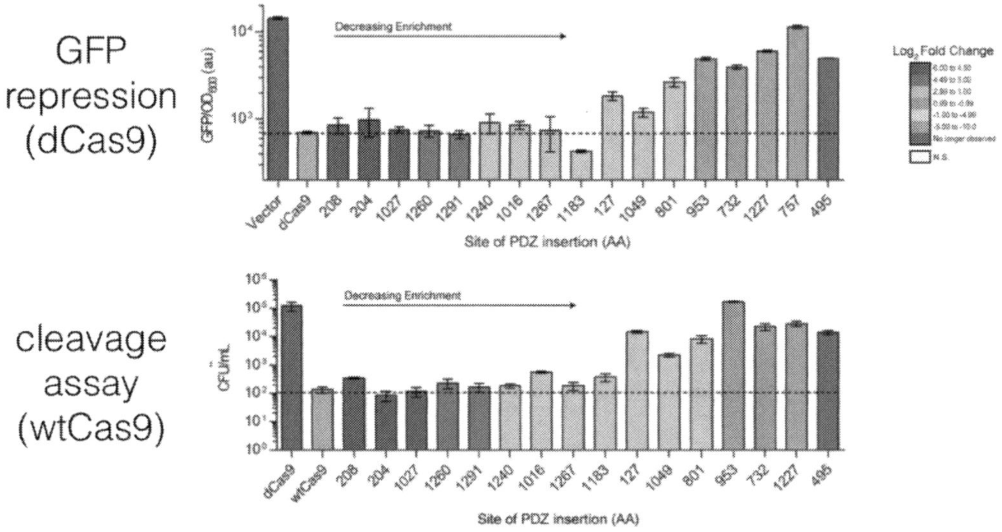

We can also screen for allosteric regulation

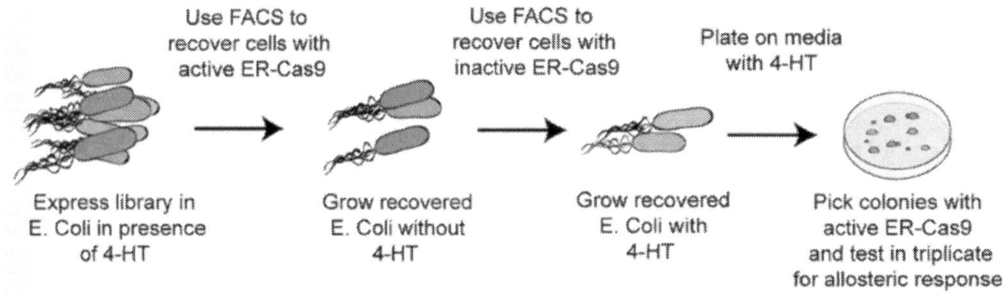

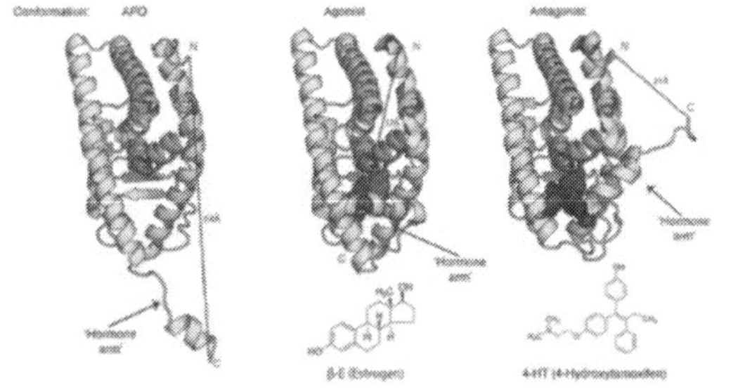

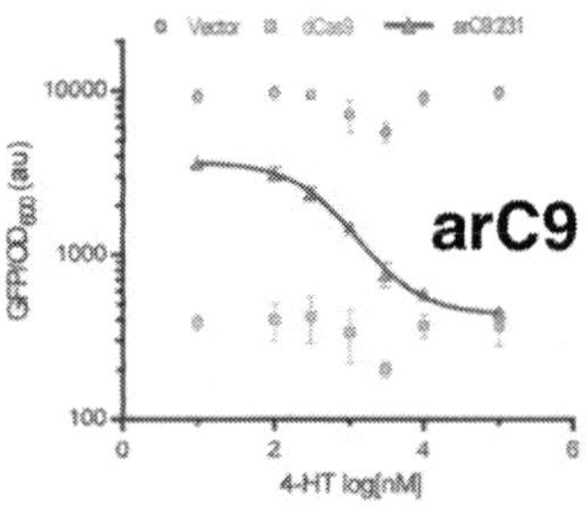

arC9 can be used to precisely control genome editing activity

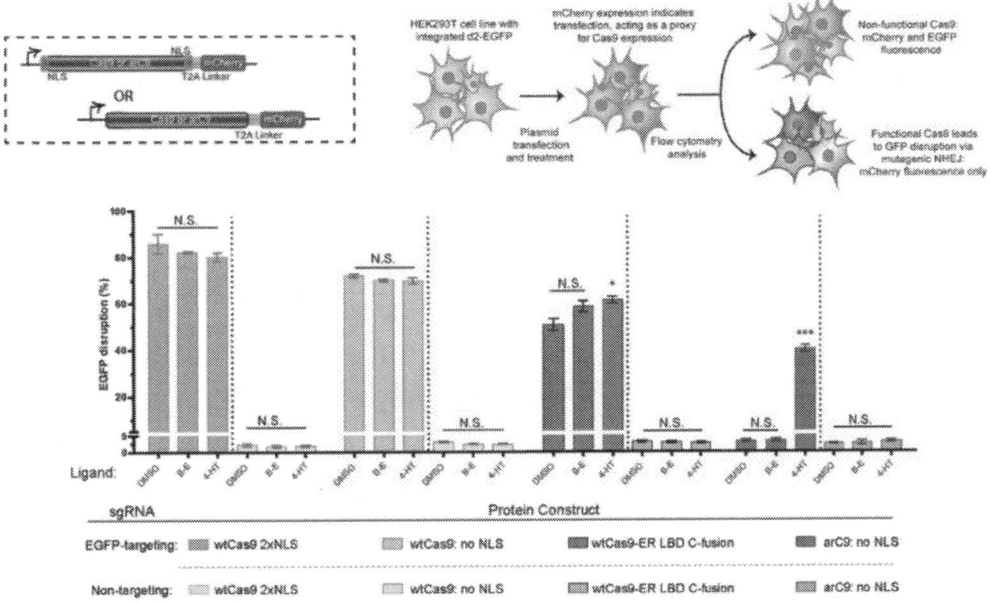

arC9 has a high dynamic range for ligand-sensing

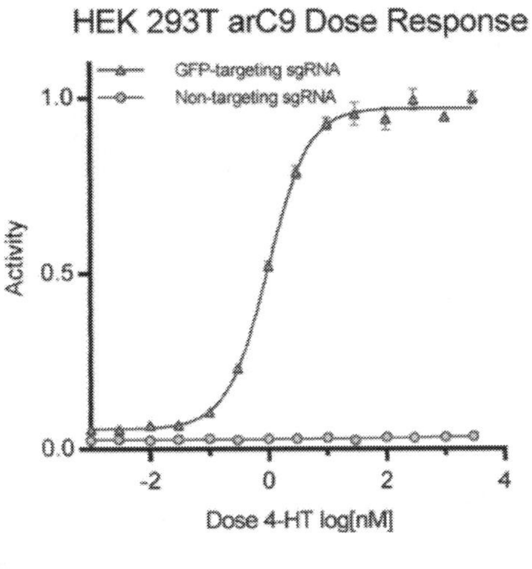

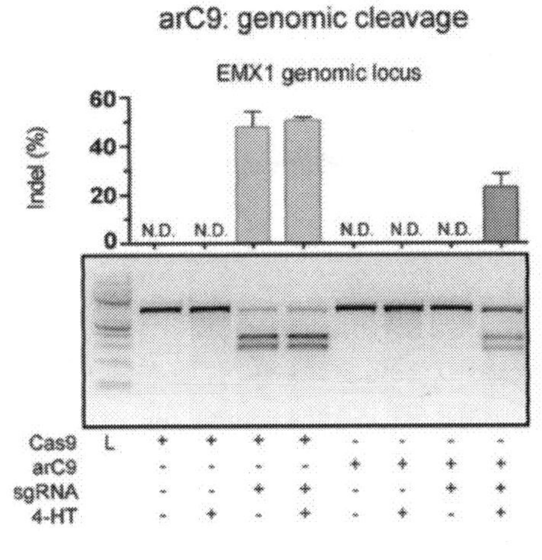

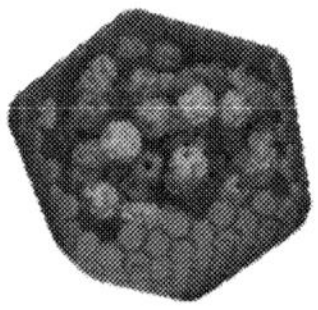

Conclusions

Domain-Insertion Profiling with sequencing (DIP-seq) is a power tool for mapping engineering hotspots in a protein of interest.

Metabolism is hard to measure but biosensors allow for high-throughput interrogation.

DIP-seq enables the rapid isolation and improvement of single fluorescent protein metabolite biosensors.

We can also engineer Cas9 with new functionality and are exploring the utility of titratable, reversible editing activity.

Acknowledgements

Savage Lab at Berkeley
University of California

contact: www.savagelab.org
code: github.com/SavageLab
reagents: www.addgene.org/David_Savage

Biosensor Team

Dana Nadler

Stacy Morgan

Katie Kortright

Benjamin Oakes

Funding: NIH/NIBIB (DP2), Energy Biosciences Institute, March of Dimes (BOC), Sloan Foundation

a

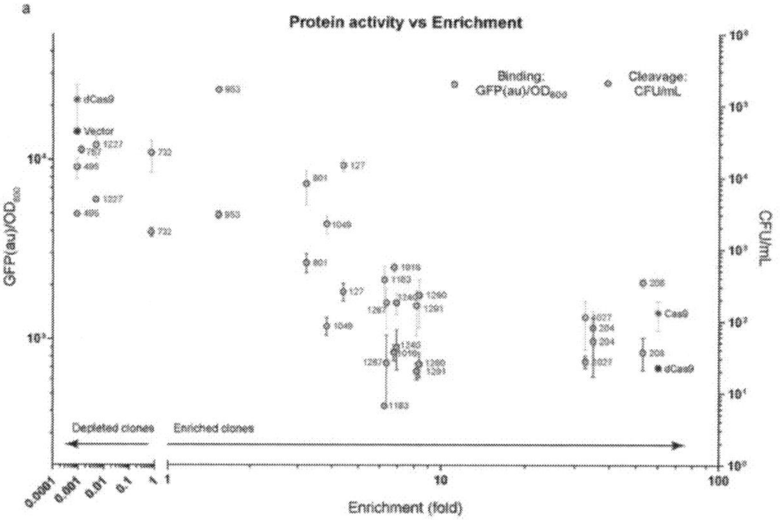

S9| **PDZ insertion clone activity vs Fold Change** (a) The data from figures 1d & e has been graphed according to the enrichment (fold change) score for each construct (Supplementary table 1), binding data based on repression of genomic GFP is graphed on the left y axis in blue. Positive and negative controls dCas9 and vector are dark blue. Cleavage data for each construct is mapped onto the right Y axis and colored in green with positive and negative controls Cas9 and dCas9 respectively in dark green. A monotonic relationship between fold change and binding/cleavage ability is readily apparent. PDZ-Cas9 activity levels plateau (at > ~4 fold enrichment) near wt Cas9 activity levels for each assay despite further increased enrichment. The spearman correlation for the cleavage data to be r= -.85, p value of <0.0001, and the binding data to be r= -.76, p value of 0.0006, indicating a strong monotonic correlation between enrichment and increased ability to both bind and cleave the genome of *E. coli*.

S6

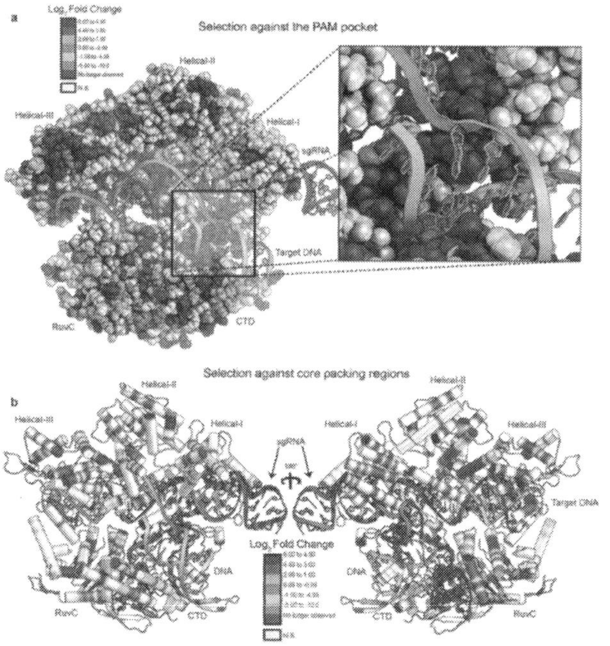

S6 | **PDZ-insertion sites avoid critical structural motifs.** (a) Mapping the log_2 fold change of the PDZ-insertions onto the RNA:DNA holo Cas9 crystal structure (PDB ID: 4UN3) demonstrates how the unbiased insertion technique can experimentally delineate regions of critical structure and function. The PAM binding pocket (PAM residues in orange) and the RNA:DNA channel are selected against (blue). (b) PDZ insertions are not readily observed in core packing regions of the RuvC and Helical-III domains and are also selected against in the sgRNA binding grooves (Fig 1C) and Arg helix.

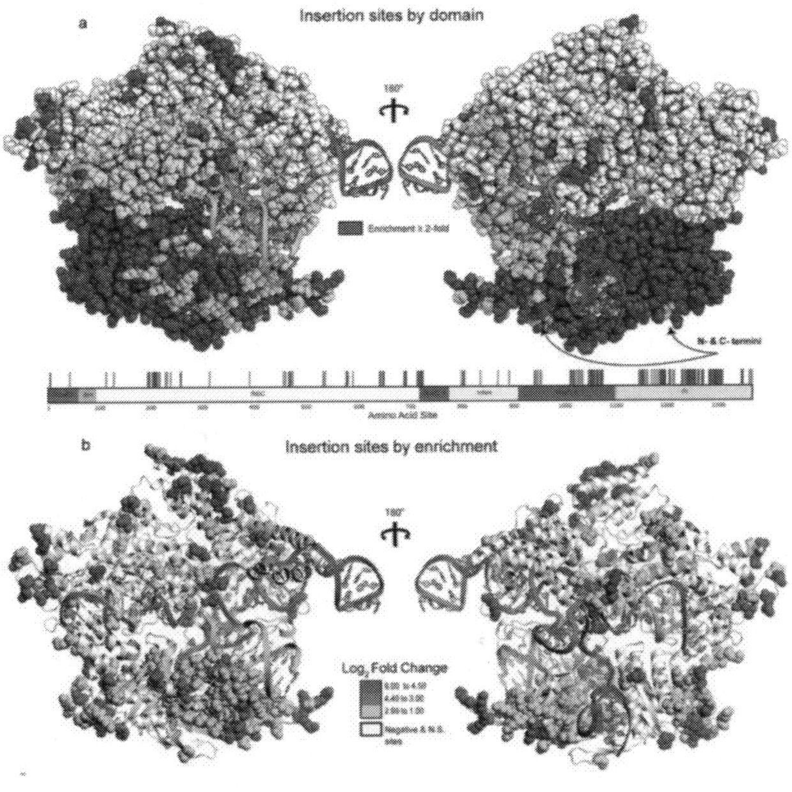

S18

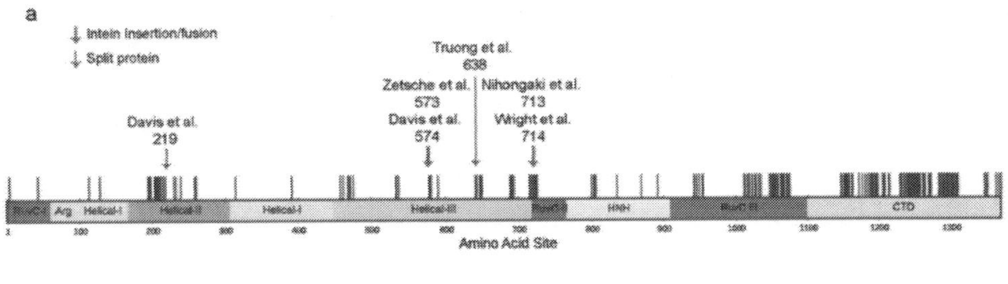

S18 | **Comparison of previously identified sites with hotspots Identified in this study.** (a) Amino acid sites rationally identified and utilized in previous studies which have split Cas9, introduced intiens or split intiens into the protein have been mapped onto the hotspots identified with the PDZ insertion.

S17

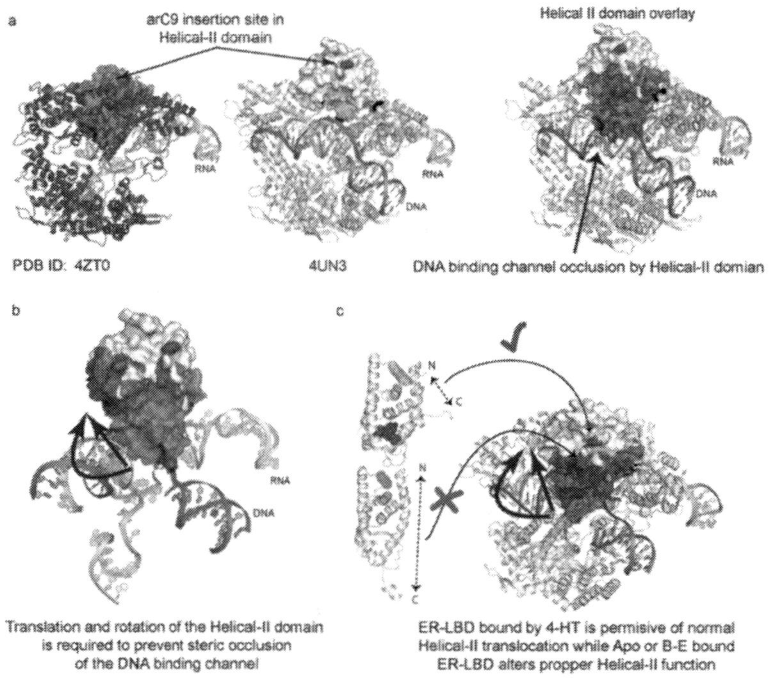

S17 | **Model for arC9 based regulation**. (a) PDB models 4ZT0 (black) and 4UN3 (white) are Cas9 structures which are respectively sgRNA only and sgRNA and DNA bound. When aligned it is possible to observe that in the RNA only bound structure the helical-II domain, highlighted

S1.

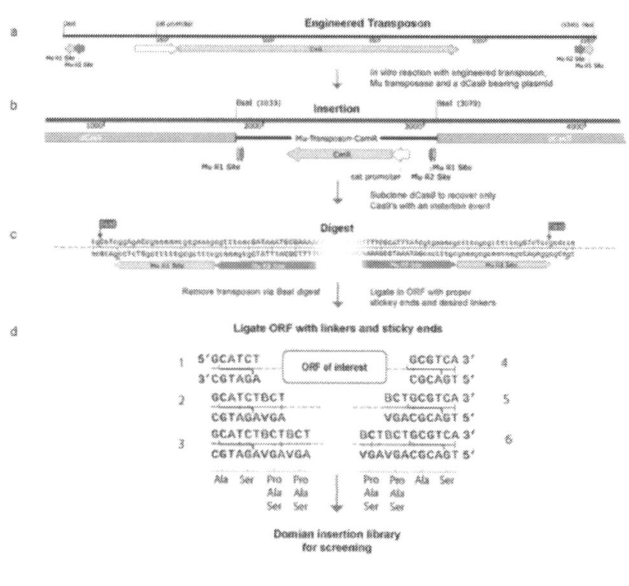

S1 | **Construction of a transposition library** (a) Schematic of the engineered Mu transposon used for transposition. (b) Schematic showing the inserted transposon. (c) ORF excision of an intra-Cas9 transposon using Type II-S restriction enzyme BsaI. Any ORF can be subsequently ligated into these sticky ends. (d) Description of the sticky ends and linkers used to ligate in the PDZ and ER-LBD domains. Ala and Ser are hardcoded on each side providing the correct sticky ends for a Golden Gate ligation and additional diversity is provided by BCT codons (encoding Ala, Ser, or Pro). In total, there are 13 possible amino acid variations per terminus

Assessing long-term and reversible activity of arC9

S16

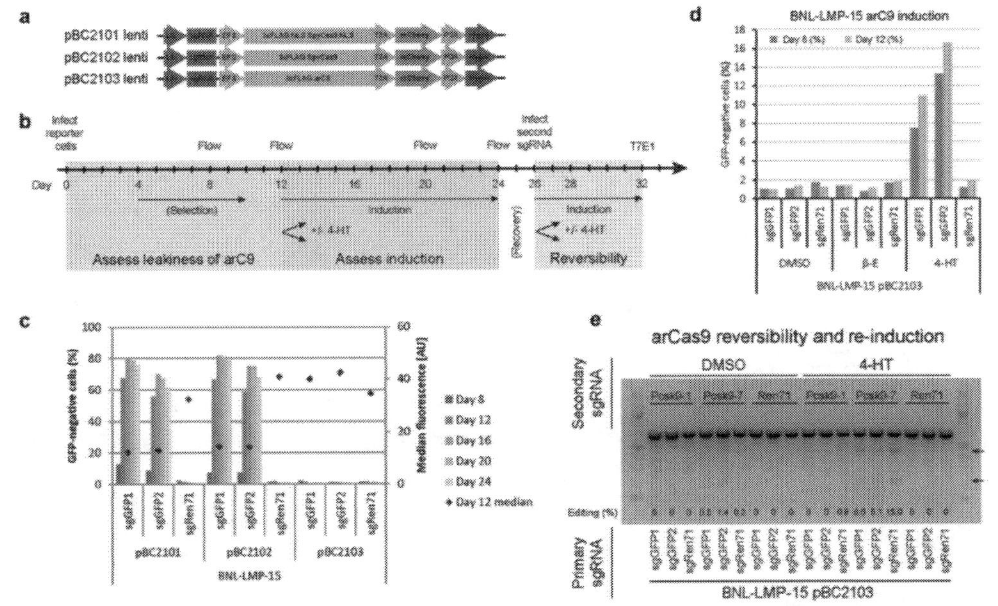

Transposons create random DNA insertions

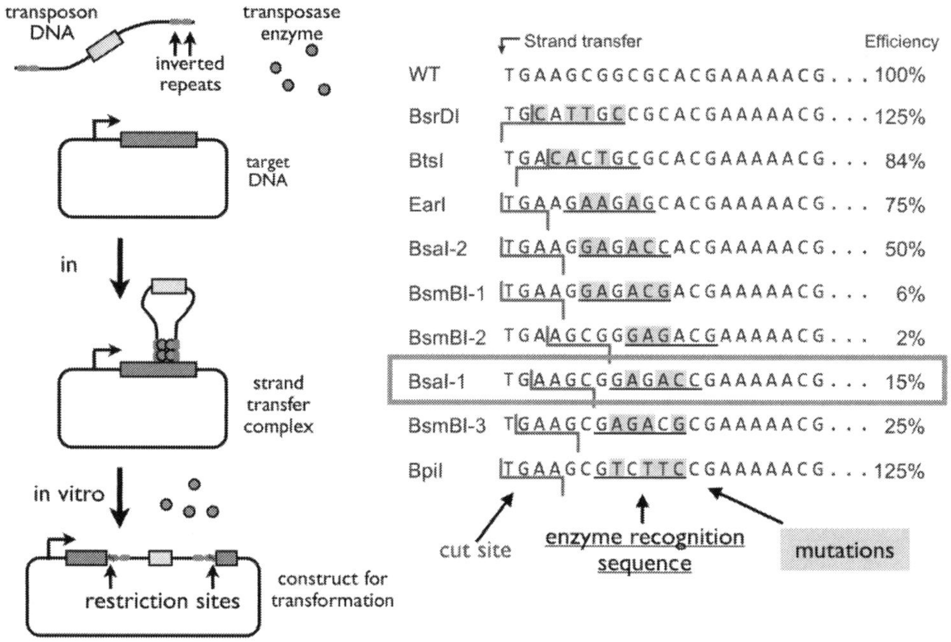

Example of productive domain insertion

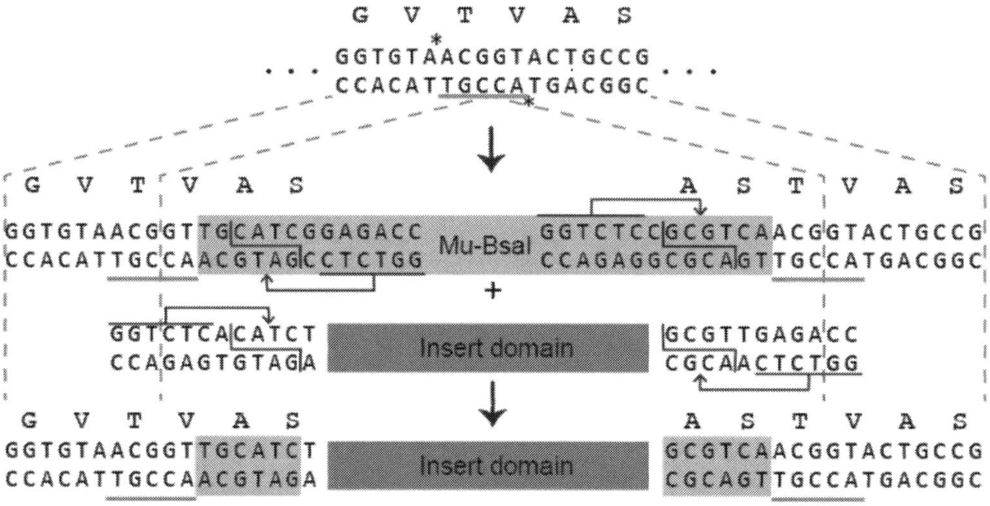

Fluorescence-activated cell sorting (FACS) enriches productive cpGFP insertions

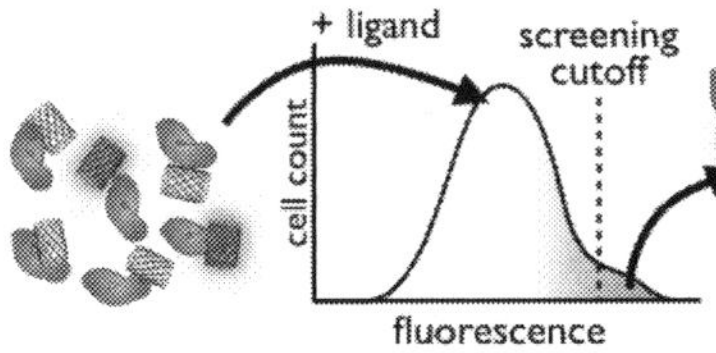

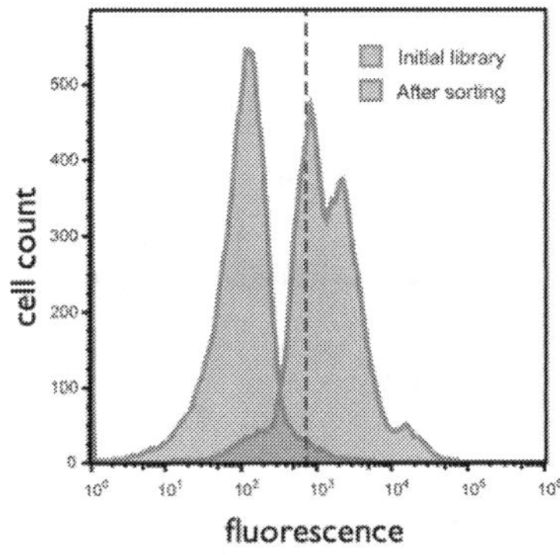

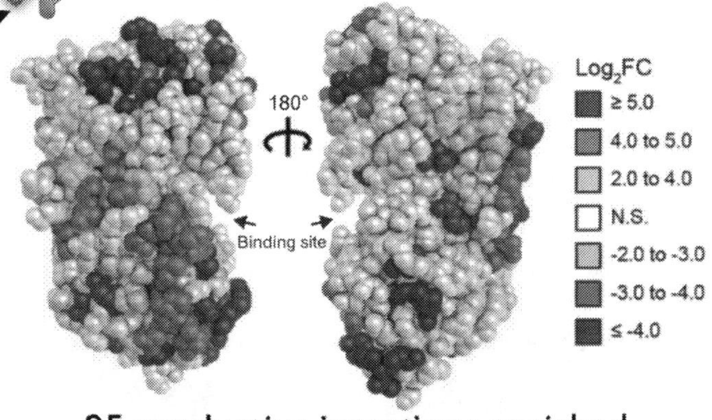

95 productive insertions enriched
(240 of 370 detected)
4 unproductive insertions enriched

PDB: 1ANF

Fluorescence-activated cell sorting (FACS) enriches productive cpGFP insertions

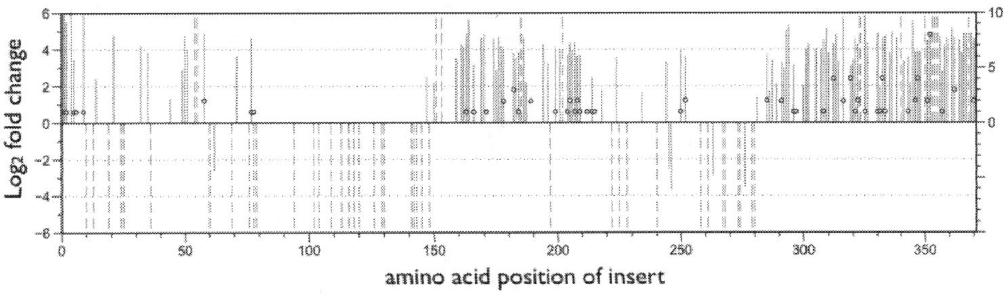

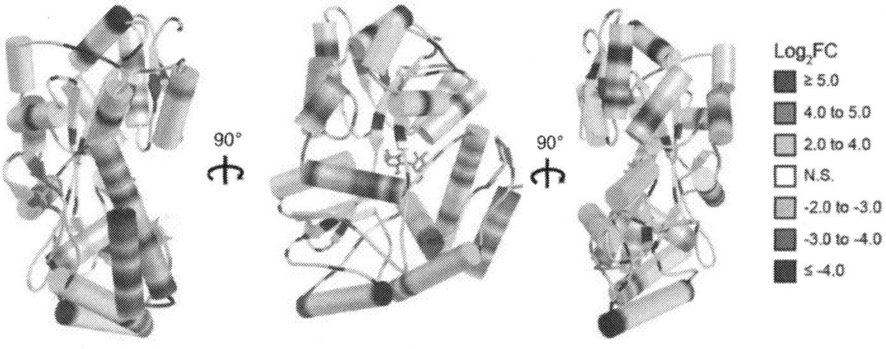

Functional biosensor sites are not isolated to the PBP "hinge" region

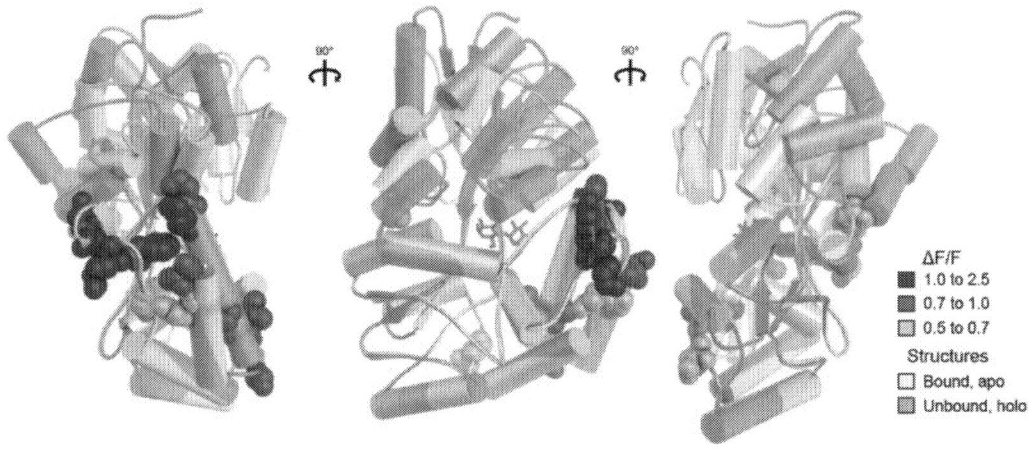

Linker-sequence optimization at a hotspot improves biosensor dynamic range

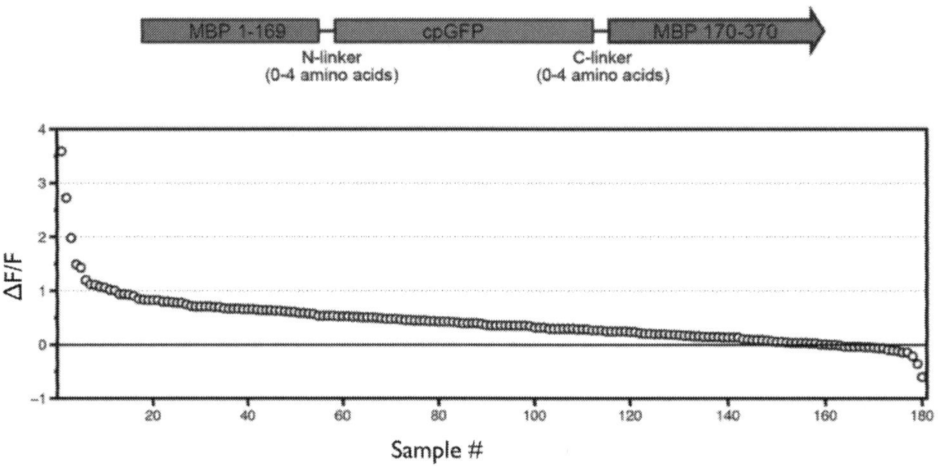

Linker-sequence optimization at a hotspot improves biosensor dynamic range

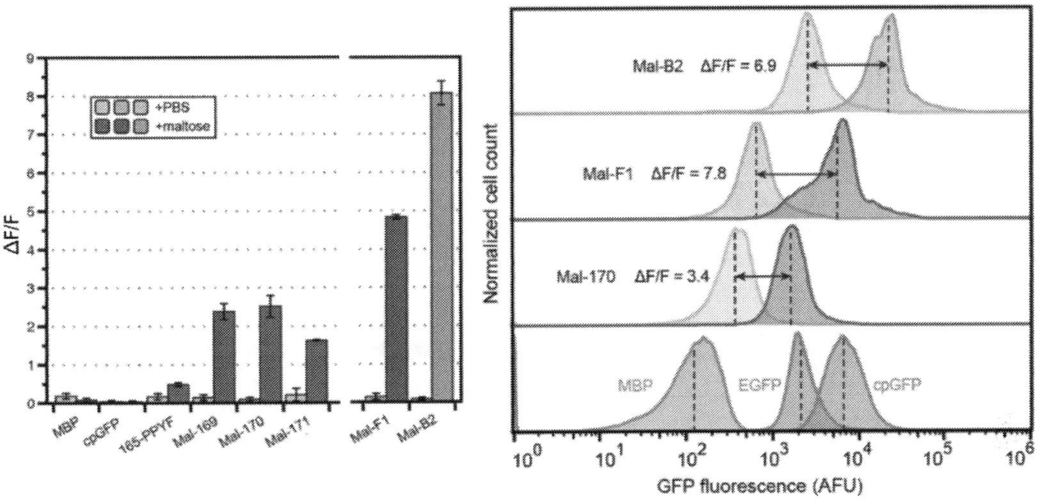

Linker sequence optimization of Mal-170

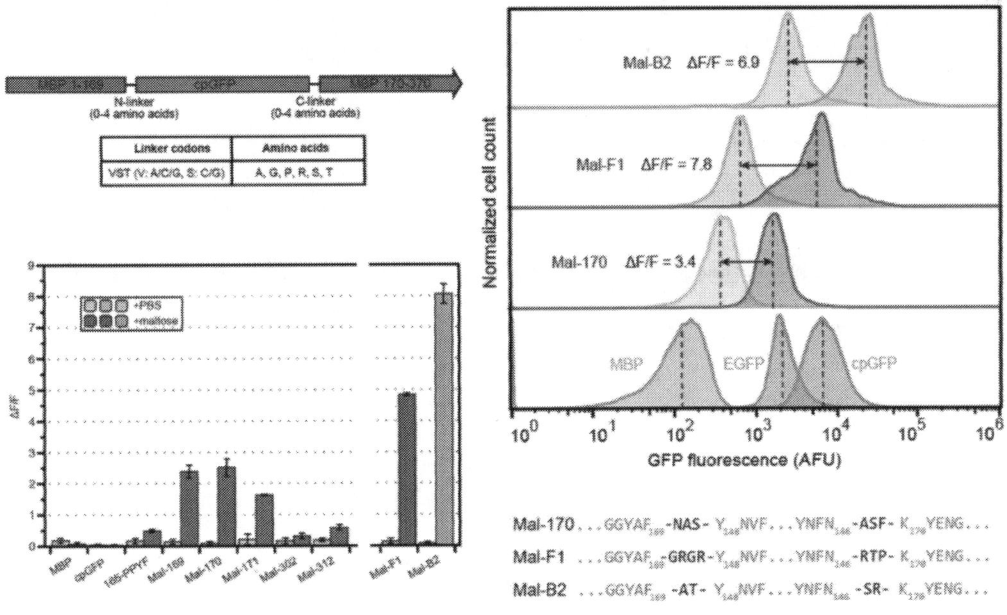

In vitro characterization of Mal-B2

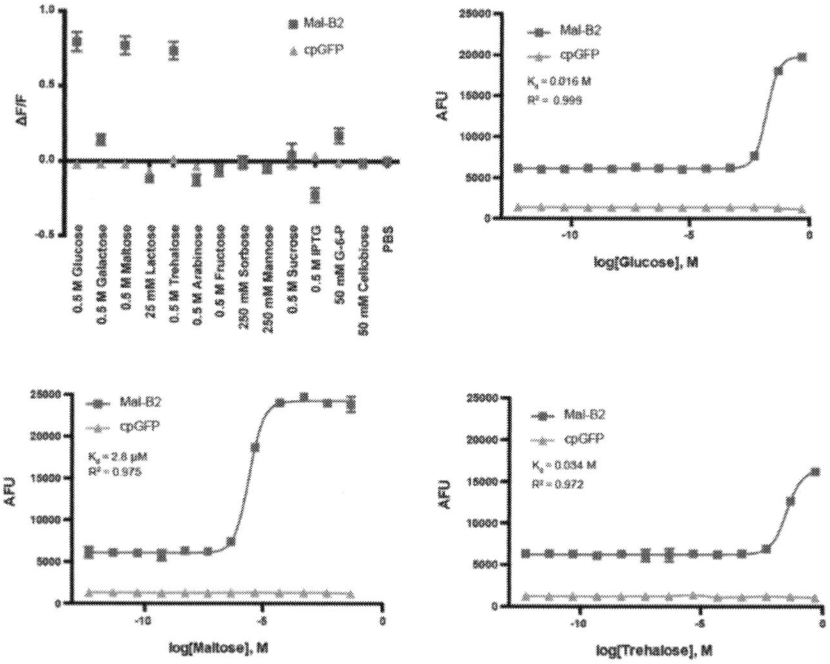

Multiple rounds of FACS screening to enrich trehalose biosensor switching behavior

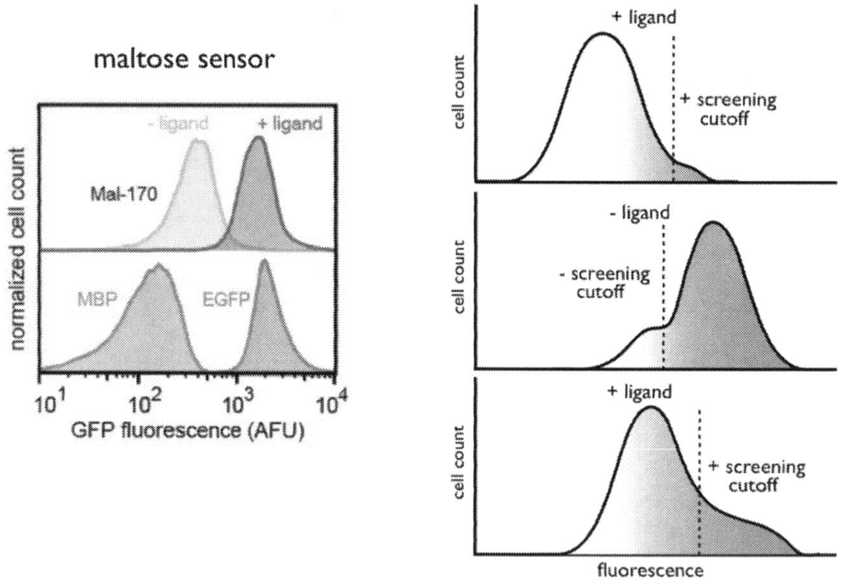

Multiple rounds of FACS screening to enrich trehalose biosensor switching behavior

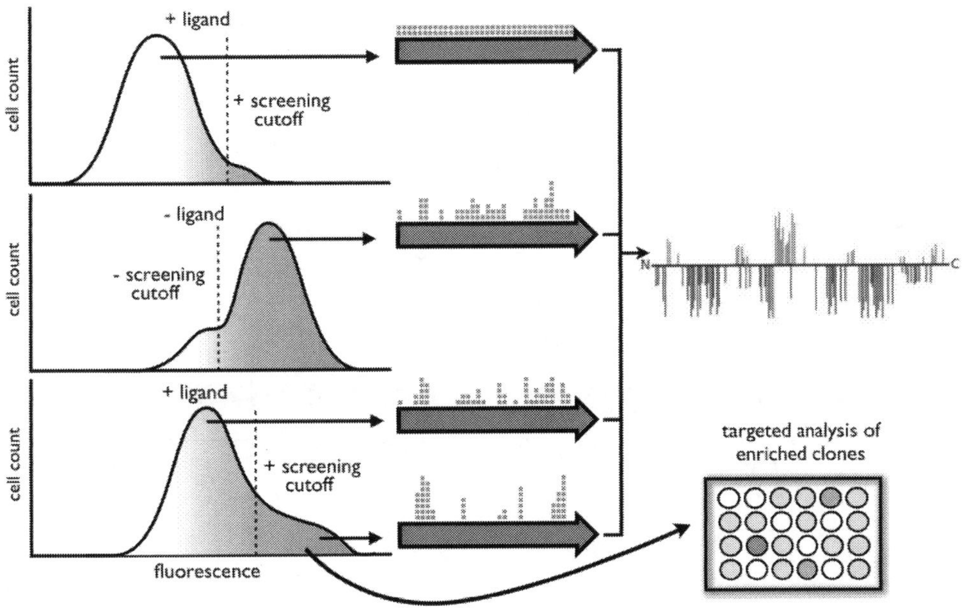

MBP and TMBP are homologs

~25% sequence identity
~42% sequence similarity

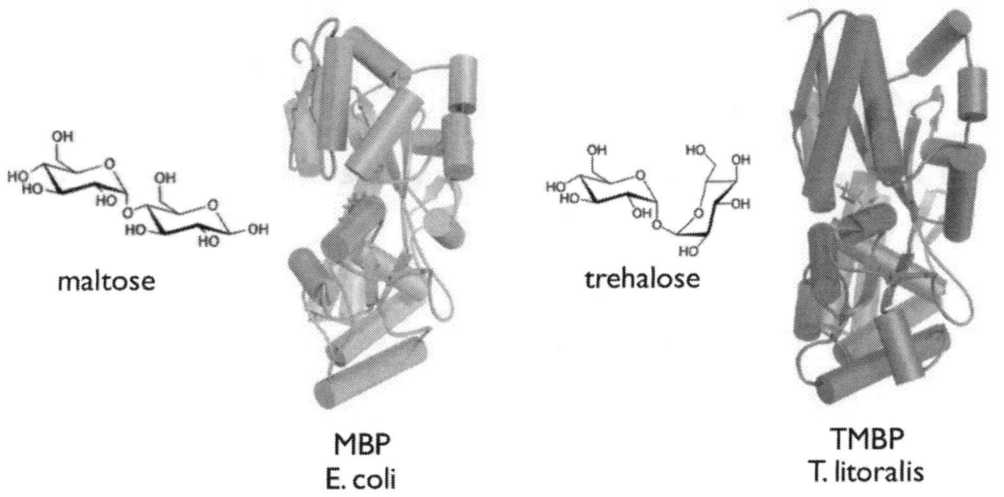

PDB: 1ANF, 1EU8

Initial TMBP domain-insertion library has extensive coverage

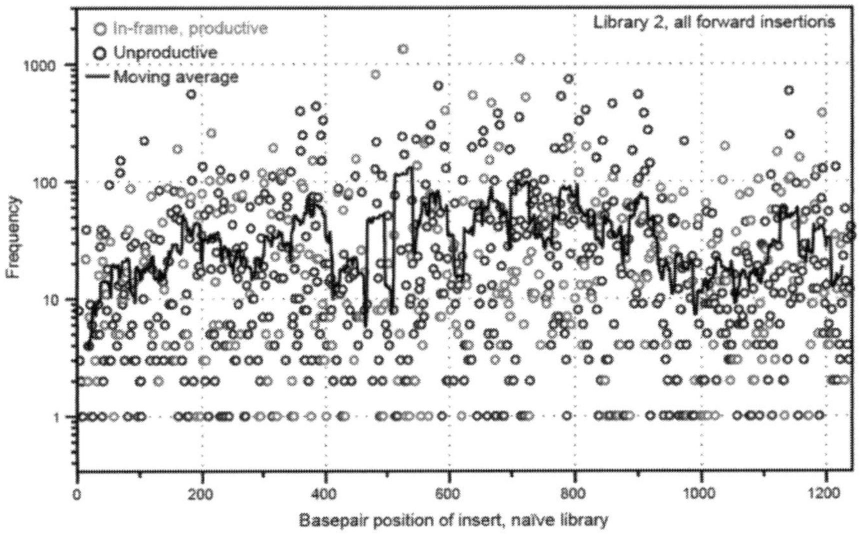

Final read count does not correlate with enrichment after 3 sorts

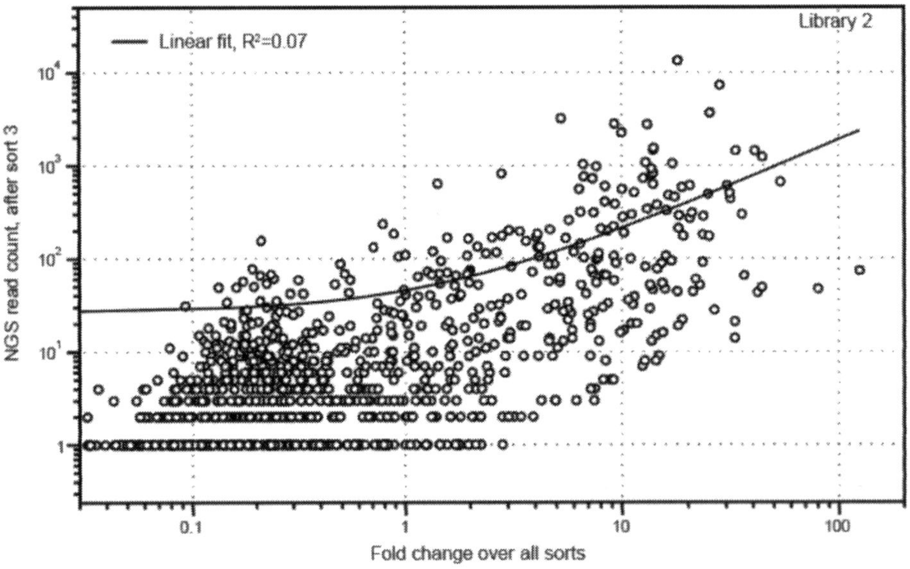

3 sorts of TMBP domain-insertion libraries enriches for productive constructs

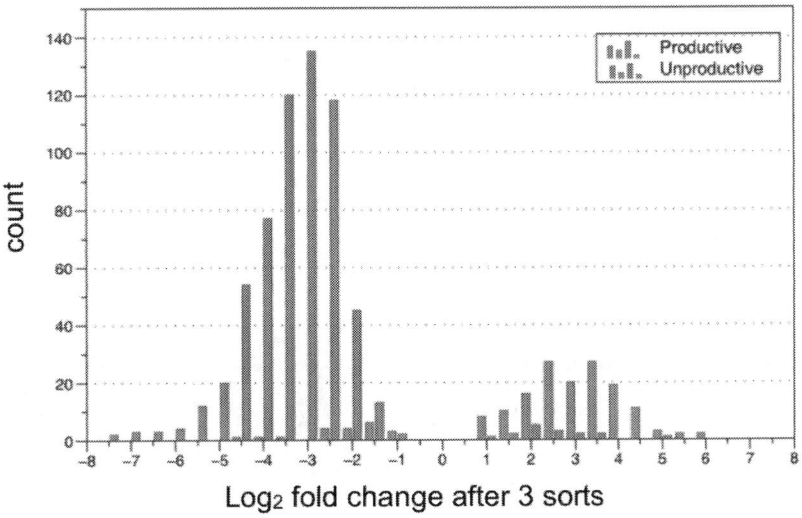

Enrichment at each round of sorting did not correlate with biosensor activity

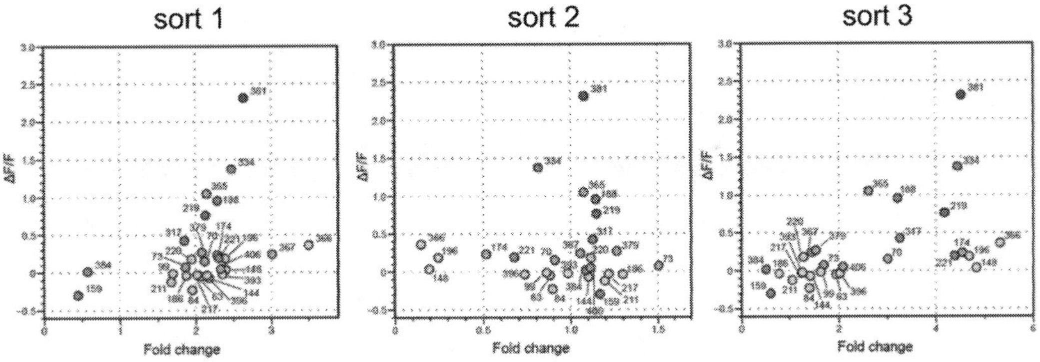

TMBP insertion sites with significant biosensor activity

TMBP insert site	ΔF/F
381	2.31
334	1.37
365	1.05
188	0.95
219	0.76
317	0.42
366	0.36
373	0.32
379	0.27
367	0.24
174	0.23

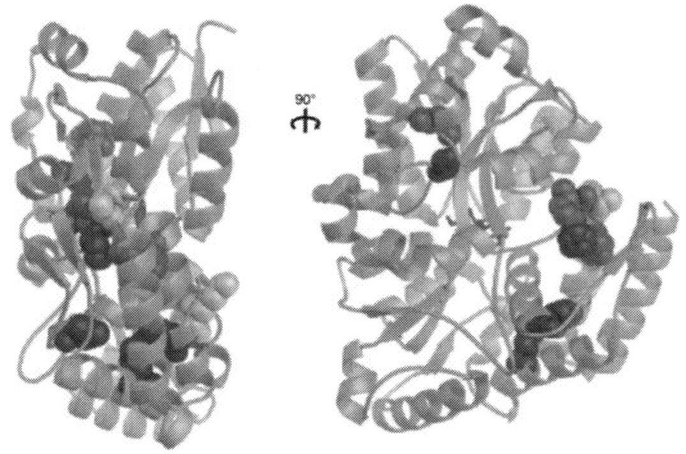

Comparison of TMBP and MBP insertion sites with significant biosensor activity

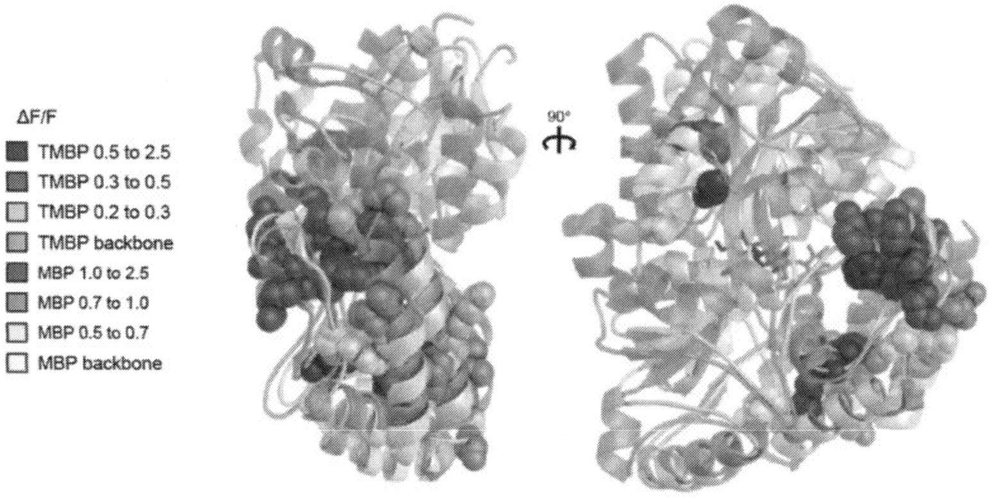

Tre-C04 in vitro characterization

Linker codons	Amino acids
VVC (V: A/C/G)	A, D, G, H, N, P, R, S, T

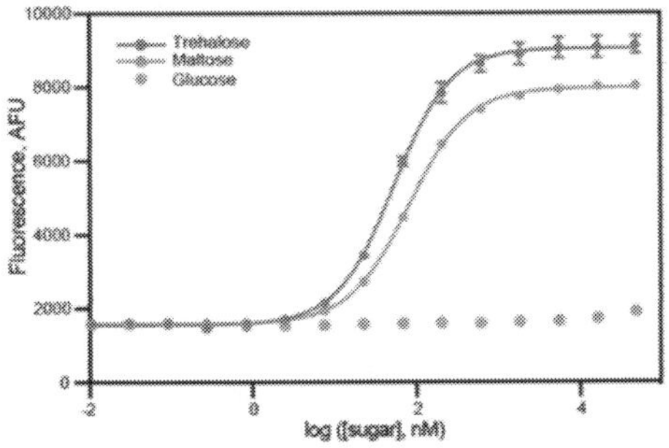

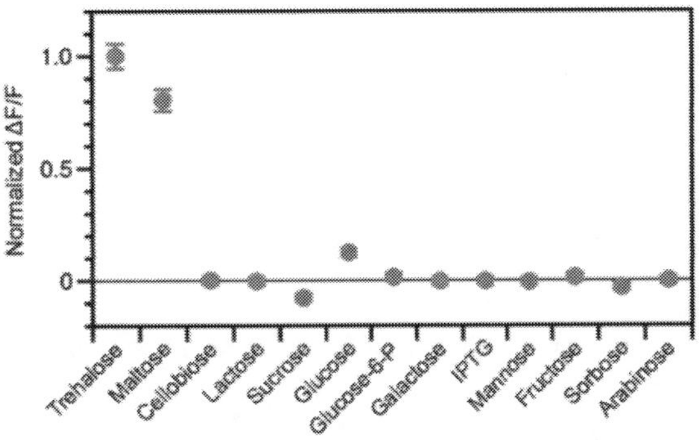

Tre-334 ...LGWNP$_{333}$ - **GAS** - Y$_{148}$NVF...YNFN$_{146}$ - **ASP** - G$_{334}$RVDV...

Tre-C04 ...LGWNP$_{333}$ - **GDR** - S$_{147}$YNV...YNFN$_{146}$ - **RSR** - G$_{334}$RVDV...

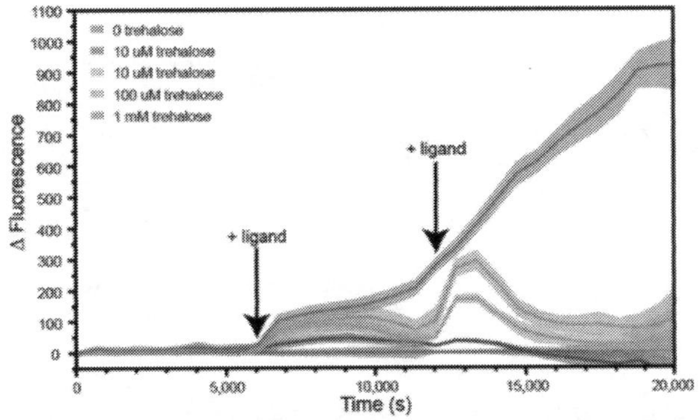

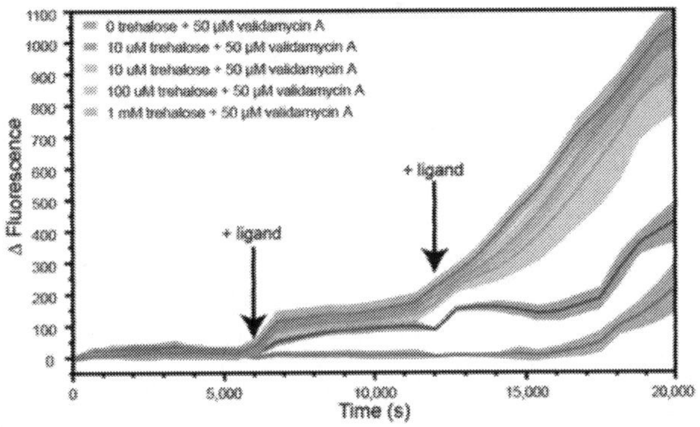

Tre-C04 reports dynamic changes in trehalose concentration *in vivo*

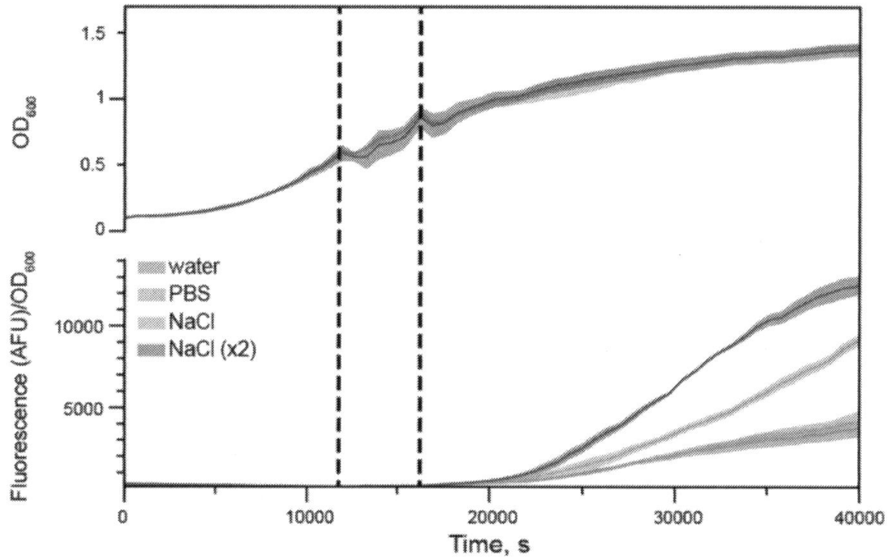

Tre-C04 in vivo response validated by extracted sugar measurements

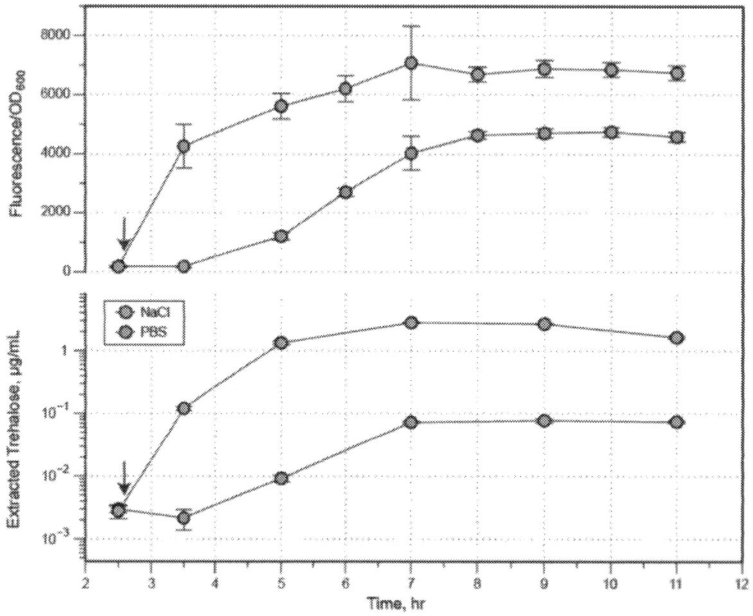

Domain-insertion profiling of Cas9 reveals hotspots for engineering

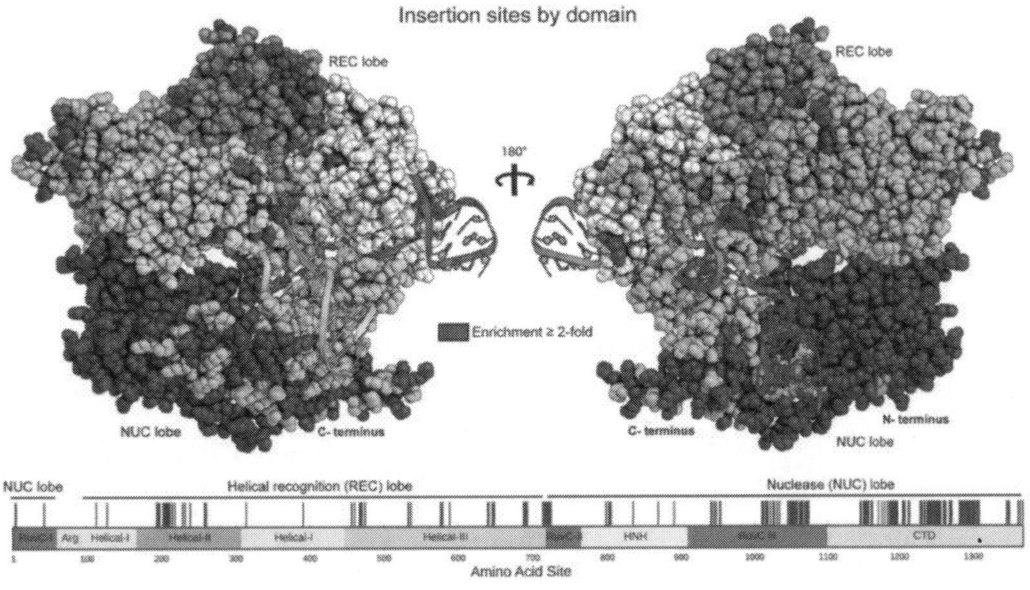

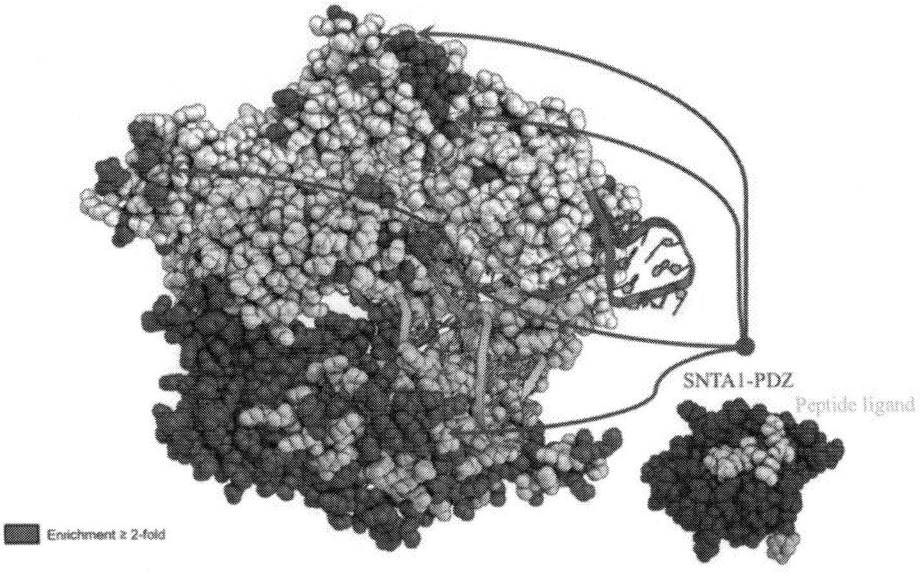

Domain-insertion profiling identifies hotspots that can be stacked

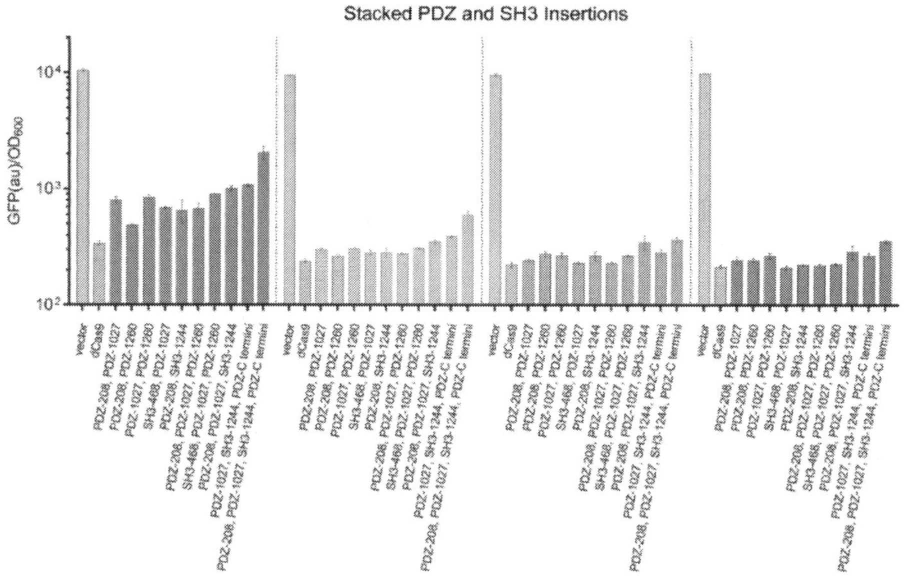

Uncovering How RNA Molecules 'Make Decisions' On the Fly: Towards Understanding and Engineering CoTranscriptional RNA Folding

Lucks, J. B., Cornell University

Originally presented on: 7/20/2016 17:00:00 - 17:25:00

RNAs are emerging as a powerful substrate for engineering gene expression and cellular behavior since they are now known to control almost all aspects of gene expression. As with all biomolecules, RNA function is intimately related to its structure, since RNA can adopt structures that selectively modulate gene expression. Central questions in biology and bioengineering then are: How do RNAs fold inside cells?; and How can we engineer these folds to control gene expression? In this talk, I will present our work at the interface of these two questions and share results that are beginning to uncover design principles for engineering RNAs for an array of applications, ranging from metabolic engineering to diagnostics.

I will start by presenting our work on engineering RNA molecular switches that control transcription. The desire to uncover design principles for engineering these RNAs motivates our development of SHAPE-Seq, a technology that couples chemical probing with next-generation sequencing and that helps characterize RNA structures on an 'omics' scale. I will then describe our exciting recent developments in using SHAPE-Seq to help break open one of the final frontiers of RNA structure-function relationships by uncovering at nucleotide resolution how RNAs fold cotranscriptionally. Specifically I will highlight new data on uncovering the ligand-dependent folding pathways of riboswitches. This new ability is allowing us to ask deep questions about how RNA molecules make regulatory decisions 'on the fly' during the dynamic process of transcription. By probing the fundamental processes of RNA folding and function, these studies are expected to greatly aid RNA engineering.

The Emerging Organism Engineering Industry
Shetty, R., Ginkgo Bioworks
Originally presented on: 7/21/2016 9:30:00 - 10:00:00

There is an emerging demand for sourcing plant-derived extracts (nutraceuticals, flavors, fragrances, sweeteners, etc.) from engineered microbes. While recent advances in synthetic biology and metabolic engineering provide feasible approaches to engineering such organisms, commercial success for developing these â??culturedâ?□ ingredients presents specific challenges. Unlike biofuels, where efforts can be focused on one particular molecule given the enormous market size, cultured ingredients require developing different organism lines in a rapid and low cost fashion. This requires a scalable solution for bio-manufacturing of organisms, which is provided by our state of the art foundry that continues to grow. I will describe how organism development at Ginkgo leverages our foundry to accelerate the design/build/test using specific examples. In particular, I will highlight the value of combining computer-aided engineering software tools, cheap gene synthesis and high resolution-accurate mass LCMS to develop engineered microbes. Finally, Iâ??ll touch on how our improvements in manufacturing organisms lend to opportunities outside of cultured ingredients.

Hardware/Wetware Co-Design: CAD for Microfluidic Synthetic Biology

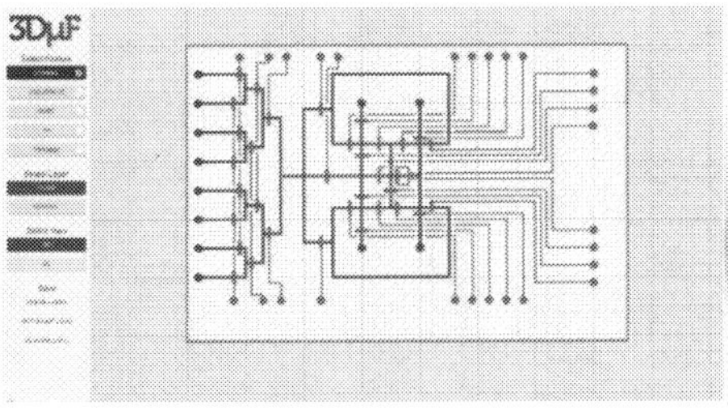

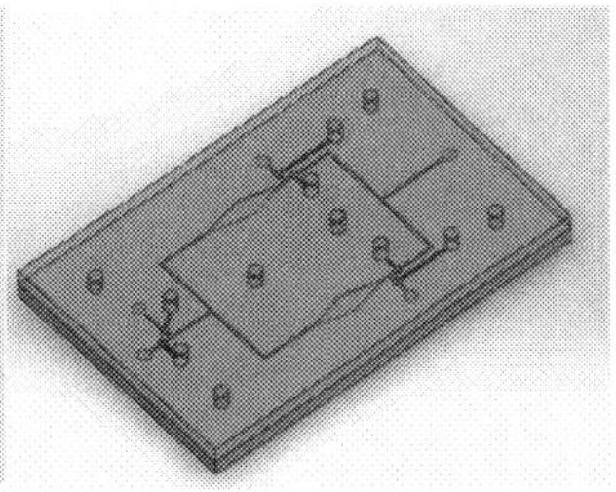

Prof. Douglas Densmore
Bio-Design and Automation
SEED 2016, Chicago
July 21st, 2016

SEED Synthetic Biology, Engineering, Evolution & Design

LIVING COMPUTING PROJECT · BOSTON UNIVERSITY · MIT · MIT Lincoln Laboratory · NSF · Fluigi

BU Electrical & Computer Engineering | BU Biomedical Engineering

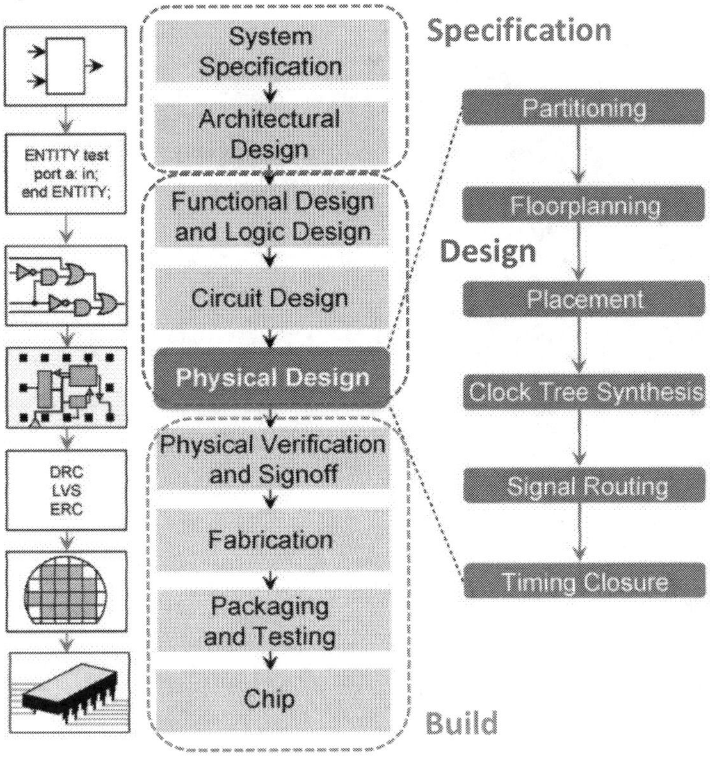

https://en.wikipedia.org/wiki/Physical_design_%28electronics%29

2

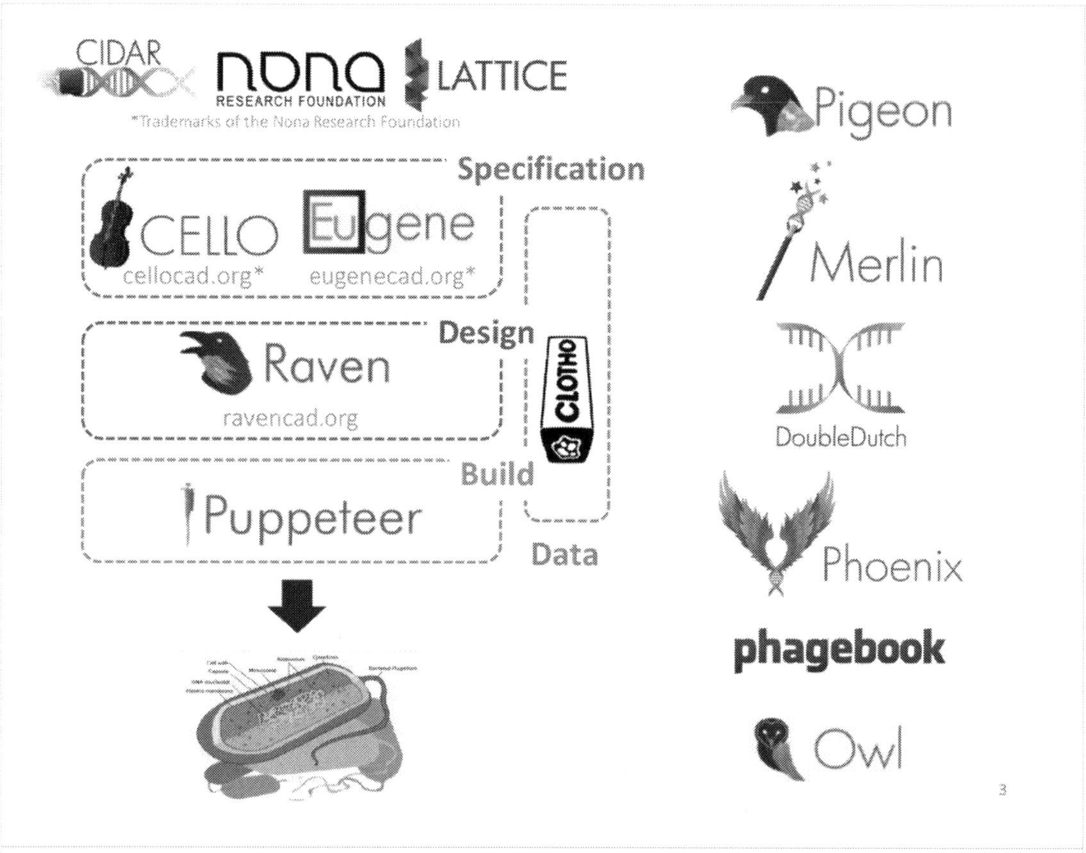
CIDAR
nona
RESEARCH FOUNDATION
LATTICE
*Trademarks of the Nona Research Foundation
Specification
CELLO
cellocad.org*
Eugene
eugenecad.org*
Design
Raven
ravencad.org
Build
Puppeteer
CLOTHO
Data
Pigeon
Merlin
DoubleDutch
Phoenix
phagebook
Owl
3

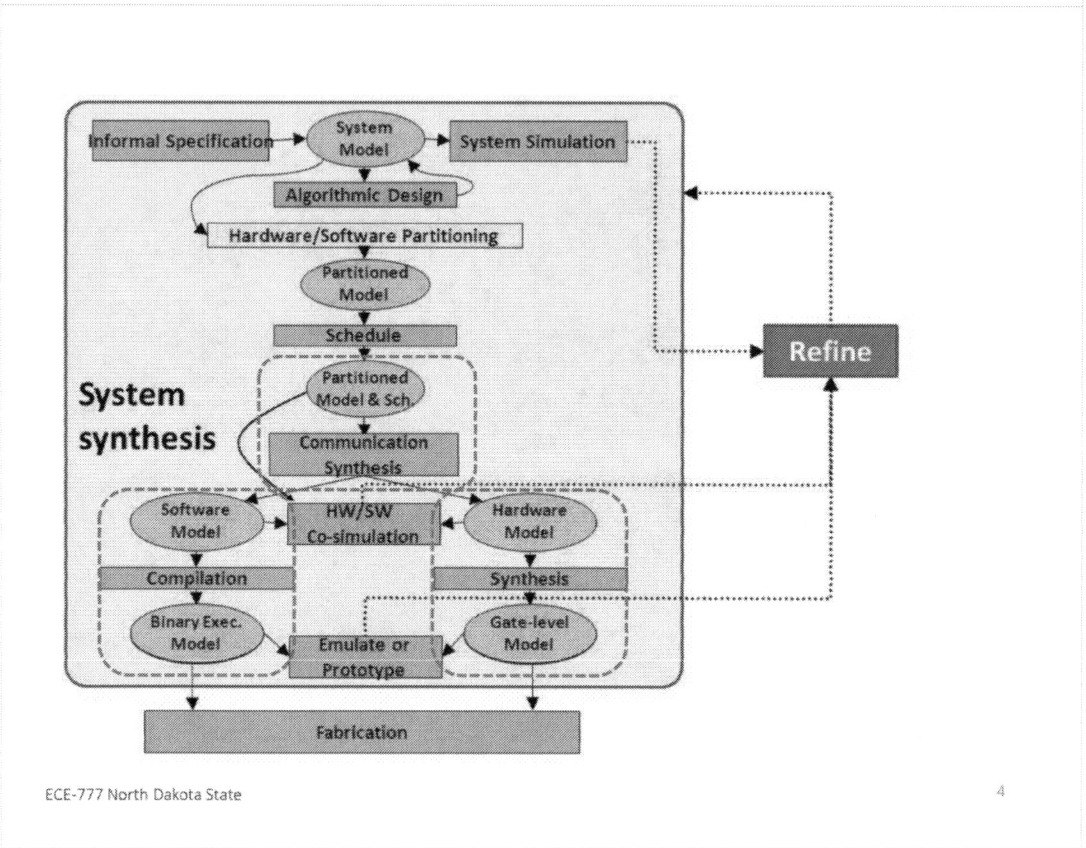
Informal Specification
System Model
System Simulation
Algorithmic Design
Hardware/Software Partitioning
Partitioned Model
Schedule
Partitioned Model & Sch.
Communication Synthesis
System synthesis
Refine
Software Model
HW/SW Co-simulation
Hardware Model
Compilation
Synthesis
Binary Exec. Model
Emulate or Prototype
Gate-level Model
Fabrication
ECE-777 North Dakota State
4

Hybrid Bio-Electronic Systems Vision

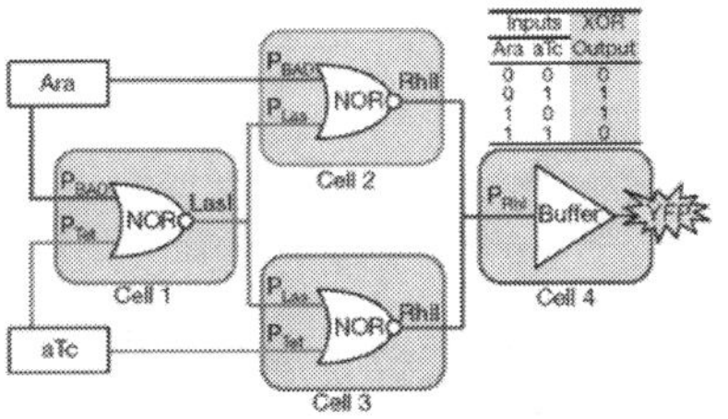

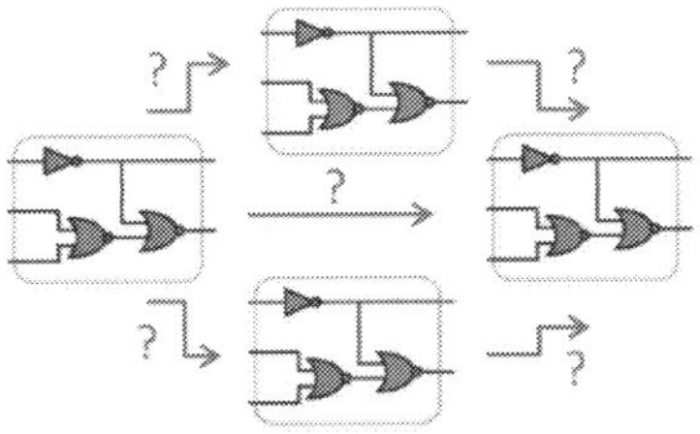

A. Tamsir, J. J. Tabor and C. A. Voigt, Nature, 2011, 469, 212–215.

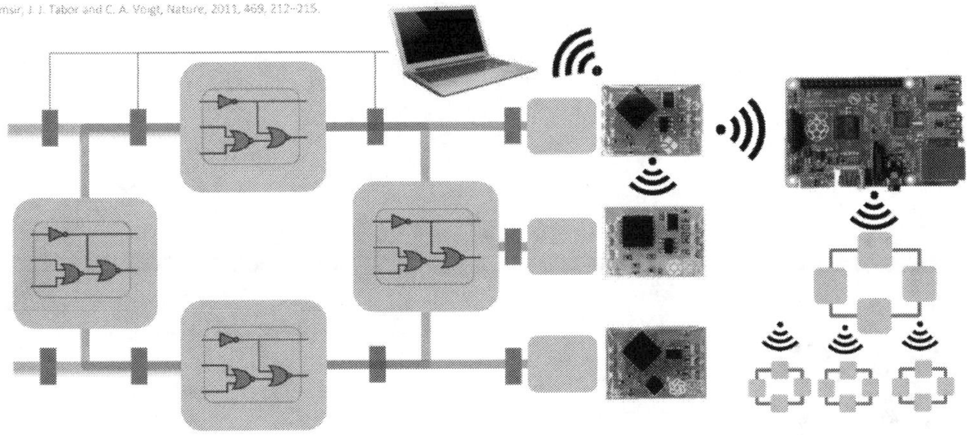

5

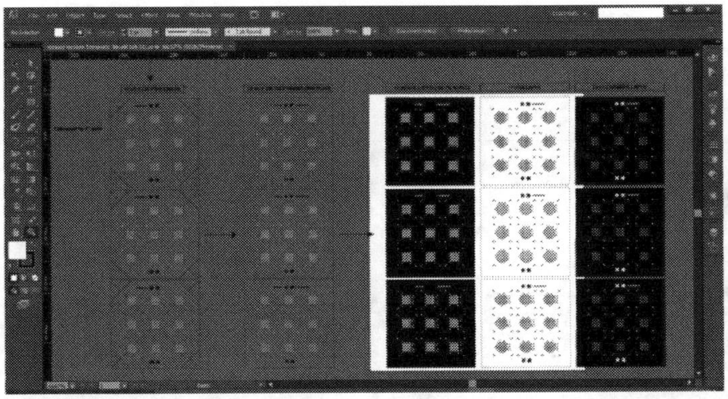

Min Feature Size/Spacing	
<1μm[1]	5μm/25μm[2]
Bonding Pressure Capacity	
50 psi[1]	5 psi[1]
Equipment Cost	
~$80,000	~$2,300
Fabrication Time	
1 Day	1 Hour
Who does most of the work?	
Human	Machine

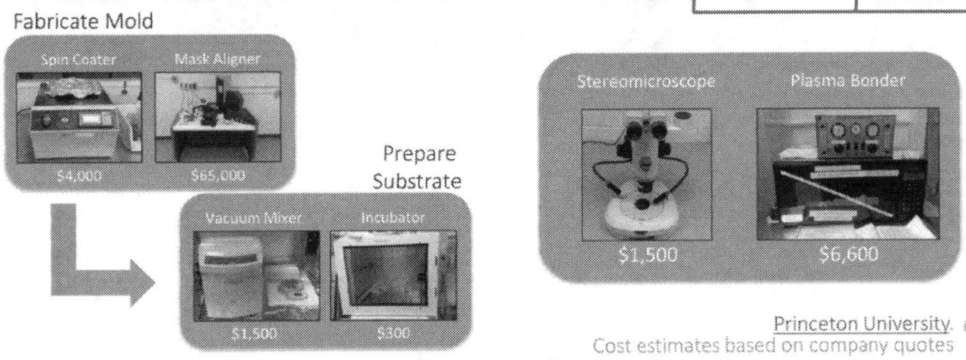

Princeton University. 6
Cost estimates based on company quotes

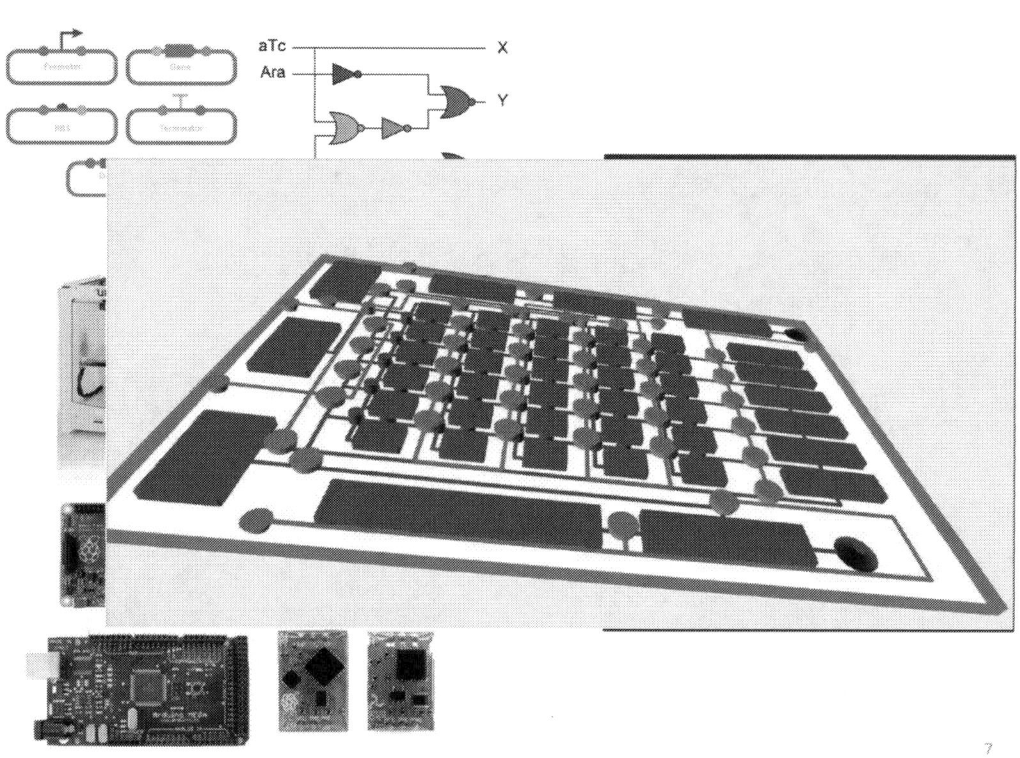

7

Fluigi

Silva | Lashkaripour | Sanka | Lippai | Ortiz | Huang

MF

MINT

3DμF

MIT Lincoln Laboratory

DRAPER

METAFLUIDICS

nona RESEARCH FOUNDATION

8

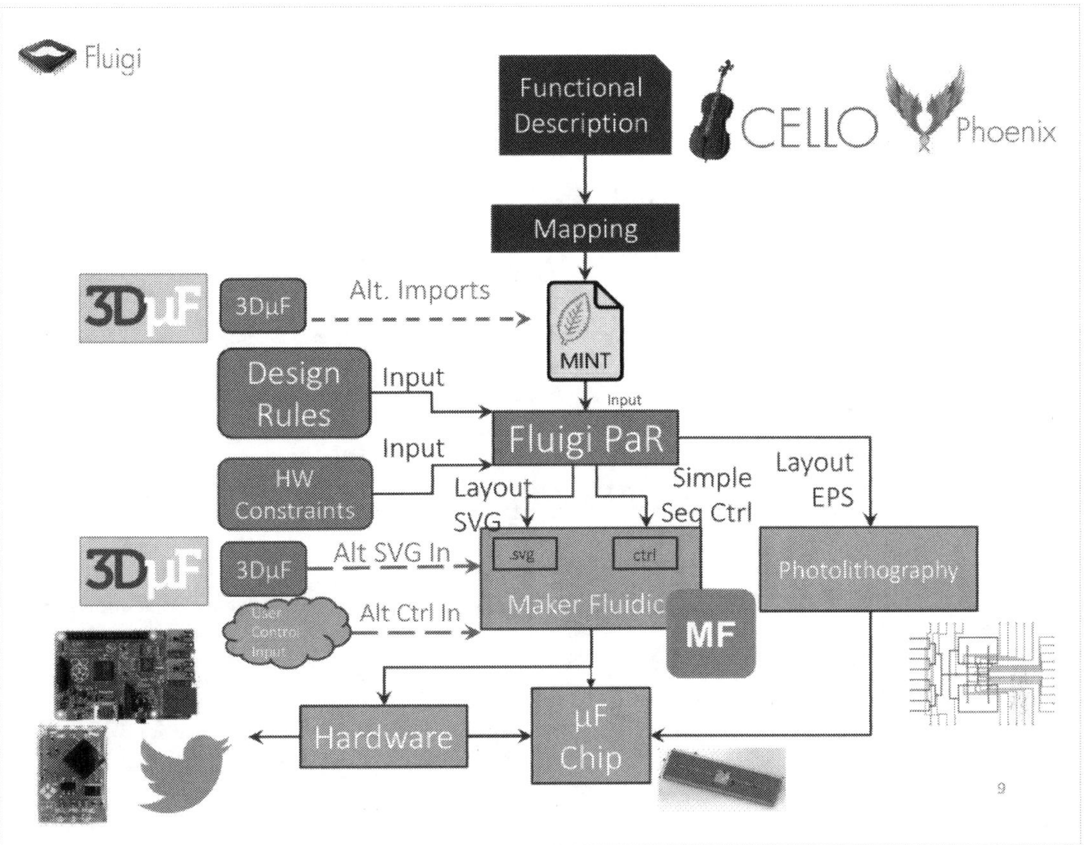
Fluigi
Functional Description
CELLO
Phoenix
Mapping
3DμF
Alt. Imports
MINT
Design Rules
Input
Input
Fluigi PaR
Input
HW Constraints
Layout SVG
Simple Seq Ctrl
Layout EPS
Alt SVG In
.svg
ctrl
Maker Fluidic
MF
Photolithography
User Control Input
Alt Ctrl In
Hardware
μF Chip
9

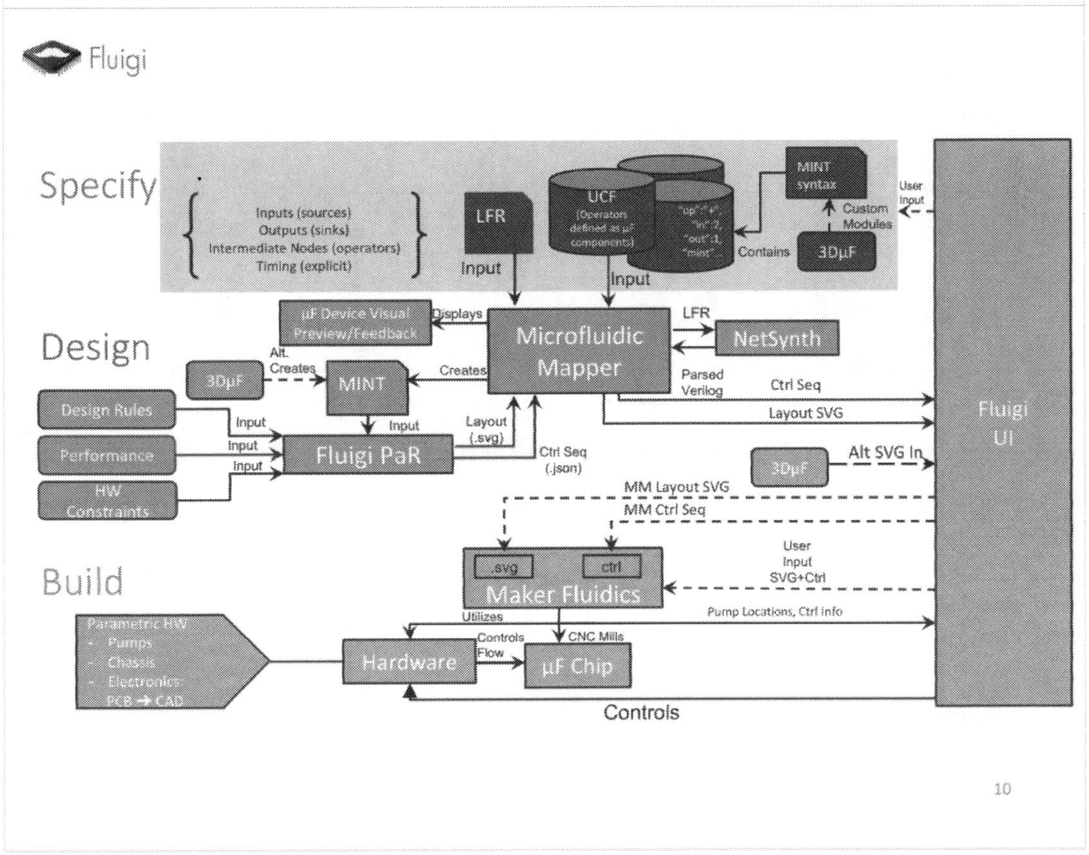
Fluigi
Specify
Inputs (sources)
Outputs (sinks)
Intermediate Nodes (operators)
Timing (explicit)
LFR
Input
UCF
(Operators defined as μF components)
Input
MINT syntax
Custom Modules
Contains
3DμF
User Input
Design
μF Device Visual Preview/Feedback
Displays
Microfluidic Mapper
LFR
NetSynth
Alt. Creates
Creates
Parsed Verilog
Ctrl Seq
Layout SVG
Design Rules
Performance
HW Constraints
Input
Fluigi PaR
Layout (.svg)
Ctrl Seq (.json)
Alt SVG In
MM Layout SVG
MM Ctrl Seq
Fluigi UI
Build
.svg
ctrl
Maker Fluidics
User Input SVG+Ctrl
Parametric HW
- Pumps
- Chassis
- Electronics: PCB → CAD
Utilizes
Pump Locations, Ctrl info
Hardware
Controls Flow
CNC Mills
μF Chip
Controls
10

Fluigi

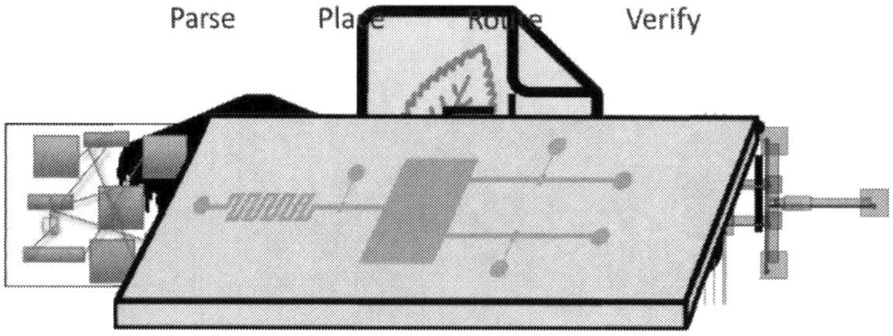

11

Defining Primitives
Composition
Hierarchy

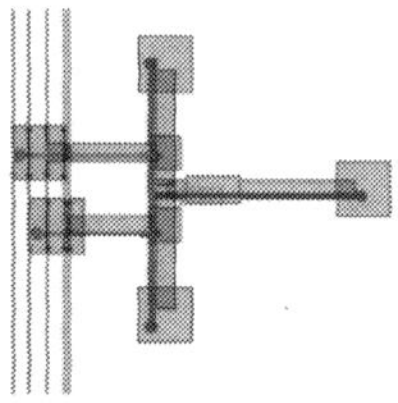

Placement
Routing
Design Rules
Control Generation

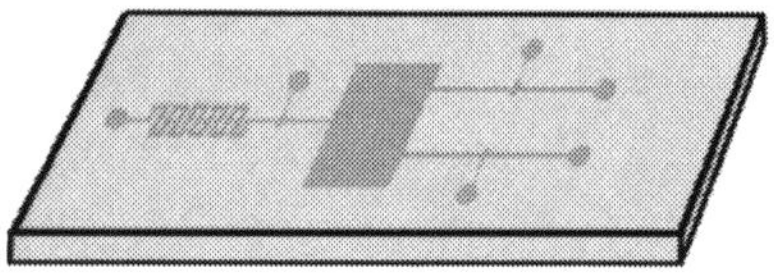

Fabrication
Deployment

12

Fluigi

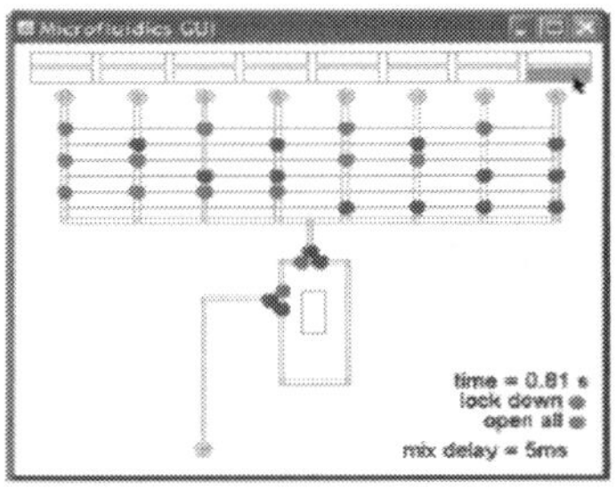

Micado/BioStream

Amin, Nada, William Thies, and Saman Amarasinghe. "Computer-aided design for microfluidic chips based on multilayer soft lithography." ICCD 2009. IEEE International Conference on, pp. 2-9. IEEE, 2009

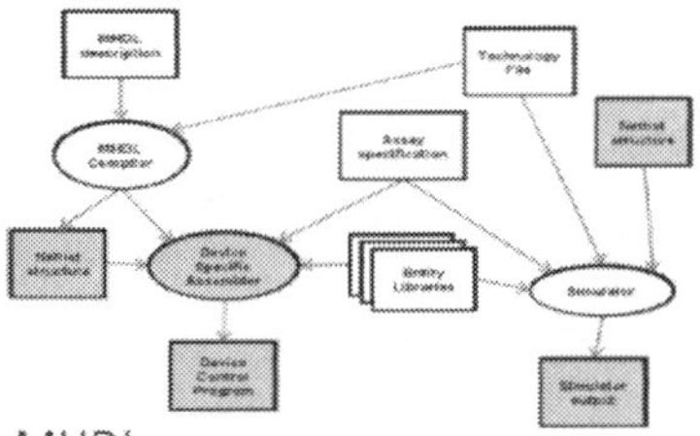

MHDL

McDaniel, Jeffrey, et al. "Design and verification tools for continuous fluid flow-based microfluidic devices." Design Automation Conference (ASP-DAC), 2013 18th Asia and South Pacific. IEEE, 2013.

MINT

MIcrofluidic NeTlist

Human Readable
Extensible
Parameterized
Hierarchical
Composable
Lightweight

13

Fluigi

```
DEVICE net_test

LAYER FLOW
PORT p1, p2 r=100;
V LONG CELL TRAP ct1 numChambers=10
    chamberWidth=100 chamberLength=100
    chamberSpacing=50 channelWidth=100;

CHANNEL c1 from p1 3 to ct1 1 w=100;
CHANNEL c2 from ct1 2 to p2 1 w=100;

END LAYER

LAYER CONTROL
PORT p3 r=100;

VALVE v1 on c1 w=300 l=100;
VALVE v2 on c2 w=300 l=100;

NET net1 from p3 1 to v1 4, v2 4
    channelWidth=50;
END LAYER
```

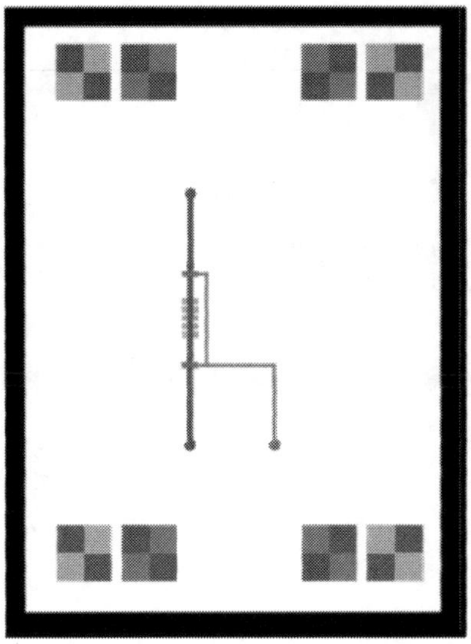

14

Fluigi

```
DEVICE grad_cells

LAYER FLOW
V BANK pb1 of 4 PORT r=100 dir=RIGHT spacing=1200 channelWidth=100;
PORT p1 r=100;
NODE n1;
H GRADIENT GENERATOR g 1 to 4 numBends=10 bendSpacing=100 bendLength=600
channelWidth=100;
V MUX m1 1 to 4 spacing=500 flowChannelWidth=100 controlChannelWidth=60;
V TREE t1 4 to 1 spacing=500 flowChannelWidth=100;
V BANK ctb of 4 CELL TRAP numChambers=20 chamberWidth=100 chamberLength=100
chamberSpacing=30 spacing=500 channelWidth=100;
CHANNEL c1 from pb1 1 to m1 1 w=100;
CHANNEL c2 from pb1 2 to m1 3 w=100;
CHANNEL c3 from m1 2 to g 1 w=100;
CHANNEL c4 from g 2 to m1 1 w=100;
CHANNEL c5 from m1 2 to ctb 1 w=100;
CHANNEL c6 from m1 3 to ctb 2 w=100;
CHANNEL c7 from m1 4 to ctb 3 w=100;
CHANNEL c8 from m1 5 to ctb 4 w=100;
CHANNEL c9 from ctb 5 to t1 1 w=100;
CHANNEL c10 from ctb 6 to t1 2 w=100;
CHANNEL c11 from ctb 7 to t1 3 w=100;
CHANNEL c12 from ctb 8 to t1 4 w=100;
CHANNEL c13 from t1 5 to p1 4 w=100;
END LAYER

LAYER CONTROL

H BANK cpb1 of 2 PORT r=100 dir=DOWN spacing=1200 channelWidth=60;
H BANK cpb2 of 2 PORT r=100 dir=UP spacing=1200 channelWidth=60;
CHANNEL cc1 from cpb1 1 to m1 7 w=60;
CHANNEL cc2 from cpb1 2 to m1 9 w=60;
CHANNEL cc3 from cpb2 1 to m1 6 w=60;
CHANNEL cc4 from cpb2 2 to m1 8 w=60;

END LAYER
```

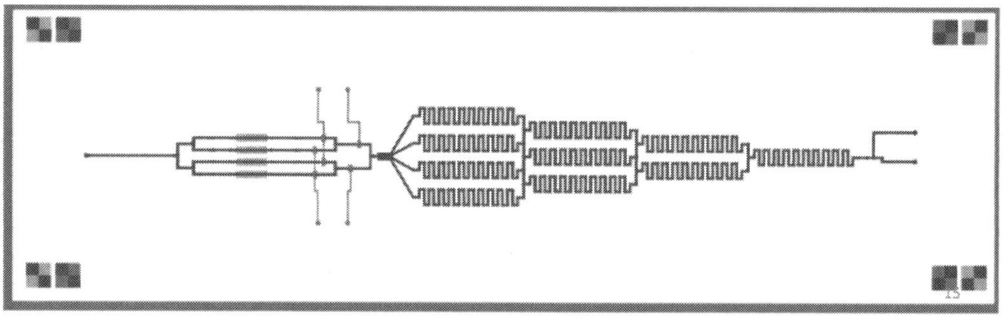

Fluigi

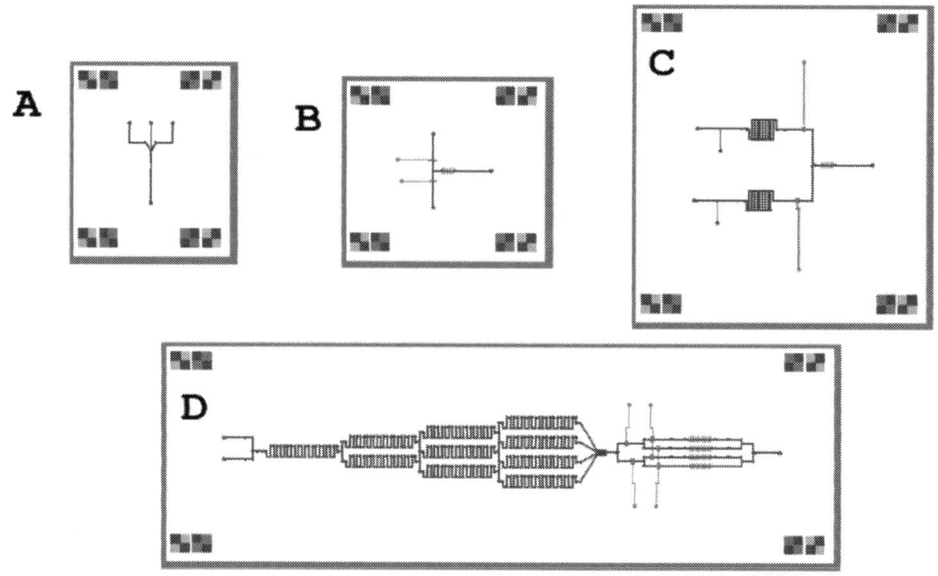

A. Flow Focusing Device

B. Cell Trap Switch

C. Cell Trap + Droplet Generator

D. Gradient Generator + Switches + Cell Traps

16

Fluigi

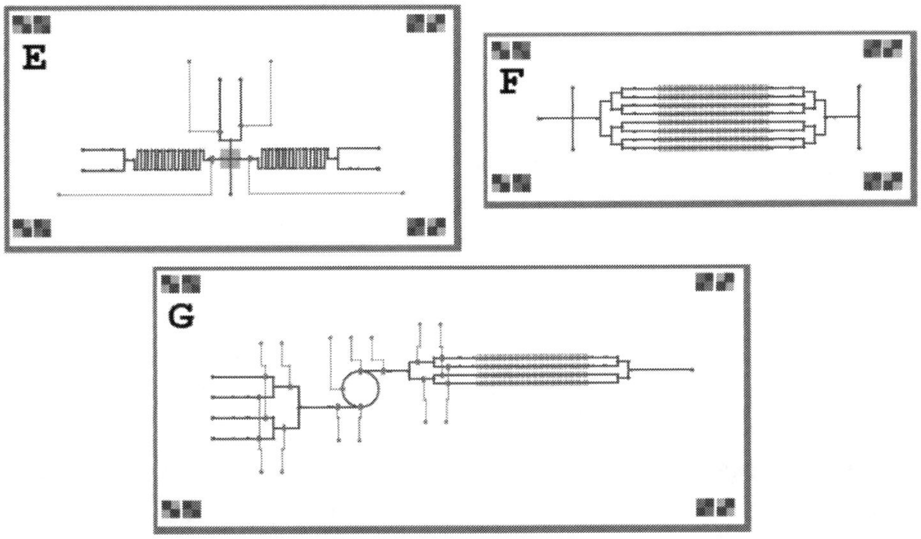

E. Four input cell trap

F. Modified Cell Grids

G. Four input rotary with traps

17

Fluigi

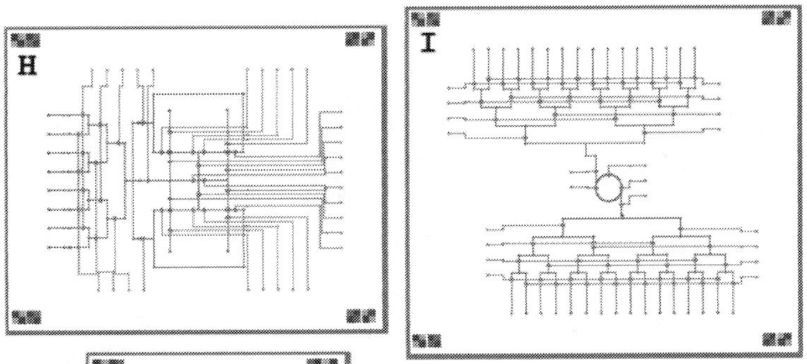

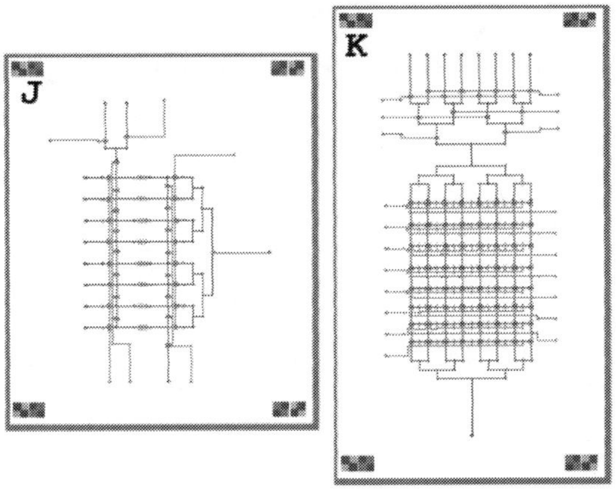

H. Four input "logic array"

I. 16 input/16 output rotary pump

J. 8 cell traps, 3 inputs each

K. 64 cell traps with 8 inputs

18

Fluigi

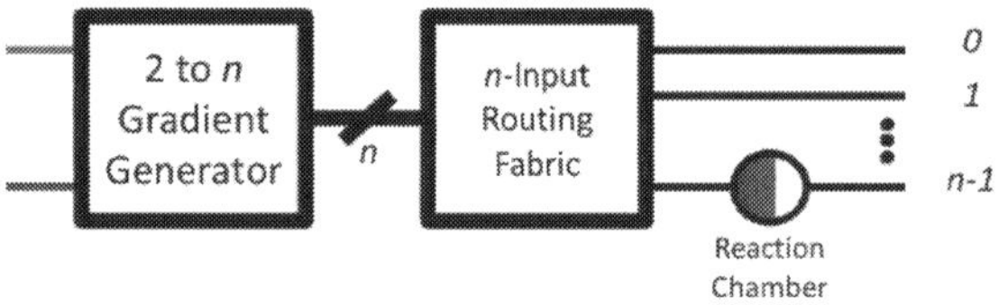

Lab on a Chip

PAPER

A reconfigurable continuous-flow fluidic routing fabric using a modular, scalable primitive†

1 Introduction

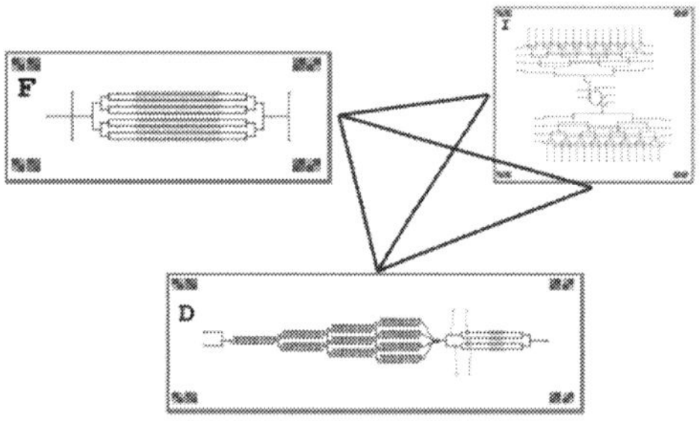

R. Silva, S. Bhatia, and D. Densmore, "A reconfigurable continuous-flow fluidic routing fabric using a modular, scalable primitive," *Lab Chip*, vol. 16, pp. 2730-2741, 2016

19

Fluigi

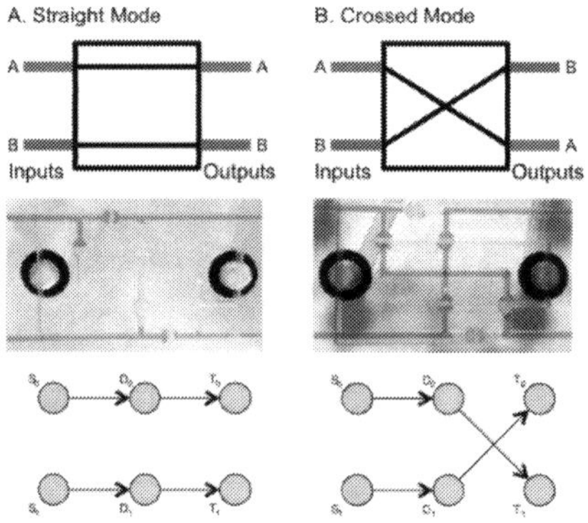

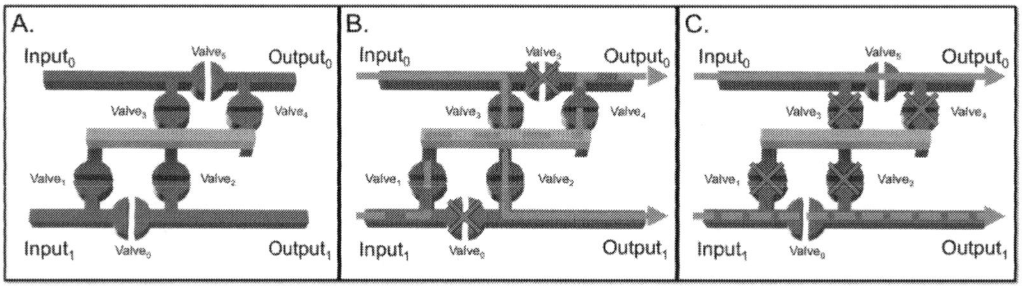

20

Fluigi

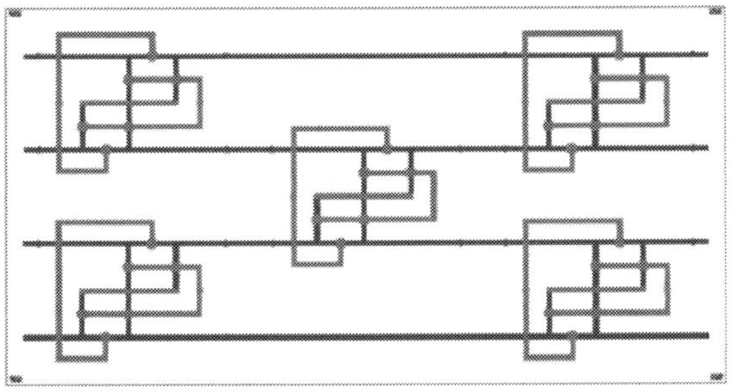

C: transposer_web

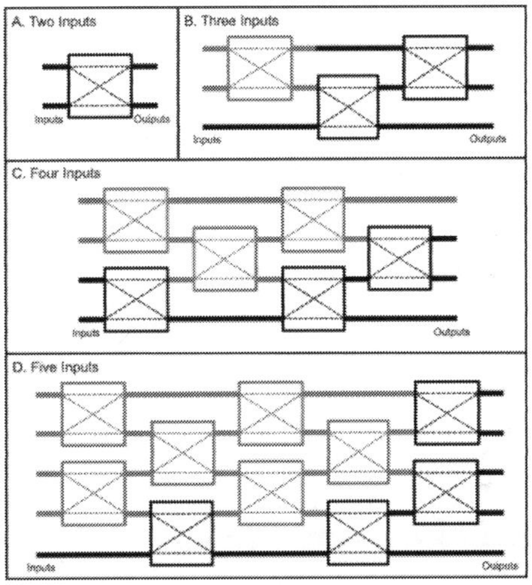

21

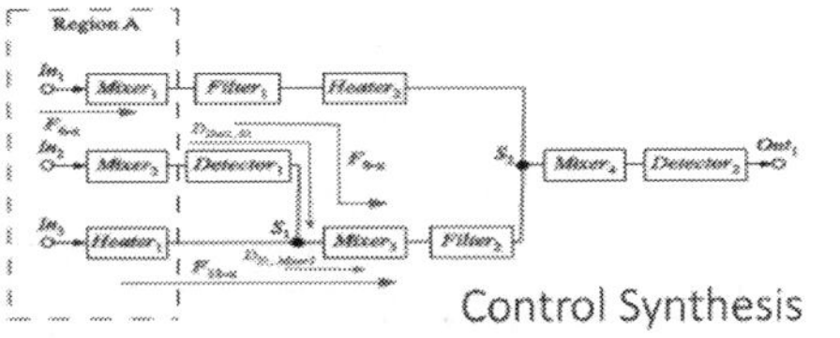

Control Synthesis

Minhass, Wajid Hassan, Paul Pop, Jan Madsen, and Tsung-Yi Ho. "Control synthesis for the flow-based microfluidic large-scale integration biochips." ASP-DAC 2013 18th Asia and South Pacific, pp. 205-212. IEEE, 2013.

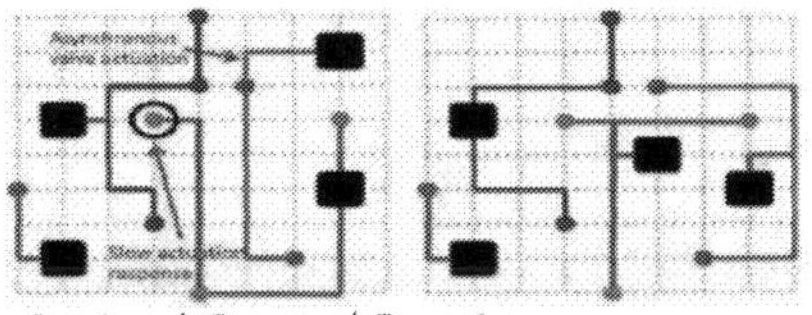

Optimal Control Routing

Hu, Kai, et al. "Control-layer optimization for flow-based mVLSI microfluidic biochips." Compilers, Architecture and Synthesis for Embedded Systems (CASES), 2014 International Conference on. IEEE, 2014.

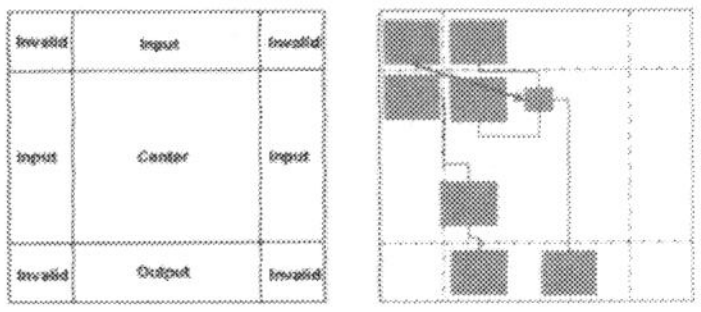

Placement Algorithms

McDaniel, Jeffrey, Brendon Parker, and Philip Brisk. "Simulated annealing-based placement for microfluidic large scale integration (mLSI) chips." Very Large Scale Integration (VLSI-SoC), 2014 22nd International Conference on. IEEE, 2014.

Fluigi

(Core)

Design Rules
Placement
Routing
Control Generation
Fabrication Plans
API

22

Fluigi

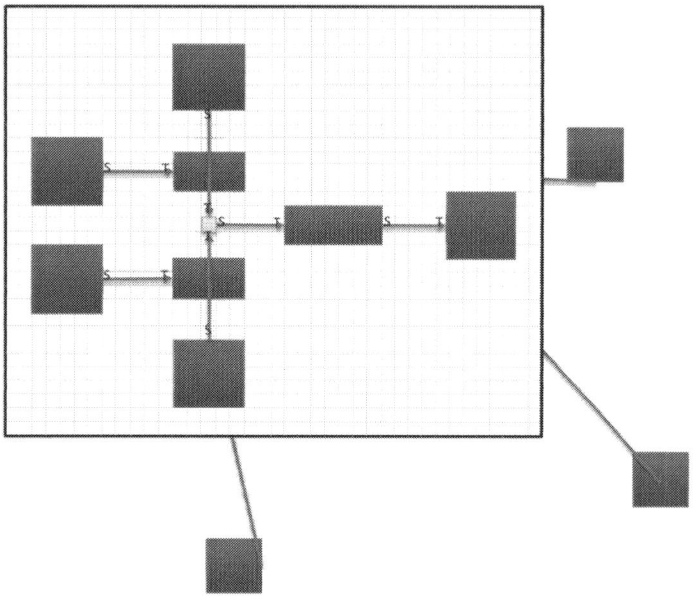

- **Step 1** - Generate Graph from MINT Design.
- **Step 2** – Use Simulated Annealing to Place the components correctly.
- **Step 3** - Route the channels between the components on the grid.

23

Fluigi

1: Add element to interval tree. Search tree for intersections.

2: Add element to interval tree. Search tree for intersections.

3: Add element to interval tree. Search tree for intersections.

4: Add element to interval tree. Search tree for intersections

5: Remove element from interval tree

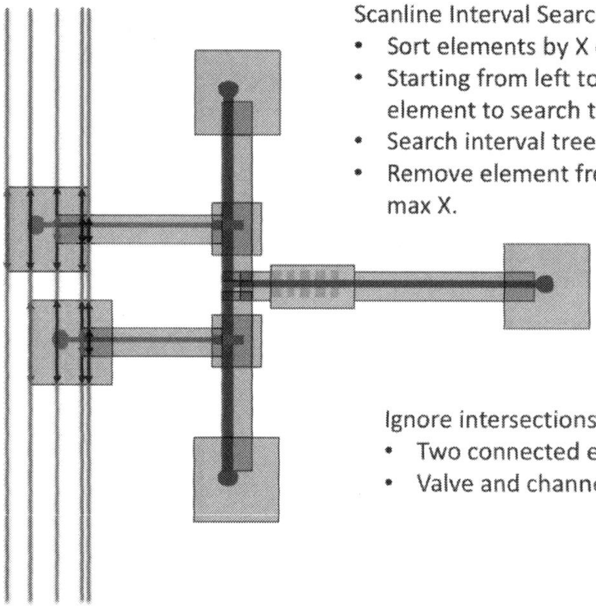

Scanline Interval Search

- Sort elements by X coordinate
- Starting from left to right, add element to search tree at min X.
- Search interval tree for intersections
- Remove element from search tree at max X.

Ignore intersections between:

- Two connected elements
- Valve and channel it controls

24

Fluigi

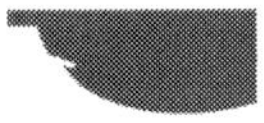

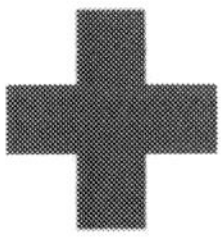

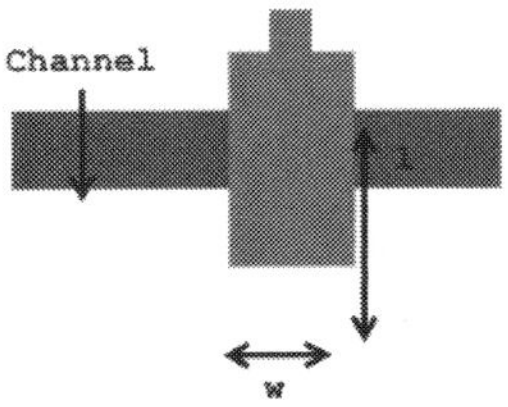

PORT <string> r=<int>; NODE <string>;

VALVE <string> on <channel> w=<int> l=<int>;

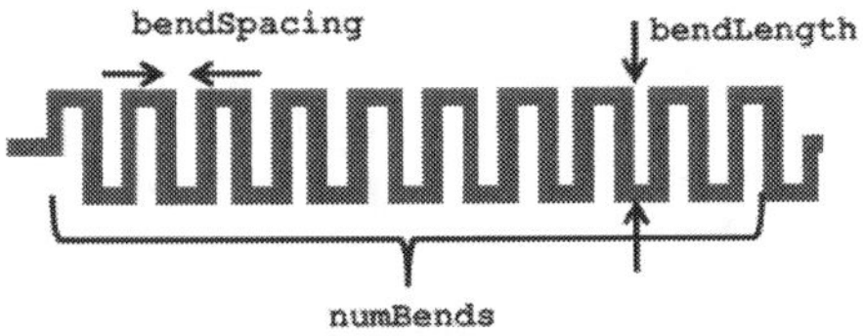

H MIXER <string> numBends=<int> bendSpacing=<int> bendLength=<int> channelWidth=<int>;

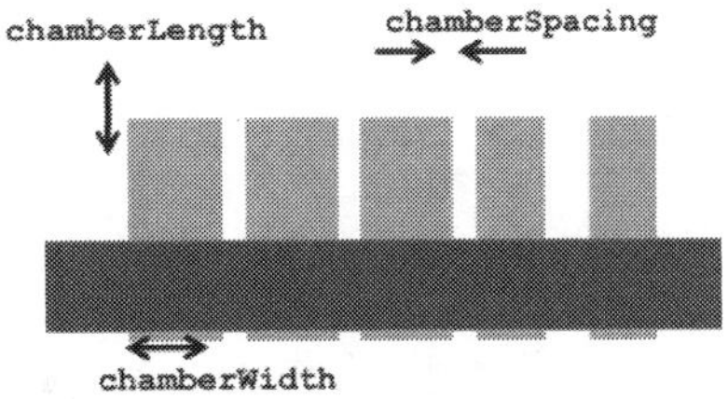

H LONG CELL TRAP <name> numChambers=<int> chamberWidth=<int> chamberLength=<int> chamberSpacing=<int> channelWidth=<int>;

25

Fluigi

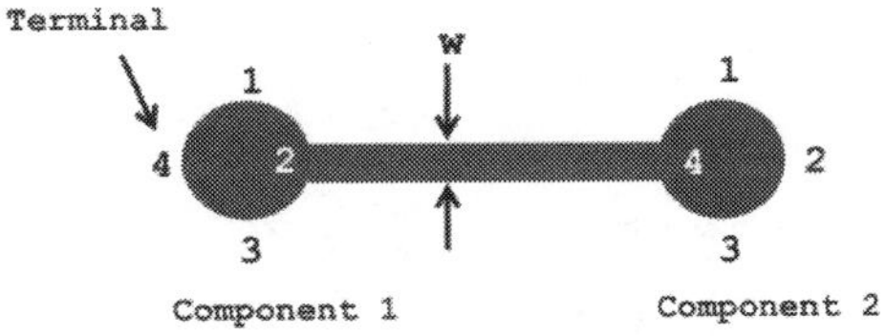

CHANNEL <string> from <Component 1> <terminal> to <Component 2> <terminal> w=INT;

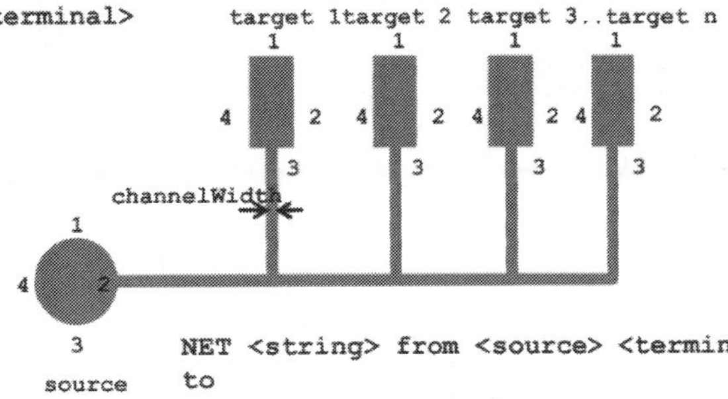

NET <string> from <source> <terminal> to <target 1> <terminal>,... <target n> <terminal> channelWidth=<int>;

26

Fluigi

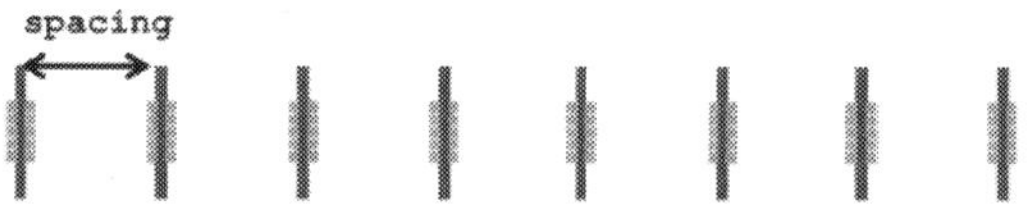

H BANK <string> of <int> CELL TRAP
numChambers=<int> chamberWidth=<int>
chamberLength=<int> chamberSpacing=<int>
spacing=<int> channelWidth=<int>;

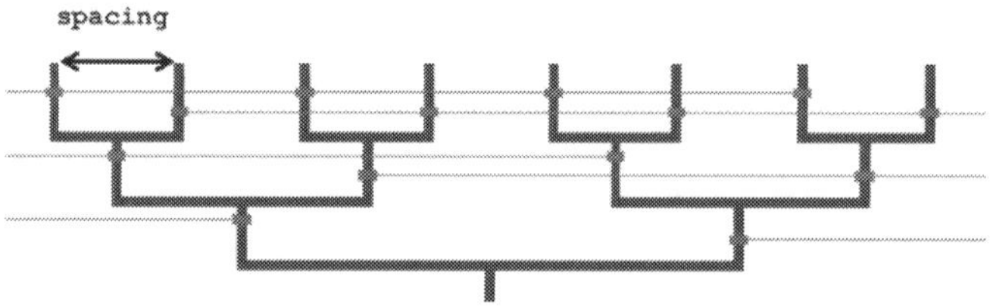

H MUX <string> <int> to 1 spacing=<int>
flowChannelWidth=<int> controlChannelWidth=<int>;

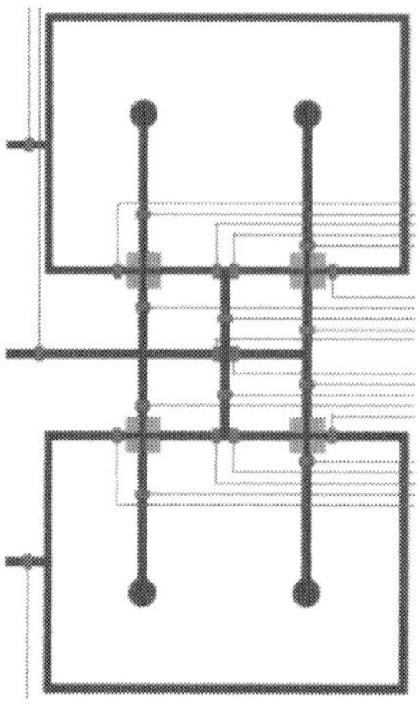

LOGIC ARRAY <string>
flowChannelWidth=<int>
controlChannelWidth=<int>
chamberLength=<int>
chamberWidth=<int>
r=<int>;

27

Fluigi

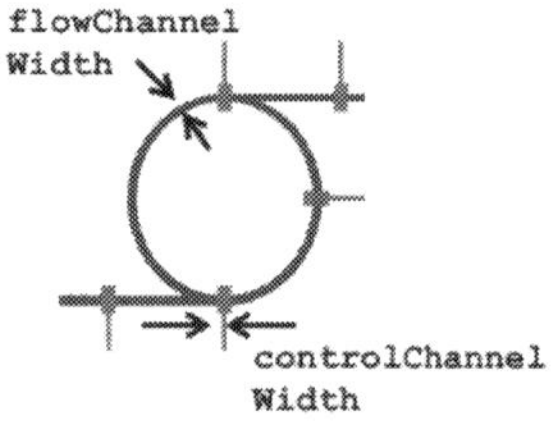

H ROTARY PUMP <string>
radius=<int>
flowChannelWidth=<int>
controlChannelWidth=<int>;

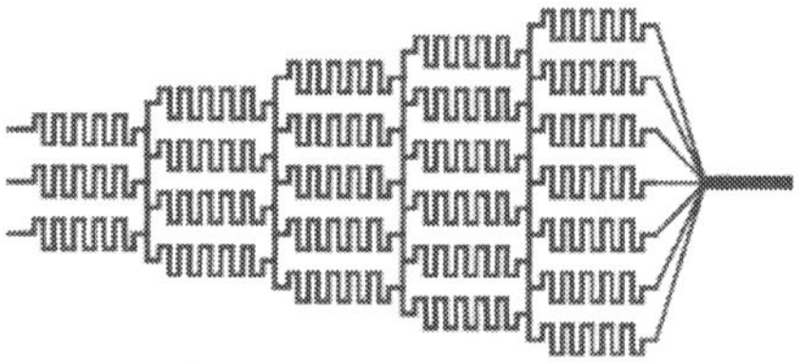

H GRADIENT GENERATOR <string>
<int (mixers in first column)> to
<int (mixers in last column)>
numBends=<int> bendSpacing=<int>
bendLength=<int>
channelWidth=<int>

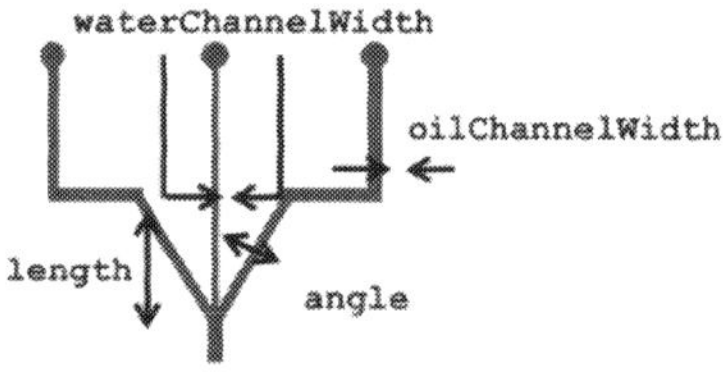

V DROPLET GENERATOR FLOW FOCUS <string>
radius=<int> oilChannelWidth=<int>
waterChannelWidth=<int>
angle=<int> length=<int>;

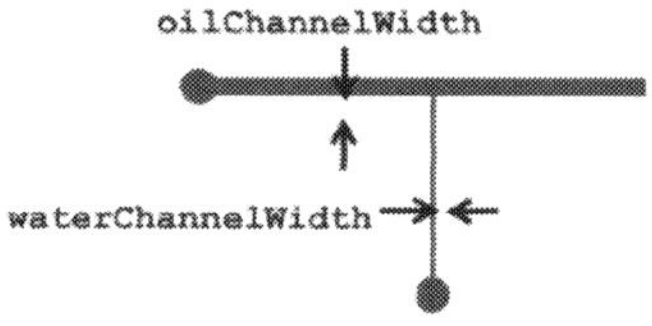

H DROPLET GENERATOR T <string>
radius=<int> oilChannelWidth=<int>
waterChannelWidth=<int>;

28

Fluigi

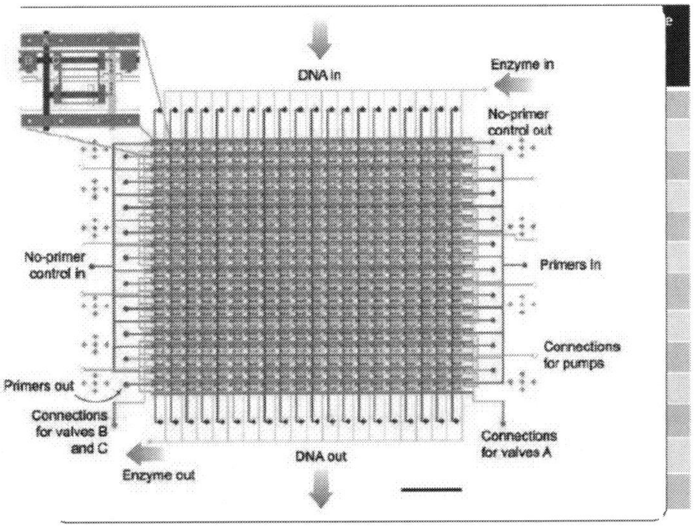

Steve Quake Stanford

29

Fluigi

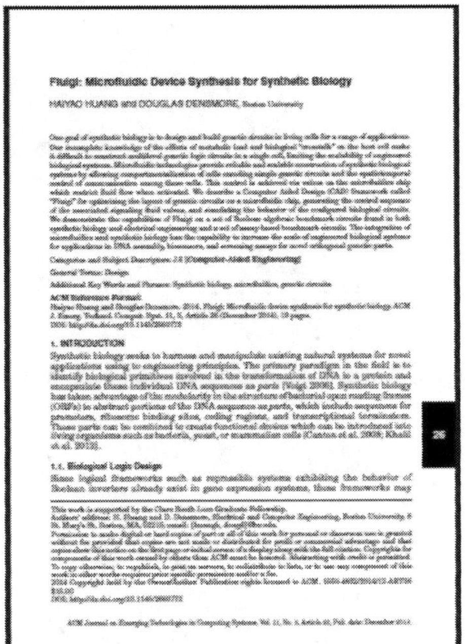

Fluigi: Microfluidic Device Synthesis for Synthetic Biology

HAIYAO HUANG and DOUGLAS DENSMORE, Boston University

H. Huang and D. Densmore, "Fluigi: Microfluidic Device Synthesis for Synthetic Biology," *J. Emerg. Technol. Comput. Syst. Special Issue on Synthetic Biology*, vol. 11, iss. 3, p. 26:1–26:19, 2014.

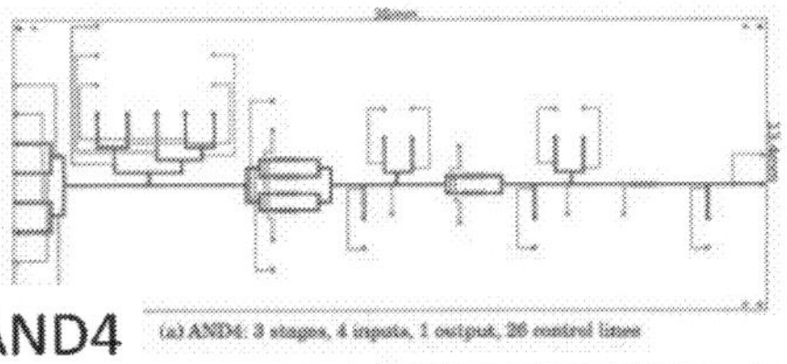

AND4

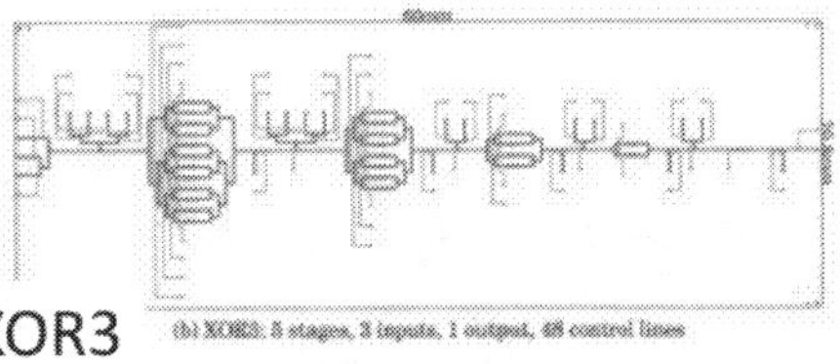

XOR3

Half Adder

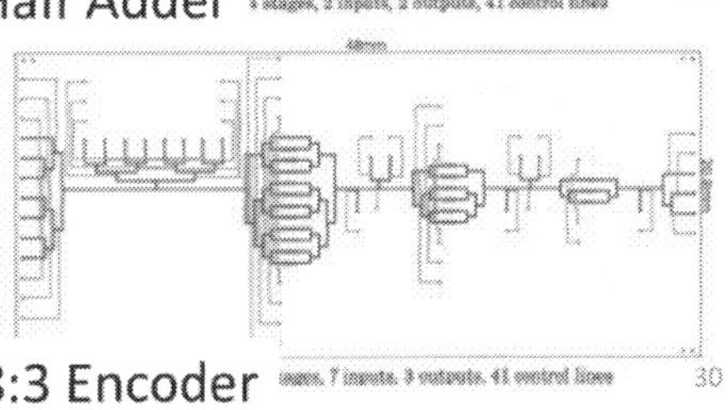

8:3 Encoder

30

Fluigi

DRAPER LABORATORY

Microfluidics for Biomedical Devices

SEMBLER

MIT Lincoln Laboratory

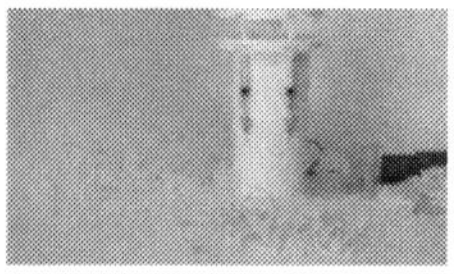

3D Printable Fluidic Hardware

Microfluidic Design Repository

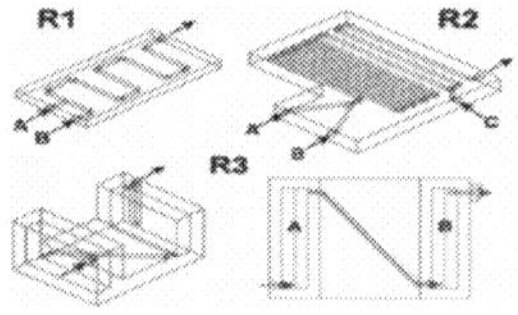

MF

MakerFluidics

- Low Cost
- Automatable
- Accessible
- Fluigi Compatible
- Integrated Electronics
- Flexible Technology

31

Fluigi

1.5 cm x 2cm

Port radius: 100um
Flow channel width : 100um
Control channel width: 50um
Cell traps: 100um x 100um chambers, 3um tall, 4 chambers per cell trap
Valves: 100um x 300um

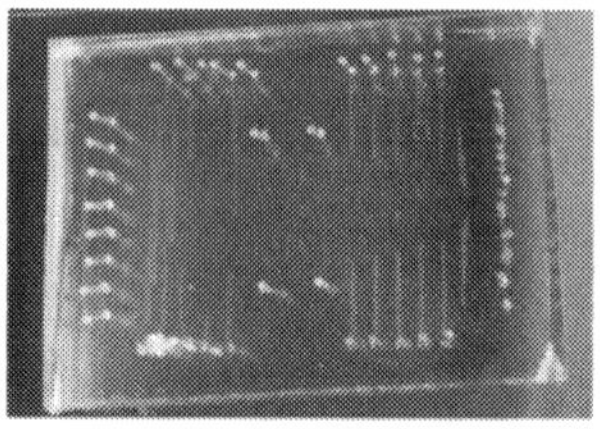

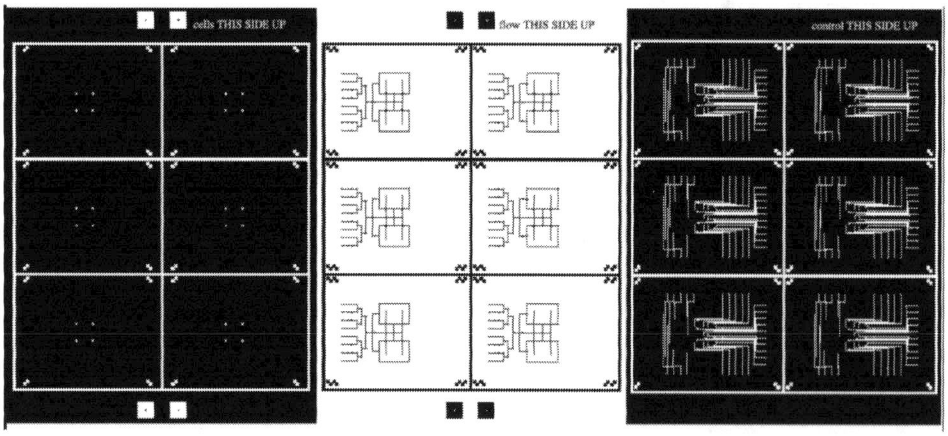

32

Fluigi

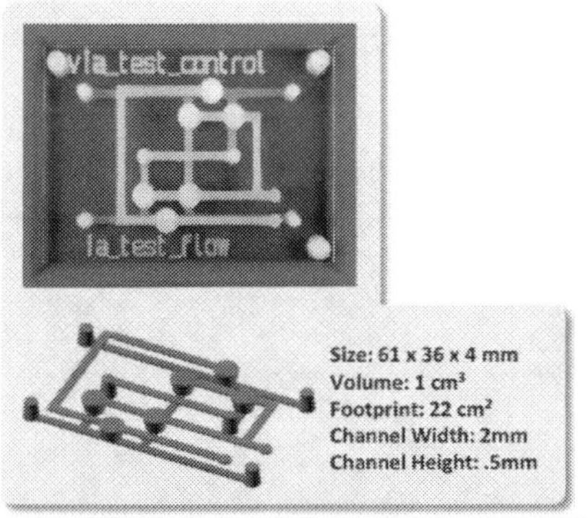

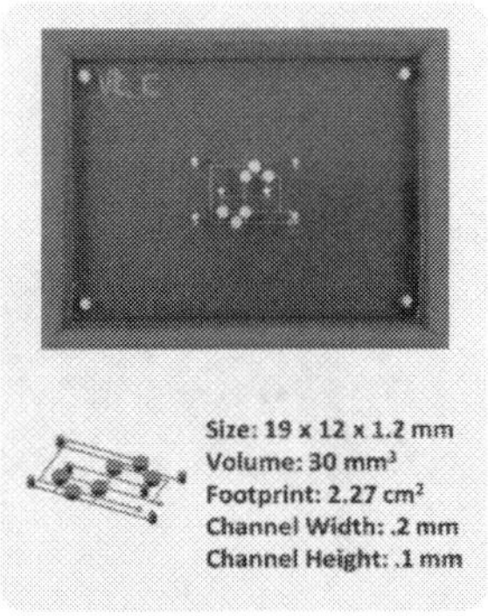

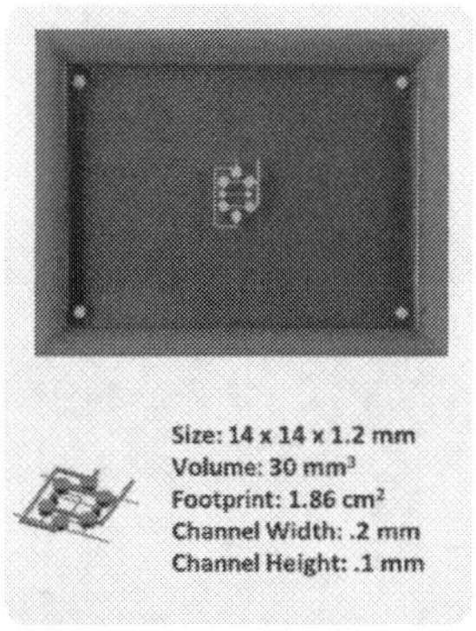

33

Fluigi

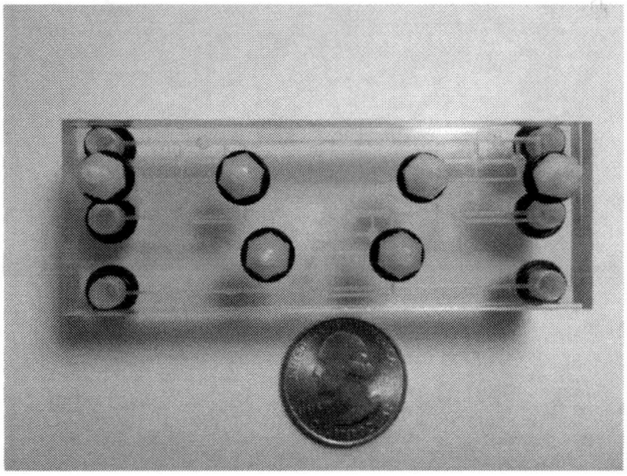

34

Fluigi

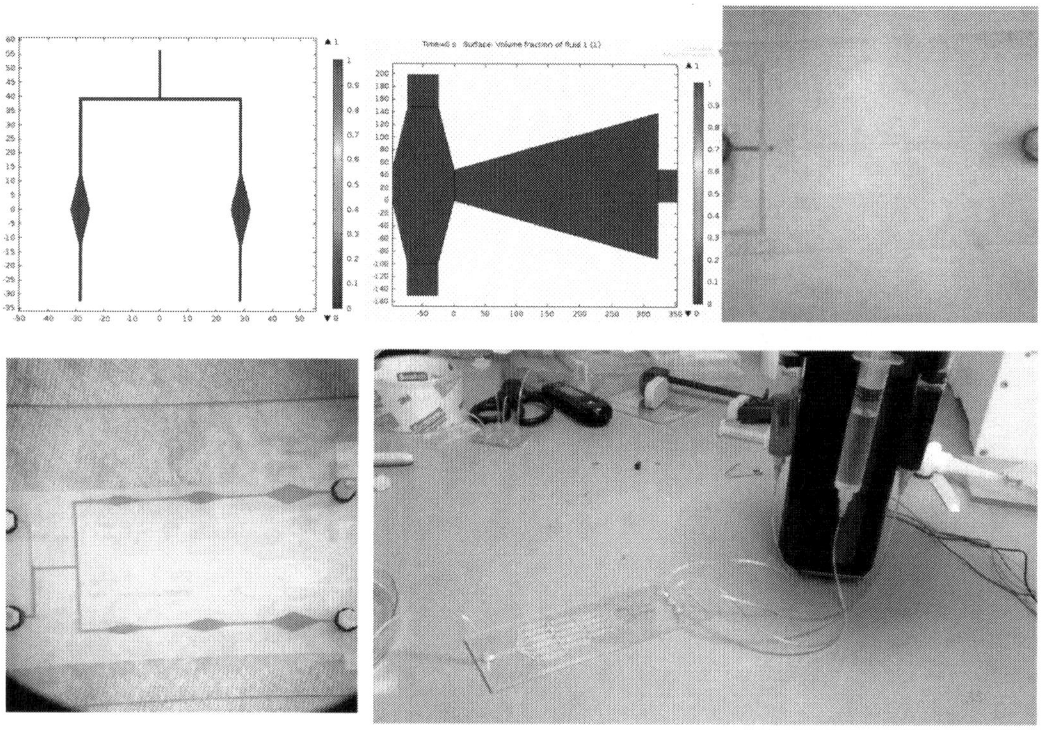

Fluigi

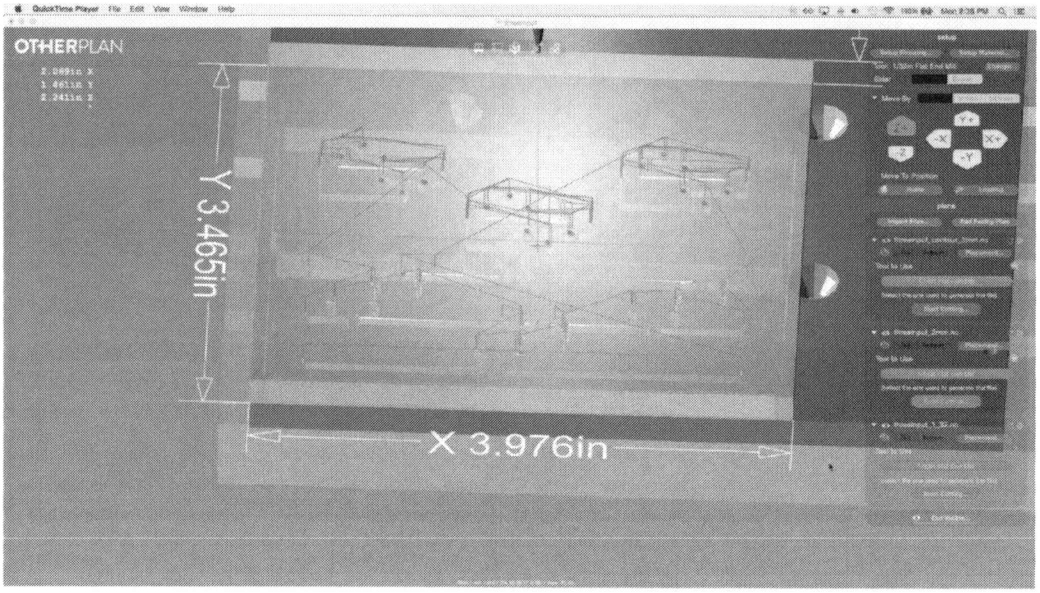

36

Fluigi

Microfluidics using an Othermill

Maximum Accuracy	25 μm[a]
Materials[a,b]	**Thermoplastics** (Polystyrene, PMMA, Polycarbonate, etc.) **Light Metals** (Brass, Copper, Aluminum) **Wax** **Cured Thermosets**
Surface Roughness[b]	0.4 – 2 μm[b]
Minimum Feature Size	Tool Dependent (end mills can be purchased as small as 5 μm in diameter)[c]
Cost per device	<1hr <$1
Machine Cost	$2,199[a]

[a] Other Machine Company (www.othermachine.co/othermill/tech-specs/)
[b] Guckenberger, David J., et al. "Micromilling: a method for ultra-rapid prototyping of plastic microfluidic devices." Lab on a Chip 15.11 (2015): 2364-2378.
[c] Performance Micro Tool (www.pmtnow.com)

37

Fluigi

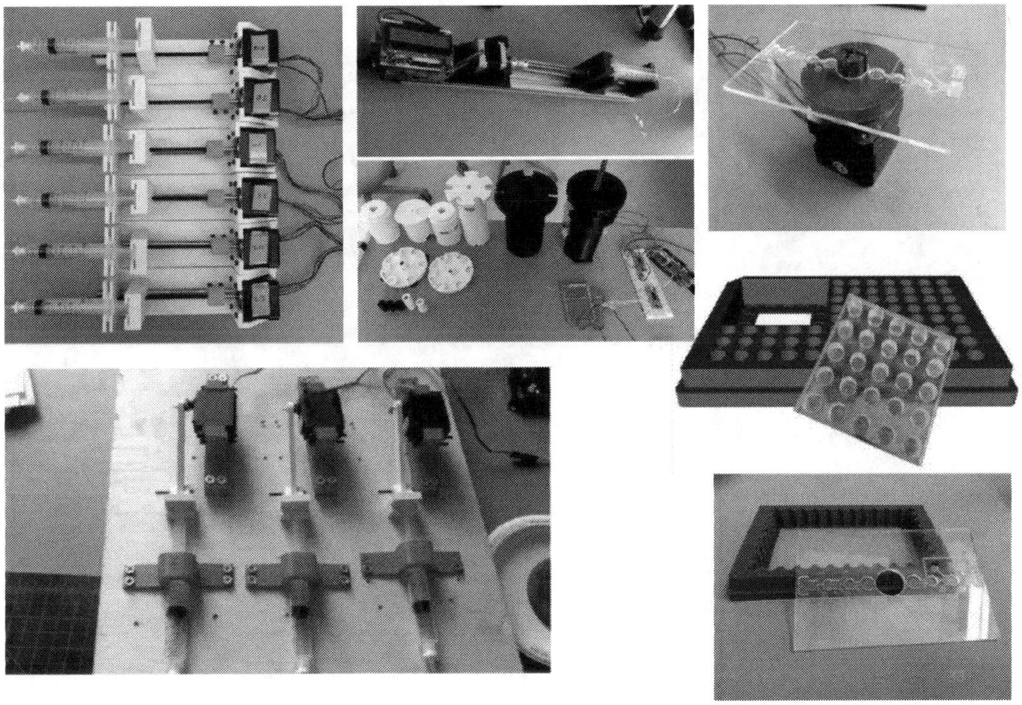

Fluigi

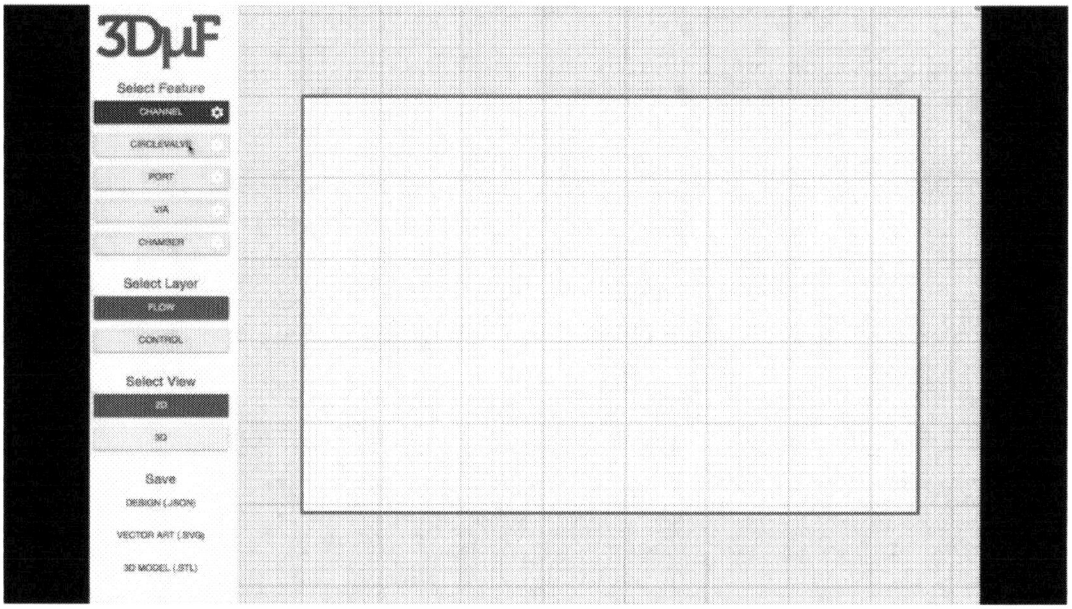

39

Fluigi

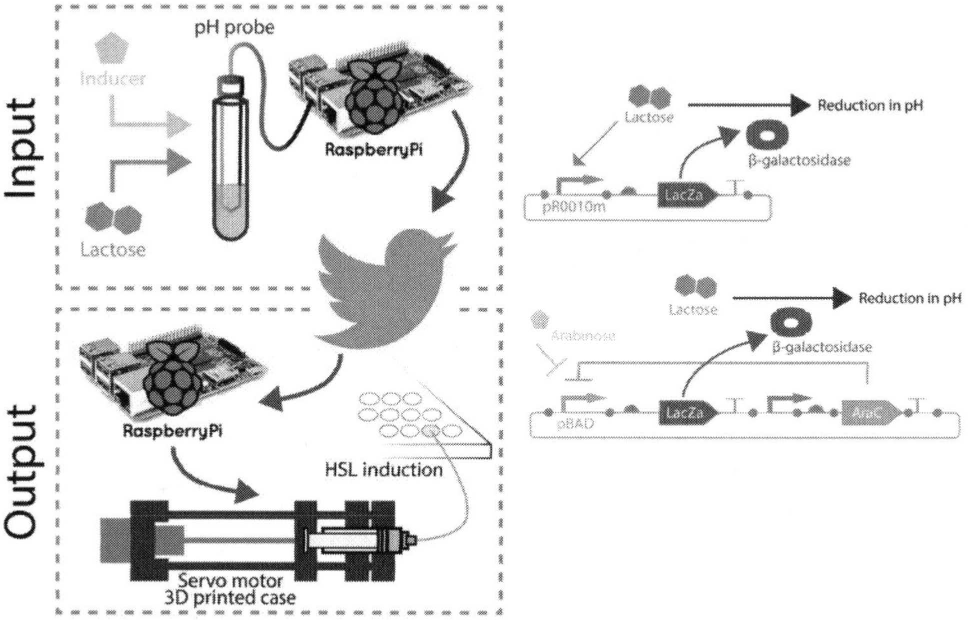

40

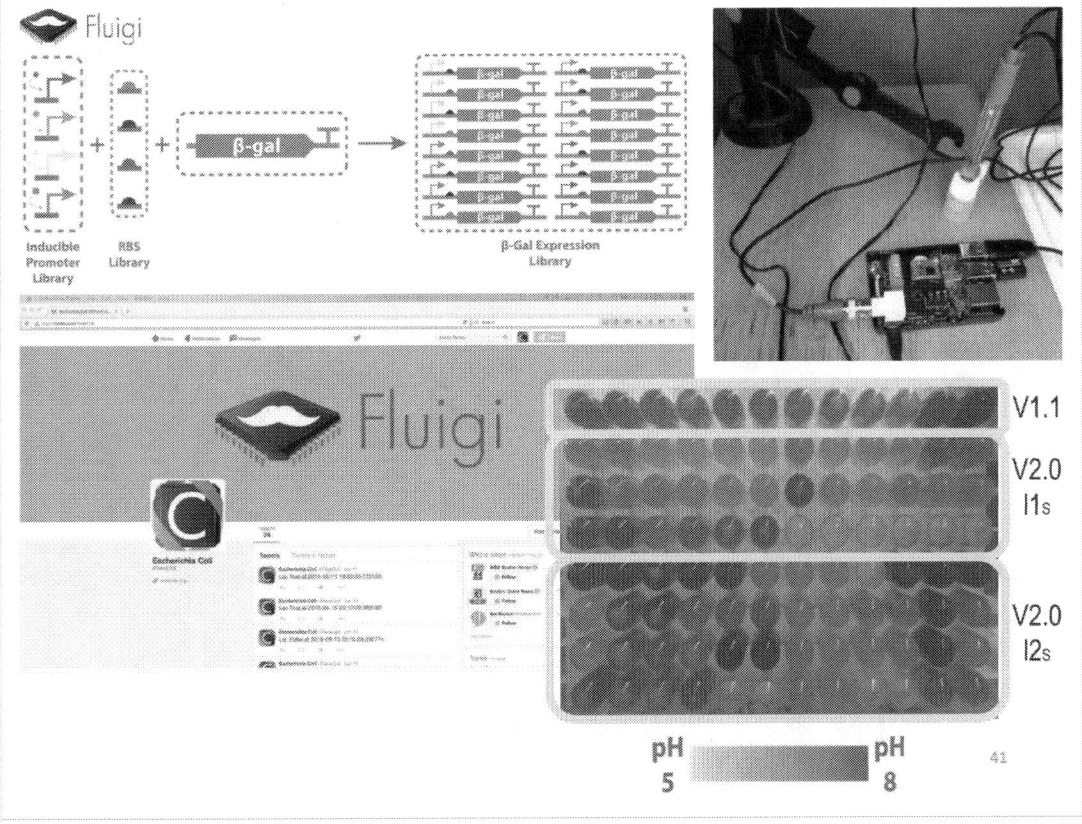
Fluigi
β-gal
Inducible Promoter Library
RBS Library
β-Gal Expression Library
Fluigi
V1.1
V2.0 I1s
V2.0 I2s
pH 5
pH 8
41

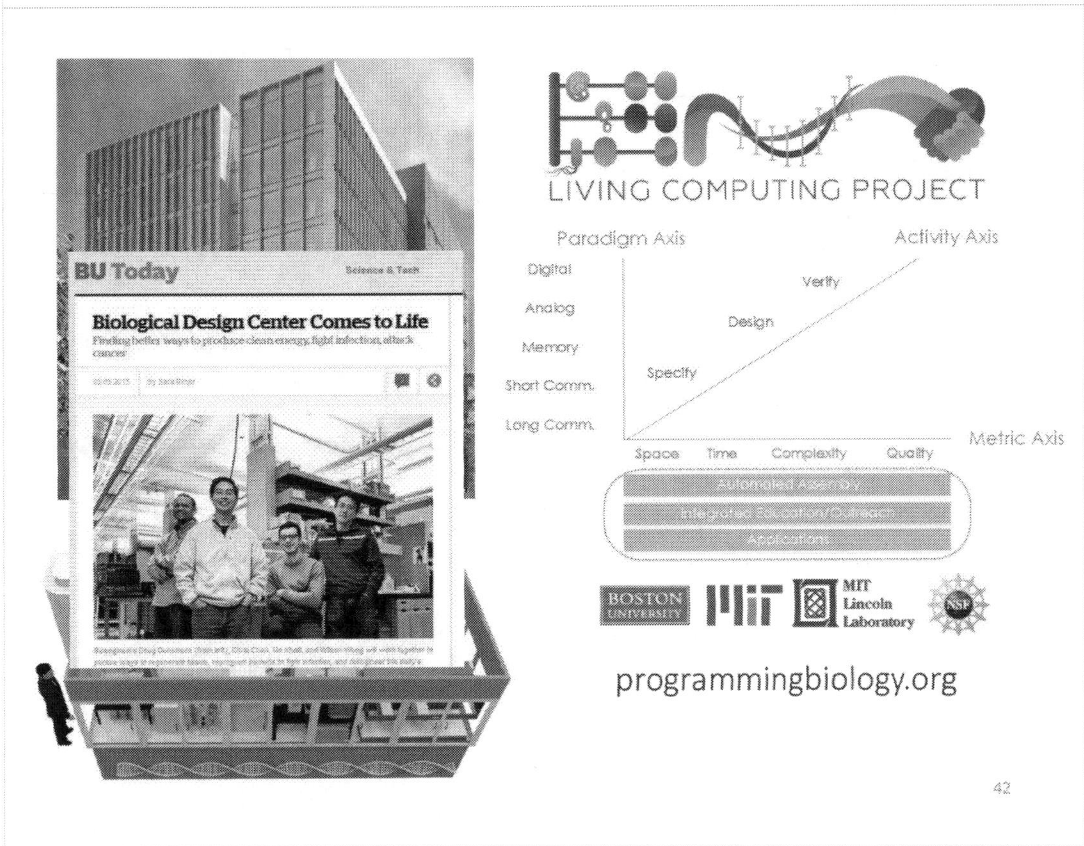
BU Today
Science & Tech
Biological Design Center Comes to Life
Finding better ways to produce clean energy, fight infection, attack cancer
LIVING COMPUTING PROJECT
Paradigm Axis
Activity Axis
Digital
Analog
Memory
Short Comm.
Long Comm.
Verify
Design
Specify
Space
Time
Complexity
Quality
Metric Axis
Automated Assembly
Integrated Education/Outreach
Applications
BOSTON UNIVERSITY
MIT Lincoln Laboratory
NSF
programmingbiology.org
42

IWBDA 2016

Newcastle upon Tyne, UK August 16-18th

www.iwbdaconf.org

nona

RESEARCH FOUNDATION

www.nonasoftware.org @nonasoftware

"Nona Works" 2016

February

2-Day, Open Source Hackathon

Special Thanks

LCP MIT Team – Voigt, Weiss, Lu, Del Vecchio

LCP BU Team – Khalil, Wong, Bhatia, Sgro

BU Wetlab – Ortiz, Costa, Pavan

Belta Lab

NSF, DARPA, ONR

Draper, LL (Carr)

Galagan Lab

BU iGEM

Vincent Noireaux

43

In Vivo continuous Evolution of Genes and Pathways in Yeast

Alper, H., The University of Texas at Austin

Originally presented on: 7/21/2016 11:00:00 - 11:30:00

Directed evolution serves as a critical bridge between sub-optimal and optimal biological components, even in light of rational design approaches. While efforts have been made to automate and improve this process for bacterial systems, less attention has been placed on fungal systems. Here, we demonstrate an *in vivo* mutagenesis approach in yeast by synthetically optimizing the retrotransposon Ty1 to enable libraries of 10^8 per liter per round. We demonstrate this approach by utilizing *in vivo*-generated libraries in the directed evolution of global transcriptional regulators, single enzymes, and multi-gene pathways. Fundamentally, this library can be used as an input to screen and/or select for any measurable phenotype, and enables *continuous* mutant generation and selection for growth-associated phenotypes. In each case described here, we obtain rapid and significant improvements in performance over days instead of weeks, a clear advantage over traditional *in vitro* approaches, especially in fungal systems. Furthermore, we demonstrate transferability of this approach to divergent yeasts such as *Kluyveromyces lactis* and alternative *S. cerevisiae* strains. Finally, we demonstrate how this approach can be coupled with microdroplet technologies to enable rapid screening and selection of novel metabolic production phenotypes.

Engineering High-Titer Heterologous Protein Secretion in Bacteria

Tullman-Ercek, D., University of California, Berkeley
Azam, A., University of California Berkeley
Wong, H. T., University of California Berkeley
Originally presented on: 7/21/2016 11:30:00 - 12:00:00

Secretion is emerging as a useful strategy for exporting proteins of biotechnological interest from bacteria. The type III secretion system (T3SS) in *Salmonella enterica* is an ideal path to protein export because it is non-essential for bacterial metabolism and allows for target proteins to cross both bacterial membranes in one step, via characteristic needle-like protein structures (1). We engineered a super-secreting strain of *Salmonella* for the high-titer production of a variety of biochemically challenging heterologous proteins, such as degradation-prone biopolymer proteins, antibodies, and toxic antimicrobial peptides. We achieve titers on the order of 100 mg/ml for a variety of proteins – a 100-fold improvement on wild type levels – at relative purity. To design the super-secreting strain, we exploited a native T3SS protein, SipD. SipD is one of the first three proteins to be secreted through the assembled T3SS. These "translocon" proteins form a complex at the tip of the T3SS needle, instigating secretion of native proteins into mammalian cells in the first stage of an infection through a poorly understood signal transduction event (2). Our experiments suggest that SipD also acts as an intracellular regulator in a tightly controlled secretion hierarchy. Interestingly, these two roles have opposite effects on protein secretion, and occur on different sides of the cell envelope: the regulatory activity of SipD serves to inhibit secretion of heterologous proteins, while exogenous addition of SipD to T3SS-expressing cultures increases secretion. Using a combination of rationally mutated and truncated variants of SipD, we demonstrate that different domains of the protein are involved in each function and pinpoint the key features responsible for conferring the titer increases. This study serves as an example of how a complicated multi-protein apparatus can be deconstructed and rationally engineered, enabling a new route to industrial protein production.

References

(1) K Metcalf, C Finnerty, A Azam, E Valdivia, D Tullman-Ercek, *Applied and Environmental Microbiology* (2014).

(2) M Lara-Tejero and JE Galán, *Infection and Immunity* (2009).

Functional Incorporation of Unnatural Amino Acids into Proteomes

Thyer, R., UT Austin
Shroff, R., University of Texas at Austin
Cotham, V., University of Texas at Austin
Maranhao, A., University of Texas at Austin
Tack, D. S., University of Texas at Austin
Gollihar, J.,
Brodbelt, J. S., UT Austin
Ellington, A. D., University of Texas at Austin
Originally presented on: 7/21/2016 12:00:00 - 12:30:00

Orthogonal translation machinery and engineered genomes can now be used to readily introduce non-canonical amino acids into cellular proteomes. This raises the question of what amino acids should be introduced, for what functionality? We have focused our efforts on two amino acids that have both structural and metabolic functionality, selenocysteine and L-DOPA.

Selenocysteine is normally incorporated via machinery that has evolved to introduce it across from opal codons programmed to interact with a so-called SECIS element in the RNA and with SelB, an EF-Tu equivalent specific for selenocysteine. We have removed these constraints by reprogramming an amber suppressor tRNA to be charged with selenocysteine. We have introduced this new, orthogonal translation machinery into the â??Amberlessâ?? E. coli, and have carried out long term evolution experiments (LTEE) to direct highly efficient incorporation of selenocysteine. These experiments were facilitated by the development of a conditionally essential protein, NMC-A beta-lactamase, that was completely dependent upon selenocysteine for function. We will detail results of the LTEEs that suggest genetic adaption of the evolved strain increases fitness and enables improved production of proteins containing the 21st amino acid.

More recently, we have engineered translational machinery for the incorporation of the redox active non-canonical amino acid L-DOPA. An evolved aminoacyl-tRNA synthetase enables highly specific incorporation of L-DOPA into proteins in E. coli. We have also shown that L-DOPA can form novel crosslinks with cysteine, which may provide the basis for the development of essential, crosslinked proteins that will â??addictâ?? strains to the amino acid. Furthermore, we have optimized an aromatic detoxification pathway for the biosynthesis of L-DOPA. These discrete components formÂ the building blocks for establishing a parallel metabolism in E. coli and will allow the development of stable and independent organisms with expanded genetic codes.

Algae Culture in a Photobioreactor NEEDS Specific Growth MEDIA, LIGHT Distribution and Operational Aspects of Chemical Engineering

Brahma, P. N. K. Sr., Institute of genetic Engineering

Cultivation of Algae has shown recently an important role in the field of Energy, Food and Health care. Macro-and Micro-algae propagations have shown two different industrial applications. Macro-algae are efficient in CO_2 capture and for energy source as bio-mass and are cost effective in open pond culture. Micro-algae on the other hand provides the source for food in presence of light and CO_2 i.e. ***Spirulina, Blue Green Algae and or Cyan bacteria*** species involved in productions of bio-fuel and food. To study biological and physico-chemical properties of Algae, differentiated by cellular properties of micro and macro-Algae, different cultural methods, were applied to optimize their growths. By changing the increase residence time $i = V_R / V^o$ (i.e. taw = volume of the reactor / flow of nutrient), proper light distribution, the growth of algae cells were monitored in PBR, as shown in figures. To increase the growth rate of algae and to reduce the lag// induction- phase, a hybrid reactor system is proposed, where spiral tube and tapered fluidized mixing column are included. To maintain the growth process continuous, a control panel based many valve systems are used. Based on algal mass productions, the sparging of air and CO_2, into reactor was optimized. Light distribution into transparent glass based reactor is designed; figure bottom left. A combined reactor model has been proposed to design the process in continuous mode, figure, bottom extreme right. Glass reactor photo-bioreactors (PBR) has been used primarily to study all such growth propagation of algae collected from local sources. To increase and to differentiate the growth potential of unicellular/ monoculture algae from multicellular algal weeds, water circulating open pond has also been used. Controlled photoreaction has been found to be essential to optimize the changing variations of lipid (oil), protein (food), carbohydrate (fertilizer) profiles. Unicellular algae are of interest for the production of polysaccharides, pigments, polyunsaturated fatty acids, and enzymes (proteins), supposed to be efficient in a monoculture of algae, aseptically grown, at fixed pH, viscosity, temperature, and salt constituents. Figure bottom extreme left shows the light distribution of one lab- PBR and the middle shows the growth of algae, at side wall of our Lab-PBR. The temperature control both by a chiller and by thermo stating water bath have been reproduced, as shown in figure, top lab-(PBR). To maintain aseptic environment inside the reactor a magnet stirrer is used. To achieve average photon flux, various hollow glass tube (heat and shell) and spiral tubes are proposed, where light will be placed inside the reactor. It has been proved from our experiment, that the changing properties of algae could also be varied both by external open sun-light and internal LED light. We will study these properties in future both by algal cellular and biochemical properties, enzymes, fats and oils.. The algae (unknown) were collected from the local ponds of Badu- Madhyamgram-Kolkata- India.

Haemoperfusive Membrane Encapsulated Activated Charcoal Packed Column Artificial Kidney in Replacing Haemodialysis

Brahma, P. N. K. Sr.,

Blood has both Newtonian and non-Newtonian characters, Using this unique natural constituents of blood, concentration gradient (DC) based hollow fiber membrane haemodialysis was conceptualized early in 16th centuries, figure extreme left top. Hollow fiber haemodialysis reactor, figure extreme right has been in recent development and is being extremely used in most of the kidney dialyzing centre, where primarily the concentration gradient(DC) is used. They are very popular in performing various toxic separations (including blood creatinine and urea of uremia patients). However it was observed that the availability and the cost of such dialyzing hollow fiber reactor (Artificial Kidney AK) are not very affordable, specially in developing countries. A packed column membrane encapsulated activated charcoal, figure top right. The pack column is made by a magazine (sack) to be placed in the empty reactor vessel covered and pressed by the upper lead through spiral thread, suitable in changing the sacks. The sacks are filled with AC (activated cells, membrane encapsulated activated charcoal and or some other non-toxic natural absorbent), and was placed inside the AK (reactor). According to the reductions of creatinine level, some absorbents show useful result in separating toxin molecules. A combined diffusive and convective activity of toxins (Uric acids) removals from blood is proposed and such haemoperfusive packed columns significantly improved the overall clearance of toxins from blood. Medical trials are need to be completed The convective component is particularly important for the transport of middle molecules in which the diffusive component becomes small due to rapid increasing of concentration polarization on the membrane surface, and decreasing nature of membrane permeability with increasing molecular weight (Mw) have been satisfied partially by pakedcolum absorptive system. The concentration of uric acid is measured by governing equations; Ln C = A. Co/ Ct . V_R, where C= Concentration and D= diffusivity and V_R= Volume of the reactor. AC (Artificial cells), A = Membrane surface area and membrane permeability. Membrane encapsulated activated charcoal are designed according to the nature of uremia MM (Middle Molecular structure), that to be absorbed through membrane to activated absorbents. In such cases the two membrane activities were considered, one is transporting from blood inside sack membrane to outside dialysis fluid, and the other moves from blood to AC, towards membrane encapsulated absorbents, which was disposed off time to time, when AC activities are found reduced.

Plasmid GENE Trasfer Invitro into Auxotroph Pheno-Genotypes, F-, Escherichia coli K-12, Generate Hybrid Prototrophs to Damage Clonal Selectivity

Brahma, P. N. K. Sr., Institute of genetic Engineering

After the discovery of DNA double helix in 1953 by Watson and Crick, the concepts of DNAà RNAàProtein, the dogma of life was more or less clear and the concepts of recipients and donor specially in microbial systems predominated to study jumping gene (transposable= Tn) element, IS1sequences, involved in plasmid recombination and antibiotic resistance. Slowly this concept was moved to design vaccine and for the productions of hormones (insulin, growth hormones, interleukin) (i.e., to develop genetically engineered ***hybrid Escherichia coli***) that would express and help many such fundamental aspects of genetic engineering and molecular biology. In such design environment with microbes, some basic important items were included, it was essential; to know the functions of the donor (i.e. the microbes, those are industrially important), whether they posses, plasmids or not; to know the recipient(i.e the universal recipient, that accepts foreign DNA, the well known ***Escherichia coli,*** K-12, the Yale strain. The strain was isolated through various mutations, caused by X-ray and UV, and was compatible to accept the foreign DNA and to help to express at K-12 auxotrophic pheno-genotypes. After that the concept of selectivity of specific DNA in transforming recipient (F-) through plasmid DNA vectors took leading part. pBR322, PSU101, and other plasmid vectors were developed. All these plasmid DNA vectors have small DNA sequence (i.e. length less than 3 Kb) and carry operon (OP) (operator and promoter sequences) and Ori(Origin of replication), support clone DNA to express at auxoytophic pheno-geno types. In all such cases the selectivity in recipient (F-) ***Escherichia coli K-12*** at auxotrophs was challenged(i.e it will grow in minimal agar, only in presence of amino acids(leucien, methionine, tryptophan). These amino acids were damaged in mutations. In addition to that some antibiotic resistant markers were added, that perhaps was not present in vector plasmid and cloned targeted DNA. Combined antibiotics selectivity allowed growing the selective clone on minimal agar plates. Beyond in ***vitro*** plasmid transformation, ***in vivo*** some self designed co-conjugation process were applied (i.e allowing the donor plasmid to express at autotrophic phenol-genotypes). However in most of all recombination, r-DNA technology of plasmids lead to prototroph. In this paper the author will show one of such experimental steps with MRHU (D-mannose resistant haemagglutination of human erythrocyte), a gene that indirectly signifies the colonization and adherence factor of fatal pathogenic activities of ***Escherichia coli.*** At the frequency of 10^{-3} a hybrid 5405 clone was selected, expressing MRHU(+) factor of the donor at auxotrophic phenol-geno types and was used for Balb/c mice immune response against adherence and fatal pathogenic activities of 026:EPEC (Enteropathogenic ***Escherichia coli***).

Development of a Multi-Omics Approach to Study Differential Gene Regulation- a Potential Workflow for the Development and Validation of Synthetic Genetic Circuits

Mandal, K., National Institute of Immunology
Bader, S. L., Institute for Systems Biology
Kumar, P., CSIR-Institute of Genomics and Integrative Biology
Malakar, D.,
Campbell, D. S., Institute of Systems Biology
Pradhan, B. S.,
Jain, V.,
Moritz, R. L., Institute of Systems Biology
Majumdar, S. S.,

The characteristic functionality of a biological system is governed by its proteome profile. This in turn is determined by the transcriptome profile which is governed by the nucleus. Transcriptionally active regions of the chromatin mainly defines the genetic circuit of a cell type and determines its function as well as identity. Any change in genetic circuit of a cell type *(principally done in any synthetic biology study)* would alter its entire transcriptome and proteome profile. Development and validation of such altered genetic circuits demands use of multi-omics approach to ensure appropriate tracking of the entire transcriptome and proteome. To this end, we developed a computationally integrated transcriptomics and Next-Generation Proteomics *(SWATH-MS)* approach for the holistic visualization of the molecular components of a particular cell type.

We used infant *(5-days old)* and pubertal *(12-days old)* rat Sertoli cell *(Sc)* as a model to establish such a workflow. Infant Sc is immature and incapable of supporting germ cell differentiation whereas, pubertal Sc is mature and can support robust spermatogenesis. Sc at two different developmental stages are functionally distinct with respect to each other and they have distinct genetic circuits.

Infant and pubertal Sc were cultured in-vitro and their RNA was used for DNA microarray. In pubertal Sc, 663 genes were upregulated and 735 genes were downregulated. The cellular abundance of these transcripts are governed by their respective promoters. Promoters *(+/- 2kb from Transcriptional Start Site)* of all these differentially expressed genes were retrieved and interrogated for the abundance of various Transcription Factor Binding Sites *(TFBS)*, as compared topromoter sets which are not expressed differentially, using vertebrate non-redundant Positional Weight Matrixes *(PWM)* from TRANSFAC database employing MATCH™ utility. The discrepancy between the frequency of individual TFBS on up and down-regulated gene-set promoters *(observed occurrence)* and that of control promoter sets *(expected by chance)* was evaluated independently by determining the statistical variable chi-square ($\chi 2$). Considering a significance level of $p<0.05$, we identified 61 TFBS on up-regulated and 56 TFBS on down-regulated gene-set promoters. Out of these 19 and 14 TFBS were found exclusive on up-regulated and down-regulated gene-set promoters respectively. Interactome analysis using string database suggested 75 genes to be associated with the TFBS in case of downregulated-gene-set and 87 genes in case of upregulated-gene-set. Nuclear and cytoplasmic fractionation of both the cell types were used for mass spectrometry for Data Dependent Acquisition *(DDA)* and SWATH-MS analysis. DDA data obtained from each sample was merged and processed to generate a comprehensive rat SWATH-assay library. This assay library was interrogated using the SWATH-MS data for comprehensive proteome relative quantification. This helped in cataloguing the list of proteins which were enriched in nucleus *(in infant and pubertal Sc)* and how many of them were associated with the enriched TFBS obtained in each case. This study revealed a holistic picture of differential gene regulation by the TFBS in coordination with the nuclear proteome of infant and pubertal Sc. The workflow can be applied for the development and validation of gene regulation in synthetic genetic circuits.

Designer Biosensors That Respond to New Small Molecules
Raman, V., University of Wisconsin-Madison

Genetic regulatory systems inducible by small molecules are indispensable tools for building genetic circuits, assaying gene function, and high-throughput metabolic engineering. Microbial allosteric transcription factors (aTFs) constitute a major class of proteins that confer such small-molecule inducible gene regulation, but aTF–inducer pairs are currently limited by those that naturally occur. Altering inducer specificity in these proteins is difficult because mutations that affect inducer binding may also destroy the allosteric behavior. Here we engineer an allosteric transcription factor, LacI from *E. coli*, to respond to four new inducer molecules, including two which are fully synthetic. We employ computational protein design and multiplex assembly of 13,600 specified gene variants to engineer the LacI ligand binding pocket for new inducer recognition. We use a two-stage enrichment screen that identifies LacI variants which allosterically regulate gene expression in response to fucose, gentiobiose, lactitol or sucralose. Following activity maturation, we identify LacI variants responsive to these new inducers that exhibit specificity and induction response values comparable to the wild-type LacI response to its synthetic inducer, IPTG. We expect this approach to be a generalized method for altering aTF inducer specificity because it relies only on sequence and structure information, which is available for dozens of additional aTFs. The ability to create designer sensors and switches will enable applications including dynamic control of cell metabolism, cell biology and synthetic gene circuits.

Engineering a Growth Sensor to Detect Antigen-Antibody Interactions in Mammalian Cells

Nguyen, T. D., The University of Tokyo
Kaku, Y.,
Inoue, S.,
Nagamune, T., School of Engineering, The University of Tokyo
Kawahara, M.,

Intracellular antibody (intrabody), an antibody that can function intracellularly, is a promising tool for analyzing protein functions and therapeutic use. However, since the reducing cytoplasmic environment generally leads to low antibody stability, it has been difficult to generate effective intrabodies.

In this study, we designed an scFv-c-kit as a growth sensor for detecting antigen–antibody interactions directly in the cytoplasm of mammalian cells. This growth sensor is constructed by fusing an scFv fragment and the cytoplasmic domain of c-kit as a receptor tyrosine kinase. The roles of the scFv fragment and c-kit are to detect an antigen and transduce a growth signal, respectively. In principle, as scFv-c-kit binds to an oligomer antigen, it will form dimers and transduce a growth signal. Therefore, detection of the growth signals allows selecting functional intrabodies. Since the selection is performed in interleukin-3-dependent Ba/F3 cells, the cells with antigen–antibody interactions can be selected as growing cells in the culture medium without interleukin-3.

Firstly, we tried to exploit the growth sensor to select an scFv clone against rabies virus nucleoprotein (RV-N) since it is known that RV-N forms oligomer complexes in the cells. We constructed two RV-N expression vectors; one contained RV-N fused to V5 and His tags, and the other had a dimer of RV-N connected by a flexible linker with the same tags. Then, we introduced them into Ba/F3 cells to obtain stable cells expressing RV-N (termed as Ba/N and Ba/N2). Next, we constructed an scFv-c-kit library expression vector, in which the scFv library is derived from the phage-displayed Tomlinson I scFv library after a single-round panning against RV-N. After the transduction of the scFv-c-kit library to Ba/N and Ba/N2, library selection was performed in 24-well plates to select cells having scFv-c-kit growth signals. The scFv sequences from growing cells were analyzed to find twelve different scFv clones. The antigen specificity evaluation revealed that three clones showed RV-N-dependent growth signals, one of which showed selective binding to RV-N.

However, during the library selection, noise of the growth sensor caused selection of false-positive clones. Therefore, it was necessary to improve the growth sensor performance. Using a structure-folding software, we recognized that scFv and c-kit were too close to each other. This may cause inherent interaction between them and subsequent cell growth noise. Hence, we decided to engineer the growth sensor structure by inserting one of three flexible linkers G_4S, $(G_4S)_2$, and $(G_4S)_3$ between scFv and c-kit. The results revealed that the insertion of the flexible linkers significantly reduced the growth noise of one of the non-specific clones, especially with the G_4S linker. Moreover, the insertion of the G_4S linker remarkably increased the signal-to-noise ratio of a specific clone for RV-

N.

Taken together, we demonstrated that the scFv-c-kit growth sensor could be feasible to select functional intrabodies, and structural engineering could ameliorate its performance. This may develop into a promising intrabody selection method in the future.

A General Strategy to Construct Small Molecule Biosensors in Eukaryotes

Feng, J.,

Biosensors for small molecules can be used in applications that range from metabolic engineering to orthogonal control of transcription. Here, we produce biosensors based on a ligand-binding domain (LBD) by using a method that, in principle, can be applied to any target molecule. The LBD is fused to either a fluorescent protein or a transcriptional activator and is destabilized by mutation such that the fusion accumulates only in cells containing the target ligand. We illustrate the power of this method by developing biosensors for digoxin and progesterone. Addition of ligand to yeast, mammalian, or plant cells expressing a biosensor activates transcription with a dynamic range of up to ~100-fold. We use the biosensors to improve the biotransformation of pregnenolone to progesterone in yeast and to regulate CRISPR activity in mammalian cells. This work provides a general methodology to develop biosensors for a broad range of molecules in eukaryotes.

Genome Engineering of Genomically Recoded Escherichia coli enables Multi-Site Non-Standard Amino Acid Incorporation in Cell-Free Synthesized Proteins at High Yield and Purity

Martin, R. W., Northwestern University
Hong, S. H., Illinois Institute of Technology
Des Soye, B. J., Northwestern University
Kwon, Y. C., Northwestern University
Davis, R. G., Northwestern University
Kelleher, N.,
Jewett, M. C., Northwestern University

Site-specific incorporation of non-standard amino acids (NSAAs) into proteins and biopolymers by amber suppression makes possible new chemical properties, new structures, and new functions. However, competition between release factor 1 (RF1) and orthogonal tRNAs during protein translation have limited the technology. Here, we describe the development of a high yielding cell-free protein synthesis (CFPS) platform from crude extracts of genomically recoded *Escherichia coli* lacking RF1. Because this recoded strain has not been previously optimized for CFPS, we exploited multiplex automated genome engineering (MAGE) to design and construct synthetic genomes that, upon cell lysis, lead to improved extract performance. We sought to stabilize DNA, mRNA, amino acids, and protein products in CFPS by targeting the inactivation of fifteen putative negative protein effectors. Twenty-seven strains were generated and tested in CFPS, allowing us to catalogue the systemic impact of making numerous gene disruptions both individually and in combinations. The protein synthesis activities of our most productive cell extracts were more than five-fold greater as compared to the extract from the parent strain, achieving synthesis of more than 1.6 g/L of active superfolder green fluorescent protein (sfGFP). We also show high incorporation efficiency (>95%) of the NSAA *p*-acetylphenylalanine (pAcF) in sfGFP at single and multiple positions. Upon optimizing the levels of the orthogonal translation system components, we further show the ability to synthesize elastin-like polypeptides containing up to 40 pAcFs with high fidelity (>95%) and yield (~100 mg/L). To our knowledge, this is the highest level of site-specific incorporation in pure proteins to date. Our work has implications for using whole-genome editing for CFPS strain development, expanding the chemistry of biological systems, and cell-free synthetic biology.

Reproducibility of Fluorescent Expression from Engineered Biological Constructs in E. coli

Beal, J., Raytheon BBN Technologies
Haddock-Angelli, T., Boston University
Gershater, M.,
de Mora, K.,
Lizarazo, M.,
Hollenhorst, J.,
Rettberg, R.,

We present results of the first large-scale interlaboratory study carried out in synthetic biology, as part of the 2014 and 2015 International Genetically Engineered Machine (iGEM) competitions. Participants at 88 institutions around the world measured fluorescence from three engineered constitutive constructs in E. coli. Few participants were able to measure absolute fluorescence, so data was analyzed in terms of ratios. Precision was strongly related to fluorescent strength, ranging from 1.54-fold standard deviation for the ratio between strong promoters to 5.75-fold for the ratio between the strongest and weakest promoter, and while host strain did not affect expression ratios, choice of instrument did. This result shows that high quantitative precision and reproducibility of results is possible, while at the same time indicating areas needing improved laboratory practices.

Random Homology-Aided Transformation for Genome and Plasmid Streamlining

Suarez, G. A., University of Texas at Austin
Barrick, J. E., University of Texas at Austin

Synthetic biologists aim to achieve predictable and controllable genetic devices, yet they have to struggle against the massive complexity already present within the simplest living cell. In this context, any synthetically implemented process will unavoidably experience some level of interference that may alter its desired outcome. One way to approach this problem is to remove all non-essential components from a genome, a feat that so far has shown both promise and limitations. Although up to 16% genome reductions have been achieved in *E. coli* without any significant loss in cellular fitness (Posfai et al., 2006), most attempts to achieve broad scale genome reductions have negatively impacted cellular growth. Dispensable genes have been found to function in stress responses and allow for efficient usage of resources to build new cells (Karcagi et al., 2016). For these reasons, it is extremely difficult to determine which genomic regions can be removed without compromising a cell's full viability.

We aimed to develop a method capable for generating combinatorial libraries of deletion variants that retain full cellular viability. Taking advantage of the high transformability of the naturally competent bacterium *Acinetobacter baylyi* ADP1, we tested the possibility of using fragmented genomic DNA (gDNA) religated to a selective marker to template homology-aided random genome deletions. By designing the marker cassette with Type IIS terminal restriction enzyme sites facing outwards, a mutant's digested genomic DNA contains fragment with the marker gene attached to 16-bp sequences flanking where it was integrated into the genome. These sequences serve to map out the precise site of deletion/integration event. We named this combinatorial methodology for generating pools of random deletion mutants random Homology-Aided Transformation (rHAT). Using rHAT we were able to retrieve fit and viable mutants with deletions averaging ~5kb (smallest 1832bp and largest 9227bp; 20 isolates assayed in total), although a surprising number of the retrieved transformants (>50%) had undergone genome re-arrangements that apparently lead to duplications in the genome while still generating a local deletion in one copy of an amplified region. Determining which genomic regions remain robust through normal or induced mutational processes can greatly benefit genome engineering projects and synthetic biologist, e.g., by helping to map out both dispensable and unstable genomic regions which would be unsuitable for genome engineering.

Furthermore, the demonstration that ADP1 can be efficiently transformed via rHAT may prove valuable for evolutionary genome reduction experiments. For example, we expect that evolving a population of cells with the continual input of randomly religated gDNA will likely accumulate deletions over evolutionary time, thereby serving to evolve fit variants with reduced genomes or deletions which may then be combined to make a minimal ADP1 genome. Future experiments will also test rHAT as a combinatorial approach to generate pools of minimized plasmids in ADP1 hosts; smaller plasmids are typically easier to transform into a cell. Overall, our preliminary results using rHAT, open the door to the use and development of this method as a tool to achieve simpler.

more efficient, and more reliable synthetic biology.

Synthetic Polyketide Enzymology: Platform for Biosynthesis of Antimicrobial Polyketides

Go, M. K., National University of Singapore
Wongsantichon, J.,
Cheung, V. W. N., National University of Singapore
Chow, J. Y.,
Robinson, R. C.,
Yew, W. S.,

We applied enzymological principles in synthetic biology and describe its use as an enabling platform in synthetic biology for the purposeful production of compounds of biomedical and commercial importance. In particular, we demonstrated the use of synthetic polyketide enzymology as a means to develop lead polyketide based compounds for antimicrobial therapeutics, as exemplified by the modular coupling of acid:CoA ligases to type III polyketide synthases in the biosynthesis and development of polyketide-based biochemicals. Using wild-type and rationally designed mutants of a type III polyketide synthase isolated from *Oryza sativa* (OsPKS), we produced a chemically diverse library of novel polyketides and identified two bioactive antimicrobials, 4-hydroxy-6-[(1*E*)-2-(4-hydroxyphenyl)ethenyl]-2*H*-pyran-2-one (bisnoryangonin) and 3,6,7-trihydroxy-2-(4-methoxybenzyl)-4*H*-1-benzopyran-4,5,8-trione (26OH), respectively, from a screen against a collection of Gram-positive and Gram-negative bacteria. The purification, crystallization, and structural resolution of recombinant OsPKS at 1.93 Å resolution are also reported. Using the described route of synthetic polyketide enzymology, a library of OsPKS mutants was generated as an additional means to increase the diversity of the polyketide product library. We expect the utility of synthetic enzymology to be extended to other classes of biomolecules and translated to various purposeful functions as the field of synthetic biology progresses.

Design Space Exploration of the Violacein Pathway in Escherichia coli Based Cell-Free System

Nguyen, P., California State University, Long Beach
Wu, Y.,
Guo, S.,
Murray, R. M., California Institute of Technology

In this study, an Escherichia coli (E. coli) based transcription translation cell-free system (TX-TL) was employed to sample various enzyme expression levels of the violacein pathway. The pathway was successfully reconstructed in TX-TL. Its variations produced different metabolites as evident from the extracts' assorted colors. Analysis of the violacein product via UV-Vis absorption and liquid chromatography-mass spectrometry (LC-MS) detected 68 ng of violacein per 10 μl reaction volume. Significant buildup of prodeoxyviolacein intermediate was also detected in the equimolar TX-TL reaction. Finally, design space exploration experiments suggested an improvement in violacein production at high VioC and VioD DNA concentrations.

Synthetic Speciation: Programmable Transcriptional Activators to Prevent Gene Flow in Sexually-Reproducing Populations

Smanski, M. J., University of Minnesota
Maselko, M., University of Minnesota

Controlling gene flow between populations of sexually-reproducing organisms has applications in public health, agriculture, and control of invasive species. We have developed a new genetic approach to prevent sexual crossing between a genetic modified organism (GMO) and otherwise sexually compatible organisms. Here we will describe the new technology, demonstrate its effectiveness using a Saccharomyces cerevisiae as a model system, and highlight applications in insect-pest control, crop production, and invasive-species control.

At-Rich Gene Synthesis

Walsh, M.,
Niles, J., MIT
Carr, P. A.,

Over the last 15 years, the costs of oligonucleotide synthesis and gene synthesis have declined sharply and the practice of ordering these from commercial entities is now common lab protocol. These companies also use synthetic oligonucleotides as building blocks for gene synthesis, making the synthesis of whole genes and larger constructs far more accessible. However, there are limitations to which double-stranded DNA sequences they can synthesize with confidence. These include ceilings on overall and local AT-content, homopolymers, palindromes, hairpins and repeats. We report here improvements in the process of gene synthesis adapted for such difficult sequences. In this work, two genes from the AT-rich organism *Plasmodium falciparum* were synthesized from designed oligonucleotides using a polymerase construction and amplification method. The component oligonucleotides were designed using DNAWorks and multiple polymerases were evaluated.

Effects of Ribosome Autocatalysis and Negative Feedback in Resource Competition

Chandra, F.,
Del Vecchio, D.,

Resource competition, and primarily competition for ribosomes, can lead to unexpected behavior in synthetic genetic circuits as well as affect the growth of host cells. Current models studying the effects of resource competition assume a constant production of ribosomes and these models describe the experimental results well. However, ribosomes are autocatalytic since they are partially made of protein which must be translated by ribosomes themselves. Autocatalysis has been shown to have detrimental effects on a system's stability and robustness. Additionally, there are known feedback regulations on ribosome synthesis such as inhibition of rRNA synthesis via ppGpp. We have developed and compared three two-state models of ribosome and protein synthesis to investigate conditions under which these regulatory actions have a significant effect in situations of increased ribosome demand, defined by the mRNA level of synthetic genes. These models incorporate: 1) constant production of ribosomes, 2) autocatalytic production of ribosomes, and 3) autocatalysis and negative feedback, respectively. Our modeling results indicate that for sufficiently low demand, autocatalysis has little or no effect. However, beyond a certain demand level, the system goes through a transcritical bifurcation at which the only non-negative steady state is at zero ribosome concentration. The presence of negative feedback, in turn, can shift this bifurcation, thus restoring the qualitative behavior observed in a model with a constant ribosome production at low demand. Additionally, autocatalysis affects the dynamics of the system, leading to a potential overshoot in the time response of the synthetic gene to changes in induction level. Our results show that ribosome autocatalysis has a significant effect on the system robustness both in the steady state concentration of ribosomes and the transient responses of the desired synthetic product, however the existing negative feedback on ribosome production compensates for the effects of the necessary autocatalytic loop and restores the behavior seen in the system with constant ribosome production.

Effects of Ribosome Autocatalysis and Negative Feedback in Resource Competition

Fiona Chandra, Domitilla Del Vecchio

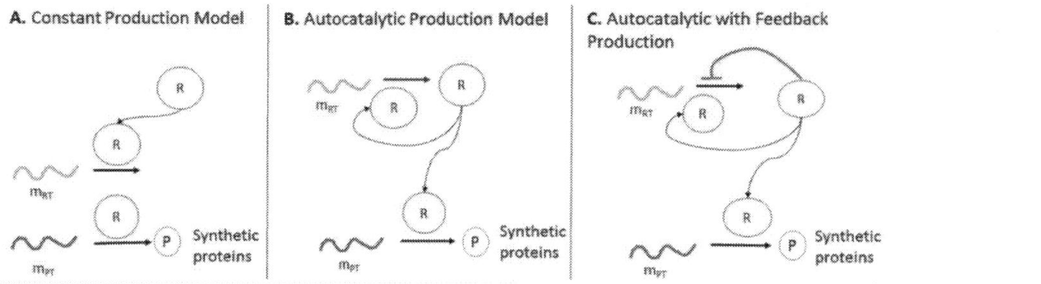

- We can mathematically prove that ribosome autocatalysis always leads to lower robustness in ribosome level to synthetic demand.
- Additionally, it leads to overshoot in the temporal response of the synthetic protein at high demand.
- Feedback can only partially mitigate, but not completely compensate for, the effects of autocatalysis.

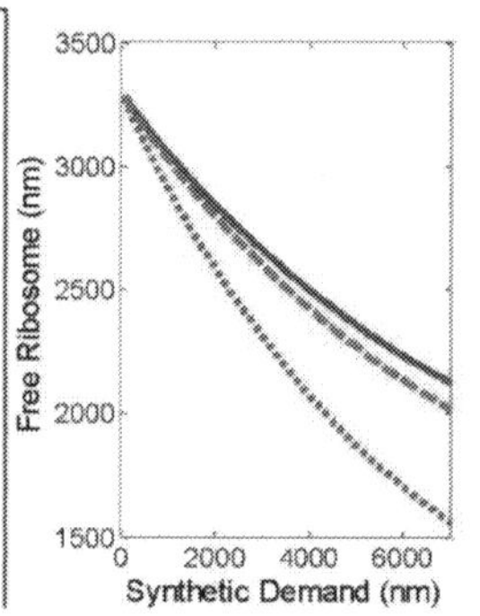

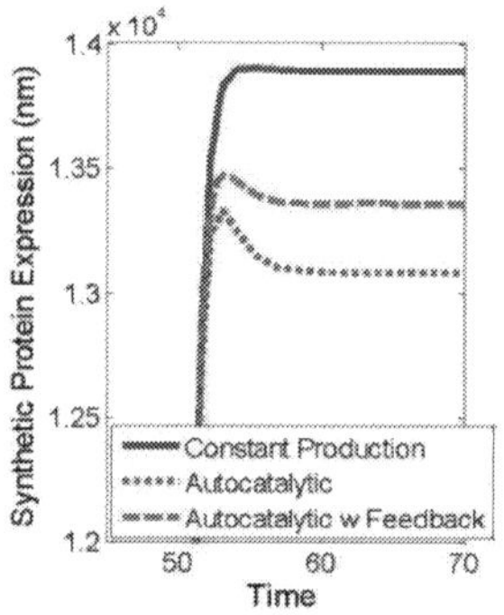

Orthogonal Gene Expression: Creating 3 Promoter Libraries Linked to Different Sigma Factors Which Are Orthogonal Between Eachother and Towards the Host

Bervoets, I., Free University of Brussels
Van Brempt, M., Ghent University
Maertens, J., Ghent University
De Mey, M., Massachusetts Institute of Technology
Charlier, D.,

Technological advances in synthetic biology, systems biology, and metabolic engineering have boosted applications of industrial biotechnology for an increasing number of complex and high added-value molecules. In general, the transfer of multi-gene or poorly understood heterologous pathways into the production host leads to imbalances due to lack of adequate regulatory mechanisms. Hence, fine-tuning expression of synthesis pathways in specific conditions is mandatory together with the decoupling of host systems.

Here we develop an orthogonal expression system in *Escherichia coli* with three heterologous sigma factors that function orthogonal between each other and towards the host organism. Furthermore, for each of these sigma factors, a promoter library will be created that retains this orthogonal feature.

First, several sigma factors originating from *Bacillus subtilis* were tested for their orthogonality in *E. coli* on the level of promoter recognition, by using a red-fluorescent reporter system. Secondly, the potential of sigma factors, and mutants thereof, from *B. subtilis* to work together with the *E. coli* core RNA polymerase was tested. This was accomplished by expressing these proteins together with their promoters; three distinct promoters for each factor. Subsequently, cross talk between the heterologous sigma factors and the promoters was evaluated by examining all combinations. Based on the results three heterologous sigma factors with one of their respective promoters were chosen for further work. For each of these factors a promoter library will be generated that remains orthogonal towards the host and towards the alternative sigma factors. The promoter libraries will be attained by randomising part of the promoter, one strategy will be to randomise regions of 5 nucleotides throughout the promoter while the second strategy will be randomisation of the complete linker sequence. Subsequently the promoter strengths will be measured through sequential rounds of fluorescent-activated cell sorting (FACS) to result in a set of orthogonal promoters for each sigma factor, covering a range of different strengths.

The use of these different orthogonal sigma factors combined with a range of promoter strengths for each factor offers the opportunity to fine-tune expression of different genes in heterologous pathways.

Protoplast Fusion with the Hydrogen-Oxidizing Bacterium, Cupriavidus Necator

Holt, E. T., University of Lousiville
Gulari, E., University of Michigan

The hydrogen-oxidizing bacteria, *Cupriavidus necator*, is prominently known for its ability to naturally produce the biodegradable plastic, polyhydroxybutyrate (PHB). Attempts to increase production of PHB have led to a number of tools enabling the bacteria's genetic tractability. A shortcoming of these tools is the ability to holistically modify *C. necator's* genome. A technique for this purpose which has been applied to other organisms is called protoplast fusion. Protoplast fusion involves the removal of the bacteria's outermembrane and subsequent fusion of the cells allowing the genes of the various genomes to be shuffled. To enable the use of protoplast fusion with *C. necator*, we developed a protocol and show the effectiveness of this technique using two strains of amino acid auxotrophs. To date, this is the first application of this technique with *C. necator*.

DNA Component Order Independence Using Insulated Bacterial Promoters

Carr, S., Boston University
Beal, J., Raytheon BBN Technologies
Densmore, D.,

Unexpected impact from flanking DNA sequences has been identified as an important cause of failed *a priori* genetic circuit design. Genetic parts characterized in one context often experience unexpected changes in expression level when moved into the context of their target genetic device. Furthermore, individual transcriptional units (TUs) or modules expected to be orthogonal to others within a device are, in practice, dependent upon the order in which they are arranged within a synthetic genetic device. Effective insulator techniques have been developed for the promoter::RBS and RBS::gene boundaries, but no systematic method for insulating promoters from their upstream context has previously proven effective.

We address the issue of upstream promoter context dependence with a novel method based on a randomized insulator library: we have developed a high-throughput, flow cytometry-based screen that randomly samples from a library of 4^{36} insulators created in a single cloning step. Screened against a reference device, our method provides precise control over genetic circuit expression. This new method of custom-insulator selection can be performed on both constitutive and transcriptionally regulated bacterial promoters, and can also be used to tune promoter expression level to a desired dynamic range. Finally, we validate the efficacy of this method for insulating promoters by showing that insulating the promoters in a genetic NOT-gate circuit using this method improves circuit performance and nearly eliminates the effect of the order in which the promoters are organized in the device.

High-Efficiency, Multi-Copy, Markerless Integration of Large Biochemical Pathways in Saccharomyces Cerevisiae

Shi, S., Agency for Science, Technology and Research
Liang, Y., Agency for Science, Technology and Research
Zhang, M. M., Agency for Science, Technology and Research
Ang, E. L., Agency for Science, Technology and Research
Zhao, H., University of Illinois at Urbana Champaign

Saccharomyces cerevisiae has many advantages as a production host due to its robustness, tolerance towards low pH, genetic tractability and long history of industrial use. However, despite recent advances in genome editing capabilities, the chromosomal integration of large multi-gene biochemical pathways for stable industrial production using *S. cerevisiae* remains challenging. To address this limitation, we developed a simple strategy for high-efficiency, single-step, markerless, multi-copy chromosomal integration of full biochemical pathways in *S. cerevisiae*. In this so-called Di-CRISPR (Delta integration CRISPR-Cas) strategy, we specifically designed guide RNA sequences to direct Cas9 cleavage at delta sites that are distributed throughout the yeast genome. The generation of double stranded breaks at the delta sites greatly enhanced the efficiency of homologous recombination at these loci and allowed simultaneous integration of multiple large DNA fragments. With our newly developed Di-CRISPR platform, we were able to attain highly efficient and markerless integration of large biochemical pathways ranging from 8 kb to 24 kb, achieve an unprecedented 18-copy genomic integration of a 24 kb combined xylose utilization and (R,R)-2,3-butanediol (BDO) production pathway in a single step, and the stable production of BDO directly from xylose with our engineered strain. The simplicity and high efficiency of the Di-CRISPR strategy would provide a superior and more industrially-relevant alternative to high copy plasmids and would render this strategy an invaluable tool for genome editing and metabolic engineering in *S. cerevisiae*.

Engineering Improved Bio-Jet Fuel Tolerance in Escherichia coli Using a Transgenic Library Approach

Tomko, T., The University of Vermont
Dunlop, M., The University of Vermont

Microorganisms are capable of producing advanced biofuels that can be used as 'drop-in' alternatives to conventional liquid fuels. However, vital physiological processes and membrane properties are often disrupted by the presence of biofuel and limit the production yields. In order to make microbial biofuels a competitive fuel source, finding mechanisms of improving resistance to the toxic effects of biofuel production is vital. This investigation aims to identify resistance mechanisms from microorganisms that have evolved to withstand hydrocarbon-rich environments, such as those that thrive near natural oil seeps and in oil-polluted waters. Using genomic DNA from *Marinobacter aquaeolei*, a hydrocarbon-degrading marine microbe, we constructed a transgenic library that we expressed in *Escherichia coli*. We exposed cells to inhibitory levels of pinene, a monoterpene that can serve as a jet fuel precursor with chemical properties similar to existing tactical fuels. Using a sequential strategy of a fosmid library followed by a plasmid library, we were able to isolate a region of DNA from the *M. aquaeolei* genome that conferred pinene tolerance when expressed in *E. coli*. We determined that a single gene, *yceI*, was responsible for the tolerance improvements. Overexpression of this gene placed no additional burden on the host. We also tested tolerance to other monoterpenes and showed that *yceI* selectively improves tolerance.

The JBEI Experiment Data Depot

Morrell, W., Sandia National Laboratory
Forrer, M., Sandia National Lab
Hillson, N. J., Joint BioEnergy Institute
Garcia Martin, H., Lawrence Berkeley National Laboratory
Birkel, G., Lawrence Berkeley National Laboratory
Eads, J.,

Ask a dozen researchers for data supporting their results, and you will get data formats numbering an order of magnitude higher. At the Joint BioEnergy Institute (JBEI), we are developing the Experiment Data Depot (EDD) to help reduce the friction in sharing data sets and collaboration. The EDD is a database and set of software tools aiming to organize and annotate actionable data with conditions and other metadata. The aim is not to be a generic laboratory information management system (LIMS); our focus is instead on collating processed results from multiple inputs, and enabling the whole to be exported to computational models, analysis tools, visualization packages, or other applications. The database framework organizes data in a heirarchcal structure for each experiment: multiple samples and conditions, each having multiple assays, each having multiple measurements, each having multiple data points. Additional metadata can be added at each level of heirarchy, and can later be queried; e.g. finding all data linked to a particular strain. Basic access controls are built in, to allow only specific individuals or groups to view or edit data.

Our original use-case for the EDD collected respiration, mass spectrometer, and optical density data for 13-C metabolic flux analysis of host metabolic models. The software is pending an open-source release, allowing the broad framework to be adapted to the needs of individual labs. We have added modules to better support the collection of RNA transcription levels, protein expression levels, flow cytometry, and chromatography data imports. Additional tools extract data to characterize genetic parts or enzymes, or build instructions for robotic measurement of samples. We hope that the framework can be used to handle more types of data, providing a common framework for sharing data.

Electricity-Generating Microbial Communities in Biotrickling Filter-Microbial Fuel Cells for Gaseous Ethyl Acetate Treatment

Lin, C. W., National Yunlin University of Science and Technology
Wu, C. H.,
Shih, J. C.,

Electricity-generating microbial communities in biotrickling filter-microbial fuel cells for gaseous ethyl acetate treatment

Chi-Wen Lin*, Chih-Hung Wu, Jing-Chen Shih

Department of Safety, Health and Environmental Engineering, National Yunlin University of Science and Technology, 123 University Rd., Sec. 3, Douliou, Yunlin 64002, Taiwan, ROC

Abstract

This study developed an integrated one-stage biofiltration system treating gaseous contaminants using biotrickling filter (BF) coupled with microbial fuel cell (MFC) by taking advantages from the MFC systems and biofiltration technologies. A unique anode design allowing fully contact between coke and graphite anode forming a three-dimensional electrode in the anode region of the biotrickling filter-microbial fuel cell (BF-MFC) system was developed to amplify anode area. Moreover, a new membrane design providing effective delivery of the proton released from microorganisms biodegrading ethyl acetate (EA), from anode to the cathode was developed by using a polyvinyl alcohol-membrane electrode assembly which separating anode from cathode in the BF-MFC. In addition, we tested for the first time the feasibility of using coke as the anode material to form a three-dimensional electrode for amplifying electricity production in MFC. The innovative BF-MFC system was then applied to investigate the performance under various inlet loads, the electricity generation, and microbial changes toward gaseous EA biodegradation. The effect of EA levels on microbial community structures was also assessed by polymerase chain reaction-denaturing gradient gel electrophoresis (PCR-DGGE) analysis. The results show that an EA elimination capacity of 83.80 $g/m^3 \cdot h$ and the maximum power density of 49.1 mW/m^2 were observed. Microbial community analysis revealed that two distinct groups of exoelectrogenic microbes and EA-degraders were existed. The former dominated in coke media, whereas the latter was present in the interior surface of the BF-MFC and in suspension in the recirculating tank. These results exhibited that the developed BF-MFC provided an alternative for the environmentally friendly removal of gaseous EA in combination with electricity production.

Keyword: biotrickling filter-microbial fuel cell (BF-MFC), denaturing gradient gel electrophoresis (DGGE) analysis, polyvinyl alcohol-membrane electrode assembly (PVA-MEA), generation electricity

Tunable Thermal Bioswitches for Noninvasive Genetic Regulation

Abedi, M., Caltech
Piraner, D., Caltech
Moser, B., Caltech
Shapiro, M. G., California Institute of Technology

Rapid advances in synthetic biology are driving the development of genetically engineered bacteria as microbial therapeutic and diagnostic agents for a host of human diseases. A critical capability for many envisioned *in vivo* applications is the ability to control the function of engineered bacteria *in situ*, as currently done using chemical inducers or visible light. However, systemic chemicals typically lack the spatial precision needed to control activity at specific anatomical sites, such as in the gastrointestinal tract or tumors, while optical approaches suffer from poor penetration in tissues. On the other hand, temperature can be controlled both globally and locally – at depth – using technologies such as focused ultrasound, infrared light and magnetic particle hyperthermia. In addition, body temperature can serve as an indicator of bacterial entry and exit from the host organism and the host's condition. Given this potential, there are surprisingly few high-performance thermal bioswitches to control gene expression in engineered bacteria. The ideal bioswitch should have a sharp thermal transition resulting in a large change in activity (> 100-fold) over a few °C, and its switching temperature should be tunable to enable a broad range of applications. Existing temperature-dependent regulators of gene expression fail to fulfill these criteria. To address this limitation, we have developed two families of tunable, orthogonal, temperature-dependent transcriptional repressors covering the biomedically relevant range of 32 to 46°C. These bioswitches feature sharp transitions (3-5°C) and a large dynamic range (> 300-fold), superior to comparable protein and RNA-based regulators. The switching temperature can be tuned using a simple high-throughput screen, and orthogonality between the two bioswitch families enables multiplexed thermal control. We demonstrate the utility of engineered thermal bioswitches in two microbial therapy applications, first restricting the growth of engineered bacteria to body temperature to limit contagion outside the intended host, then using focused ultrasound to thermally activate engineered microbes at a specific targeted location. Given the multitude of ways that temperature can be controlled *in situ*, these tunable thermal bioswitches will enable a wide range of applications in bacterial synthetic biology and could also be adapted for use in eukaryotic systems.

D. Piraner and M. Abedi contributed equally to this work.

Minimum Flow Cytometry Calibration Information (MinFlowCal)

Beal, J., Raytheon BBN Technologies
McLean, P., NIST
Haddock-Angelli, T., Boston University
Teague, B.,
Wang, L.,
Munson, M. S., National Institute of Standards and Technology
Munro, S., NIST
Salit, M., National Institute of Standards and Technology

Flow cytometry is a powerful method for inspection of the behavior of large numbers of individual cells in a population. However, in the absence of appropriate data processing, calibration, and sufficient reporting, measurement results obtained with a flow cytometer are difficult to interpret or compare to existing datasets. MinFlowCal is a Synthetic Biology Standards Consortium (SBSC) project aimed at addressing these issues by delineating a minimum-information standard and calibration method that will facilitate appropriate reporting of flow cytometry data. We describe the use of commercially available fluorescent beads to calibrate fluorescence signal values to a common fluorescence unit, e.g., mean equivalent fluorochrome (MEF). Adopting and reporting calibrated flow cytometry measurement results will enable comparability and evaluation of reproducibility across instrument platforms and laboratories.

MinFlowCal: Minimum Flow Cytometry Calibration Information

- **Need:** absolute unit measurements from large numbers of single cells
- **Technology:** Calibrated Flow Cytometry
- **Products:**
  1. Minimum standards for reporting on unit conversion
  2. Accessible protocols for unit calibration

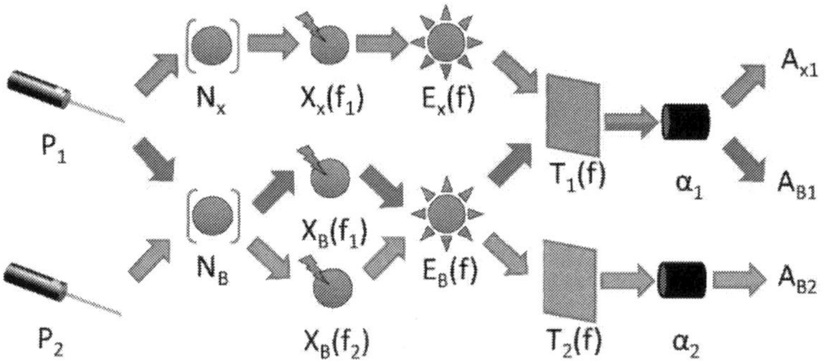

POSTER #9, Beal et al.

An In Vivo mutation Prevention System
Chavez, A.,

Alejandro Chavez[1,2,3,8], Benjamin W Pruitt[1,8], Marcelle Tuttle[1], Rebecca S Shapiro[1,4,5,6,7], Ryan J Cecchi[1], Jordan Winston[1], Michael Tung[1], James J Collins[1,4,5,6,7,9], George M Church[1,3,9]

[1]Wyss Institute for Biologically Inspired Engineering, Harvard University, Cambridge, Massachusetts, USA.

[2]Department of Pathology, Massachusetts General Hospital, Boston, Massachusetts, USA.

[3]Department of Genetics, Harvard Medical School, Boston, Massachusetts, USA.

[4]Institute for Medical Engineering & Science, Massachusetts Institute of Technology, Cambridge, Massachusetts, USA.

[5]Synthetic Biology Center, Massachusetts Institute of Technology, Cambridge, Massachusetts, USA.

[6]Department of Biological Engineering, Massachusetts Institute of Technology, Cambridge, Massachusetts, USA.

[7]Broad Institute of MIT and Harvard, Cambridge, Massachusetts, USA.

[8]First two authors contributed equally

[9]Corresponding author

Point mutations have the remarkable ability to drastically alter organismal phenotype, conferring antibiotic resistance to bacterial pathogens and driving oncogenesis in healthy human cells. Despite a wealth of research into the mechanisms and implications of point mutations, there is no method to directly prevent their occurrence. To date we can only document when mutations occur and utilize this information to predict their effect on disease progression and chemotherapeutic outcome.

Through careful tuning of the RNA-guided endonuclease Cas9, we developed an *in vivo* "mutation prevention" system that can prevent the occurrence of targeted point mutations with little to no latent toxicity to the host organism. We demonstrate the efficacy of the mutation prevention system by simultaneously inhibiting up to four targeted mutations that confer antibiotic resistance within the model organism *E. coli*.

Our results provide a general framework upon which to develop highly specific Cas9-gRNA complexes and make possible entirely new avenues of discovery in fields as diverse as molecular ecology and infection control.

Characterizing RNA Thermometers through Shape-Seq

Meyer, S. I., Cornell University
Lucks, J. B., Cornell University

Whether harvested from nature or designed synthetically, RNA regulators offer synthetic biologists a valuable way to control gene expression and engineer cells that can sense and respond to their environment. As part of our on-going efforts to expand the toolbox of RNA regulators available for use, we report the functional and structural characterization of a series of RNA thermometers derived from *Salmonella*. These thermometers regulate the translation of a downstream gene via a temperature-sensitive hairpin that occludes the RBS at low temperatures but not at high temperatures. Using a combination of fluorescent reporter gene read-outs and RNA structural probing via SHAPE-Seq, we dissect the underlying mechanism of these thermometers in order to allow for their use and modification in synthetic biology applications.

"Passcode" Kill Switch: Modular Hybrid Transcription Factors for Programmable Biocontainment

Chan, C. T. Y., Massachusetts Institute of Technology
Collins, J. J., Broad Institute of Harvard and MIT

With the widespread use of genetically engineered microbes in medical and industrial applications, it is increasingly important to develop security measures that prevent the release of these microorganisms into the natural environment. Biological containment systems use an environmental signal sensing strategy in which survival of target cells requires the presence of specific small molecules in their assigned environments. Alteration of survival requirements may be necessary when the engineered microbes are used for different applications, which is difficult for many pioneering biocontainment systems. Here we develop a safeguard genetic circuit with programmable biocontainment conditions in E. coli. To create modular environmental sensors for building this genetic circuit, we first used a biochemistry approach to identify a conserved boundary between environmental sensing modules (ESMs) and DNA recognition modules (DRMs) from transcription factors (TFs) in the LacI/GalR family, in which these discrete ESMs and DRMs can be mixed and matched to create hybrid TFs that link different environmental inputs to the control of a specific promoter for gene expression. Then, we harnessed these hybrid TFs to build a transcriptional repression-based genetic circuit that blocks toxin gene expression under only one of the eight potential combinations generated by three environmental inputs—the "passcode" for cell survival. Exposure of the engineered cells to any of the other seven combinations induces toxin gene expression and cell death. Importantly, we can reprogram the survival "passcode" by reconnecting input detection to promoter activity regulation with different hybrid TFs. Our development of a genetic circuit-based kill switch demonstrates a promising approach to generate biocontainment systems with complex and customizable environmental requirements for cell survival and proliferation.

Piace: Parallel Integration and Chromosomal Expansion of Biofuels Pathways in E. coli

Goyal, G.,
Alonso-Gutierrez, J., Joint Bioenergy Institute (JBEI)
Kang, A.,
Lee, T. S., Joint BioEnergy Institute
Keasling, J. D., Lawrence Berkeley National Laboratory
Hillson, N. J., Joint BioEnergy Institute

Robust fermentation of biomass-derived sugars into biofuels demands the reliable microbial expression of underlying metabolic pathways. Previous studies have shown that plasmid-based expression systems, which may suffer from instability issues (despite selective pressure), can result in inconsistent and highly variable biofuel production rates and titers. An established mitigation strategy, Chemical Induced Chromosomal Expansion (CIChE; Tyo et al. 2009), exploits RecA-mediated copy number expansion of a pathway post integration at a single chromosomal locus, and offers the gene copy numbers of plasmids while maintaining pathway stability even without the addition of antibiotics (a significant process cost-savings). Innovating beyond CIChE, we are developing PIACE (Parallel Integration and Chromosomal Expansion) to facilitate the simultaneous and independentlyadjustable expansions of multiple pathways at separate chromosomal loci. PIACE offers a set of BglBrick-compatible suicide vectors, each of which site-specifically integrates into a distinct chromosomal locus via a single homologous crossover and has high efficiency for transformation and chromosomal integration of long (> 3kb) pathways that reside in super-coiled sequence-validated plasmids. As a proof of principle for PIACE, we have integrated rfp and gfp at two separate chromosomal loci, expanded rfp and gfp independently in the presence of heat-curable plasmid pRedi2RecA, and demonstrated that both expansions are stable over 50 cellular generations in the absence of antibiotic selection pressure. We are now applying PIACE to the chromosomal stabilization and optimization of the isopentenol fuel pathway in E. coli DH1. PIACE affords an efficient, tunable method for improving the robustness, productivity, and titers of engineered metabolic pathways.

Integrated Biocad Toolchain Enables Search for Experimentally Validated Components

Plahar, H., Joint BioEnergy Institute, Lawrence Berkeley National Laboratory
Forrer, M., Sandia National Lab
Coble, J., Joint BioEnergy Institute, Lawrence Berkeley National Laboratory
Morrell, W., Sandia National Laboratory
Keasling, J. D., Joint Bioenergy Institute
Adams, P. D., The Joint BioEnergy Institute
Garcia Martin, H., Lawrence Berkeley National Laboratory
Hillson, N. J., Joint BioEnergy Institute

Biological computer-aided design (bioCAD) tools are gaining widespread adoption in commercial, academic, and government research settings. As more and more tools emerge, there is an increasingly urgent need to integrate them into a larger toolchain that aggregates functionality and data. We have recently integrated several of our bioCAD tools (DeviceEditor, DIVA, ICE, EDD Analytics, and EDD) to enable researchers designing biological systems to query for components (e.g., promoters) that have been experimentally validated to meet particular design criteria (e.g., high transcriptional levels in E. coli). In our bioCAD toolchain, the user first specifies the desired component's properties in DeviceEditor, a visual bioCAD canvas. The user's specifications form the basis of a query that, via the DIVA software platform, is sent from DeviceEditor to ICE. ICE replies to the query with the list of components that meet the user's criteria. The user then selects from the list those desired components that should be added to the DeviceEditor design canvas. Importantly, EDD Analytics derived the specifications for the components in ICE from a combination of experiment data in the EDD (Experiment Data Depot) and context meta-data (e.g., microbial host) in ICE. As such, during the component selection process, the user has access to the primary experimental data in EDD supporting the design selection decision. Our initial demonstration of this integrated bioCAD toolchain consists of a researcher searching for low, medium, or high strength promoters with flow cytometry data supporting their strength classifications.

Molecular Membrane Engineering for Nanoreactors

Di Bartolo, N., University of Bristol
Komar, J., University of Bristol
Jones, M., University of Bristol
Collinson, I., University of Bristol
Swainsbury, D., University of Sheffield

Membrane bound compartments are at the heart of biological complexity, allowing nature to control and drive biochemical reactions. The aim of this project is to create artificial bionanoreactors and linked vesicle networks that mimic this natural compartmentation. The approach is to engineer synthetic energy-transducing biomembrane vesicles that incorporate engineered membrane proteins and soluble components and to use light to generate electrochemical gradients to power and gate transport of materials between the nanoreactor and its environment. The ultimate goal is to control the flux of specific molecules into or out of the nanoreactor and the chemistry within them, enabling catalytic control for synthesis, degradation, delivery, sensation etc.

A key aspect of this work is to gain control over the composition of individual nanoreactors by developing component membrane proteins that self-assemble into hetero-oligomeric complexes with defined stoichiometry and orientation. A range of photosynthetic reaction centre membrane proteins modified with extra-membrane components such as coiled-coil bundles, the SpyCatcher/SpyTag system and the maltose binding protein have been engineered and are currently being assessed for expression, stability and their ability to form high affinity interactions. Their impact on reconstitution and sidedness in assembled proteoliposomes is also being studied through the development of a simple, convenient and generic assays based on the His-tags used to aid the purification of the various integral membrane proteins being used in the project.

We are also developing the use of digital holographic microscopy, which measures refractive index changes, as a non-invasive measurement of transport. When coupled with conventional spectroscopic techniques, this will allow us to engineer nanoreactors with optimised influx and efflux of reactants and products in response to chemical and optical stimuli. We have established that the technique can be used to image individual liposomes of 1000 nm diameter and currently exploring methodologies for immobilising liposomes on surfaces for the application of measurements on single liposomes.

Interactions Between Non Homologous End Joining Repair and Type II Crispr-Cas Systems

Bernheim, A., Institut Pasteur
Touchon, M.,
Rocha, E. P., Institut Pasteur
Bikard, D., Institut Pasteur

CRISPR-Cas systems confer bacteria and archaea an adaptative immunity against phages and other invading mobile genetic elements. Type II CRISPR-Cas systems target foreign DNA by generating double strands breaks, which can be repaired either by homologous recombination (template-based) or by Non Homologous End joining (NHEJ) that does not require any template. These systems will limit the efficiency of CRISPR-Cas systems if they repair the double strand breaks in invading DNA. The interference of homologous recombination in this process is expected to be negligible due to the lack of a template for repair. We therefore hypothesized that the NHEJ system could interfere with CRISPR immunity. To test this idea, we used comparative genomics to study the interactions between CRISPR-Cas systems and NHEJ. We report a negative association between NHEJ and Type IIA CRISPR Cas systems. We then investigated experimentally the possible molecular interactions causing this negative association. Our results suggest that NHEJ has no effect on CRISPR interference. However, the introduction of the Bacillus subtilis NHEJ system in Staphylococcus aureus abolished spacer acquisition by a Type IIA CRISPR-Cas system.Our findings give insights on the complex interactions between CRISPR-Cas systems and repair mechanisms in bacteria.

High-Throughput DNA Assembly Pipeline for Natural Product Discovery in Streptomycetes

Hsu, S. Y., University of Minnesota-Twin Cities

Disease suppressive soil has the ability to inhibit disease development even in the presence of pathogens and susceptible plant hosts. Studies suggest that the interaction of soil microbiota, especially their secondary metabolites production, is one of the primary factors of disease suppressiveness. Research approaches combining culture-based assessments and DNA-based techniques have expanded physiological and molecular knowledge about the operative mechanisms of soil microbial communities. However, these procedures could produce biased representation of the microbial community due to extraction difficulties, differential culturing requirements and relative differences of amplification of target sequences among different organisms. Here, we report a high-throughput genomic approach to identify putative biosynthetic gene clusters and their corresponding secondary metabolites from microbes isolated from suppressive soil.As a proof-of-concept, three new soil Streptomycetes strains from disease suppressive soil were isolated and sequenced. The *de novo* assembly by PacBio® SMRT sequencing pipeline produced three high-quality draft genomes and were able to capture large read, extreme GC content and highly repetitive regions which the second-generation sequencing was incapable of. Using draft genome as an input, bioinformatics analysis by antiSMASH for prediction of novel natural products identifies corresponding putative gene clusters to be cloned and characterized. To simplify the design of targeted multigene constructs for high-throughput cloning, we developed a DNA fabrication pipeline which includes primer design automation algorithm and a standardized, modular vector system based on Type IIS restriction enzymes. In light of the sequence data and an alternative modular assembly method, we hope to use this genome-guided method to elucidate the biochemical properties of novel secondary metabolites found in the microbiota of disease suppressive soil. In addition, we provide a tool to rapidly convert sequence data to diverse secondary metabolite information.

Novel Lactate Biosensor for Use in Low Oxygen Conditions

Dekker, L., Imperial College London
Goers, L., Imperial College London
Polizzi, K. M., Imperial College London

Lactate is of interest in a variety of different industrial and biotechnological applications. In *Escherichia coli, t*he *lldPRD* operon is induced in the presence of L-lactate and is involved in lactate transport and metabolism. The BioBrick standard assembly method was used to construct a whole cell lactate biosensor which contains a gene encoding a flavin mononucleotide-based fluorescent protein (FbFP) under the control of the *lldPRD* operon promoter. The biosensor also contains the *lldR* regulatory gene which acts as a repressor in the absence of lactate, and an inducer in the presence of lactate. To enable the biosensor to function in anaerobic conditions, the ArcA-P binding site on the *lldPRD* operon was modified to prevent repression of the operon, allowing the biosensor to function in both aerobic and anaerobic conditions. The biosensor exhibited a concentration dependent response to lactate in the range of 0.05–5 mM lactate. This biosensor could be used in low oxygen environments to detect lactate. Due to the growing demand for lactate, this biosensor could play a valuable role in industrial processes.

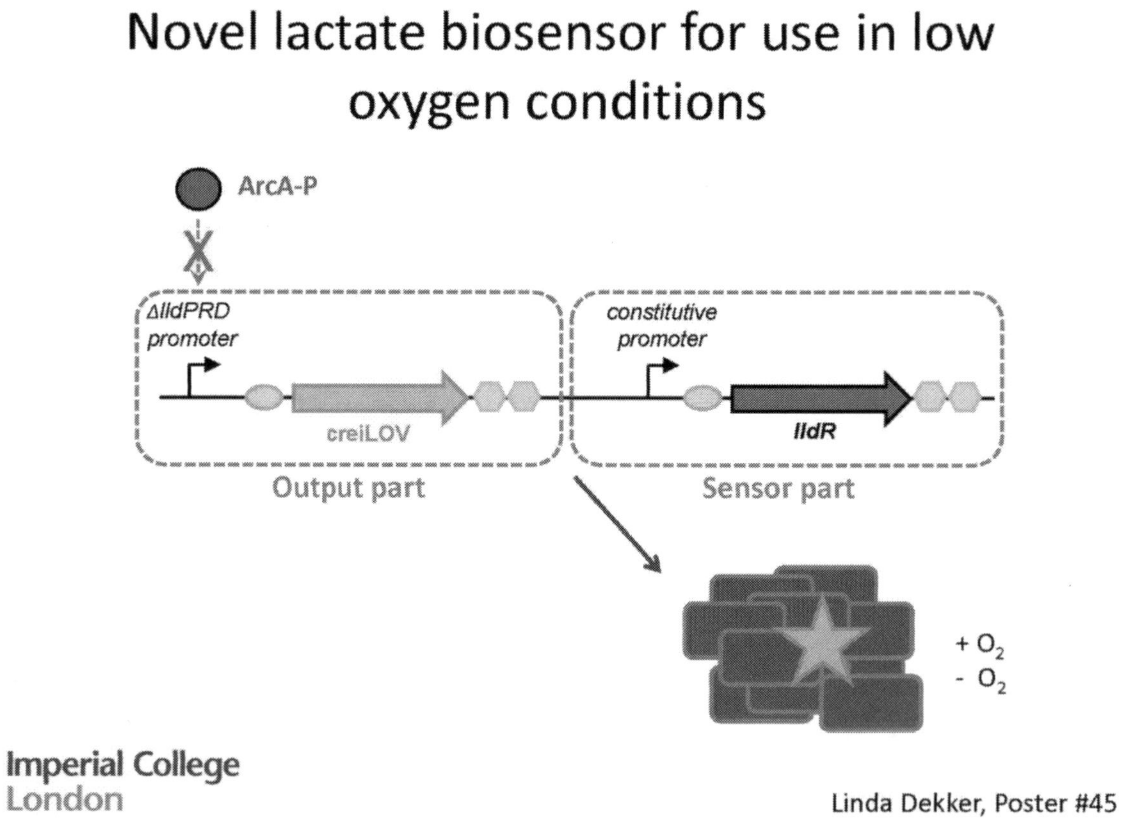

Whole Genome Regulator Mutant Library for Expedited Identification of Beneficial Mutations for Enhanced Tolerance Towards Environmental Stress

Choudhury, A., University of Colorado, Boulder
Wang, Z., Tianjin University
Kaar, J. L., University of Colorado
Gill, R. T., University of Colorado Boulder
Garst, A., University of Colorado Boulder

Use of *E coli* as an industrial strain entails tolerance to multiple environmental stresses. Resistance towards such stresses is a complex phenotype requiring modification of expression of multiple genes across the genome. In the past, mutations in regulatory proteins have enabled such complex modifications. Universal resources for rapid identification of such beneficial mutations are lacking. We propose the development of a library with single point mutations on functional sites of regulatory proteins across the *E coli* genome. The library was subject to multiple environmental stresses like presence of ethanol and isobutanol, low pH, high temperature and presence of antibiotics. From the selections, multiple novel beneficial mutations were identified with enhanced tolerance to these stresses. These mutations might enable better understanding of tolerance mechanisms towards these stresses. The library can act as a platform for expedited strain engineering for applications in metabolic engineering.

Synthetic Speciation: New Technology for Transgene Biocontainment and Pest Biocontrol

Maselko, M., University of Minnesota
Smanski, M. J., University of Minnesota
Heinsch, S., University of Minnesota
Kim, S.,

We present here the proof of concept for a generalizable method which can be used to generate reproductivly isolated 'synthetic species'. This technology has numerous potential applications including engineering a genetic incompatibility between crops which have been modified to produce medications or industrial compounds and those which are grown for human consumption. The use of synthetic animal species may be used for the biocontrol of pests such as insect disease vectors or invasive species.

Sequence to Activity Mapping: Computational Approaches and Experimental Tools

Pines, G., University of Colorado
Garst, A., University of Colorado Boulder
Bassalo, M.,
Halweg-Edwards, A., Colorado University, Boulder

We developed a method for rapidly investigating sequence-to-activity relationships in a large scale, at the genomic level of *E. coli*. This method combines phage-based recombineering with CRISPR genome editing: site-specific gRNAs and editing templates are synthesized in parallel on a microarray and later pool-cloned into a dedicated plasmid. The gRNA, together with an inducible Cas9 digest the genome in a sequence-specific manner, while the editing template is integrated genomically by the recombineering machinery. The editing template harbors two mutations: the first is the desired modification and the second silently mutates a neighboring PAM sequence, which is necessary for CAS9 cleavage. This approach allows to build genome-wide libraries and complete saturation of whole genes. Since the plasmid editing templates highly correlates with the genomic edits, simple plasmid sequencing replaces whole genome interrogation, making this system easily trackable. We demonstrate CREATE by completely saturating the essential E. coli *folA* gene, generating its full sequence to activity map. Moreover, we also challenged these library cells with trimethoprim, a FolA specific inhibitor, and found two mutations within the same site that confer resistance to this drug.

We previously reported on a new approach for saturation mutagenesis. In contrast to the commonly used NNK codon, we compute the minimal set of degenerate codons to exactly cover the desired set of amino acids. Importantly, the compressed codons do not include stop codons, and cover every amino acid only once, eliminating redundancy. Moreover, the codons selected are with high usage scores, as defined by the target organism. We now report a dedicated website that includes all these and more functionalities, improving accessibility and ease of use for the synthetic biology and bioengineer's community.

Precise Control of Lycopene Production to Enable a Fast-Responding, Minimal-Equipment Biosensor

McNerney, M., Georgia Institute of Technology
Watstein, D., Georgia Institute of Technology
Styczynski, M., Georgia Tech

Zinc deficiency contributes to hundreds of thousands of childhood deaths annually, but current detection methods are cost-prohibitive for use in the regions most affected. Our group recently developed a low-cost bacterial biosensor that can produce different visible outputs based on the concentration of zinc in which the cells grow. However, these initial sensor cells always produce some pigment, which necessitates that they be grown from a very small inoculum so that initial coloration is minimal – thus requiring overnight assay times. To make an easy-to-use, field-deployable assay, we are engineering cells that completely repress pigmentation during growth and quickly produce the red pigment lycopene upon induction. This would allow sample addition to concentrated pre-cultured cells, decreasing assay time from about 24 hours (the time required for growth) to 3 hours (for visible pigment production without growth).

A sensor that produces a pigment instead of fluorescent proteins is advantageous in resource-limited settings, since sensor readout can be interpreted without advanced equipment. However, complete repression of pigment production requires more precise control than what is necessary for most synthetic biology applications, since small amounts of enzyme can produce visible amounts of pigment. Inducing high levels of pigmentation poses an additional challenge, as overexpression of *crtEBI*, the genes that produce lycopene, can cause cell toxicity and prevent accumulation of lycopene. Since maximum pigment production does not correspond with maximum protein production, precise control of the induced expression level is also necessary to express the appropriate amount of pigment-producing proteins.

Three commonly used inducible promoter systems—the *lac*, *ara*, and T7 systems—were tested for precise control of lycopene production. The arabinose-inducible promoter *pBAD* proved most repressible but still could not fully repress lycopene production during uninduced growth, and in all systems lycopene production decreased upon induction because of *crtEBI*-related toxicity issues. Modification of polymerase binding sites in the promoters significantly decreased induced expression level (enabling overexpression of lycopene upon induction), but only slightly decreased the repressed level (which was insufficient to repress lycopene production during uninduced growth). Translational modifications proved much more effective in controlling lycopene production. When the original ribosomal binding sites for the *crtE*, *crtB*, and *crtI* genes were replaced with weak RBSs, the *lac* and *ara* regulatory systems completely repressed pigmentation during uninduced growth and produced visible amounts of lycopene at full induction. Upon induction, the *lac* and *ara* systems produced visible pigment in about 3 and 6 hours, respectively – a large improvement over the 18-24 hours required in the original assay. Zinc-responsive elements and other pigments can be added to these systems to create circuits that produce different pigments in response to different zinc concentrations.

As the applications of synthetic biology expand to systems that require precise control of metabolic state, transcriptional control alone will likely be insufficient. Even the most repressible promoters could not fully repress lycopene production, suggesting that post-transcriptional modifications may be a better starting point in combatting unwanted protein expression.

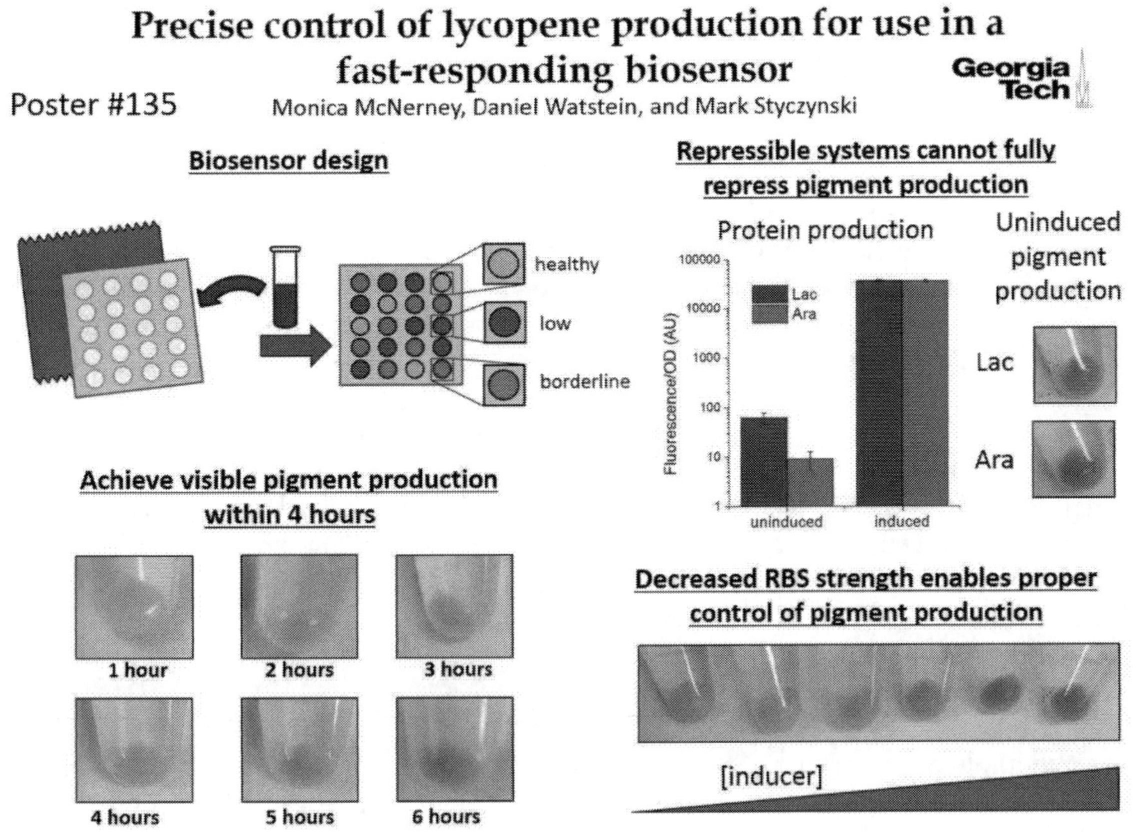

Biophysical Limits of Control Implementations in Synthetic Gene Networks

Dy, A. J., MIT
Collins, J. J., Broad Institute of Harvard and MIT
Del Vecchio, D.,

Synthetic biology has made progress in applying engineering principles to forward-engineer cellular behavior, but there are still knowledge gaps for moving from high-level design to functional systems. One such gap is choosing the correct parts for a given network topology and ensuring optimal performance in a cell that has limited resources. While network motifs have been evaluated for different functions, the precise implementations have not been studied to determine differences in robustness and efficiency.

Here, we analyze the biophysical limits imposed by different molecular mechanisms when engineering bio-molecular control. As an example, we focus on the ability of incoherent feedforward loops (IFFLs) to achieve adaptation and maintain a set-point despite changing inputs. Unlike the simplest models of the IFFL motif, our model includes biophysical parameters that allow comparison of system error and resource usage over a range of input values.

For the same IFFL network topology we model three implementations of the negative interaction: TetR transcriptional repressor, T7 RNA polymerase (RNAP) that transcribes a small RNA (sRNA) that blocks translation, and *mf*-Lon protease that degrades target protein. Each implementation takes a general input u (e.g. copy number, transcriptional machinery, or translational machinery) that activates transcription or translation of both the target protein and each of the three regulating nodes. All can be shown to exhibit adaptation in the simplest mathematical model, but have tradeoffs in error and resource usage in our mechanistic model.

All three implementations were tuned to the same starting expression and steady-state error for a transcriptional input change of 1 to 10, then input changes were simulated over 3 orders of magnitude. Over this range, the percent error for all three are within one percent of each other but ribosome usage can differ by orders of magnitude. The T7 RNAP-sRNA circuit uses the least ribosomes, but has monotonically increasing error as the input increases beyond the transcriptional input for which the circuit was tuned. The TetR repressor circuit uses more ribosomes but the error goes to a fixed amount as the input goes to infinity. Similar adaptation is achieved for input changes to translational capacity but ribosome usage still differs by orders of magnitude.

Our work can be used to choose biological realizations of gene networks that satisfy design constraints on error and resource usage. Such a theoretical framework can demonstrate quantitative tradeoffs among transcriptional, translational, and post-translational control that are not readily apparent. For instance, T7 RNAP-sRNA feedforward may be better for ribosome limited applications while TetR feedforward may be better for cases in which a small error is needed over a large transcriptional input range. We use these insights to guide our experimental realizations of feedforward or feedback control that can be tested and compared to the mathematical model. This analysis provides guidelines for modeling and designing circuits that are robust in resource limited cells.

Assembly of Membrane Encapsulated Nanoreactors

Komar, J., University of Bristol
Di Bartolo, N., University of Bristol
Swainsbury, D., University of Bristol
Jones, M., University of Bristol
Collinson, I.,

This project aims at engineering synthetic self-contained energy-transducing biomembrane vesicles that can serve as nanoreactors, incorporating both natural and bespoke membrane proteins. The ultimate goal is to control the flux of specific molecules into or out of the nanoreactors and the chemistry within them.

In order to achieve this, a range of component membrane proteins have been produced, which will act as gateways to control the influx and efflux of molecules and ions in a contained phospholipid environment. These include multidrug transporters, e.g. AcrB and Sav1866, sugar transporters, e.g. SglT, the light-driven sodium pumping rhodopsin NaR (KR2) and the light-driven proton pump – bacteriorhodopsin (bR). Various methods are being developed to control the orientation, stoichiometry and self-assembly of the components upon reconstitution into liposomes.

In addition to a traditional method of reconstitution directly into liposome vesicles, nanoreactors can be also assembled using a supported lipid bilayer. SAGE peptide nanocages are being explored for that purpose as a potential scaffold that would increase the robustness of synthetic liposomes.

Moreover, various assays are being explored for the purpose of monitoring membrane transport in assembled nanoreactors using spectroscopic methods, including assays from antibiotic and sugar transport.

This system will be further developed for the purpose of catalysis, synthesis of chemicals and degradation of toxic molecules.

Characterization of Bacillus Subtilis Colony Biofilms Via Mass Spectrometry and Fluorescence Imaging

Si, T., University of Illinois at Urbana Champaign
Zhao, H., University of Illinois at Urbana-Champaign
Sweedler, J. V., University of Illinois at Urbana-Champaign

Colony biofilms of *Bacillus subtilis* are a widely used model for studying cellular differentiation. Here, we adapt matrix-assisted laser desorption ionization mass spectrometry imaging (MALDI-MSI) to examine the cellular and molecular heterogeneity in *B. subtilis* colony biofilms. From *B. subtilis* cells cultivated on a biofilm-promoting medium, we detected two cannibalistic factors, which were not found with the same strain in previous MALDI-MSI studies under different culturing conditions. Given the importance of cannibalism in matrix formation of *B. subtilis* biofilms, we employed a transcriptional reporter to monitor matrix-producing cell subpopulations using fluorescence imaging. These two complementary imaging approaches were applied to characterize three *B. subtilis* strains, including the wild type isolate NCIB3610 and two mutants with defective and enhanced biofilm phenotypes (Δ*spo0A* and Δ*abrB*). Upon deletion of key transcriptional factors, correlated changes were observed in biofilm morphology, signaling and cannibalistic factor distribution, and matrix-related gene expression, providing new insights on cannibalism in biofilm development. This work underscores the advantages of using multimodal imaging to compare spatial patterns of both chemicals and protein expression patterns, obtaining information on cellular heterogeneity and function not available when single characterization methods are used. We are adapting this method to high-throughput screening for synthetic biology and metabolic engineering applications.

Integration of an Inducer-Producing Metabolic Pathway in an Artificial Cell

Coutable, A., BrisSynBio, a BBSRC/EPSRC-funded Synthetic Biology Research Centre
Tang, T. Y. D.,
Booth, R., University of Bristol
Perriman, A. W., University of Bristol, Bristol
Mann, S., BrisSynBio, a BBSRC/EPSRC-funded Synthetic Biology Research Centre
Anderson, J. L. R., University of Bristol

Synthetic cell-like system are common architectures employed to study biological processes (enzyme-catalysed reactions, simple metabolic pathways, transcription/translation) and inter-cellular communication (between artificial cells or between artificial and natural cells) or as drug-delivery and gene therapy agents. Conventional artificial cells are based on the self assembly of lipid components (liposomes, water-oil lipid droplets) providing an amphiphilic environment essential for integrating membrane proteins. Alternative synthetic cell-like systems such as polymersomes were developed to offset lipid systems which can be limited by the impermeability of the lipid membrane or by an environment that is not sufficiently crowded. Another promising alternative to lipid-based protocellular micro-compartments are proteinosomes, which are delineated by a semi-permeable, enzymatically active membrane consisting of protein-polymer nano-conjugates. In this work, we demonstrate the versatility of such proteinosomes as multi-functional protocellular micro-compartments. The enzyme-polymer membrane functions as an integral part of a multi-enzyme pathway, facilitating the production of a natural inducer which subsequently initiates fluorescent protein expression from an *in vitro* transcription-translation system encapsulated within the artificial cell.

Optimized Genetic Tools and Overexpression Strategies in the Development of Efficient Production Systems in Photosynthetic Cyanobacteria

Nagy, C. I.,
Kallio, P., University of Turku
Thiel, K., University of Turku
Dandapani, H., University of Turku
Patrikainen, P.,
Aro, E. M., University of Turku

Because of the gradually increasing need of fossil fuels and because of the inherent ecological and geopolitical concerns, the search for alternative/renewable energy sources is imminent. Cyanobacteria can serve as cell factories for the production of fuels or other commodity chemicals because of the many benefits that their physiology and genetics offer: photosynthesis, CO_2 fixation, fast growth rate, relatively simple genetic manipulation, and feasible culture conditions. Although cyanobacteria have many advantages over heterotrophic microorganisms like *E. coli* or *S. cerevisiae*, their biotechnological potential is less explored because the relative scarce information about their genetic and metabolic processes. However, by means of synthetic biology techniques cyanobacterial strains can be effectively modified to produce valuable compounds.

Our research focuses on the construction of synthetic pathways in the cyanobacterium *Synechocystis* sp. PCC6803 to be used for the production of a wide range of industrially valuable compounds. Since genetic alterations can cause perturbations and metabolic imbalance, in order to establish stable and efficient heterologous metabolic pathways, fine-tuned genetic elements and related synthetic biology tools needs to be evaluated at transcriptional and translational level. A library of 18 integration sites (targeting the chromosome and the native plasmids), a library of 13 ribosome binding sites (7 cyanobacterial and 6 from *E. coli*), 8 inducible promoters (5 native and 3 from *E. coli*), enhancer-, insulator- and transcriptional terminator elements have been designed, constructed and are being evaluated for an optimized metabolite production in this strain. The aim is also to explore the effect of the order of the coding genes on the translation efficiency; therefore we have chosen 3 different fluorescent reporter genes for evaluation. In the assembly process we use a set of plasmids and a specific gene-brick system; subsequently these constructs will be integrated in the genome of the cyanobacterium or alternatively maintained on an autonomously replicating artificial plasmid.

On metabolic level, our goal is to establish an efficient production of desired end-products directly from CO_2 and water using sunlight as energy. In practice this means targeted modification of the host to maximize the photon conversion efficiency and metabolic flux towards the production of pyruvate. Since pyruvate is one of the central metabolites, several metabolic modifications have been accomplished to direct and enhance the metabolic flux towards this chemical. Two native reporter compounds have been chosen to serve as markers for the pyruvate production efficiency: lactate and ethanol. Besides these, we are monitoring sucrose production to evaluate the effect of the alteration of the photosynthetic electron transport pathway. In order to divert the metabolic flux from non-essential photosynthetic- and biochemical processes towards target products, several genes and subsequently pathways have been altered that are related to alternative photosynthetic electron transport, glucose metabolism and Calvin cycle.

With this comprehensive and extensive study we aim to design an efficient cyanobacterial strain with an improved biotechnological potential.

Improving RNA Regulator Dynamic Range with Dual Transcriptional Translational Control Mechanism

Westbrook, A., Cornell University
Lucks, J. B., Cornell University

RNA regulators are emerging as powerful tools for building transcriptional cascades and logic circuits. However, the dynamic range of RNA transcriptional regulators still doesn't match that of their protein counterparts. This can lead to propagation of incomplete repression and cause circuit leak. Here we demonstrate how transcriptional regulators can be modified to regulate both transcription and translation, increasing repression from 85% to 98% and activation from 11 fold to over 900 fold, using transcriptional termination and RBS sequestration in a single compact RNA molecule. In order to build sophisticated RNA-only circuitry, orthogonal regulators are necessary. Initial attempts to engineer multiple regulators resulted in major crosstalk. However, by engineering the antisense to be more selective, the orthogonality was greatly improved. Additionally, we show that these modified, dual control regulators can be used in transcriptional cascades to mitigate circuit leak. As synthetic biology moves beyond parts engineering to networks, we anticipate these regulators be useful for engineering sophisticated circuits.

Accelerating Genome Engineering in E. coli: From Single Step Pathway Integration to Trackable Genomic Diversity

Bassalo, M.,
Garst, A., University of Colorado Boulder
Pines, G., University of Colorado
Gill, R. T., University of Colorado Boulder

Engineering the *E. coli* genome for a broad range of biotechnology applications commonly require two tasks: the ability to import heterologous functions stably into the host genome and to search the genome space efficiently for beneficial mutations affecting the phenotype of interest. We facilitated the process of importing new functions by developing a highly efficient CRISPR-based strategy that allows integration of entire metabolic pathways. This strategy accomplishes integration in a single step without requirement of selectable markers, maintaining high efficiency in distinct genomic positions. Using this strategy, we integrated a 10 kb construct to implement isobutanol production in a single day in *E. coli* with 50% efficiency. Such capability will increase the throughputs of testing metabolic pathways for interesting biological functions. To efficiently search the genome for beneficial mutations, we previously developed a technology (CREATE) that allows generation of thousands of rationally designed genomic mutations using multiplexed oligonucleotide synthesis and cloning. The fitness of each mutant can then be tracked in parallel using the CREATE plasmid as a trans-acting barcode. We applied this technology to introduce more then 50,000 mutations across the genome and identified enriched mutants in libraries spanning from a single protein to whole genome scale. We are now improving the tracking strategy by introducing a barcode recording element into the CREATE plasmid. As a result, each designed genomic edit is accompanied by the integration of a small barcode into a neutral site in the genome, maintaining a log of previously introduced mutations in the cell. We demonstrated high correlation between genomic edits and barcode recording. This technology allows trackable generation of combinatorial mutants as well as generation of portable libraries that can be shuttled to other organisms. Together, both of these technologies will facilitate genome engineering in *E. coli* and accelerate the process of generating interesting strains from both a scientific and biotechnology perspective.

Rational Membrane Engineering Increases E. coli Robustness and Production of Biorenewables

Tan, Z., Iowa State University
Yoon, J. M.,
Chen, Y., Iowa State University
Lombardo, J., University of Missouri
Black, W.,
Nielsen, D. R., Arizona State University
Shanks, J. V., Iowa State University
Jarboe, L., Iowa State University

Production of fuels and chemicals by using renewable material is limited by toxicity from inhibitors in feedstock and end-products, many of which generally cause membrane damage to host strains. Herein, three different strategies were performed to construct robust strains with increased production of biorenewables by engineering membrane of *E. coli*.

Firstly, novel trans unsaturated fatty acids (TUFA) were produced and incorporated into the membrane of *E. coli* MG1655 by expression of cis-trans isomerase (Cti) from *Pseudomonas aeruginosa*. While the engineered strain was found to have no increase in membrane integrity, a significant decrease in membrane fluidity was observed. As a result, tolerance to exogenously added octanoic acid and octanoic acid production were both increased relative to the wild-type strain. This membrane engineering strategy to improve octanoic acid tolerance was found to require fine-tuning of TUFA abundance. Besides improving tolerance and production of carboxylic acids, TUFA production also enabled increased tolerance in *E. coli* to other bio-products, e.g. alcohols, organic acids, aromatic compounds, a variety of adverse industrial conditions, e.g. low pH, high temperature, and also elevated styrene, another versatile bio-chemical product. TUFA permitted enhanced growth due to alleviating bio-product toxicity, demonstrating the general effectiveness of this membrane engineering strategy towards improving strain robustness.

Secondly, since a phospholipid molecule generally consists of a hydrophilic phosphate head and two hydrophobic fatty acid tails and most previous engineering studies attempted to engineer membrane fatty acid tails, little was known about effect of phospholipids head engineering. To this end, we proposed the idea that engineering phospholipid head to booster the robustness of microbes. Specifically, key enzymes candidates which are responsible for different phospholipids biosynthesis were chosen and engineered in *E. coli* MG1655. Results showed that increase of phospholipid A significantly increased membrane integrity and membrane electrochemical potential. In the case of fatty acids, this membrane engineering was also found to increase both tolerance to exogenously added fatty acids by 29.1% at most and fatty acids production by 42% relative to the wild-type strain. There was a positive relationship between phospholipid A content and tolerance increase to fatty acids. Besides increasing tolerance and titer of carboxylic acids, phospholipid A production also enabled increased tolerance of *E. coli* to other important bio-products, e.g. ethanol, hexanol, toluene, styrene,

Moreover, tolerance to several important inhibitors in hydrolysate of lignocellulose, e.g. furfural, hydroxymethylfurfural (HMF), acetate and a variety of adverse industrial conditions, e.g. low pH, high temperature, were also elevated by phospholipid A production.

Thirdly, besides phospholipids, membrane bound proteins are also important components of cell membrane, while we know little of their engineering effects on membrane physical properties and strain robustness. Herein, two membrane protein candidates, X and Y were engineered in *E. coli* MG1655. Results showed that, individual disruption of X and overexpression of Y significantly increased membrane integrity during fatty acids production. Besides its individual effect, the two engineering manipulations have a synergistic role in further increasing fatty acids production. These engineering expands the understanding of membrane proteins in strengthening membrane properties and strain robustness.

Automated Design for Synthesis and Assembly

Oberortner, E., Lawrence Berkeley National Laboratories
Cheng, J. F., Lawrence Berkeley National Laboratories
Hillson, N. J., Joint BioEnergy Institute
Deutsch, S., Lawrence Berkeley National Laboratory

Not every DNA sequence that one could conceive can be effectively manufactured using state-of-the-art DNA synthesis technologies. Over the last years and through the synthesis of many million base pairs, the DNA synthesis community has acquired knowledge regarding the feasibility of manufacturing synthetic DNA. Empirical analysis have enabled to identify common sequence features that impact the success/failure rates of synthetic DNA. Such features include, for example, repeating sequences, sequences with low or high GC content, or sequences that contain restriction sites.

A current major bottleneck for an efficient cost and time effective DNA synthesis process is the lack of incorporation of the DNA synthesis knowledge into the design process. Non-compliant designs result in severe process inefficiencies as: (i) sequences are often rejected by DNA synthesis vendors, (ii) when not rejected, they are synthesized at a higher cost or longer cycle time.

One major collaborative goal of DOE JGI, JBEI, and FutureBio is to develop scalable solutions to design and build complex DNA constructs for strain characterization in parallel with a high success rate. Therefore, we have developed the Sequence Polishing Library (SPL) that consists of various software tools that enable the automation of a "Design for Synthesis and Assembly" workflow. The SPL enables the verification of DNA sequences against specified synthesis constraints, such as repeats or GC content. In case of violations, careful modifications can be suggested that alter the DNA sequence without changing its functionality, such as swapping codons in the DNA's coding regions. The SPL partitioning tool decomposes DNA sequences that exceed the maximum length of synthesis into synthesizable building blocks. The challenge here is to find appropriate overlapping sequences between properly sized building blocks in order to assemble the synthesized building blocks effectively. Thus, the partitioning tool enables the user to configure the characteristics of the overlaps, which are mostly determined by the utilized assembly protocol, such as length or melting temperature.

Every tool of SPL supports the exchange of sequence information using common data exchange formats, such as FASTA, GenBank, and SBOL. The SPL Web Application enables users to learn and utilize each tool interactively in an easy manner. In addition, every tool provides a RESTful API in order to programmatically invoke its functionality and to integrate it into automated workflows.

Functional Coordination of Iron-Sulfur Clusters by Selenocysteine

Miller, C.,
Broecker, M.,
Ho, J.,
Söll, D., Yale University

Iron-sulfur clusters are widely recruited by proteins throughout nature, playing a critical role in oxidative phosphorylation. Canonically, iron-sulfur clusters are coordinated by cysteine residues and (rarely) by histidine residues. Here we describe replacement of four cysteine residues required for iron-sulfur cluster coordination with selenocysteine (Sec), the 21st amino acid. This substitution was accomplished using an engineered tRNA designed for EF-Tu mediated Sec insertion through stop codon suppression.

In *E. coli* protein Formate Dehydrogenase H (FDH_H), residues Cys8, Cys11, Cys15, and Cys42 coordinate a [4Fe-4S] cluster essential for protein function. We show that single substitution all four residues, and simultaneous substitution of up to three residues, results in production of functional protein *in vivo*. Kinetic assays of proteins with a single residue replaced show no significant change in activity.

Watching the Ribosome Making Peptide Bonds Between Unnatural Amino Acids

Melnikov, S., Yale University
Söll, D., Yale University

Here we describe a strategy, based on high-resolution ribosome crystallography, which allows one to catch the ribosome in the act of protein synthesis with unnatural amino acids. This approach paves the way to understanding how the ribosome discriminates (or fails to discriminate) natural from unnatural substrates, and how the active site of the ribosome can be remodeled to accommodate unnatural building blocks and produce artificial protein-like structures.

Open Vectoreditor: Freely Open-Source Javascript Web Application for DNA Visualization, Annotation, and Editing

LaFrance, S., Lbnl
Hillson, N. J., Joint BioEnergy Institute
Plahar, H., Joint BioEnergy Institute, Lawrence Berkeley National Laboratory
Coble, J., JBEI
Rich, T.,
Nixon, J., UC Berkeley

Open VectorEditor is a web application for the visualization, editing, and annotation of DNA sequences. It offers features typical of DNA sequence editors, including virtual restriction enzyme digest gel-electrophoresis visualization, melting temperature calculation, and support for (and inter-conversion between) several DNA sequence file formats (including FASTA, Genbank, and SBOL).

Recently, we have rewritten VectorEditor from the ground up, starting from an Adobe Flash implementation, to a modern javascript web application framework (React / Cerebral / Baobab), ensuring VectorEditor's longevity and enabling its maintenance and continuous development. Additional benefits of the new javascript implementation include touchscreen-friendly features and accessibility for mobile device users, which are increasingly relevant as researchers begin to more commonly use tablets in the wet lab for experiment protocol display and lab book note-taking.

The new VectorEditor javascript code-base remains freely available as an open source project (under a liberal BSD license), and any user can customize the interface for their own organization's needs, use VectorEditor as a standalone tool, or embed it into other web applications.

VectorEdit or:

Open-Source Javascript Web Application for DNA Visualization, Annotation, and Editing

Sarah A LaFrance, Nathan J. Hillson, Jacob Coble, Thomas Rich, Hector Plahar, Joshua Nixon

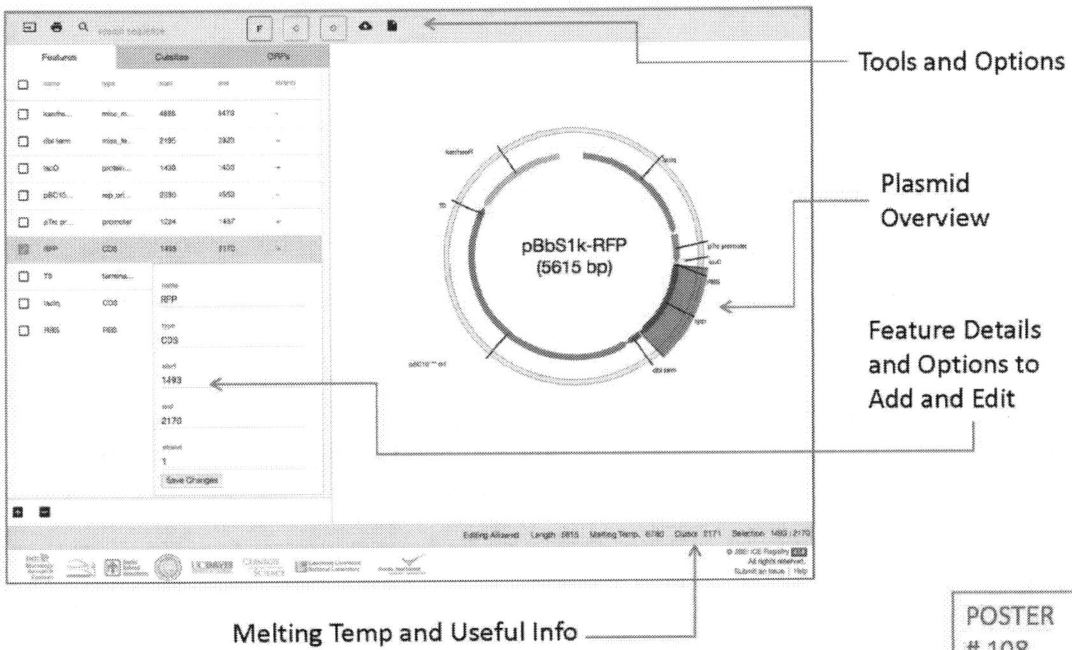

POSTER
108

Mathematical Modelling Determines Post-Segregational Killing Efficacy of Toxin-Antitoxin Systems in Commensal Escherichia coli Nissle 1917

Fedorec, A. J. H., University College London
Ozdemir, T., University College London
Barnes, C., University College London

Plasmids are commonly used as vectors for introducing synthetic circuits into bacterial hosts. These circular pieces of extra-chromosomal DNA are easily editable and provide the modularity that is one of the central goals of synthetic biology. However, ensuring plasmids are stably inherited can be problematic. Uneven distribution of plasmids within a cell at the time of division leads to the creation of plasmid-free cells which often outgrow the plasmid-bearing population. Antibiotic selection, commonly used in the laboratory, cannot be used as a plasmid stability mechanism in environments such as the gastrointestinal tract, where the disruption of other bacterial communities is undesirable. Naturally occurring plasmids have evolved several mechanisms for enhancing plasmid stability, such as active partitioning and post-segregational killing. Here we implement deterministic and stochastic models describing the population dynamics of plasmid-bearing bacteria. Using flow cytometry and automated clustering with mixture-models, we quantify the change in the plasmid-bearing fraction of populations of the commensal bacteria *Escherichia coli* Nissle 1917 in the presence of different toxin-antitoxin systems. Bayesian methods are used to fit our models to the data, with a GPU accelerated tau-leaping implementation enabling simulation of the stochastic model. Our methods allow for the separation of growth rate differences, between plasmid-bearing and plasmid-free bacteria, from the plasmid stability parameters. We also demonstrate the difficulty of separating the rate of plasmid loss from the efficacy of an employed post-segregational killing mechanism. Further, our results show that the non-native Axe/Txe toxin-antitoxin provides stability beyond the period for which native, and more commonly used, Hok/Sok was effective. As synthetic biological applications progress to more complex environments consideration for the stability of synthetic circuits becomes essential. We anticipate that the use of mathematical and statistical modelling will become increasingly useful to rationalise the design of synthetic circuits as more components become available, and to predict their function as the complexity of the target environment increases.

An Update on Synbis â€“ the Synthetic Biology Information System

Bultelle, M. A., Imperial College
Sainz de Murieta, I., Imperial College
Kitney, R. I., Imperial College

Systematic rational design — also known as design-build-test cycle — has become key in the development of synthetic biology towards an engineering discipline. Typically, the goal is to build a new pathway or modify an existing one, to endow it with a set of qualitative or quantitative properties (for instance a response function to a given chemical species that falls within an acceptable range) , while reliably operating under a list of constraints (host, environmental conditions etc).

A common concept in engineering is that systems can be rapidly produced from the combination of standardized components. For this to happen in synthetic biology, there is a strong need of developing large catalogues of fully characterised parts and make them available in public registries, so they can be picked and reused for design.

Ideally, BioParts should be characterised in a range of contexts (plasmids, hosts, medium…) to the highest standards - using validated protocols ensuring a high level of reproducibility. Also, the catalogues must be available online, offer programmatic access to their data, and support widely-used data standards to integrate with existing and future CAD tools.

We present here the latest developments on SynBIS. SynBIS is an information system developed at the Centre for Synthetic Biology and Innovation (CSynBI) at Imperial College to support its multistage characterisation pipeline of biological parts, effectively:

- supporting the collection of information on constructs, construction and experimental protocols
- supporting and monitoring data acquisition
- automating data analysis, curation and the generation of Biopart datasheets
- compiling all collected and generated data into the SynBIS catalogue

As part of the design of SynBIS, we have striven to support and develop best practices in term of data information standards and reproducibility. SynBIS supports established data standards (SBOL for construct description; SBML for modelling) and nascent standards for protocol description. To enable data acquisition at scale, we have developed the first data-acquisition standard in synthetic biology (DICOM-SB) – basing it on the highly successful DICOM standard used in medical informatics.

Finally, we have developed a flexible datasheet model – based on the workflow of a canonical characterisation pipeline and serialised in XML – linking a BioPart to all the data generated in a characterisation experiment: raw data, analysis results (presenting qualitative and quantitative information describing the part's behaviour) and the metadata required to ensure reproducibility of the experimental and analysis results.

The SynBIS parts catalogue (www.synbis.bg.ic.ac.uk) is an open access resource. It is organised around datasheets (and Biopart performance), and can be searched (beside standard parameters such as name or sequence) by placing conditions on the input-output function of a part.

Characterisation information is offered in human readable formats —HTML5 and pdf— as well as a machine readable formats —SBOL, SBML,XML and DICOM-SB. Direct programmatic access is available through a public API to all analysis, simulation and CAD software tools.

Towards the First Data Acquisition Standard in Synthetic Biology

Sainz de Murieta, I., Imperial College
Bultelle, M. A., Imperial College
Kitney, R. I., Imperial College

One of key drivers for the transformation of synthetic biology into a fully-pledged engineering discipline, will be the explosion of data captured and analyzed— thanks in no small part to the availability of new imaging modalities (generating ever larger files) and the development of high throughput platforms. At present there is no established standard supporting data acquisition. Developing such a standard is therefore of the utmost importance for the field.

We present here DICOM-SB, an extension to the Digital Imaging and Communications in Medicine (DICOM) standard and *de facto* standard in biomedicine. Developing such an extension of DICOM is advantageous, as DICOM has properties that would benefit any data acquisition standard, namely:

1. It is built on a modular, extensible data model for the experimental process.
2. Its binary representation optimizes data storage.
3. It provides services oriented to automate the exchange of information between modalities and repositories.

Also, it has long term success and proven industry orientation. Therefore any extension can be expected to enjoy swift adoption by hardware manufacturers and the engineering community.

Originally developed at Imperial College to support its multistage characterisation pipeline of biological parts (called SynBIS) and work by partners such as NTU Singapore, DICOM-SB has now matured to the point a formal submission to the DICOM committee is in process. The main achievements regarding DICOM-SB are as follows.

1 – Data model. The DICOM-SB Hierarchical Information Model was constructed by modifying DICOM's model and adding synthetic biology-specific Information Entities, including *Tranformation* (of a host by a set of components/bioparts), *Experiments* (performing the procedures required to analyze the change of behavior that the integrated components produce in the host), *Compartments* (wells), *Stimuli* , *Equipment* and *Series*.

2 – Supporting new modalities. A new Synthetic Biology module (called the SBRD) has been constructed to model some of the modalities used in synthetic biology, but not already present in DICOM – specifically plate reader data and flow cytometry. The SBRD was constructed both by reusing the Waveform Information Module of DICOM, and defining new objects for the metadata.

3- Supporting communication. DICOM offers services to allow the communication of data objects between different Application Entities. In a data acquisition context, there is a need for a service that stores the acquired object into a data repository - we have developed a web service that implements a *DICOM Message Service Element* (DIMSE): the *Store* service.

It is important to note that far from competing with other synthetic biology standards, DICOM-SB actually actively complements them. This coexistence is demonstrated within our SynBIS platform where SBOL, protocol, SBML and DICOM-SB files are bound by our datasheet model.

Medical imaging has benefited from this automation for over the last decade thanks to DICOM being embedded in most of their commercial modalities. We believe the adoption of a similar standard by the synthetic biology modality suppliers would be a key milestone for synthetic biology — particularly in relation to interoperability and industrial translation.

Syntrack: Streamline Complex DNA Assembly Processes

Meng, J., DOE Joint Genome Institute
Tarver, A., DOE Joint Genome Institute
Hamilton, M., DOE Joint Genome Institute
Bendall, M., DOE Joint Genome Institute
Simirenko, L., Joint Genome Institute
Hillson, N. J., Joint BioEnergy Institute
Cheng, J. F., Lawrence Berkeley National Laboratories
Deutsch, S., Lawrence Berkeley National Laboratory

The assembly of multiple DNA fragments into final constructs is a complex multi-step process that requires close integration with laboratory automation systems and needs to incorporate QC/QA outcomes at multiple steps. In addition when dealing with large batches of DNA parts, which result in many different deliverable constructs, assembly steps often require re-work. Hence, tracking the required laboratory steps, QA/QC inputs, the location of the intermediates, and the status of the deliverables is a complex process. To facilitate all critical DNA construction operations we have developed the SynTrack system at JGI. Syntrack is a dynamic, workflow-driven data management system that tracks the DNA build process starting from a standardized design file that defines the final construct deliverables to the hierarchical relationships of the DNA parts and intermediate fragments. Syntrack first generates DNA ordering forms, tracks when DNA parts are received, and then proceeds through all the subsequent PCR, assembly and cloning steps. The system captures QC outcomes, defines location in plates, and creates interactive plate maps that operators can query. Driven by DNA assembly workflows, the system leverages all construct design and production operations, and integrates with laboratory automation equipment throughout the DNA construction process. For example, it first generates instructions for setting up assembly and cloning reactions with the Labcyte ECHO. Then, it integrates outputs from plate readers, colony pickers, and nextgen sequencers. The system supplies a flexible method to allow for changes in the assembly process and the addition of re-work at any stage, allowing the ability to move backwards and forwards throughout the workflow. During all assembly steps, the status of the hierarchal DNA fragment structures is closely monitored and updated accurately with a date/time stamp of every fragment status change or action.
The source code is written in Django web framework. The code is open, modular, and optimized for rapid data entry with CSV or Excel inputs. The complex requirements of all DNA assembly data tracking have been condensed into more simple, secure interfaces that provide laboratory operators with the ability to see the status of their DNA constructs, anywhere in the process. The application integrates data from multiple sources within the assembly line with automated generation of plate map and well instructions. It handles DNA Assembly requests at a variety of levels, from an individual fragment to a batch of constructs. SynTrack also presents visual warnings or can deny access, when proper event sequence is not followed. We show that SynTrack supports ad hoc and modular application design. Without significant efforts, Syntrack produces a superior maintainable application that can adapted to changes in the workflow as technology evolves. Syntrack creates productivity enhancements and better process visibility to support a scalable DNA foundry operation.

Nanoparticle Delivery Biofactory Using E. coli as the Synthetic Host

Pasula, R. R., Nanyang Technological University
Lim, S., Nanyang Technological University

Nanoparticles exhibit unique physical, mechanical, optical and electromagnetic properties owing to the increase in surface area to volume ratio as compared to the bulk material. The unique characteristics pave ways to a wide array of applications such as detection of proteins, tumor destruction, tissue engineering and MRI contrast enhancement. Biological synthesis of these nanoparticles using bacterial hosts is being explored as an alternative to chemical synthesis as it is environmentally friendly. Protein nanocages, when employed as templates facilitate production of size controlled, monodispersed nanoparticles. Bacterial hosts harboring these protein nanocages could be engineered with surface ligands for tissue-specific accumulation for enhanced imaging in MRI applications.

Archaeoglobus fulgidus Ferritin (*AfFtn*) a self-assembling, 24-subunit iron storage protein which maintains iron homeostasis, is used as a template for iron nanoparticle synthesis in *E. coli*. To enhance iron uptake and reduce iron efflux, *E. coli* has been further engineered to express iron influx transporter protein (*feo*B) while knocking out iron efflux protein (Δ*fie*F). In vivo iron loading in *AfFtn* was achieved by incubating *E. coli* with excess iron in the growth medium. The iron loaded host is further targeted towards colorectal cancer cells for nanoparticle delivery.

Catalytic Enhancement of Enzymatic Pathways on a Nanoparticle Surface

Vranish, J., NRC Fellow at Naval Research Lab
Oh, E., US Naval Research Laboratory
Walper, S.,
Susumu, K.,
Medintz, I., US Naval Research Laboratory

Enzymes are of tremendous interest for the industrial production of numerous products due to their incredible ability to carry out chemical transformations that are highly specific and often quite difficult, if not impossible to replicate using synthetic means. However, their use has been largely limited due to their instability, purification costs, and difficulty in separating the enzymes from products for re-use. Efforts at enhancing the performance of enzymes and their viability in industrial syntheses have often focused on the areas of protein engineering and enzyme immobilization. Protein engineering, while successful in certain cases, is often quite difficult and often produces only modest improvements in catalytic activity or preferences for somewhat analogous substrates. Immobilization of enzymes on the other hand often leads to decreased catalytic activity relative to free enzymes, thus limiting its overall effectiveness. Recently our lab has demonstrated that immobilization of enzymes on quantum dot (QD) surfaces is often able to enhance the activity of enzymes, including a "perfect" diffusion-limited enzyme. Contrary to many other methods of enzyme immobilization, this technique is quite facile, requiring only a histidine tag to attach the protein to the inorganic surface. We have now investigated the effect of QD immobilization within a multienzyme cascade, using freely diffusing glucose oxidase (GOX) coupled to QD-immobilized horseradish peroxidase (HRP). The immobilized HRP displays enhanced enzyme kinetics, and appears to be activated either by substrate accumulation near the QD surface or improvement in the catalytic properties of HRP. By optimizing the ratio of GOX to HRP, we were able to preserve the catalytic enhancement of HRP, demonstrating that the presence of additional catalytically-coupled solution phase enzymes with the QD immobilized enzyme doesn't appear to interfere with the QD-dependent enzymatic enhancement. We are now extending the power of enzyme immobilization on QDs by binding multiple enzymes from glycolysis to a single surface to determine whether we can further enhance the activity of the system by minimizing diffusion between enzymes. As part of this study we are also investigating whether QD-dependent enzymatic enhancement of a histidine-tagged protein can be achieved without first purifying the enzyme from the cell lysate. Additional efforts are underway to attempt to organize QDs on larger inert surfaces in an attempt to create additional surface area for enzyme binding and colocalization. It is our hope that these efforts will lead to a widely applicable method for immobilizing enzymes while preserving and enhancing their activity within complex enzymatic pathways. By doing so, we hope to create better "green" catalysts that can be used for the industrial synthesis of numerous products.

Engineering E. coli for Dual Color Detection of 2,4,6-Trinitrotoluene

Harbaugh, S., Air Force Research Laboratory
Martin, J.,
Salis, H., Pennsylvania State University
Goodson, M.,
Weinstein, J.,
Kelley-Loughnane, N.,

Genetically engineered microorganisms have been applied for accurate and sensitive detection of target analytes in a diverse variety of fields such as medicine, environmental monitoring, defense, food processing and safety. Here we combine *in vitro* selection and computational design to identify a novel RNA regulatory sensing element: a synthetic riboswitch that initiates a recombinase-based memory switch in *E. coli* cells in response to 2,4,6-trinitrotoluene (TNT). TNT is one of the most commonly used explosives for military and industrial applications; its detection is important from a security perspective and due to its high toxicity.

Riboswitches are typically composed of two functional domains: an aptamer domain that binds to a specific analyte; and an expression platform that controls the expression of a downstream gene via conformational changes that are induced by analyte binding to the aptamer domain. First, we identified a TNT-binding aptamer using a magnetic bead-based SELEX (systematic evolution of ligands by exponential enrichment) technique specifically developed for selection of structure-switching aptamers for small molecule analytes. This method is advantageous because it does not require modification of the analyte. Modifying a small molecule can block groups that are critical for binding to an aptamer, which can affect the selectivity of identified aptamers. Next, we applied computational RNA design to facilitate and expedite converting the resulting TNT aptamer into a functional riboswitch. This was based on a recently developed statistical thermodynamic model that predicts the sequence-structure-function relationship for translation-regulating riboswitches that activate gene expression inside cells. In contrast to commonly used *in vivo* selection and screening approaches that require analysis of a large number of riboswitch clones, computational design resulted in two functional riboswitches selected from the top five predicted sequences when tested in *E. coli* cells. These two riboswitches, TNTrs14 and TNTrs15, demonstrated activation ratios of 19.2 (± 5.0) and 18.8 (± 0.3), respectively. In the cell-based sensor, TNTrs14 and TNTrs15 were placed upstream of FimE recombinase coding sequence in one plasmid, and a second plasmid carried a fluorescent reporter composed of two fluorescent protein genes (mKate2 and GFPa1) flanking an invertible DNA segment (*fimS*) containing a constitutively active promoter. Without an analyte, the fluorescent reporter constitutively expresses green fluorescent protein (GFPa1). Addition of TNT initiates translation of *FimE* causing unidirectional inversion of the *fimS* segment and constitutive expression of red fluorescent protein (mKate2). This dual color cell-based sensor allows not only the detection of the presence of the target analyte, but also determination of the viability and localization of an uninduced sensor. By leveraging parallel computational modeling with construction and testing we were able to expedite the design, build, test cycle to develop a sensor-circuit device with ‘real-world’ applications.

This research is funded by the Air Force Office of Scientific Research.

A Genetically Encoded Reporter for Diffusion Weighted Magnetic Resonance Imaging

Mukherjee, A., California Institute of Technology
Wu, D., California Institute of Technology
Davis, H., California Institute of Technology
Shapiro, M. G., California Institute of Technology

Background: The ability to image gene expression within the context of intact, living mammalian organisms is critical for facilitating the eventual clinical translation of engineered genetic circuits and cell□based therapeutics. However, most genetically encoded reporters, based on fluorescent and luminescent proteins have limited utility in this respect due to the poor penetration of light into deep tissues. In contrast to optical techniques, magnetic resonance imaging (MRI) enables the acquisition of *in vivo* images with excellent depth penetration and high spatial and temporal resolution. However, existing MRI reporter genes, based primarily on metal-binding proteins or chemical exchange saturation transfer probes, are limited by their reliance on metal ions or relatively low sensitivity. In this work, we introduce an entirely new class of genetically encoded reporters for MRI that works by altering water diffusivity in cells. Specifically, we show that overexpression of the human water channel aquaporin 1 (AQP1) produces robust contrast in diffusion weighted MRI by increasing effective water diffusivity without affecting cell viability. Low levels of AQP1 expression (< 1 μM), or mixed populations comprising as few as 10% AQP1-expressing cells produce contrast that is readily observable using MRI. Finally, we demonstrate the utility of AQP1 *in vivo* by imaging gene expression in intracranial tumor xenografts.

Results: To evaluate AQP1 as a genetically encoded reporter for diffusion weighted MRI, we used lentiviral transfection to generate several mammalian cell lines stably overexpressing this channel or GFP. AQP1-expressing CHO, U87 and Neuro2A cells showed 187%, 82% and 95% increases in water diffusivity, respectively, compared to GFP controls. Next, using doxycycline-controlled AQP1 expression, we demonstrated that AQP1 can report on varying degrees of gene expression in a dose□dependent fashion. Notably, AQP1 expression at submicromolar concentrations was found to be sufficient for enhancing water diffusivity by 53%, thereby, placing AQP1 among the most sensitive MRI reporters. Furthermore, to evaluate AQP1 as a viable reporter for imaging small subsets of labeled cells in heterogeneous populations, we developed Monte Carlo models for water diffusion in mixed cell populations, which suggested that AQP1-based increase in water diffusivity should be evident even in low (~ 10%) AQP1□labeled cell fraction scenarios. Consistent with these predictions, our experiments revealed that as few as 10% AQP1 expressing cells is sufficient to enhance overall water diffusivity by 21%. Finally, we demonstrated that AQP1 can produce induction□dependent contrast in mouse tumor xenografts, with the average intensity in AQP1 tumors changing by 41% following doxycycline injection, compared to control tumors.

Conclusion: Our results establish AQP1 as the first genetically encoded reporter for diffusion weighted MRI. Importantly, as a sensitive reporter that can be readily imaged in heterogeneous cell populations and low cell-fraction scenarios, AQP1 appears ideally suited to fulfil the need for a sensitive, nontoxic reporter for labeling and tracking immune and stem cell-based therapeutics. Overall, the high performance, biocompatibility, and engineering capacity of aquaporin-based reporters will enable a multitude of applications ranging from basic biological studies to the noninvasive evaluation of cell-based devices and engineered gene circuits in the context of preclinical animal models.

A Hacked 3D Printer Towards the Printing of Materials through Biological Organisms

González, L. M., Carnegie Mellon University
Nielsen, A.,
Wallace, A., Massachusetts Institute of Technology
Voigt, C. A., University of California, San Francisco

Over the last couple of years, 3D printing technology has received a lot of attention and this is not a surprise as the technology has an unlimited number of applications from a materials perspective. Addictive manufacturing surpasses subtractive manufacturing, as printing of intricate structures is now possible with a simple 3D CAD model. With this project, we are seeking to print 3D composite materials with multifunctional capabilities using synthetic gene networks to drive the printing process via encoding the materials to print at the DNA level in microbial organisms. To do the printing part, we modified a commercially available 3D printer, the MakerBot Replicator 2X. We first modified the printer's extruder head to enable printing of a blend of agarose and live cells from two individual reservoirs instead of the standard Acrylonitrile butadiene styrene (ABS) or Polylactic acid (PLA) plastics reel from a spool. Using an optical interrupter, we detect when the piston in the slider-crank configuration is moving up and fill the chamber with the agarose/cells blend that eventually carves into the structures. A solenoid valve, normally closed, controls the agarose input and a DC liquid pump (and a second solenoid) controls the cell input via an Arduino microcontroller. When the piston is retracting (moving up), we applied a long enough time delay on the solenoid and DC liquid pump to compensate for the loss in pressure in the mixing chamber. In addition, a set of check valves prevents the flow of these liquids back into their respective reservoirs. The piston displaces the agarose and cell blend through the nozzle as it is moving down. This design permits us to use an established Skeinforge Gcode generator with minor modifications. In this case the sensing of the piston position occurs concomitantly with the machine-inherited 3D printing process. For the agarose feeding system, we have a hermetically sealed chamber with one inlet for air and one outlet for the agarose to reach the printing head. To maintain the agarose in the molten state (upon exposing to 60°C) a magnet wire is wound around the tubing. The house air fills the chamber and because of the built-up pressure the agarose is propelled to this printing head. Using a rheometer, we characterized the gelation time, the melting temperature and the shear thinning properties of this agarose to ensure the agarose can support itself without distortion. From a synthetic biology perspective, we are developing a broad host CRISPRi system (based on light sensing and recombinases tools) to control the production of cellulose in *Gluconacetobacter xylinus* during the printing process.

Production of a Modular Toolkit for Creating Prosthetic Gene Networks within Yeast

Shaw, W.,
Ellis, T., Imperial College London
Wigglesworth, M., AstraZeneca

As the field of synthetic biology is maturing, it is becoming clear that it will have a major impact in clinical therapy, particularly with the design of prosthetic gene networks, which are being specifically built for the autonomous treatment of human disease without the limitations of traditional pharmacotherapies. These closed-loop biosynthetic sensor/actuator devices can functionally integrate and interface with the host metabolism in order to detect and manage disease states with the production of an appropriate therapeutic *in situ*. Using yeast as an engineerable testbed, I am generating a platform for developing prosthetic gene networks in a rational, bottom-up manner by creating a highly-characterised and standardised set of modular components, along with a well-defined workflow for rapidly iterating the design/ build/ test cycle. It is believed this new framework will yield systems with improved predictability for tuning to patient idiosyncrasies, provide a cheaper and faster strategy for creating prosthetic gene networks, and generate a variety of parts and devices that can be reused for future projects. To achieve this, I am refactoring major components of the native yeast pheromone response pathway under synthetic regulation, in an effort to identify the best strategy for linking medically relevant G-protein coupled receptors to an appropriate therapeutic response. Initial work has revealed that regulation of individual components can be used to alter the dose response characteristics of the signalling pathway, which can be used as an approach to network tuning.

Protease-Based Feedback for Overcoming Growth Rate and Enzymatic Queueing Limitations

Tomazou, M. III, Imperial College London
Stan, G. B., Imperial College London

An increasing number of synthetic genetic circuits are relying on degradation tags for achieving fast dynamics that are not rate-limited by dilution due to cell division. In bacteria, the SsrA tags are commonly used to target proteins to the native-to-E. coli ClpXP protease. However, the over-expression of tagged proteins interferes with the physiological activity of ClpXP by overloading it and consequently reducing its degradation activity. This 'enzymatic queuing' effect has an impact on the anticipated dynamics of the expressed synthetic genetic circuits as tagged transcription factors become post-translational coupled. More importantly, this can lead to pathological host phenotypes, where cells grow slower or even fail to divide. In this work we are employing biomolecular feedback control in order to a) sense the load imposed on proteases, such as the native ClpXP or *mf*-Lon from *mycoplasma*, by increasing amounts of tagged proteins and b) automatically adjust the total level of proteases to compensate and alleviate the pathological effects caused by the post-translational coupling between the tagged species. The basic architecture of the network comprises a self-regulated protease. This is realized by either introducing a self-targeting tag on the protease, or through a protease-degradable transcriptional activator circuit. From an engineering design viewpoint, our in silico analysis has highlighted unique and interesting properties of this feedback architecture. First, the varying levels of tagged proteins, in the closed loop network, exhibit negligible sensitivities with respect to each other. In addition, the system is significantly more robust and shows faster recovery from transient perturbations, but more importantly it has reduced sensitivity to dilution rate changes. This suggests that our protease feedback mechanisms can be used to maintain protein levels at a stable, almost constant level during and between cell growth phases. Undergoing experimental work aims to implement, demonstrate, and inform further the modelling aspects of this work in order to optimize it's 'buffering' capacity. Potential applications include delivering a biomolecular controller that can significantly increase robustness in synthetic circuits in terms of perturbations and changes in cell growth rates and allow for the predictable fine-tuning of the dynamics of synthetic biology systems.

Efflux Pump-Mediated Antibiotic Resistance in Dynamic Environments

Langevin, A., University of Vermont
Dunlop, M., The University of Vermont

Antibiotic resistance has become a major public health concern as bacteria produce strategies to evade drugs, leading to recurring infections and overuse of antibiotics. Understanding the complex strategies bacteria employ to resist treatment could provide insight into different methods of targeting these evasive cells. There are many methods that bacteria use to achieve antibiotic resistance including the production of antibiotic-inactivating enzymes, the transfer of genetic material encoding for resistance genes, and the use of efflux pumps. Here, we focus on efflux pumps, which are membrane transporters that can export undesired compounds such as drugs out of the cell. Multidrug resistance pumps, such as AcrAB-TolC from *Escherichia coli*, are especially well known for their ability to export a wide variety of antibiotics. Although these cell-membrane proteins provide antibiotic resistance for cells, there are potential trade-offs, as there is a substantial cost to cells when pumps are expressed. An excess of pumps can alter the fluidity of the cell membrane and slow growth. We hypothesize that the ideal pump expression level involves an environmentally-dependent trade-off between the benefit of pumps and the cost of their expression. For example, if antibiotics appear frequently then high pump expression levels are required. In addition, the frequency and duration of antibiotic appearance will affect this trade-off. In this work, we evaluate the benefit of the AcrAB-TolC pump under a range of induction conditions and under variable antibiotic stresses. We assess the performance of cells with and without pumps when the antibiotics are presented in different forms – step, ramp, pulse – and when the pumps are induced with respect to the antibiotic levels. Our overall goal is to quantitatively determine the trade-off between resistance and hindered growth for cells containing these pumps. To achieve this we are co-culturing cells with and without the efflux pumps. The relative fraction of cells with and without the pumps can change with time under variable antibiotic and inducer conditions. This work will provide insight into how bacteria optimize the use of efflux pumps as strategic mechanisms for surviving in stressful conditions.

Secretion and Self-Assembly of Protein-Protein Conjugates from Bacillus Subtilis

Gilbert, C., Imperial College London
Ellis, T., Imperial College London

The design and construction of biological materials is a burgeoning field that promises to deliver novel nanostructures and polymers for diverse applications. Proteins have attracted considerable interest as substrates for biomaterial synthesis owing to their potential to bring together both favourable structural and functional biological properties. Using tools and techniques from synthetic biology, it is hoped that functional protein modules (e.g. catalytic activity, elasticity, mechanical strength) can be recombined to create novel biomaterials tailor-made for diverse applications. Further, through the use of genetic circuits complex polymer architectures can be pre-programmed into cells. We aim to develop a cell-based co-culture system for protein polymerisation, in which secreted recombinant protein monomers assemble into protein polymers and nanostructures outside the cell. Here we describe a method for the secretion and conjugation of various recombinant proteins from the Gram positive bacterium *Bacillus subtilis*. In addition, we show the applicability of this technology to the creation of thermo-tolerant enzymes and for the self-assembly of multi-enzyme complexes.

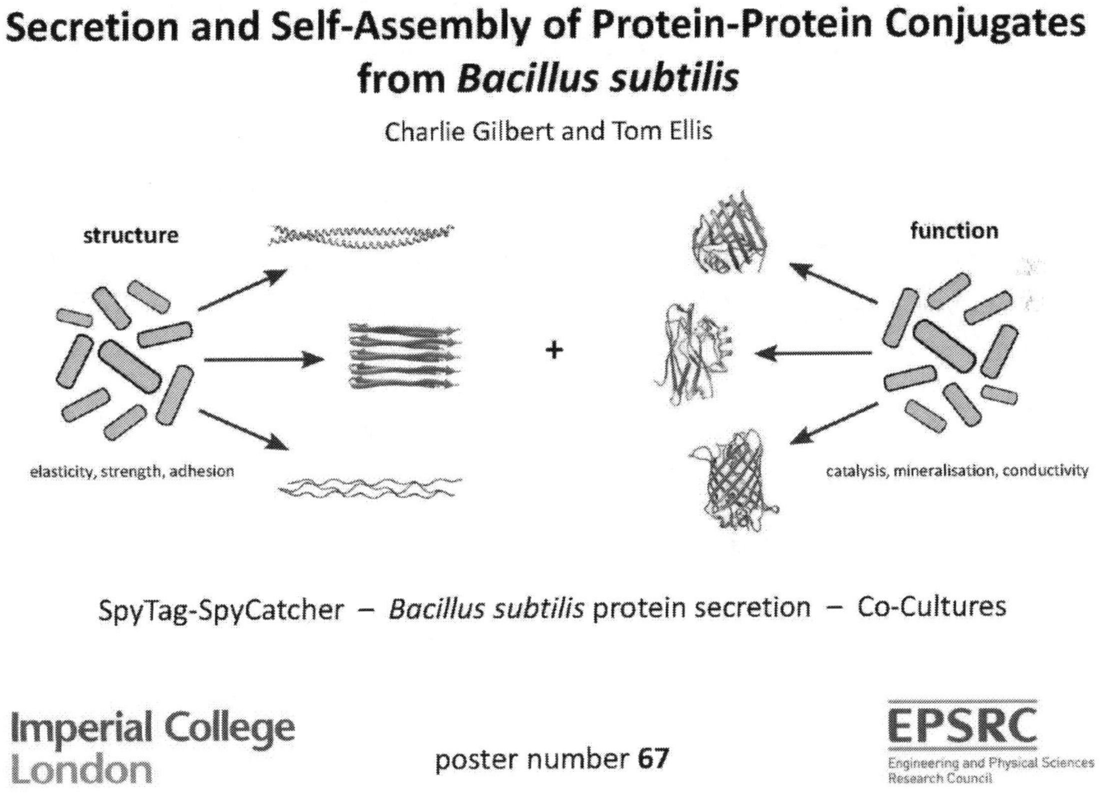

Engineering Efficient Specialist Strains for a Synthetic Multi-Species Microbial Consortium

Saleski, T., University of Michigan
Lin, X., University of Michigan

Use of synthetic microbial consortia can help to address some of the limitations of monoculture bioprocessing, including high metabolic burden on the cells and the need to optimize multiple pathways or functions in the same species. Design of synthetic microbial consortia, however, brings the challenge of identifying or engineering strains that can perform optimally within a mixed population and regulating the population composition. Here we present our work on developing methods to optimize microbes for performance in co-culture.

We are working on a synthetic co-culture of the cellulolytic fungus *Trichoderma reesei* and an *Escherichia coli* strain that has been designed to produce the drop-in biofuel isobutanol directly from cellulosic substrate in one consolidated bioprocess [1]. The *E. coli* strain in this system shows suboptimal performance compared to monoculture conditions, demonstrating lower yield and titer, higher byproduct formation, and high plasmid loss. We are developing new ways to generate *E. coli* strains optimized for co-culture performance. The first is a directed evolution approach involving screening for improved growth with a selection agent in co-culture with *T. reesei*. Second, we are taking a systems biology approach to compare the *E. coli* metabolic state in monoculture and co-culture in order to guide further genetic modification and improvement. Third, we integrate and optimize the copy number of the isobutanol production pathway in the *E. coli* genome to enable more stable performance.

[1] Minty JJ, Singer ME, Scholz SA, Bae C-H, Ahn J-H, Foster CE, Liao JC, & Lin XN (2013) Design and characterization of synthetic fungal-bacterial consortia for direct production of isobutanol from cellulosic biomass. *Proc Natl Acad Sci USA* **110**(36): 14592-14597.

Engineering *Escherichia coli* for more stable isobutanol production in co-culture with *Trichoderma reesei*

Tatyana Saleski, Xiangyu Zhao, Adam Krieger, Xiaoxia "Nina" Lin
Department of Chemical Engineering, University of Michigan, Ann Arbor

Consolidated Bioprocess Co-culture

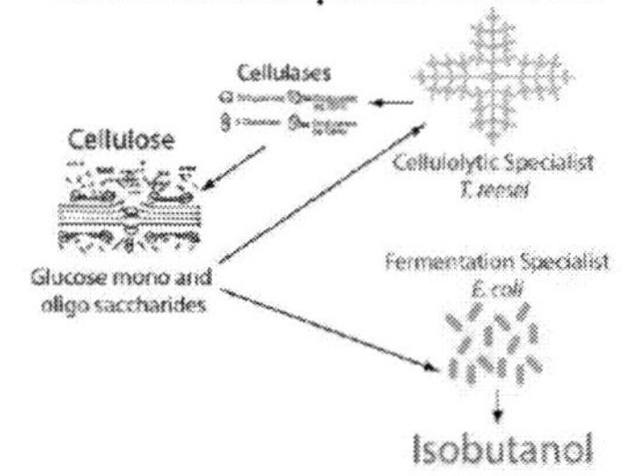

Conditions are non-ideal for *E. coli*:

- pH 6
- Low glucose
- Potentially inhibitory compounds (on real biomass)

Suffers from plasmid loss (Minty JJ et al., *PNAS*, 2013)

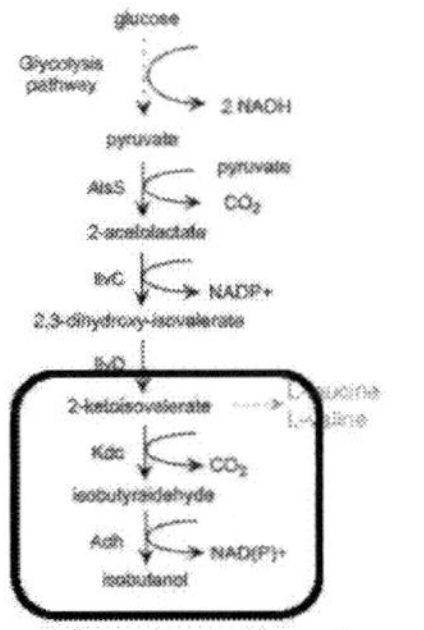

(Adapted from Astumi S et al., *Appl Microbiol Biotechnol*, 2010)

Goal: Integrate and optimize copy number by chemically inducible chromosomal evolution (Tyo KE et al., *Nat Biotechnol*, 2009) to improve co-culture performance

(Adapted from Tyo KE et al., 2009)

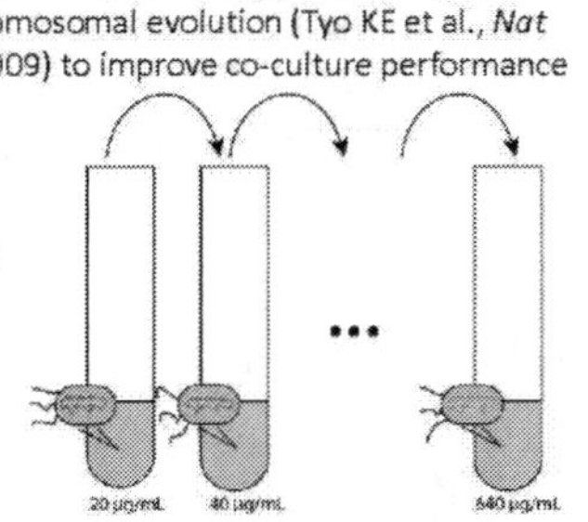

Increasing chloramphenicol concentration -> increasing copy number
Stabilize by deleting *recA*, then transform with pSA69 (*alsS*, *ilvCD*)

Copy number measured by qPCR of *cat* gene

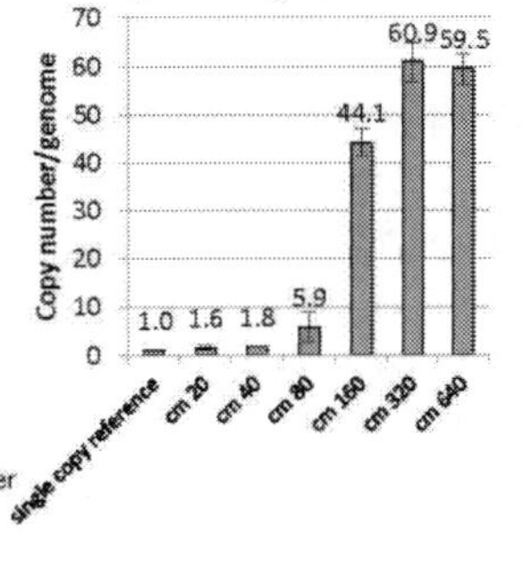

Integrated strain performance:

***E. coli* monocultures**

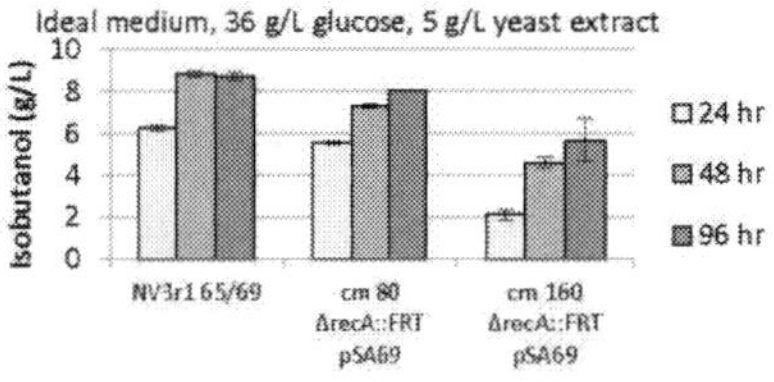

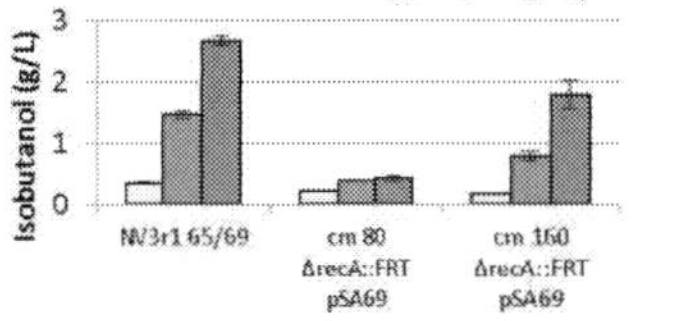

***T. reesei/E. coli* co-cultures**

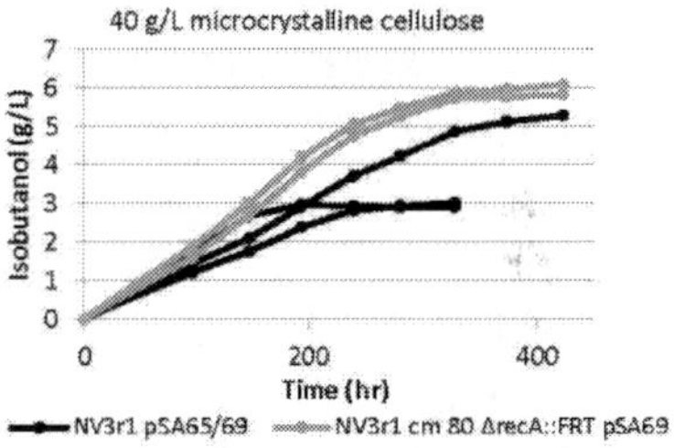

(further testing in progress)

Engineering a Leather Biomaterial

Jakab, K., Modern Meadow
Marga, F.,
Kaesser, R.,
Cassingham, D.,
Brown, E.,
Lee, J.,
Williamson, D.,
Budoff, S.,
Borden, J.,
Forgacs, G.,

Tissue engineering combined with synthetic biology aims to build controlled, functional, and multicellular biological structures. Here we demonstrate the application of these principles to produce novel biomaterials of animal origin, namely leather. A leather-like material does not require the complete anatomical skin of the animal; rather it requires a collagenous extracellular matrix (ECM). To construct this matrix in vitro, bovine fibroblasts are grown over 18 days, producing self-assembled ECM tissue sheets. Upon harvesting, the sheets are briefly seeded with another cell layer, then wrapped around a plastic mandrel with five additional sheets to form a multilayered tissue structure approximately 15 layers deep. This thick tissue construct is cultured for an additional three weeks to allow for layer-to-layer attachment and fusion via additional collagen synthesis and assembly. The resulting tissue is an approximately 20 x 5 cm, 0.5-1 mm thick sheet. Harvested samples proceed through proprietary tanning chemistry to produce finished cultured leather. The final tensile strength is 10-15 MPa, elongation at break approximately 30%, and samples are hydrothermally stable up to 95C, which is comparable to animal-derived leather. Our novel method of leather production has several unique features compared to traditional leather: products are prepared with minimal inorganic waste; leathers are not size-constrained; the entire process is performed in a closed and controlled environment; and novel material properties can be engineered.

High-Throughput Evolution of in Vitro Orthogonal Translation Systems Via an Integrated Platform Merging Cell-Free Protein Synthesis and Microfluidics

Kwon, Y. C., Northwestern University
Tran, T. M., California Institute for Quantitative Biosciences, University of California, San Francisco
Gan, R.,
Abate, A. R., California Institute for Quantitative Biosciences, University of California, San Francisco
Jewett, M. C., Northwestern University

The extraordinary power of protein biosynthesis in conjunction with site-specific non-standard amino acid (nsAA) incorporation into a growing peptide chain has opened a new frontier in the chemistry of life. Site-specific nsAA incorporation provides not only new protein function and structure, but also endows unique chemical properties. However, nsAA incorporation efficiency is about 1000x lower than that of natural amino acids in translation due to the challenge of reengineering the protein translation apparatus, a complex biochemical orchestra composed of nearly 80 macromolecules including proteins, RNAs, and cofactors. To address this challenge, we have developed a high-throughput platform for the evolution of an *in vitro* orthogonal translation system (OTS) that improves nsAA incorporation efficiency. For the synthesis of proteins containing nsAAs, we utilized the crude-lysate cell-free protein synthesis (CFPS) system. Coupling this protein biosynthesis platform to efficient microfluidic devices capable of droplet generation, merging, and sorting, we established a high-throughput evolution platform capable of screening a large library of translation machinery variants. We evolved one residue of the tRNA binding interface and two residues of the amino acid binding pocket of an aminoacyl-tRNA synthetase (aaRS) from *M. jannaschii* for *para*-azido-L-phenylalanine (pAzF) incorporation. Each droplet containing a variant of the aaRS was merged with a CFPS droplet. The resulting droplet synthesized super folder green fluorescence protein with 5 amber codons (sfGFP 5xUAG). Droplets were then sorted based on fluorescence. We achieved high incorporation efficiency (>95%) by screening 5 randomized consecutive amino acids for each residue hypothesized to be critical active sites for the aaRS. We expect this work to meet the growing demands for OTS evolution as well as accelerate broader efforts in protein engineering.

Using Business Process Management to Enhance Synthetic Biology Workflows

Reynolds, C., Imperial College London
Holdbrook, J., Imperial College London
Ainsworth, C. M., Imperial College London
Bultelle, M. A., Imperial College
Exley, K., Imperial College London
Sainz de Murieta, I., Imperial College
Kitney, R. I., Imperial College

Synthetic biology is based on engineering principles, and a key principle for the characterisation of parts is the consistency and replicability of workflow processes. This requires a method of specifying processes, modelling data structures, defining user interfaces and the integration with existing standards and with technologies, both computational and robotic. The method needs to allow the collection and analysis of as much data as possible.

Business Process Management (BPM) encompasses a field of techniques used to optimise workflows. In this study, BPM methods are used to model synthetic biology processes to standardise experiments, increase the ease of automation and allow analysis of workflow data. This study demonstrates an implementation of BPM for synthetic biology using the Bizagi BPM Suite, with the data structure stored in a SQL Server Database, and custom ASP.NET websites used for additional data handling and analysing.

As part of this integration, the workflow processes were first modelled using industry-standard Business Process Model Notation (BPMN), a widely accepted format for specifying processes in graphical form. The processes were then automated and combined to produce an application that could be accessed by an end user via a web-browser. Tasks in the process can be undertaken by an experimentalist or integrate directly with other technologies.

The benefits of using BPM include:

- Ease of constructing and altering processes with minimal programming.
- Defining simple user interfaces.
- Visibility of all sub-processes and individual tasks that form a process.
- Reproducibility of processes due to design of processes in standard BPM notation.
- Standardised representation of protocols.

Design and Implementation of Whole-Cell Biosensors for Micronutrient Detection in Low Resource Environments

Watstein, D., Georgia Institute of Technology
McNerney, M., Georgia Institute of Technology
Styczynski, M. P., Georgia Institute of Technology

Micronutrient deficiencies are a significant healthcare concern across the globe, affecting more than 2 billion people and causing billions of dollars in economic damage. These conditions, though straightforward to treat, remain a problem because they are often difficult to recognize and diagnose, requiring lab tests that are prohibitively expensive in both material and human resources for those in developing or remote areas. As obligate consumers of the same micronutrients, bacteria possess cellular machinery to control intracellular micronutrient levels and have corresponding regulatory mechanisms to respond to varying concentrations in their environment. We are developing a novel diagnostic tool based on whole-cell bacterial sensors using designed genetic circuitry to direct existing or minimally engineered cellular machinery to trigger specific changes in color in response to defined micronutrient levels. Such a test would be cheap, require minimal equipment and minimal training to administer and interpret, allow on-site diagnosis of micronutrient deficiencies in the populations most at risk, and facilitate epidemiological studies of such deficiencies and assessment of the efficacy of intervention strategies on a population scale.

Towards this goal, we present work on a whole-cell biosensor for the detection of zinc, a key micronutrient; zinc deficiency is responsible for around 1.7% of global all-cause child mortality. We have demonstrated a multi-output circuit using two native zinc-responsive *E. coli* transcription factors, ZntR and Zur, and we present a pigment-based reporter strategy based on the violacein and carotenoid pathways which obviates the need for additional detection equipment (which would be required for bioluminescent or fluorescent reporters) while maintaining unambiguous sensor output. We have demonstrated changes in cell color easily visible to the naked eye around medically relevant zinc concentrations and will discuss the strategies used to tune the sensor, including optimizing regulator production, vector selection, altering translational efficiency, and promoting protein degradation. Using multiple mechanisms in tandem allowed the circuit to change output over a small fraction of the natural dynamic range of the transcription factors used and overcome the inherent challenges of using sequential pigmented metabolites in a single pathway as different circuit outputs. The reporting strategy is modular and is in principle adaptable to reporting other analytes for which differentially regulated transcription factors are available. The resulting whole-cell biosensor is capable of distinguishing conditions that are high, intermediate, or low in zinc, and is currently being engineered for optimal response in human serum and reduction to practice in a format ready for use in the field. This work represents the first *E. coli* whole-cell biosensor for zinc with human-readable output to minimize cost per assay and facilitate epidemiological study of zinc deficiency and evaluate the efficacy of nutritional intervention.

Dynamic Responses of Laci/Galr Chimera-Based Transcriptional Logic Gates

Hirning, A.,
Gomez, M., UC Berkeley
Bennett, M.,

Within the synthetic biology field, various logic gates have been constructed with broad applications in mind such as biological computing and sensors in a dynamic environment. Additionally, these logic gates may be integrated into larger networks, mediating response or coupling dynamics of modular genetic circuits based on environmental conditions. However, less work has been done to understand the limitations of transcriptional logic gates in these varied applications. It is clear that reliable logic gates necessitate responses on a faster time-scale than time-varying environmental conditions. Therefore, we aim to understand the fundamental limitations on the dynamic range of transcriptional logic gates. Transcriptional logic gates function as low-pass filters, responding faithfully to low frequency (long period) signals and unfaithfully to high frequency (short period) signals.

The focus of this study is chimera-based transcriptional logic gates. Recently, Shis et al. (2014) showed that chimeric proteins derived from the LacI/GalR family of transcriptional repressors can be used to create transcriptional AND gates *in vivo*. These chimeric proteins have the same operator (DNA) binding domain but different ligand binding domains; hence, they will bind to the same operator site but are induced by different sugars. Only if all repressors are induced will repression of the reporter be relieved, creating a multi-input AND gate.

We consider chimera-based transcriptional logic gates in this study for two reasons: 1) we predict their dynamic range will be larger than repressors with transcriptionally mediated inputs and 2) they provide us with a multitude of orthogonal inputs. By utilizing different inducible or constitutive promoters, and different combinations of inducer ligands, we have created two different transcriptional logic gates– IMPLY and AND. Using microfluidic devices capable of delivering either one or two input signals, we have measured the response of these two gates to different periodic inducer input frequencies and created frequency response plots.

We use a mathematical model to predict the critical frequency at which the response of the logic gates begins to degrade. Using control theory tools, we find that the critical frequency is dependent on the degradation rate of the fluorescent reporter. Furthermore, we find that modeling enzymatic degradation can play a critical role in predicting dynamic range depending on the range of protein expression. This nonlinearity can further complicate the frequency drop-off rate of the logic gate response.
Experimentally, at the higher frequencies we see a large delay in initial response of the logic gates despite a minimal phase shift at steady state response. These delays are much larger than transcriptional delays and require further investigation.

Additionally, we are currently constructing and testing NOT and NOR logic gates. These are similar to the IMPLY and AND gates with an added inverter, increasing the number of genes that need to be transcribed for "computation". We predict that the 'direct induction' gates (IMPLY, AND) will have a much lower delay and act on timescales more relevant to synthetic circuits, while the 'indirect induction' gates (NOT, NOR) have higher delay and respond on slower time scales than IMPLY and AND.

Development and Implementation of an Automated Data Analysis Workflow for High Throughput Part Characterisation Data

Ainsworth, C. M., Imperial College London
Bultelle, M. A., Imperial College
Sainz de Murieta, I., Imperial College
Exley, K., Imperial College London
Kitney, R. I., Imperial College

Synthetic biology relies on constructing devices from small genetic parts which are being catalogued by various registries and databases. However, the characterisation data associated with these parts is often brief or missing and so knowing how these parts will function is unclear. The Centre for Synthetic Biology and Innovation at Imperial College has begun to develop automated characterisation pipelines for biological parts. This can create large amounts of characterisation data (for example, an automated characterisation platform can collect data for up to 24 constitutive promoters a day) and so the data analysis methods must be high throughput to match. We describe an analysis pipeline used to handle the data stream which demonstrates a high level of automation in the procedure. Once analysed, the data is presented on the SynBIS database to be used by other scientists when designing synthetic constructs. We have standardised the data input: the raw files are time series data from plate readers and fcs files from flow cytometers. The R programming language is used to read the files andÂ is where the analysis takes place. The data is stored according to the â??tidy dataâ?□ principles (Wickham, 2014) to ensure that the data objects are both human readable, but also amenable to a modular and flexible approach of analysis. Just as modularity is becoming recognised as important when building biological devices in Synthetic Biology, it is also important for the analysis of the characterisation data. Each step, accounting for backgrounds, fitting growth curves, calculating synthesis rates, etc. is treated as a separate module of analysis to make the pipeline as flexible as possible. This means that the scripts can handle data that may already have backgrounds accounted for etc. We present an algorithm developed for the analysis of constitutive and inducible biological part characterisation data, be it from time series or flow cytometry sources. The approach has allowed us to quickly analyse large datasets and calculate the key metrics, as well as extract the associated metadata so that the output can be presented on SynBIS. We have been able to compare far more data simultaneously which allows for a more big data approach; excluding data that is anomalous in cluster analysis for example. As a result, we can analyse large amounts of characterisation data quickly and accurately and output the results straight onto SynBIS where the data is disseminated. This characterisation data will aid a Synthetic Biologist in building new devices with predictable properties, reducing the time that the design, test, build cycle takes. Wickham, H. (2014). Tidy Data. Journal of Statistical Software, Vol. 59 (10).

Imperial College
London

Development and Implementation of an Automated Data Analysis Workflow for High-Throughput Part Characterisation Data

Catherine Ainsworth, Matthieu Bultelle, Kelan Exley, Iñaki Sainz de Murieta, Prof Richard Kitney

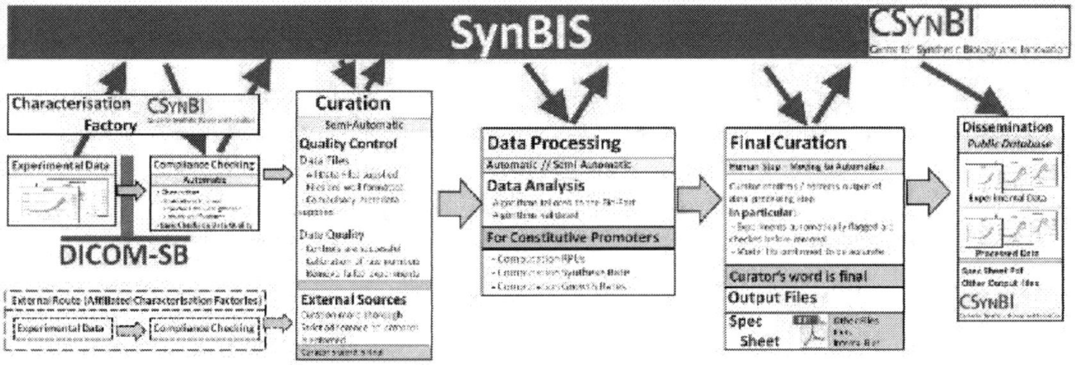

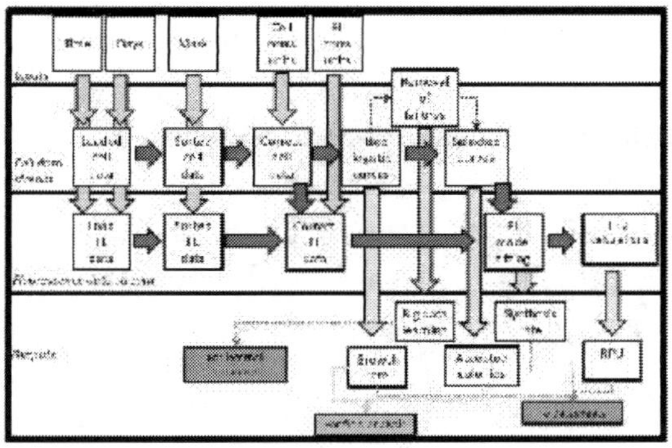

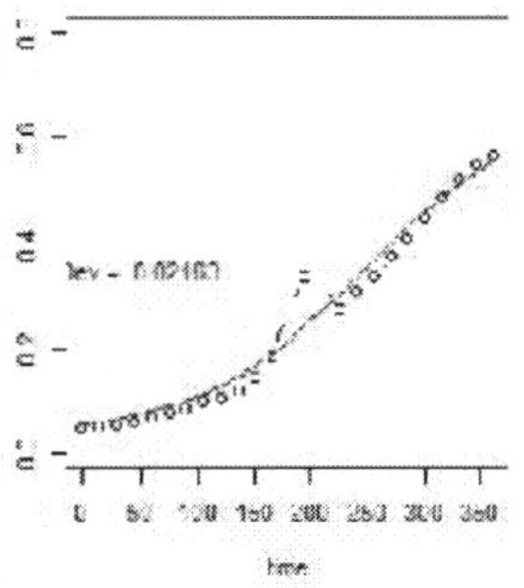

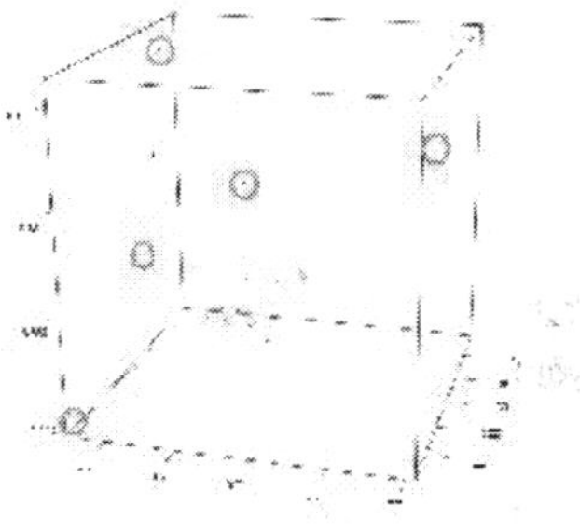

Loading As a Design Parameter for Genetic Circuits

Kumar, N., Massachusetts Institute of Technology
Huang, H. H.,
Del Vecchio, D.,

A major goal in synthetic biology is to modularly design, build and interconnect biomolecular circuits that perform as originally devised. However, the dynamics of circuits are sensitive to their context, and implementing circuits *in vivo*could potentially degrade their performance. Retroactivity, a form of context-dependence, affects the dynamics of an upstream system upon interconnection with a downstream system akin to impedance in electrical networks.

We study the effects of retroactivity, modeled as DNA promoter sites, on a genetic oscillator. The DNA promoter sites sequester transcription factors from the upstream system thereby changing the dynamics of the oscillator. The activator-repressor (A-R) oscillator consists of two proteins that transcriptionally activate and repress themselves and one another. Conditions for a stable limit cycle include sufficiently high transcriptional leakiness, relatively strong activation, and weak repression. Our analysis indicates that increasing the load on the repressor protein can activate a quenched clock and give rise to stable oscillations. Conversely, increasing the load on the activator protein dampens oscillations and causes the system trajectories to converge to an equilibrium point. Load, therefore, provides a method to tune the relative strength of the activation and repression branches for stronger or weaker oscillations, without resorting to degradation tags or changing the promoter regions.

Cell-Free Mixing of Escherichia coli Crude Extracts to Prototype Isoprenoid Biosynthesis

Dudley, Q., Northwestern University
Jewett, M. C., Northwestern University

Cell-free metabolic engineering (CFME) is advancing a powerful paradigm for prototyping and designing biochemical synthesis pathways. To date, most CFME efforts use either purified enzymes or a crude extract from a source strain containing the entire enzymatic pathway. In this work, we used mixing of crude lysates, or extracts, from *Escherichia coli* to reconstitute a portion of the isoprenoid pathway; isoprenoids are a diverse class of compounds with applications as pharmaceuticals, fragrances, pesticides, biofuels, and chemical feedstocks. Specifically, we individually overexpressed three enzymes, mixed the resulting lysates, and utilized the native glycolytic enzyme cascade to convert glucose to mevalonate, an intermediate of isoprenoid synthesis. Avoiding the synthesis of cytotoxic HMG-CoA in the source strain, this plug-and-play approach was used to screen enzyme variants and optimize enzyme ratios which facilitated the construction of an all-in-one strain. The open nature of CFME allowed exploration of cofactor landscapes at varying ATP, NAD^+, and CoA concentrations. Genomic modification to the source strain confirmed that metabolism of acetate salts via the *ackA-pta* pathway is critical for cofactor recycle and high titers ($>10\ g{\cdot}L^{-1}$). Our results highlight the high cofactor turnover of cell-free lysates and the ability to rapidly prototype and debug enzymatic pathways *in vitro*.

Application of a Transient Viral Vector for Efficient Genome Editing in Plants

Cody, W., Texas A&M Agrilife Research
Scholthof, H., Texas A&M University
Mirkov, T. E., Texas A&M University

CRISPR-based gene editing systems have been adapted as rapid transient screening devices in many model systems, but a completely transient genetic screening tool has not been developed for plants. An *Agrobacterium* launched *Tobacco mosaic virus* coat protein (CP) deletion mutant viral vector, TRBO, was evaluated for its ability to deliver effective single guide RNAs (sgRNAs) *in planta*. A Cas9 expression vector was developed to be delivered through agroinfiltration along with TRBO-sgRNA. The GFP expressing 16c *Nicotiana benthamiana* plants were used as a host to measure indel percentages of a sgRNA target that was designed within *GFP* and was flanked by either two, one or no ribozymes. Furthermore, a replicase mutant was tested to ensure mutagenesis was based on TRBO-dependent replication and transcription. Observed indel rates were highest for the replication-competent sgRNA construct carrying no ribozymes, producing transient indel rates of nearly 70% with the majority of mutations occurring 3 days post-inoculation (dpi), and peaking between 6 and 7 dpi. Reduced levels of GFP protein expression and decreased green fluorescence in imaging at 12 dpi coincide with high indel percentages. Additionally, endogenous *N. benthamiana* paralogs *Argonaute 1-H* and *Argonaute 1-L* were targeted using a singular sgRNA, which effectively created indels within both genes. The TRBO-sgRNA system was then tested for its ability to deliver multiple sgRNAs. Surprisingly similar indel efficiencies were reported for the delivery of two sgRNAs expressed through the same viral CP subgenomic promoter, without a separating linker sequence, as a single sgRNA. This demonstrates, to the best of our knowledge, the highest and quickest plant somatic cell mutagenesis reported. Delivery of sgRNAs using the TRBO system has the ability to reduce the time and cost of producing non-transgenic gene edited plants while also offering a transiently delivered gene knockout screening method for functional genetic studies as a more precise alternative screening tool compared to viral induced gene silencing screens. Possibly the most exciting and promising capabilities for Cas9/TRBO-sgRNA, as demonstrated by the ability to deliver multiple guides, is its potential use for plant metabolic engineering processes.

Computational Colloids: Engineered Bacteria As Computational Agents in the Design and Manufacture of New Materials and Structures

Dade-Robertson, M.,
Guyet, A.,
Rodriguez Corral, J.,
Wipat, A.,
Mitrani, H.,
Zhang, M.,

We report on a novel project to integrate a civil engineering application with Synthetic Biology through the development of a mechanical sensing bacteria system. The project integrates a team of molecular biologists, civil engineers and an architect. The aim of the project is to create a bio-based system to strengthen soils in response to mechanical forces. In this system, synthetic bacteria would act as a biological grout enabling, for example, self-constructing foundations. Aside from the geotechnical applications such a technology would push well beyond the current state of the art and challenge a new generation of engineering designers to think at multiple scales from molecular to the built environment and to anticipate civil engineering with living organisms.

Microbial activity is important to many geotechnical process and bacteria can be responsible for cementing soils. Bacteria, such as *Sporosarcina pasteurii* and *Bacillus megaterium* produce ammonia which, in turn, causes calcium carbonate to precipitate. Microorganisms also have a significant impact on the fertility of soils in the upper layers. Until recently it was thought that bacteria activity decreases substantially with depth. However, bacteria are known to move freely through many types of soils, and they can also attach to soil particles to form biofilm. This buildup of biofilms can, in turn, change the geotechnical characteristics of soils.

The growth, survivability, adaptation and associated genetic response of bacteria to pressure have been studied in relation to low-pressure environments (sub-0.1 MPa), and ocean bacteria in high-pressure environments (from 10 MPa to over 100 MPa). However, little work has been done at moderate pressures associated with, for example, pore pressures in geotechnical processes (between 0.1 and 10 MPa).

In our project, we have initially studied differential gene expression in response to moderate pressure using *Escherichia coli* K-12 as model organism. Selected gene promoters are being cloned upstream of reporter genes to test sensitivity to pressure (at 0.1 to 1 MPa) and engineer a mechanical sensing bacteria. In parallel experiments are being carried out to test that the promoters' response was pressure-stress specific and not related to other biological stresses. We are also developing a method to build a prototype system using hydrogels as a proxy for soil volumes and to visualize the response of the bacteria to pore pressure changes in three dimensions. In addition, we developed a computational model to link geotechnical modelling with gene expression data which integrates a finite element model of soils under load with diffusion models to calculate effective stress and pore pressure. The resulting visualisations give us an indication of how our proposed system may perform and support the process of model driven design.

Effects of Resource Sharing on Gene Expression Dynamics

Caveney, P. M., University of Tennessee Knoxville
Norred, S. E.,
Chin, C., University of Tennessee Knoxville
Retterer, S. T., Oak Ridge National Laboratory
Collier, C. P., Oak Ridge National Laboratory
Simpson, M. L., Oak Ridge National Laboratory

Gene expression is a noisy process[1], and a large portion of expression noise is bursting, where many proteins are made in short bursts punctuated by periods of no expression[2]. Many sources of bursting have been investigated (e.g. supercoiling[3] and transcription factors[4]), but the possible utility of bursting vis-à-vis resource sharing is not well understood. Here we explore how recyclable resources (e.g. polymerase and ribosomes) are temporally shared among actively expressing genes. By confining reactions to cell-sized PDMS chambers and POPC vesicles, we compare how resources are shared when they are either partitioned experimentally or naturally by the genes themselves. We find that resources are shared in a bursty manner (i.e. few genes have many resources for brief periods of time), and surprisingly, chambers with higher protein abundance increase the size bursts instead of initiating more frequent bursts. This is likely due to the difficulty in assembling the ~100 components necessary for protein expression[5]. Once assembled, the components are spatially correlated and thus have a higher probability of rebinding than if the solution were well mixed[6]. We corroborate our results with a simple model in which genes that transcribe mRNA early gather a disproportionate share of the protein expression resources resulting in a larger burst of protein expression. Our results further the understanding of burst frequency saturation[7] and important cell-fate decisions[8].

1. Blake, W. J., Kaern, M., Cantor, C. R. & Collins, J. J. Noise in eukaryotic gene expression. *Nature* **422,** 633–637 (2003).

2. Dar, R. D. *et al.* Transcriptional burst frequency and burst size are equally modulated across the human genome. *Proc. Natl. Acad. Sci.* **109,** 17454–17459 (2012).

3. Chong, S., Chen, C., Ge, H. & Xie, X. S. Mechanism of transcriptional bursting in bacteria. *Cell* **158,** 314–326 (2014).

4. van Zon, J. S., Morelli, M. J., Tănase-Nicola, S. & ten Wolde, P. R. Diffusion of transcription factors can drastically enhance the noise in gene expression. *Biophys. J.* **91,** 4350–4367 (2006).

5. Shimizu, Y. *et al.* Cell-free translation reconstituted with purified components. *Nat. Biotechnol.* **19,** 751–755 (2001).

6. Takahashi, K., Tănase-Nicola, S. & Ten Wolde, P. R. Spatio-temporal correlations can drastically change the response of a MAPK pathway. *Proc. Natl. Acad. Sci.* **107,** 2473–2478 (2010).

7. Sanchez, A. & Golding, I. Genetic determinants and cellular constraints in noisy gene expression. *Science* **342,** 1188–1193 (2013).

8. Dar, R. D., Hosmane, N. N., Arkin, M. R., Siliciano, R. F. & Weinberger, L. S. Screening for noise in gene expression identifies drug synergies. *Science* **344,** 1392–1396 (2014).

Analysis of Yeast Metabolism to Overproduce Isobutanol and n-Butanol

Tian, M., University of Wisconsin–Madison
Sato, T., Great Lakes Bioenergy Research Center
Reed, J. L., University of Wisconsin

Cells can synthesize cellular building blocks from different substrates using a large network of reactions. Metabolic engineering utilizes this reaction network to re-wire the cell to produce chemicals of interest. Traditionally, metabolic engineers made mutations in individual or competing pathways. Constraint-based modeling is a very powerful tool to understand and engineer metabolism. It provides a deep and systematic way to analyze the metabolic network and thus leads to rational strain designs that differ from those found using traditional approaches. SimOptStrain is a strain design algorithm that uses constraint-based models to suggest gene knockout and gene addition strategies to enhance biochemical production.

SimOptStrain was applied to a model of *Saccharomyces cerevisiae* to identify strategies that are predicted to increase isobutanol and n-butanol production with glucose or xylose as the substrate. We also computationally evaluated how different enzyme location (mitochondria or cytosol) and cofactor (NADH or NADPH) usage would impact the production rate. By analyzing the metabolic network, we provided gene knockout and gene addition strategies to enhance the biochemical production and insights into how these alterations would impact the metabolism in yeast. The proposed strategies are currently being tested experimentally.

ICE: A Distributed and Interconnected Biological Part Registry

Plahar, H., Joint BioEnergy Institute, Lawrence Berkeley National Laboratory
Nnadi, O.,
Chen, J., Joint BioEnergy Institute, Lawrence Berkeley National Laboratory
Ham, T.,
Dmytriv, Z.,
Adams, P. D., Joint BioEnergy Institute
Keasling, J. D., Lawrence Berkeley National Laboratory
Hillson, N. J., Joint BioEnergy Institute

Advances in synthetic biology research and genetic design automation, along with the availability of rapid and reliable DNA sequencing has resulted in a steady increase in the number of complex engineered biological devices. Maintaining information about the constituent parts of these devices, along with the ability to share them with other members in the synthetic biology community as well-characterized components with the goal of reusability is essential. To this end, we have extended the Inventory of Composable Elements (ICE), an open-source web based platform for managing information about synthetic biological parts, to include the ability to efficiently share these elements in a distributed manner. Referred to as "Web of Registries", this feature aims to increase the efficiency of bioengineering by enabling scientists to publish and share datasets across multiple ICE installations across labs in the synthetic biology research community and in potentially different geographic locations. Thus, users from one ICE instance (e.g. the JBEI public repository) can query and access parts, sequences, strains and Arabidopsis seeds of interest stored in other connected ICE instances (e.g. the Synberc repository). Also, through the REST API, information in ICE can be accessed using a web browser or through third party BioCAD tools. Support for multiple data formats (e.g. Genbank, Fasta and the Synthetic Biology Open Language) also opens up the data and platform to these higher level synthetic biology design tools. These third party tools can also access the shared data in the "web of registries" through the available APIs.

Cell-Free Platform for Natural Product Discovery

Heinsch, S., University of Minnesota
Smanski, M. J., University of Minnesota

The discovery process for natural products is often hindered by time consuming operations including strain engineering and tuning of genetic regulation. Here, we present a platform for the high-throughput discovery of natural products. Cell-free transcription and translation systems allow for the simultaneous evaluation of several genetic architectures within a context that is largely divorced from host metabolic and regulatory processes. As a proof of concept, this platform is used to produce the ribosomally synthesized and post-translationally modified peptide patellamide.

A Droplet Microfluidic Platform for Automating Genetic Engineering

Gach, P., Sandia National Laboratories
Shih, S. C. C., Joint BioEnergy Institute
Sustarich, J.,
Keasling, J. D., Lawrence Berkeley National Laboratory
Hillson, N. J., Joint BioEnergy Institute
Adams, P. D., Joint BioEnergy Institute
Singh, A. K., Joint Bioenergy Institute

We present a water-in-oil droplet microfluidic platform for transformation, culture and expression of recombinant proteins in multiple host organisms including bacteria, yeast and fungi. The platform consists of a hybrid digital microfluidic/ channel-based droplet chip with integrated temperature control to allow complete automation and integration of plasmid addition, heat-shock transformation, addition of selection medium, culture, and protein expression. The microfluidic format permitted significant reduction in consumption (100-fold) of expensive reagents such as DNA and enzymes compared to the benchtop method. The chip contains a channel to continuously replenish oil to the culture chamber to provide a fresh supply of oxygen to the cells for long-term (~5 days) cell culture. The flow channel also replenished oil lost to evaporation and increased the number of droplets that could be processed and cultured. The platform was validated by transforming several plasmids into *E. coli* including plasmids containing genes for fluorescent proteins GFP, BFP and RFP; plasmids with selectable markers for ampicillin or kanamycin resistance; and a Golden Gate DNA assembly reaction. We also demonstrate the applicability of this platform for transformation in widely used eukaryotic organisms such as *S. cerevisiae* and *A. niger*. Duration and temperatures of the microfluidic heat-shock procedures where optimized to yield transformation efficiencies comparable to those obtained by bench-top methods with a throughput up to 6 droplets/min. The proposed platform offers potential for automation of molecular biology experiments significantly reducing cost, time and variability while improving throughput.

A Droplet Microfluidic Platform for Automating Genetic Engineering

Philip C. Gach, Steve C.C. Shih, Jess Sustarich, Jay D. Keasling, Nathan J. Hillson, Paul D. Adams, and Anup K. Singh

- The temperature-controlled hybrid microfluidic platform automatically implements heat-shock transformation, antibiotic selection, culture and assay in discrete droplets.

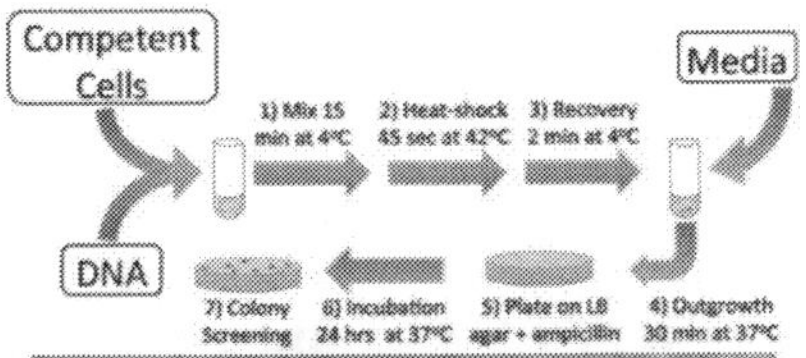

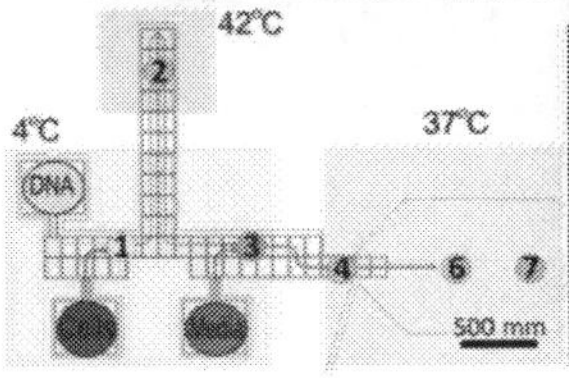

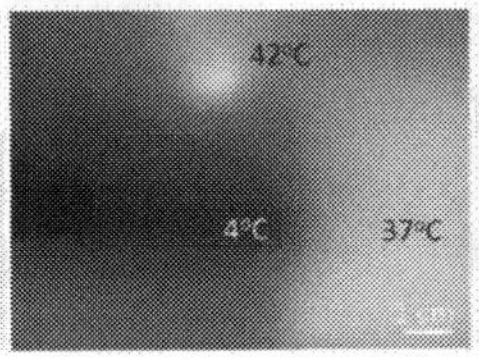

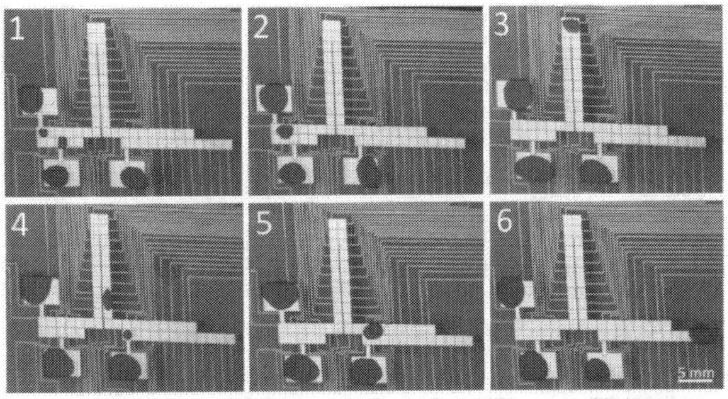

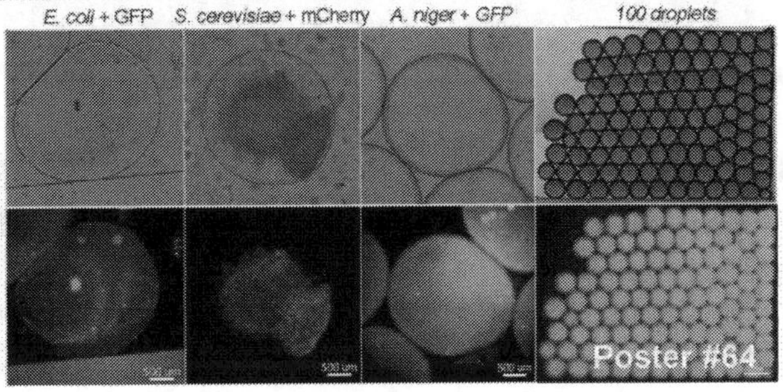

Deep Sequencing-Based Protein Engineering to Optimize Functional Enzyme Expression in Synthetic Metabolic Pathways

Klesmith, J., Michigan State University
Whitehead, T., Michigan State University

Synthetic biologists primarily optimize enzyme expression levels in designed metabolic pathways by modifying transcription and translation. However, many enzymes can not be rescued by such methods as they are intrinsically unstable, thus requiring changes to the protein sequence itself. In my presentation I will detail different deep sequencing-based approaches to enable the rapid and parallel determination of sequence effects on protein function for complete gene-encoding sequences. The first approach, called Fluxscan, is able to resolve the sequence-flux relationship of an enzyme in a growth-based selection. I will show application of this method to determine the function of 93% of all possible single point mutants (over 8,000) in a levoglucosan utilization pathway implanted in *Escherichia coli*. I will describe how computational protein design augmented by these experimental datasets resulted in an designed metabolic pathway that supported 15-fold improvement in growth rate on levoglucosan as the sole carbon source, and describe how this advance improves hybid thermochemical/biochemical processing of renewable biomass. The main shortcoming of FluxScan is that the method relies on a coupling of enzyme activity to growth rate, yet many pathways cannot be so coupled. In the second portion of my talk I will describe an alternative deep sequencing-based tool to improve enzyme expression, for which I will supply datasets from several model enzyme systems.

Improving Genomically Recoded Escherichia coli for the Production of Proteins Containing Non-Standard Amino Acids

Perez, J., Northwestern University
Carlson, E. D., Northwestern University
Weisser, O.,
Jewett, M. C., Northwestern University

The expansion of the genetic code to co-translationally incorporate non-standard amino acids is a rapidly growing field in synthetic biology. This is due to its potential to transform how we synthesize materials and therapeutics, investigate protein structure, and understand the evolution of the translation system. A pioneering effort has recently developed an Escherichia coli strain lacking all TAG codons and release factor 1 which allows for genetic encoding of an additional, non-standard amino acid. This strain however has not previously been optimized for protein production. Here we describe the construction of a series of genomically recoded organisms that are optimized for protein production and have the additional ability to tune protein expression through a T7 RNA polymerase (T7RNAP). Using multiplex automated genome engineering (MAGE), we targeted the individual deletion of 5 putative negative effector genes as well as all double and several triple mutants. We found that reduction of nuclease and protease activity increases wildtype sfGFP production by 260% and sfGFP containing 2TAGs production by 225% and 560% with p-azidophenylalanine and N6-(propargyloxycarbonyl)-L-Lysine, respectively. Next, we designed and optimized an IPTG and a Tetracycline inducible T7RNAP cassette for reduced background expression and high final cell density. These cassettes were genomically inserted into the top mutant strains. We envision that our library of strains will provide the community with multiple strain options for expression of proteins containing non-standard amino acids with increased protein yield and cell density.

Definition and In Vivo Evaluation of Mathematical Models to Predict the Effects of Copy Number Variations and Cell Burden in Interconnected Synthetic Circuits

Pasotti, L.,
Zucca, S.,
Casanova, M.,
Bellato, M.,
Massaiu, I.,
Arbuschi, M.,
Murgiano, M.,
Serra, A.,
Cusella, G.,
Magni, P.,

The predictability of bottom-up designed synthetic circuits is a major issue in biological engineering. Several aspects currently prevent the quantitative prediction of circuits output, e.g., retroactivity, crosstalk among parts, limited cell resources, noise and context-dependent variability of components. For these reasons, the potential of synthetic biology cannot be fully exploited in high-impact application fields, thus requiring trial-and-error approaches, and time-consuming re-characterization or debugging of the constructed biological systems.

Mathematical models are usually adopted to support the design of synthetic circuits. Empirical and mechanistic models of gene regulatory networks have been used in different works to capture the quantitative behaviour of circuits, predict the output of new interconnected systems, and support early design or debugging steps.

The objective of this work is to define and test predictive models of interconnected synthetic circuits, considering copy number variation of components and cell burden. Such models can be exploited for the quantitative prediction of circuits output when individual pre-characterized components providing a burden to the chassis are connected, and the copy number of some of their subparts is changed.

Copy number of biological components is known to dramatically affect the quantitative behaviour of synthetic circuits with different network motifs. Empirical models of gene networks, usually based on Hill functions to describe gene regulations, are not able to capture copy number variation effects of promoters or regulatory genes. Mechanistic models can be adopted to describe the molecular interactions in circuits. Although they are able to capture the copy number change of every model species, they often have a higher number of parameters than empirical models. Here, we define different models and provide a steady-state algebraic solution for the output of commonly adopted systems, e.g., tetR-, lacI- and luxR-based regulations, considering different assumptions. For each model, identifiability analysis is also performed and the results are used to drive the design of suitable in vivo experiments on model systems in which regulatory proteins and/or promoter copy number are varied.

Cell burden derives from limited resource availability (e.g., polymerases and ribosomes) in living organisms. For this reason, when synthetic circuits are incorporated they can alter cellular resources and, on the other hand, the functioning of the circuit could be significantly affected. Models to describe the effects of resource limitation have been recently proposed, together with a reporter gene-based experimental method to quantify cell burden. In this context, a major challenge is to quantitatively predict the burden and the effects on circuit output from the knowledge of individual genetic modules connected in the circuit. To carry out this task, existing models were selected and applied to study model systems. Transcriptional regulators cascades were used to test in vivo a rigorous bottom-up approach in which individual components are characterized, model parameters are estimated, simulation are carried out on a model of interconnected circuit and predictions are compared with experimental output.

This work provides a modelling framework and in vivo testing to support the quantitative prediction of synthetic circuits output considering two factors currently limiting the bottom-up design in biological engineering.

A Combinatorial Assembly Pipeline for Gene Expression in Mammalian Cells

Stach, C., University of Minnesota
Smanski, M. J., University of Minnesota

Mammalian synthetic biology holds promise in developing cell based therapies, natural product synthesis, protein therapeutics, and defining the role of complex signaling networks in cell development. Progress towards these goals has been hampered by the limited availability of efficient and straightforward assembly pipelines in mammalian cells. While assembly pipelines exist for the generation of multi-gene constructs, they are limited in their utility. Currently, protein expression in mammalian cells leads to heterogeneous populations of the same protein due to differences in post-translational modifications. Here I present a combinatorial assembly pipeline developed for use in mammalian cells that can be used to identify genes important to expression of homogeneous protein populations. It is based on the use of Type IIS enzymes to assemble a series of defined parts such as promoters, terminators, and genes, into expression constructs. This system permits the expression of genes in a predictable manner and the ability to assemble and express genes from various pathways to examine their role within the cell.

Topologies of Phosphorylation Cycles As Insulation Devices

Shah, R.,
Del Vecchio, D.,

One of the problems faced in building large circuits from small parts or modules in synthetic biology is that of retroactivity. Retroactivity is the change in the dynamics of a system when it is connected with another system or load. An insulation device is a system that can be connected in between such modules to attenuate the effect of retroactivity, enabling the design of larger circuits. A single phosphorylation cycle has been theoretically and experimentally shown to have the ability to attenuate the effect of retroactivity applied by downstream systems on genetic circuits due to its fast reaction rates. However, it has been found that this retroactivity attenuation property comes at the expense of an increased retroactivity to the input, wherein the cycle slows down the signal it receives from the upstream system. In this work, we provide a library of insulation devices with different insulation properties, built from various topologies of phosphorylation cycles, utilizing its fast timescale of operation. We find that cascades of phosphorylation cycles break the tradeoff faced by a single cycle, allowing the attenuation of retroactivity applied by downstream systems, while applying a small retroactivity to the input. We show that there is an optimal number of cycles that maximally extends the linear operating region of the insulation device while keeping the desired retroactivity attenuation properties, when a common phosphatase is used. Double phosphorylation cycles can show similar retroactivity attenuation properties to a single cycle, and provide either an ultrasensitive response to the input, or a linear response with a sign-sensitive delay. These findings provide insights into cellular signaling systems, where such phosphorylation cycles are ubiquitous. In addition, they provide a variety of insulation device designs for synthetic biology applications.

Discovering the Function of a Vibrio Fischeri Metabolic Gene Using a Newly Constructed Genome-Scale Model and Genomic Library

Pan, S., University of Wisconsin-Madison
Nikolakakis, K., University of Wisconsin-Madison
Ruby, E., University of Hawaii, Manoa
Reed, J., University of Wisconsin-Madison

Vibrio fischeri is a bioluminescent marine bacterium that forms a symbiotic relationship in the light-emitting organ of the Hawaiian bobtail squid, *Euprymna scolopes*. Its metabolic capabilities are representative of many other marine bacteria, including both beneficial and pathogenic members of the genus *Vibrio* and, thus, are of great interest to researchers. To better understand the metabolism of *V. fischeri* and, in particular, to search for metabolic genes without any experimentally characterized functions, we constructed a *V. fischeri* genome-scale metabolic model and developed a novel gene-finding method using genomic libraries.

Our *V. fischeri* model is a structured systems-level representation of most known *V. fischeri* biochemical reactions and their associated genes. It is built upon both the annotated genome and literature information, and is parameterized by the biochemical composition of *V. fischeri*, as well as its growth rates and metabolite consumption and production rates. The current *V. fischeri* model consists of 1725 reactions, 1019 unique metabolites, and 822 metabolic genes. To validate the model, we tested the growth phenotypes of *V. fischeri* under 181 different sole carbon-source growth conditions. The model correctly predicted 86.2% of the experimentally validated growth phenotypes. It also accurately categorized 83.4% of the 822 metabolic genes into either essential or non-essential genes for growth in complex-nutrient medium, as based on a previous report [1].

We also developed a method to search for genes that were missing from the model. In our experiment, *V. fischeri* was able to grow in a minimal medium without β-alanine, a precursor to the synthesis of the required metabolite coenzyme A. However, *V. fischeri* is missing an aspartate decarboxylase gene (encoded by *panD* in *E. coli*), which converts aspartate to β-alanine and is required for growth in minimal medium. However, no ortholog to the *E. coli panD* could be identified in *V. fischeri*. We located the *V. fischeri* aspartate decarboxylase gene by complementing an *E. coli* Δ*panD* mutant (an auxotroph for β-alanine) with a plasmid library of *V. fischeri* genomic DNA. The library had an average insert size of 2.4 Kb, and a total size of 53 Gb. Sequencing a plasmid that rescued growth of the *E. coli* Δ*panD* mutant found the *V. fischeri* aspartate decarboxylase gene. We cloned just the coding region of the gene into a new plasmid vector, and this plasmid was again able to rescue *E. coli panD* mutant in minimal medium. These results strongly suggest that the gene functions as an aspartate decarboxylase in *V. fischeri*, despite the sequence dissimilarity between the gene and *panD* from *E. coli*. We are currently doing *in vitro* enzymatic assays and phylogenetic analyses to further characterize this gene.

[1] J. F. Brooks, M. C. Gyllborg, D. C. Cronin, S. J. Quillin, C. a Mallama, R. Foxall, C.

Whistler, A. L. Goodman, and M. J. Mandel, ?Global discovery of colonization determinants in the squid symbiont *Vibrio fischeri*.,? *Proc. Natl. Acad. Sci. U. S. A.*, 111: 17284-89, Nov. 2014.

Characterizing the Role of 23S rRNA Peptidyltransferase Center Interactions in Translation Fidelity

d'Aquino, A., Northwestern University
Liu, Y., Northwestern University
Buechel, E., Northwestern University
Jewett, M.,

The ribosome, a 2.5MDa molecular machine that polymerizes α-amino acids into proteins, is the catalytic workhorse of the translation apparatus. The catalytic capacity of the translation machinery has attracted extensive efforts to repurpose it for novel functions. One key idea is that the natural translation machinery can be harnessed to synthesize polymers containing non-natural building blocks. Expanding the repertoire of ribosome substrates and functions is a difficult task, however, because the requirement of cell viability severely constrains the alterations that can be made to the ribosome, a catalyst that sustains the life of a cell. In practice, these constraints have made the natural ribosome nearly unevolvable and, so far, no generalizable approach for modifying the catalytic peptidyl transferase center (PTC) of the ribosome has been advanced. We propose to address this grand challenge by using cell-free systems that harness the biosynthetic potential of cellular machines without using intact cells, thus removing cell viability constraints. Here, we use our *in vitro* ribosome synthesis, assembly, and translation system (termed iSAT) to generate variant ribosomes with mutations in the 23S rRNA (especially those that are dominantly lethal in the PTC and exit tunnel) and inquire how these modifications affect protein synthesis. Using iSAT, we assembled more than 50 mutant ribosomes possessing single and some double base substitutions at 23S rRNA nucleotides in the PTC. By successfully quantifying full-length protein synthesis kinetics of iSAT-assembled WT and mutant ribosomes, we unexpectedly found many key PTC mutations, which were expected to abolish ribosome activity, still permitted full-length protein synthesis. We also assessed fidelity. Our work provides a comprehensive mapping of the systems effect of multiple ribosomal mutations on protein synthesis. The understanding gained facilitates efforts to engineer ribosomes for synthetic biology.

A Scalable Peptide/GPCR Communication Language

Shen, M.,
Agmon, N., New York University School of Medicine
Billerbeck, S., Columbia University
Brisbois, J., Columbia University
Jimenez, M.,
Cornish, V.,
Boeke, J., Institute for Systems Genetics

As synthetic biology transitions from single cell engineering to the creation of multi-cell consortia, a scalable signaling language for communication within diverse communities will become necessary for information transfer. Bacterial quorum sensing and electrical communication are two such examples of engineering cell-cell communication. However, the number of specific ion channels and unique small-molecule signals available to use as a signaling language is small, and not easily expanded or diversified. Fungi, however, have already evolved a scalable signaling system in the form of their small peptide pheromone/GPCR pairs used for mating. Through fungal genome mining we have identified hundreds of extant peptide/GPCR pairs that can be further diversified through directed evolution. Additionally, we have chosen the well-studied baker's yeast *Saccharomyces cerevisiae* to serve as a genetically tractable host with an extremely well-characterized mating pathway. Using a common downstream transduction MAPK pathway in *S. cerevisiae*, we have developed a "plug-and-play" system whereby data-mined GPCR and peptide ligand modules may be easily substituted into the yeast mating and peptide secretion pathways. We have also engineered *S. cerevisiae* strains with GPCR-controlled outputs including fluorescence and colored pigment production in order to provide a quantitative readout of receptor activation. We are working to couple GPCR activation to the expression of an essential gene in our engineered strains, allowing us to establish a cyclic communication system whereby interdependent growth is controlled by peptide signaling between strains and have developed a model to describe the behavior of our engineered consortia and predict consequences of culture perturbations. As we expand our repertoire of fully characterized peptide/GPCR pairs, we aim to establish this system in additional output species including *Saccharomyces boulardii* and *Schizosaccharomyces pombe* as well. Such consortia could serve as a new biosensor paradigm and as independent hosts for the biosynthesis of modular metabolites such as polyketides.

Maintaining Plasmid Stability in Commensal Escherichia coli Nissle 1917 without Antibiotic Selection through Orthogonal Toxin-Antitoxin Systems

Ozdemir, T., University College London
Fedorec, A. J. H., University College London
Barnes, C., University College London

The rapid progress observed within the field of synthetic biology over the last decade has facilitated the design and creation of more clinically relevant and complex genetic circuits. The human microbiota consists of the various microorganisms associated with the human body and it is increasingly being implicated as a regulator of health and disease. The commensal nature of the intestinal microbiota and its constituents provide a number of well-tolerated microorganisms such as *Escherichia coli* NISSLE 1917 (EcN) that with synthetic biology could potentially exploit the intestines as an entry point and be used for monitoring or engineering the host microbiota. Most of the circuits designed thus far are based on the use of bacterial plasmids. These small circular DNA molecules replicate independently of the chromosome and are easy to edit and propagate with common molecular tools such as restriction enzymes and antibiotic selection markers. However, the use of antibiotic selection in the circuits would eventually hinder the investigative power of these approaches as the native microbiota would be severely impacted upon antibiotic administration. Toxin-Antitoxin (TA) post-segregational killing systems are a naturally occurring mechanism developed by bacteria to maintain plasmid stability. This mechanism consists of a self-regulated fragment that ensures only daughter cells that possess the plasmid survive after a cell undergoes division. Here we show that an orthogonal TA system can be used in EcN to maintain the stability of a reporter plasmid in the absence of antibiotic selection for up to 13 daily passages. We characterised three TA systems from both gram-negative and gram-positive bacteria using a plasmid-born GFP fluorescent marker, flow cytometry (FCM) analysis, mathematical modelling and Bayesian statistics. Interestingly, the Axe/Txe TA system from the gram-positive *Enterococcus faecium* was found to be more effective in EcN than the more widely characterised and used Hok/Sok system derived from *E.coli*. Our results demonstrate that TA systems can be successfully used to maintain the stability of synthetic genetic circuits in commensal strains like EcN without the need for antibiotic selection. This approach provides a foundation for the construction of synthetic circuits that perform precise in vivo investigations of the intestinal microbiota within a variety of commensal bacterial strains. Synthetic biological tools like this can help further elucidate the complex role of the microbiota in human health and disease.

Bliss: The Black List Sequence Screening Pipeline

Simirenko, L., Joint Genome Institute
Deutsch, S., Lawrence Berkeley National Laboratory
Cheng, J. F., Lawrence Berkeley National Laboratories
Hillson, N. J., Joint BioEnergy Institute

Concerns have been raised that individuals with malicious intent could exploit DNA synthesis technology to obtain genetic elements from pathogenic organisms that would otherwise be difficult to obtain. In response, the U.S. Department of Health and Human Services (HHS) issued the *Screening Framework Guidance for Providers of Synthetic Double-Stranded DNA*. This voluntary Guidance outlines the U.S. government's recommendations for screening double-stranded DNA to ensure that existing Select Agent Regulations (SAR) and Export Administration Regulations (EAR) are followed, to encourage best practices in addressing biosecurity concerns, and to reduce the risk that individuals with ill intent could exploit the application of DNA synthesis technology to obtain genetic material derived from or encoding Select Agents or Toxins, or agents on the EAR's Commerce Control List (CCL).

In accordance with the HHS guidance, the U.S. Department of Energy Joint Genome Institute (JGI) has developed a DNA screening pipeline (BLiSS – Black List Sequence Screening) to screen all sequences that it synthesizes through its DNA Synthesis Science program. BLiSS detects "sequences of concern" of at least 200 nucleotides in length on either DNA strand, including polypeptide translations using the three alternative reading frames on each DNA strand (six-frame translation). Sequences to be synthesized are aligned to GenBank's non-redundant nucleotide and protein databases. A "Best Match" approach is used to determine whether any of the sequences of concern identified within the sequence to be synthesized are unique to Select Agents or Toxins, or CCL-listed agents, toxins or genetic elements, to minimize false positives from closely related organisms or highly conserved "house-keeping genes" which do not pose a biosecurity threat.

Using the lessons learned by screening over 10,000 sequences submitted to the JGI for synthesis, we are modifying the pipeline to detect when a "sequence of concern" has a high likelihood of being a gene that is not involved in the pathogenicity of the Select Agent. This will greatly aid in the follow-up required when sequences fail the screening process. This follow-up requires human interpretation so it cannot be automated, and has been identified as the most costly aspect of implementing the *Guidance* by double-stranded DNA providers.

Library and Model Derived Synthetic Terminators for Yeast

Morse, N., University of Texas at Austin
Alper, H., The University of Texas at Austin

Control of transcription and translation are critical parts of yeast synthetic biology and metabolic engineering. While significant efforts have historically focused on the engineering of promoter elements, more recent attention has shifted toward terminator elements as a means to influence mRNA half-life and resulting protein expression. Our group has recently demonstrated that short, synthetic terminator sequences (ranging from 35 to 70 bp in length) can result in superior synthetic parts for the yeast *Saccharomyces cerevisiae*. We show that these more minimal elements (nearly an 80% reduction in sequence space) increase transcript levels nearly 4-fold over commonly used terminators such as CYC1. We then show that these short, minimal terminators can function across different fungal host systems. In an effort to move toward the model-based design of these elements, we have extracted a functional model effective at *de novo* design of these terminators. We demonstrate the utility of this model by establishing a library of synthetic terminator elements with a range of activity. In doing so, we create some of the first model-specified terminator elements for yeast. This work both expands the set of synthetic parts available for fungal systems as well as provides insights into the sequence properties and underlying mechanisms of yeast transcriptional termination.

The Joint Genome Instituteâ€™s Synthetic Biology Internal Review Process

Hillson, N. J., Joint BioEnergy Institute
Simirenko, L., Joint Genome Institute
Harmon-Smith, M.,
Visel, A.,
Rubin, E., Joint Genome Institute

Serious concerns have been raised over the possible intentional use of synthetic biology approaches to engineer pathogenic organisms as well as the possible accidental environmental release of genetically engineered organisms. Scientists pursuing synthetic biology research must diligently consider issues such as these. As such, the U.S. Department of Energy (DOE) Joint Genome Institute (JGI) has developed a Synthetic Biology Internal Review process to assess, beyond technical and scientific merit, certain broader aspects (e.g., environmental, biocontainment, biosafety, or biosecurity) of the research proposals associated with the JGI's DNA Synthesis Science program. The purpose of this internal review process is two-fold: 1) to assess the broader aspects of the research, request proposal modifications if issues of concern are not sufficiently addressed in the proposal, reject research proposals where issues of concern are not or can not be satisfactorily addressed, and output a paper-trail audit of the review process; and 2) to encourage and educate researchers to more extensively consider the broader aspects of their research, including beyond the immediate research itself.

All JGI DNA Synthesis Science proposals (including those from the JGI Community Science Program and from the DOE Bioenergy Research Centers) contain a broader implications section dedicated to a brief discussion of these broader aspects. This broader implications statement should address not merely the possible rewards but also a considered statement of the risks associated with the work. These statements serve as a useful tool to protect not only the public, but the Investigators (and their institutions), as well as JGI itself. These statements are examples of consideration and deliberation of the responsible application of science. As members of the research community, we must consider risks, and be able to show our consideration of those risks - even if they are demonstrably small.

This poster will include an update of the JGI's Synthetic Biology Internal Review process statistics to-date, including the number of reviewers (and distribution between internal and external), the number of proposals reviewed, the number of reviewer comments across proposals (and distribution across comment categories), as well as review process outcomes across proposals. These review process statistics, along with lessons learned and process improvements made at the JGI over the past 2 and a half years, provide funding agencies insight into the resource requirements and benefits of such an institutional review process, and facilitate and support the successful deployment of similar review processes at other research institutions.

The JGI's Synthetic Biology Internal Review Process

Lisa Simirenko, Miranda Harmon-Smith, Axel Visel, Edward M. Rubin, and **Nathan J. Hillson***

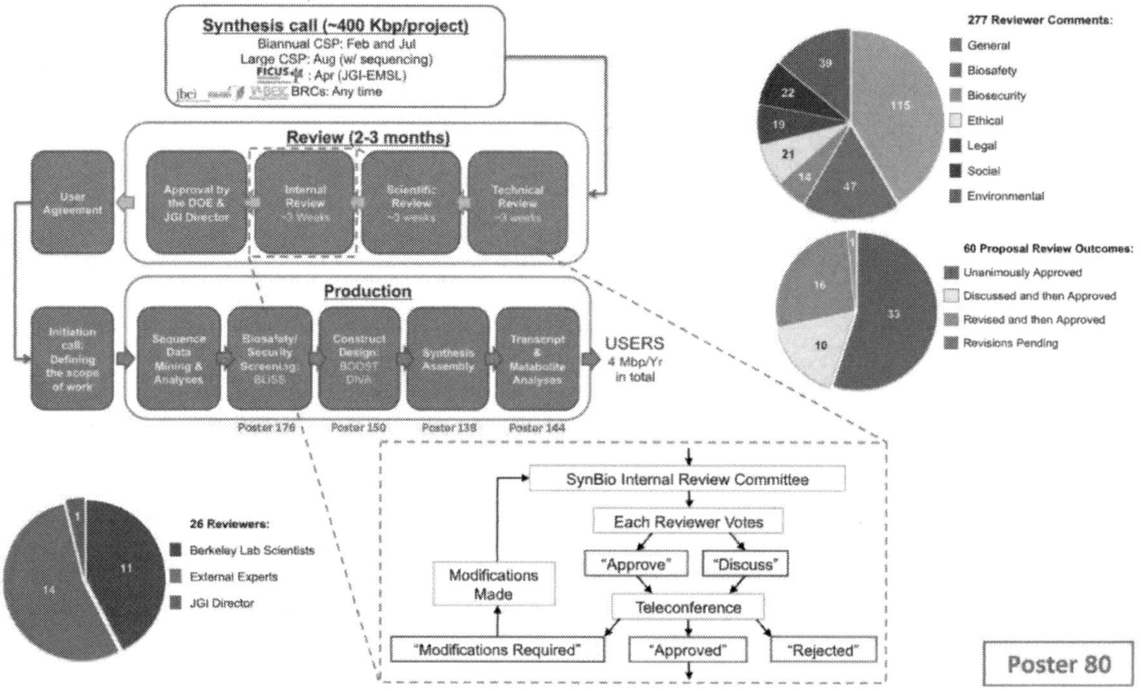

Poster 80

Engineering Synthetic Systems Inspired By Anaerobic Fungi

Gilmore, S. P., University of California, Santa Barbara
Haitjema, C. H., University of California, Santa Barbara
Henske, J. K., University of California, Santa Barbara
Sexton, J. A., University of California, Santa Barbara
Theodorou, M. K., Harper Adams University
Valentine, D., University of California, Santa Barbara
O'Malley, M. A., University of California, Santa Barbara

Anaerobic fungi in the hindgut of large herbivores are among the most robust organisms at degrading crude lignocellulose. Their remarkable cellulolytic capabilities have great potential for use in biomass breakdown and biofuel processing. Anaerobic fungi achieve cellulolytic efficiency through the production of large, multi-enzyme complexes called fungal cellulosomes. In isolation, anaerobic fungi metabolize some of the released sugars and convert them into fermentation products. In nature, however, they exist in a community with archaea, bacteria, and protozoa, which drastically alter the behavior of the fungi. By elucidating the parts responsible for efficient biomass degradation at both the protein and cellular level, we seek to replicate this efficiency in synthetic systems.

Previously, anaerobic fungi have been shown to interact closely with methane producing archaea (methanogens). Methanogens reduce CO_2 with H_2 for growth and methanogenesis, which maintains a low partial pressure of H_2 in their environments, allowing the fungi to more efficiently metabolize sugars. To further investigate this mechanism, native fungal/methanogen consortia were isolated from herbivore fecal materials. ITS profiling and genomic sequencing revealed the presence of one fungus, two methanogens, and one bacterium in one consortium, which was stable under continuous passage for over 20 months. The consortium demonstrated faster and more complete degradation of cellulosic substrates, as well as a wider range of utilized substrates compared to the monocultured fungus alone. By introducing the methanogens into cultures of other well-characterized anaerobic fungi, stable synthetic co-cultures were established. These stable synthetic consortia demonstrated similar efficiency, and suggest a promising option for conversion of crude biomass into sustainable chemicals.

Anaerobic fungi create large enzyme complexes called fungal cellulosomes, which they utilize for biomass degradation. Fungal cellulosomes are similar to bacterial cellulosomes in that the protein-protein interactions are mediated through parts termed the dockerin and cohesin, however the exact sequence for the cohesin module has yet to be established. Through a combination of –OMICs approaches and traditional biochemical assays, a large putative scaffoldin molecule was identified. The scaffoldin was heterologously expressed and screened for interaction with recombinant dockerin through an ELISA. The $K_{D,app}$ was determined using Equilibrium Surface Plasmon Resonance. A transcriptomic survey of dockerin domain-containing proteins revealed some degree of conservation in dockerin location on classes of CAZymes. Using this observation, the dockerin domains were adapted to thermostable cellulases, demonstrating its applicability as a novel protein scaffolding systems and suggesting the possibility of synthetic cellulosomes for biomass degradation.

Engineering Novel Modular Biosensors to Confer Metabolite-Responsive Regulation of Transcription

Younger, A. K. D.,
Dalvie, N. C., Northwestern University
Leonard, J. N., Northwestern University

Our ability to engineer microbial factories has accelerated as new technologies have enabled rapid mining of biological diversity and high throughput synthesis of novel ensembles of enzymatic pathways. To date, however, it remains challenging to monitor the presence of metabolites in individual cells, outside of the situations in which a natural metabolite-responsive transcriptional regulator has evolved and been identified. Such naturally occurring biosensors have proven useful in applications including both screening of high-dimensional libraries and engineering feedback control to optimize biosynthesis. To help address this need for novel biosensors, we have developed a modular and general approach for engineering novel transcriptional regulators that are actuated by the presence or absence of metabolites of interest – the Small molecule Responsive Transcriptional Regulator (SmaRTR) biosensor platform. The central goal of this approach is to develop a systematic strategy for converting proteins that bind small molecules into proteins that bind a target DNA sequence in a manner that depends on whether those small molecules are present. In our initial evaluation of this strategy, we utilized the BCR-ABL1 zinc finger protein as a model DNA-binding domain. We built and characterized a library of promoters including BCR-ABL1 binding sites at various positions to elucidate which promoter design features are most important for achieving effective zinc finger mediated-transcriptional repression. Then, we evaluated several strategies for converting the well-characterized maltose binding protein (MBP), a model of a ligand-binding protein, into a novel maltose-actuated transcription factor that regulates gene expression in a ligand-dependent manner. Our investigations using this model system resulted in several robust and functional novel biosensors, establishing the fundamental feasibility of the SmaRTR approach. We also elucidated protein and promoter engineering strategies that may ultimately be broadly applicable to developing a range of engineered biosensors. Ultimately, this technology may be harnessed to generate quantitative insights into metabolite dynamics in living cells, provide new tools for high throughput screening, and enable improved biomanufacturing through implementation of metabolic feedback control.

Rewiring Cellular Input-Output to Engineer Cell-Based Therapies That Interface Robustly with Human Physiology

Schwarz, K. A., Chemical and Biological Engineering, Chemistry of Life Processes Institute, R.H. Lurie Comprehensive Cancer Center, Northwestern University
Dudek, R. M., Northwestern University
Dolberg, T.,
Daringer, N., Northwestern University
Leonard, J. N., Northwestern University

Cell-based therapies have proven to be useful diagnostic and therapeutic tools for treating a wide variety of diseases such as arthritis, infectious diseases, and cancer. Treating cancer with genetically engineered T cells is particularly promising, and for patients with certain types of "liquid" cancers, this approach has already delivered astounding clinical benefits. However, current clinical experience indicates that extending these benefits to other types of cancer, as well as to other diseases, will require new therapeutic capabilities. In particular, there yet exists a need for technologies enabling one to program cells to sense and respond to their environment in a defined fashion. Toward this goal, our lab has developed a platform for engineering novel protein biosensors for extracellular cues, which we have termed modular extracellular sensor architecture (MESA). MESA comprises a self-contained receptor-signal transduction system in which ligand binding induces receptor dimerization, which then releases a sequestered transcription factor to regulate expression of an "output" gene or genes.

Here we present two expansions of the basic MESA capabilities, which make key strides towards meeting the needs identified above: 1) integrating ligand-specific MESA receptors with intracellular gene circuits to enable the cell to "process" multiparametric environmental cues, and (2) rewiring cellular input-output by coupling the sensing of biologically relevant extracellular signals to modulation of endogenous gene expression. To engineer biosensing of novel input ligands, we demonstrated that both camelid antibody analogs termed "nanobodies" and scFv antibody fragments could be incorporated as novel MESA ligand-binding domains. To achieve MESA-regulated induction of endogenous target gene expression, we integrated dCas9-based transcription factors as receptor "outputs". Altogether, these key proof-of-principle studies identified readily generalizable strategies for rewiring cellular input-output behavior in a user-defined fashion. This work lays the foundation for leveraging MESA receptors for a range of applications in both fundamental research as well as advanced cell-based therapies

Rewiring cellular input-output using modular extracellular sensors

Poster #171

Kelly A. Schwarz, Nichole M. Daringer, Taylor B. Dolberg, Joshua N. Leonard

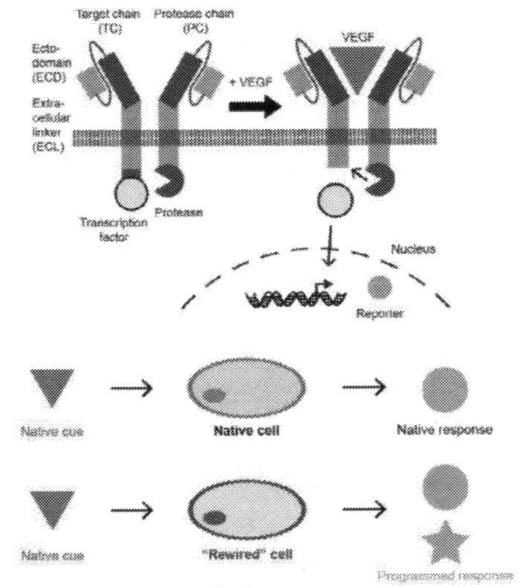

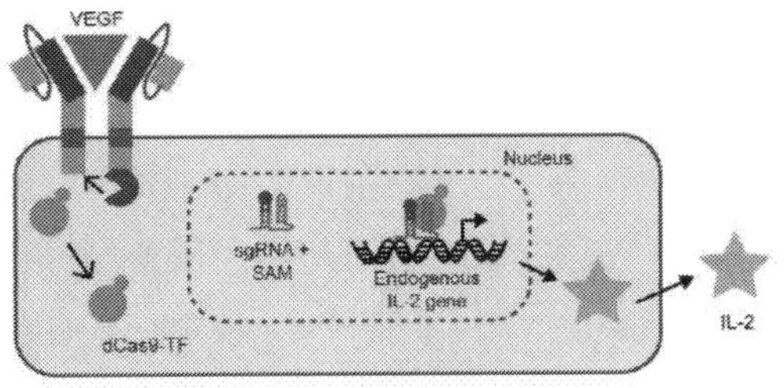

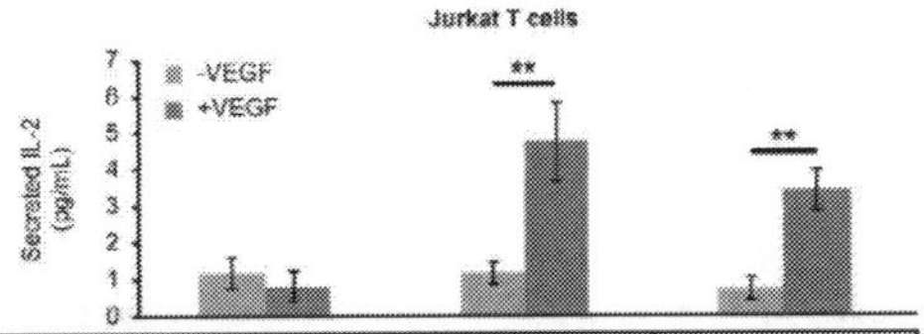

Fold	-	4.7	4.0
scFv	Vector only	V1-MESA	V2-MESA
sgRNA		sgRNA 7	sgRNA 7
TC:PC Ratio		10	10

MESA comprises a generalizable synthetic biology tool for engineering cells to manifest novel, programmable, input-output functions

CENTER FOR SYNTHETIC BIOLOGY DARPA NIH Northwestern McCORMICK SCHOOL OF ENGINEERING

Defining the Anti-Shine-Dalgarno Sequence Interaction and Quantifying Its Functional Role in Regulating Translation Efficiency

Hockenberry, A. J., Northwestern University
Pah, A. R.,
Jewett, M. C., Northwestern University
Amaral, L. A., Northwestern University

Studies dating back to the 1970s established that binding between the anti-Shine-Dalgarno (aSD) sequence on prokaryotic ribosomes and mRNA helps to facilitate translation initiation. However, nearly all of these results are based on over-expressed recombinant protein constructs, so the importance of this sequence interaction over evolutionary time for endogenous genes remains difficult to quantify. Due to their large number and and varying contexts, endogenous genes provide a rich dataset that can shed light on ill defined parameters such as the precise location of aSD binding relative to the start codon that maximizes translation efficiency, the full extents of the aSD sequence, and the functional form of the relationship between aSD binding and translation efficiency. Here, we leverage genome-wide estimates of translation efficiency to determine these parameters and show that anti-Shine-Dalgarno sequence binding increases the translation of endogenous mRNAs on the order of 50%. Our findings highlight the non-linearity of this relationship, showing that translation efficiency is maximized for sequences with intermediate aSD binding strengths. These mechanistic insights are highly robust; we find nearly identical results in ribosome profiling datasets from 3 highly diverged bacteria, as well as independent genome-scale estimates of translation efficiency. Further, we show evidence that intermediate-strength binding sequences maximize translation efficiency in controlled experimental data using recombinant GFP expression. By analyzing the rules that shape and constrain the evolution of endogenous genes, our findings highlight the important role that systems biology and genome-scale analysis methods can play in guiding synthetic design.

Modeling of Cell-Free Glycoprotein Production in Escherichia coli

Horvath, N. G., West Virginia University
Vilkhovoy, M., University of Massachusetts
Calhoun, K. A., Stanford University
Swartz, J., Stanford University
Varner, J. D., Purdue University

Cell-free systems offer many advantages for the study, manipulation and modeling of metabolism compared to in vivo processes. Many of the challenges confronting genome-scale kinetic modeling can potentially be overcome in a cell-free system. For example, there is no complex transcriptional regulation to consider, transient metabolic measurements are easier to obtain, and we no longer have to consider cell growth. Thus, cell-free operation holds several significant advantages for model development, identification and validation. In this study, we constructed an ensemble of genome-scale kinetic models of E. coli cell-free synthesis of the chloramphenicol acetyl transferase (CAT) protein. The model, which described 234 species and 276 interactions in central carbon and amino acid biosynthesis, was identified using dynamic CAT, energy, amino acid, glucose and organic acid measurements. Allosteric interactions were described using a novel rules-based approach which integrated complex allosteric regulation with traditional kinetic pathway modeling using non-linear transfer functions. Next, we used global sensitivity analysis to estimate which experimentally controllable parameters (e.g., metabolite or enzyme initial conditions) most influenced culture performance (e.g., rate of CAT production and CAT yield). From this analysis we identified a subset of metabolite and enzyme initial conditions that improved CAT protein production. Taken together, we developed a genome-scale kinetic model describing the cell-free production of a model protein. This model was identified using a comprehensive dynamic data set, and analyzed using global sensitivity analysis to determine experimentally adjustable factors which influenced culture performance. This study represents a first step toward the rational design of the dynamic operation of cell-free protein production.

Modulating Arginine Biosynthesis in E. coli nissle to Consume Ammonia in a Mouse Model of Hyperammonemia

Kotula, J. W., Synlogic, Inc.
Tucker, A., Synlogic, Inc.
Isabella, V. M., Synlogic, Inc.
Falb, D., Synlogic, Inc.
Miller, P., Synlogic, Inc.
Silva, A., Synlogic, Inc.
Millet, Y., Synlogic, Inc.
Anderson, C., Synlogic, Inc.

Urea cycle disorders (UCD) are a collection of genetic conditions in which the patient is unable to convert nitrogen to urea, resulting in the toxic accumulation of ammonia systemically (hyperammonemia). Up to 70% of the excess blood ammonia originates from the gastrointestinal tract, making a gut microorganism-based therapeutic a promising platform to absorb the excess ammonia.

Towards this goal, we have engineered strains of the probiotic bacterium *Escherichia coli* Nissle to consume ammonia by upregulating expression of the arginine (Arg) biosynthetic genes. Here we compare two strategies for optimizing expression of the Arg pathway in Nissle: deletion of the ArgR transcriptional repressor that regulates arginine biosynthesis (Nissle Δ*argR*) and mutation of ArgR binding sites in each *arg* locus (Nissle ΔArgBox). We demonstrate that both strategies yield increased arginine biosynthesis relative to wildtype Nissle, with the Nissle Δ*argR* stain producing the highest levels of arginine. We used metabolite profiling and qPCR to characterize the arginine biosynthetic pathway performance in the Nissle Δ*argR* and Nissle ΔArgBox strains. The metabolic profile of the Nissle ΔArgBox strain revealed several important differences when compared with the Nissle Δ*argR* strain. Overexpression of ArgG (argininosuccinate synthase) increases arginine yield in the Nissle ΔArgBox, but arginine production was still below that of the Nissle Δ*argR* strain. We have performed qRTPCR and RNAseq experiments to identify other potential genetic modifications that may improve arginine production in the Nissle DArgBox strain.

Concurrent with Nissle ΔArgBox strain optimization, we assessed the performance of the Nissle Δ*argR* strain as a therapy in a genetic mouse model of UCD (spf-ASH). The Nissle Δ*argR* strain effectively reduced blood ammonia levels in these mice, validating the Arg biosynthesis approach for reducing blood ammonia levels from the gastrointestinal tract.

Unravelling RNA Structure / Function Relationships Via Novel Next-Generation Sequencing Techniques

Loughrey, D.,
Lucks, J. B., Cornell University

The functionality of RNA in the cell is deeply related to its structure, and methods that merge chemical probing with next generation sequencing have created breakthroughs in the throughput and scale of RNA structure characterization. In this work we use Selective 2'-Hydroxyl Acylation Analyzed by Primer Extension Sequencing (SHAPE-Seq) to begin to resolve the RNA structure / function relationship on a scale necessitated by the omics-level questions being asked today. The flexibility and accuracy of the SHAPE-Seq technique, as part of the toolbox of powerful related NGS methods, has the ability to provide us with a reference point to both validate computational RNA structural prediction algorithms and observe RNA structures as they exist and interact in their natural environment. Here, in this work, we develop and apply SHAPE-Seq to study several model RNA systems, each of which offers an opportunity to study a different type of RNA structure / function relationship, or a different biological system. This includes providing structural analysis into protein-mediated small RNA regulatory mechanisms in E. coli and further insight into the exciting field of eukaryotic lncRNAs. We anticipate that our work with the SHAPE-Seq technology will provide the scientific community with a broad yet precise set of tools, aimed at providing unprecedented characterization of cellular RNA regulatory mechanisms. This will hopefully serve to be illuminating in offering deeper insight into a whole manifest of wide-ranging biological questions.

Using Volatile Gas Production As an Output for Genetic Circuits to Report on Programmed Microbial Behaviors in a Soil

Cheng, H. Y.,
Silberg, J. J., Rice University
Masiello, C. A., Rice University
Bennett, G. N., Rice University

Microbes drive processes in the Earth system far exceeding their physical scale, mediating significant fluxes in biogeochemical cycles. Microbial behavior also affects soil development, water quality, and crop yields. The tools of synthetic biology have the potential to significantly improve our understanding of the roles that microbes play in these processes, as well as the effects that heterogeneous environmental materials have on microbial behaviors. However, synthetic biology has not yet seen wide use within environmental materials (soils, sediments, and biomass) because synthetically programmed microbes are hard to deploy into these materials, the vast majority of which are not transparent and are heterogeneous. Traditional visual reporting strategies used to read out circuit status are challenging to use in Earth materials because environmental matrices display high absorbance and auto-fluorescence at wavelengths of light used for traditional visual reporters like GFP. These technical limitations have made it challenging to use programmed microbes to study how variation in soil environmental parameters (moisture, nutrient status, mineralogy, structure, and temperature) affect real-time biological behaviors. To help synthetic biology make the petri-dish-to-soil transition, we are using enzymes that produce small volatile gases as gene expression reporters in soils. We will present data showing that this approach can be used to report on microbial gene expression dynamics in bulk soils at and below the water holding capacity. We will also describe the signal-to-noise for this reporting strategy within a Typic Hapludalf soil from Kellogg Biological Station that consists of a mixture of sand, silt, and clay. Furthermore, we will describe our studies using gas reporting to monitor two dynamic soil microbial processes within this soil, horizontal gene transfer and quorum sensing, and our efforts to benchmark the signal from a volatile gas against standard reporting approaches. Unlike previously described reporters of gene expression, these gas reporters should be useful for studies of microbial processes within soil samples at a range of scales, including sealed vessels widely used by soil scientists to monitor the dynamics of greenhouse gas production.

Tuning the Electron Transfer Activity of a 2Fe2S Ferredoxin Using Homologous Recombination and an Escherichia coli Selection

Atkinson, J. T., Rice University
Campbell, I. J., Rice University
Bennett, G. N., Rice University
Silberg, J. J., Rice University

Iron-sulfur (Fe-S) cluster containing ferredoxins (Fd) are low potential (high energy) metalloproteins that distribute electrons across diverse metabolic pathways by leveraging interactions with an array of partner proteins (the Fd interactome). *In vitro* studies have revealed Fd paralogs with high sequence identity (>90%) can vary drastically in their functional profiles when coupling to the same e- donors and acceptors. However, these studies have not yet provided sufficient insight into Fd to tune their redox proportioning to different partner proteins for synthetic biology applications. To improve our understanding of the underlying sequence-structure-function relationships that dictate Fd redox proportioning and cycling efficiencies, we are recombining *Mastigocladus laminosus* and *Prochlorococcus* cyanomyophage P-SSM2 Fd, plant-type 2Fe2S Fd that display ~53% sequence identity, and analyzing the cellular functions of chimeras. Ferredoxin function is being quantified by analyzing the complementation of an *Escherichia coli* sulfide auxotroph, which requires electron transfer through a linear redox pathway that is made up of Fd-$NADP^+$ reductase, Fd, and sulfite reductase. To quantify the relative activity of chimeras, Fd expression was placed under control of the promoter $P_{LtetO-1}$ and bacterial complementation has been measured in the presence of varying anhydrotetracyline (aTc) concentrations. Our preliminary experiments have revealed that some chimeras require low levels of aTc (2.5 ng/mL) to complement bacterial growth half maximally, like *M. laminosus* Fd, while other chimeras only partially complement bacterial growth at high levels of aTc (250 ng/mL). Ongoing efforts are investigating whether these differences in cellular activity arise because of changes in cellular expression, protein stability, cofactor binding, redox potential, partner binding, and partner allosteric interactions. By comparing the complementation of dozens of Fd, our results will allow us to calibrate how Fd redox activity relates to structural disruption calculated using biophysical models. In addition, they will provide fundamental insight into Fd sequence-structure-function relationships, which are currently needed to inform the construction of orthogonal redox circuits that can operate alongside host cell redox circuits.

Recursive Cloning for the Assembly of Biosynthetic Pathways and Artificial CRISPR Arrays

Van Hove, B., Ghent University
De Wannemaeker, L., Ghent University
Guidi, C., Ghent University
Maertens, J., Ghent University
De Mey, M., Massachusetts Institute of Technology

Advances in the field of Synthetic Biology have given rise to a multitude of new molecular cloning techniques with improved flexibility, efficiency and accuracy. While these methods have enabled multi-part assembly of complex constructs, they are often lacking in several respects: How can additional parts be added after assembly? What about standardisation to custom specifications? Here we present Protected Oligonucleotide Duplex Assisted Cloning (PODAC), a novel cloning technique that eschews large, one-off multi-part assemblies in favour of a more iterative workflow.

PODAC is a Golden Gate derivative whereby a cloning site for future iterations is included together with the genetic part to be cloned. To protect the new cloning site from being destroyed itself during the one-pot restriction and ligation reaction, it is supplied in the form of a synthetic oligonucleotide duplex carrying methylated cytosine residues. Passage through E. coli reactivates these protected sites, resulting in a construct that can accept another part during a next cycle of PODAC.

Each cycle is performed using the same endonuclease and sticky ends. This not only reduces overall complexity but also maximises the reusability of part storage vectors. Because the restriction enzyme used (BsaI) cleaves DNA outside of its recognition site, the sequence of the sticky ends and resulting assembly scars can be chosen at will. This allows for easy standardisation to the specifications imposed by the experimental context. As a first proof-of-principle, PODAC was used to assemble the violacein biosynthesis pathway.

Moreover, because PODAC does not rely on PCR or long sequence homologies, it can safely be applied to designs containing repeated sequences. One application where this is the case is the creation of artificial CRISPR arrays. To demonstrate its broad applicability, PODAC was used to construct guide RNA arrays that target various catabolic pathways in E. coli for dCas9 mediated transcriptional roadblock silencing.

In conclusion, we believe that the wide applicability and amenability to standardisation make PODAC a useful addition to the existing Synthetic Biology toolkit.

The Development, Implementation and Transfer of a Reproducible Methodology for the Generation of High Quality Data for Part Characterisation

Exley, K., Imperial College London
Hirst, C. D.,
Ainsworth, C. M., Imperial College London
Poh, C. L., Nanyang Technological University
Kitney, R. I., Imperial College

Synthetic Biology aims to apply an engineering approach to bio design. This involves breaking down biological function into components, Parts or modules. One of the key elements of this approach is that it uses standard Parts. Is relies on the ability to accurately characterise parts and to reproduce the characterisation process. Reproducibility is at the heart of engineering biology, but this can be challenging. Consequently a collaborative project has been undertaken between Imperial College London and Nanyang Technological University (NTU) in Singapore to determine the transferability and reproducibility of our characterisation process.

The Kitney Lab at Imperial College has developed a robust and automated method to measure the function of Parts. Because we are interested in reproducibility, the joint project with NTU has focused primarily on constitutive promoters. The methodology has been designed and optimised to ensure reproducibility, with the workflow comprising a number of stages. This includes an initial sample preparation with an overnight growth, dilution step in a chemical defined media, followed by a 90 minute outgrowth and with a second dilution step to a specified cell concentration (absorbance). The cell absorbance and fluorescence data are collected at regular (15 minute) intervals during a total of 6 hours - to calculate the promoter activity. The method was employed in triplicates and over 3 different characterisation runs (i.e. 3 different days). The promoter activity was shown to be highly reproducible using this approach.

The methodology was originally developed using an automated liquid handling system (TheOnyx;ALS, Germany). To test the robustness of the workflow, the same set of promoters were also characterised on a different automated liquid handling system (a FeliX; AnalytikJena) in our centre at Imperial College. The two automated systems have differences in their pipetting capabilities, but by tightly following the workflow the characterisation data was demonstrated to be closely matched.

The end goal of this project was to show the transferability of the methodology and that difference in the site, users and/or hardware would not significantly impact the comparability of the results. To test this method was utilised in the Poh Lab at NTU. Using the workflow developed in the Kitney Lab (but performing the dilution steps manually), similar results were obtained at NTU for the characterisation of an identical set of promoters. The next stage in the project, which will also be reported, is to undertake automated characterisation in Singapore on the same set of promoters.

In conclusion, we have developed an automated, reproducible and transferable characterisation methodology. This has initially been applied to the characterisation of a set on constitutive promoters. The workflow is now being modified and applied to other Part families. A lack of reproducibility is often cited as weakness in biological experimentation more generally. In synthetic biology it is essential. The development of our automated workflow is a step towards the standardisation of characterisation protocols and the production of large numbers of Parts with associated high quality, reproducible characterisation.

Strategies to Enhance the Yield of Biocellulose in Gluconacetobacter Xylinus

Kumar, P.,
McClay, K.,
Fuller, M.,
Reed, J., University of Wisconsin-Madison

Gluconacetobacter xylinus is known to produce high quality biocellulose. The biocellulose obtained from *Gluconacetobacter xylinus* is devoid of the cell wall components which makes it very pure. This highly pure cellulose can be used for many applications including medical and defense applications. This project seeks to enhance biocellulose yields from *Gluconacetobacter xylinus* strains. Optimizing metabolic and regulatory changes to enhance chemical production is a time-consuming and expensive process. However, this process can be expedited using a genome-scale metabolic model. Metabolic models can evaluate multiple strain design strategies to identify those with the greatest strain improvements. We sequenced the genome of *Gluconacetobacter xylinus* ATCC 53582 and used this information to build a genome-scale metabolic model of this strain. The growth predictions of the genome-scale metabolic model were consistent with the experimental growth observations performed using Biolog phenotype microarrays. We applied the SimOptStrain algorithm to the metabolic model to identify gene deletion and reaction addition strategies to enhance the yield of bio-cellulose from glucose in the minimal media. The proposed genetic changes are predicted to enhance the biocellulose yields by almost three fold compared to the yields reported for the wildtype strain. Future work will focus on implementing and validating the effects of the metabolic changes experimentally.

Deciphering the Regulation of Biomass Degradation By Anaerobic Fungi

Henske, J. K., University of California, Santa Barbara
Solomon, K. V., University of California, Santa Barbara
Theodorou, M. K., Harper Adams University
Grigoriev, I.,
O'Malley, M. A., University of California, Santa Barbara

To strengthen existing renewable technologies, more efficient methods to extract sugars from crude plant biomass (lignocellulose) are required. Biomass breakdown represents a significant barrier towards the production of fuels and chemicals. However, some microbes in extreme environments have evolved to degrade lignocellulose efficiently. Anaerobic gut fungi are the primary colonizers of plant material in the guts of herbivores, like sheep and goats. Encoded within their genomes is an incredible variety of biomass degrading enzymes including cellulases, hemicellulases, esterases, and polysaccharide deacetylases. However, little is known about the overall suite of biomass degrading enzymes produced, and even less is known about their regulation.

Sequencing technologies have improved and become more accessible offering a useful method to examine regulation of biomass degrading genes and gene regulators in gut fungi. We have used RNA sequencing technologies to elucidate regulation patterns in anaerobic fungi in response to carbon source of varying complexity. Two species of fungi, *Anaeromyces robustus* and *Neocallimastix californiae*, were grown on glucose, cellobiose, cellulose, and lignocellulose. Expression counts of transcripts were compared to the base case of glucose, where expression of biomass degrading enzymes was lowest represented the smallest percentage of total transcriptome expression. In both cases gene set enrichment analysis demonstrated enrichment of cellulases and hemicellulases on more complex substrates. However, in the case of *A. robustus* enrichment appears to be primarily triggered by cellobiose, but for *N. californiae* enrichment occurs more gradually, which some GH classes enriched during growth on cellobiose and cellulose while others were enriched only on biomass. These results suggest that different species of gut fungi maintain varying global regulation networks. For *A. robustus*, while there may be multiple integrated regulatory mechanisms, the regulation of biomass degrading machinery may be primarily controlled by simple break out molecules, like cellobiose.

Further, we are integrating these results with recently obtained genomic information to identify common regulons and promoters that govern transcriptional regulation by substrates. By examining the upstream regions of genes with similar predicted function as well as similar regulation patterns, we may identify promoters used to control the expression of these enzymes. Collectively, this information will provide important insight into gut fungal metabolism, and present new strategies to amplify their breakdown abilities.

Cell-Free Synthesis and Assembly of Protein Nano-Machines on a Chip

Daube, S. S.,
Bar-Ziv, R. H., Weizmann Institute of Science

Towards the development of synthetic bio-devices that mimic basic cellular processes, we are using a novel photolithographic approach to carry out controlled gene expression reactions on a silicon chip [1]. Our approach uses surface-immobilized, highly dense gene compartments, reaching in vivo DNA densities, and localized antibodies that serve as traps for gene products [2, 3]. Addition of a cell-free extract results in a complete transcription-translation reaction that can be modulated by the orientation and density of immobilized DNA [4]. Protein products are localized on the antibody traps next to their gene source forming membrane-free compartments.

The cell-free biochip approach provides conceptual advantages for studying bio-assembly processes of large macromolecular complexes such as T4 bacteriophage and the ribosome: (i) Cell-free expression renders coupling between synthesis and assembly, better mimicking assembly in vivo (ii) Surface anchoring generates structural scaffolds for assembly [5]. We have already shown that expression of bacteriophage T4 tail protein resulted in correct self-assembly into long nanotubes on surface-patterned antibody traps, as imaged directly by transmission electron microscopy [2]. We are currently investigating the outcome of expanding the number of T4 genes that can be added to cell-free on-chip reactions and that would result in correct assembly. In addition we are developing tools to decipher viral and ribosome assembly pathways and explore the possibility to affect assembly outcome by the initial patterns of localized genes and antibody traps.

[1] Small

[2]Nature Nanotech 7, 374-378, 2012

[3] MSB 4, 181, 2008

[4]PNAS 107, (2010) 2836-2841

[5]WIREs Nanomed Nanobiotechnol, 5, (2013) 613–628

Engineering Novel Functionality into Red Blood Cells

Meinders, M., Bristol University
Satchwell, T.,
Hawley, B.,
Pellegrin, S., University of Bristol
Frayne, J.,
Toye, A.,

The red blood cell (erythrocyte) has a well-characterised life cycle and biophysical properties that make them the ideal vehicle for therapeutic use; they are biocompatible, biodegradable, are already designed to traverse the vascular system and lack nuclei. Research to date on the use of erythrocytes as cellular delivery agents has generally focused on loading mature erythrocytes with drugs, encapsulation using isolated erythrocyte membranes or functionalization of the red cell surface, alterations that can be detrimental to the biophysical properties of the erythrocyte.

Our approach is to use an ex vivo erythroid culture system that allows us to culture immature erythrocytes (reticulocytes) from either CD34+ cells isolated from peripheral blood or alternatively by using novel human immortalised erythroid cell lines. We have shown that these erythroid cells can be manipulated by altering their protein expression during erythroid development to engineer erythrocytes with desirable properties.

To utilize the erythrocyte as an effective novel cellular chassis, requires a better understanding of membrane assembly and the rules for the introduction of exogenous membrane proteins. We have determined when many key membrane proteins in the erythroid membrane assemble, and have overexpressed 9 selected endogenous and exogenous single and multipass membrane proteins. These exhibited a wide variation in the level of membrane expression. We are now engaged in a second phase of design to increase protein retention by reducing loss of proteins due to remodelling during differentiation and enucleation. To achieve this we are altering protein sorting and attachment to the underlying cytoskeleton. The eventual goal is to develop a "human red blood cell engineering" toolkit to manipulate erythrocytes during their development to obtain novel features and characteristics.

Extending Amyris' Genotype Specification Language (GSL) for Modular and Combinatorial Genome Engineering

Wilson, E. H.,
Platt, D.,

As synthetic biologists continue to push the boundaries of genome engineering techniques and strategies, the infrastructure to implement these new strategies must similarly expand. Although individual engineers have been able to produce beautiful artisanal constructs using direct DNA editing techniques, they cost significant time to intricately align raw sequences and are thus error-prone and unscalable. These shortcomings have led to the development of a host of Computer Aided Design (CAD) tools to enable engineers to more efficiently engineer genomes. These CAD tools have ranged from symbolic drag-and-drop interfaces to text-based language tools, often with the goal of moving users away from editing raw DNA and towards more general and reusable genetic elements. Genotype Specification Language (GSL) is an instance of a language-based genome engineering tool that has been developed and tested within Amyris' strain design platform for the past 6 years, and has recently been made available as open source. GSL balances the needs for quick and efficient design creation against the needs for design complexity and specificity; it abstracts functional genetic parts—such as gene, promoter, terminator, and homology parts—into a simple syntax while providing precision editing syntax to allow users to manipulate the sequences within these parts. Although GSL's code-like interface may feel most natural to those with some background in programming, its syntax style was intentionally designed to reflect notational conventions from genetics, which has facilitated easier adoption of the language among biologists as well. Today, as Amyris' pipeline expands to include the rapid and parallel scale up of hundreds of molecules, so too has GSL's syntax expanded to include new features such as variables and functions. During initial testing, these features have enabled easier specification of modular and combinatorial designs and in turn have improved engineers' ability to submit high-quality suites of designs to Amyris' Automated Strain Engineering service for construction.

Prediction of Host-Circuit Interactions Using Cell-Free Protein System

Borkowski, O., INRA

The design of synthetic circuits rarely accounts for the impact production of orthologous proteins have on the host cell. However, in bacteria the burden triggered by the production of synthetic parts can lead to significant decreases in protein production efficiency, to growth rate decreases and even cell death. Moreover, alteration of cell physiology considerably impacts the behaviour of synthetic circuits complicating our abilities to produce reliable predictions and efficient designs. To address this, we combined Cell-Free Protein Systems (CFPS), measurements in *E. coli* and modelling to enable prediction of the impact synthetic circuits have on cells.

CFPS has been used by others to rapidly obtain protein expression measurements and predict protein production in engineered *E. coli*. Here, we extend this further and use CFPS as a platform to measure competition for resources between different synthetic circuits, allowing us predict *in vivo* burden. We measured the expression and resource-use of different genes of interest, combined with different parts in both CFPS and *in vivo* with *E. coli*. The competition for resources between our previously described capacity monitor and the different genes of interest revealed a strong correlation between CFPS and *in vivo* measurements thus reinforces the utility of CFPS measurements for predicting complex behaviours in cells.

In parallel, we also developed a whole cell model to extend predictions of both *in vivo* circuit performance and burden in different growth conditions and also in cells producing multiple genes of interest. By combining these efforts, we will provide a novel method for rapid *in vitro* screening of synthetic parts that then allows robust prediction of synthetic circuit behaviours, efficiencies and their impacts on the host cell.

Directing Inward Electron Transport to the Cytosol in Mtr-Escherichia coli Using Fumarate Reductase

Baruch, M., Lawrence Berkeley National Laboratory
Ajo-Franklin, C.,

The use of exoelectrogens in microbial electrosynthesis offers the benefit of high Faradaic efficiency due to the electron transfer through direct contact between their transmembrane electron transfer (ET) system and the electrode. Currently, the only molecularly-defined inward ET pathway is found in *Shewanella oneidensis,* wherein the electrons passing through the Mtr-CymA ET system are been urilize by the periplasmic fumarate reductase FccA for fumarate reduction. Nonetheless, the reducing equivalents in this system do not pass through the inner membrane barrier to cytosolic carriers, where they would be of most value for microbial electrosynthesis. Our group has transplanted the *S. oneidensis* inward ET system into the host organism *Escherichia coli*, enabling outward ET through a heterologous system. However, it was unknown if these proteins are sufficient to confer inward electron transfer as well in *E. coli*. Using a three-electrode electrochemical system, we monitored electron flow from cathodes to unmodified and Mtr-expressing *E. coli* in response to fumarate addition. We found only the Mtr- *E. coli* consumed significant levels of current and sustained this consumption over 5 days. By using a knockout mutant and complementation, we showed that Mtr-modified *E. coli* can use the native fumarate reductase FrdABCD for cathode-driven fumarate reduction. We also found that over-expression of FrdABCD significantly increases current production in Mtr-*E. coli*. In contrast to FccA, FrdABCD is localized to the cytoplasmic face of the inner membrane and the reduction reaction occurs in the cytoplasm, making this a complete inward ET pathway. Thus, we have described for the first time a molecularly-defined inward ET pathway to the cytosol which opens the door for the delivery of electrode-derived reducing equivalents to cytosolic metabolic pathways.

Cell-Free Metabolic Engineering of Carotenoid Biosynthesis

Mohr, B., University of Tennessee
Doktycz, M. J., Oak Ridge National Laboratory

Engineering heterologous metabolic pathways is a time consuming endeavor, even in model microbial systems like *E. coli* and *S. cerevisiae* that have benefited from recent advances in genetic assembly and pathway optimization techniques. Cell-free metabolic engineering, or the use of genetically engineered biochemically active cell extracts for the optimization and synthesis of biochemical products, is an accompanying strategy that can reduce engineering times, explore biosynthetic pathways in diverse organisms and produce novel biochemicals. Isoprenoids are plant and microbial secondary metabolites of interest to metabolic engineers due to their varied ring structures and degrees of unsaturation. They have applications in pharmaceuticals, biocontrol agents and fuels. Carotenoids are a predominantly linear member of the isoprenoids that play important roles in membrane structure and resistance to oxidative stress in microbes; they have also been shown to be beneficial to human health. Here, we are examining the use of a cell free system for the development of a multi-step biosynthetic pathway for carotenoids. Crt I and Crt B from *Pantoea* sp. YR343 are being produced through cell-free protein synthesis. These protein products constitute the biosynthetic pathway for lycopene starting with the substrate geranylgreranyl pyrophosphate. Without the limitations of a cellular membrane and cell viability, gene variants and enzyme stoichiometry can be manipulated through the addition of heterologous DNA for increased product yield without repeated genetic engineering. Recent progress will be presented.

Expanding Biosensing Abilities through Computer-Aided Design of Metabolic Pathways

Libis, V., CNRS
Delépine, B., CNRS
Faulon, J. L., University of Manchester

Detection of chemical signals is critical for cells in nature as well as in synthetic biology where they serve as inputs for designer circuits. Important progress has been made in the design of signal processing circuits triggering complex biological behaviors but the range of small molecules recognized by sensors as inputs is limited. The ability to detect new molecules will increase the number of synthetic biology applications but direct engineering of tailor-made sensors takes time. Here we describe a way to immediately expand the range of biologically detectable molecules by systematically designing metabolic pathways that transform non-detectable molecules into molecules for which sensors already exist. We leveraged computer-aided design to predict such sensing-enabling metabolic pathways and we built several new whole-cell biosensors for molecules such as cocaine, parathion, hippuric acid and nitroglycerin.

Consensus Parameter Inference and Optimal Circuit Design in Cascade Circuits

Huynh, L., UC Davis
Rai, N., UC Davis
Pereira, B., UC Davis
Tagkopoulos, I., University of California, Davis

Accurate prediction and design of synthetic circuit behavior is crucial if synthetic biology aspires to scale up in circuit complexity. A central challenge towards this goal is the lack of parameter values for fundamental parts and modules used in circuit design. Experimental measurements of these values while possible, are difficult, expensive, context-dependent and time-consuming. At the same time, parameter inference from single circuits is more descriptive than predictive due to over-fitting and model sloppiness, given the vast parameter space. To address this challenge, we developed an approach where parameter values are simultaneously inferred from datasets measured in synthetic circuits that have overlapping parts. For this, we curated and integrated experimental data from literature in a Part And Module library (PAMLib). We used a consensus method and a universal model to estimate parameter values for all parts, as well as their confidence intervals of prediction. We then used a computer aided design (CAD) tool together with optimization techniques to find optimal designs for several cases of 3-gene cascades. We evaluated both *in silico* and *in vivo* the capacity of this method to predict circuit behavior and guide the design of new cascade circuits with the desired dynamic characteristics.

TNT-Cloning: A New Platform for Flexible and Universal Usage of DNA Parts

De Paoli, H., Oak Ridge National Laboratory
Tuskan, G., Oak Ridge National Laboratory
Yang, X., Oak Ridge National Laboratory

Synthetic biology has developed rapidly due to our ability in performing deep-sequencing and synthesizing/assembling large fragments of DNA. However, the testing of several clusters for genetic circuits and pathway engineering is still costly and time consuming in the absence of a flexible, universal and all-inclusive methodology for joining DNA parts. Countless methods for *in vivo* and *in vitro* cloning were developed and different syntaxes were proposed, however, the natural course led the diverse synthetic biology community to adopt similar but yet distinct methodologies, suggesting the current available methods still lack coverage. Recently a modified typeIIS-based method called TNT-cloning was described showing unprecedented flexibility and universality for cloning DNA parts (De Paoli et al, 2016, Scientific Reports). Here we present a overview and comparison of the TNT-cloning versus other typeIIS-based methods and homology-based methods along with improvements performed in the published method as the inclusion of a lethal reporter (ccdB), for faster and more efficient cloning, as well as the putative expansion of the system to yeast and mammals aside E. coli and plants. The improved TNT-cloning system represents a step forward for supplying the synthetic biology community with a clean, efficient and malleable cloning system that has the potential to establish a quick and transferable platform necessary for determination of qualitative and quantitative gene fragment interactions that involves gene sets and gene networks in several fields.

Transcription Analysis and Metabolite Profiling of the Terpenoid Indole Alkaloid Pathway in Engineered Catharanthus Roseus Hairy Root Lines

Zhao, L., Iowa State University
Sun, J., Colorado State University
Li, C.,
Gibson, S.,
Peebles, C., Colorado State University
Shao, Z., Iowa State University
Shanks, J. V., Iowa State University

Catharanthus roseus (*C. roseus*) produces more than 130 different terpenoid indole alkaloids (TIAs), and so far is the only source for obtaining vinblastine and vincristine, which are currently widely applied in chemotherapies for many types of cancers. Those TIAs exist at very low concentrations in *C. roseus*, making direct extraction from plant laborious and costly. Hairy root culture is one of the alternatives to produce valuable secondary metabolites with high biochemical and genetic stability and capacity of large-scale fermentation.

This poster summarizes the results of several *C. roseus* hairy root lines that were engineered with the goal of enhancing flux towards the TIAs. Quantitative PCR of several transcription regulators and enzymes involved in TIA pathway was done to evaluate the engineered hairy root lines in transcription level. In addition, high performance liquid chromatography (HPLC) systems coupled with either mass spectrometer or photodiode array (PDA) detectors were applied to identify and quantify the key metabolites in the TIA pathway.

Incorporating Phenotype Data to Improve Parental Strain Flux Estimations

Long, M., University of Wisconsin – Madison
Reed, J. L., University of Wisconsin

Genome-scale constraint-based models have been widely used to predict extracellular and intracellular fluxes. Many of these methods, such as Flux Balance Analysis (FBA), depend upon the assumption of maximum growth (or some other objective function). When cells have not evolved to maximize growth or where other processes (*e.g.*, regulation) limit metabolic fluxes, these methods may fail to accurately describe the flux distribution in a cell. Methods have been developed which predict a perturbed cellular state from a previously estimated parental state; however, these methods (*e.g.* Minimization of Metabolic Adjustment) are heavily dependent upon the accuracy of the parental state flux distribution.

In order to improve the accuracy of the parental state flux prediction, various methods have been developed which incorporate various types of experimental data including extracellular flux, transcriptomic, and proteomic data. However, extracellular fluxes are not sufficiently constraining to substantially improve flux predictions, gene and protein expression levels do not strongly correlate with flux or help improve flux estimations, and intracellular fluxes are difficult and costly to measure using ^{13}C metabolic flux analysis. Therefore, we propose a new method called Relative Phenotypes for Parental Strain estimation (REPPS) that uses multiple sets of extracellular flux measurements. By utilizing extracellular data for a number of perturbed reference strains (*e.g.* single gene knockouts), as well as the parental strain (*i.e.* wildtype), we can substantially improve the descriptive flux estimation of the parental strain as well as flux predictions in subsequent mutant strains derived from the parental strain. For example, using REPPS in *Escherichia coli* with data from five single gene knockout reference strains, the mean squared error between model-estimated and measured fluxes is 53% less than the equivalent parsimonious FBA prediction for the parental strain. This improved parental flux estimate further improves predictions of fluxes in derived strains, where using MOMA with the REPPS estimated parental flux distribution yielded a 58% decrease in mean squared error between predicted and measured fluxes, compared to using MOMA with the pFBA predicted parental flux distribution.

REPPS is also powerful in its ability to be less affected by omissions in a model. Since REPPS attempts to fit a wide variety of experimental data, it is less sensitive to missing or incorrect reactions that may be in a model. Therefore, REPPS can also be used for lesser studied organisms whose models are less developed or for organisms where optimal growth is not their principal objective.

These improvements in accuracy with REPPS result in both better descriptive flux distributions in the parent strain, as well as better prediction of derived-strain phenotypes and fluxes. The improvement in descriptive modelling can improve our understanding of the key mechanisms underlying observed behaviors in a strain. Furthermore, the improved predictive capabilities can reduce the number of experiments necessary to create a chemical production strain, thus improving our ability to engineer strains with desired phenotypes.

Engineering Bacterial Two Component Sensors to Study Gut Inflammation

Daeffler, K.,
Hartsough, L. A., Rice University
Sheth, R. U., Rice University
Tabor, J. J., Rice University

The vast majority of bacteria in a healthy gut produce their energy by fermentation of complex polysaccharides and amino acids. However, a minority of bacteria found in the gut, including pathogenic enterobacteria such as *E. coli* and *S. typhimurium*, have a suite of enzymes, called reductases, which enable them to produce energy by coupling ATP production to the transfer of electrons from a donor molecule to the specific electron acceptor used by the reductase. Reactive oxygen and nitrogen species produced during intestinal inflammation result in the production of oxidized species that can be used by bacteria as electron acceptors during anaerobic respiration. These energy sources, only accessible to bacteria with reductases, allow enterobacteria to thrive during inflammatory conditions and out-compete the normal microbial flora, resulting in an altered composition of the microbiome. Enterobacterial blooms have been shown to exacerbate inflammation and are associated with inflammatory bowel diseases and colorectal cancer. To identify inflammatory conditions that can lead to enterobacterial blooms, we seek to detect inflammation-generated electron acceptors directly at the site of inflammation. Enterobacteria can use many terminal electron acceptors during anaerobic respiration including nitrate, nitrite, tetrathionate, TMAO, DMSO, and thiosulfate, all of which are produced in the gut during inflammation. Bacteria have naturally evolved to sense and respond to these molecules in the environment using two-component systems (TCSs), based upon membrane-associated histidine kinases. We have developed a computational and experimental TCS mining methodology and used it to identify and engineer the first ever tetrathionate and thiosulfate sensors. Our sensors show a strong response to and a high specificity for the desired ligand with no unwanted cross-regulation in contrast to a previously reported tetrathionate sensor, which is cross-regulated by oxygen. We have optimized our sensors to work *in vitro* in a non-pathogenic lab strain of *E. coli* as well as a probiotic gut-colonizing strain of *E. coli* for *in vivo* testing in mouse models of intestinal inflammation that lead to the development of colon cancer.

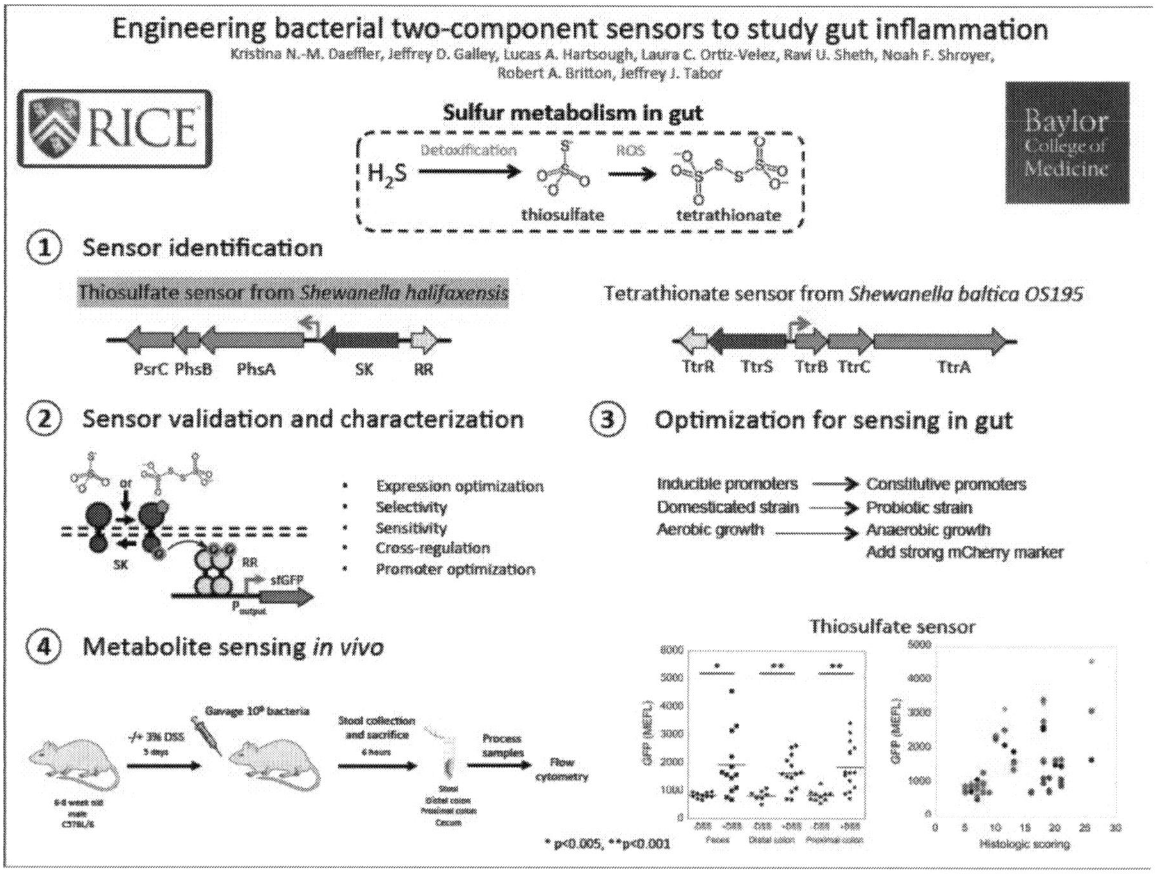
Engineering bacterial two-component sensors to study gut inflammation
Kristina N.-M. Daeffler, Jeffrey D. Galley, Lucas A. Hartsough, Laura C. Ortiz-Velez, Ravi U. Sheth, Noah F. Shroyer, Robert A. Britton, Jeffrey J. Tabor
RICE
Baylor College of Medicine
Sulfur metabolism in gut
H2S
Detoxification
ROS
thiosulfate
tetrathionate
① Sensor identification
Thiosulfate sensor from Shewanella halifaxensis
PsrC PhsB PhsA SK RR
Tetrathionate sensor from Shewanella baltica OS195
TtrR TtrS TtrB TtrC TtrA
② Sensor validation and characterization
or
SK
RR
sfGFP
Poutput
Expression optimization
Selectivity
Sensitivity
Cross-regulation
Promoter optimization
③ Optimization for sensing in gut
Inducible promoters → Constitutive promoters
Domesticated strain → Probiotic strain
Aerobic growth → Anaerobic growth
Add strong mCherry marker
④ Metabolite sensing in vivo
-/+ 3% DSS
5 days
Gavage 10⁹ bacteria
Stool collection and sacrifice
6 hours
Process samples
Flow cytometry
Stool
Distal colon
Proximal colon
Cecum
Thiosulfate sensor
GFP (MEFL)
Feces
Distal colon
Proximal colon
Histologic scoring
* p<0.005, **p<0.001

Bacteriocin Production and Screening Using Cell-Free Protein Synthesis

Hong, S. H., Illinois Institute of Technology

Extensive use of antibiotics in the community and hospitals has resulted in the occurrence of multidrug-resistant bacteria that have been significant threats. In the United States, at least 2 million people are suffered from the infection from bacteria which have resistance to one or more antibiotics, and 23,000 people die by the results of direct infection of these pathogens in each year. As antibiotics are treated, bacteria rapidly develop the ways to protect themselves by regulating intrinsic antibiotic resistance genes or acquiring new genes from other bacteria that are already equipped the antibiotic resistance mechanisms. Therefore, novel strategies are required to effectively eradicate antibiotic resistant bacteria. Bacteriocins are antimicrobial peptides produced by bacteria and exhibit narrow or broad spectrum of killing other bacteria. These peptides have great potential as alternatives of traditional antibiotics, since the killing mechanisms of bacteriocins are different from those of antibiotics: bacteriocins kill cells via pore-forming on the membrane, nuclease activity, and inhibiting peptidoglycan production, while antibiotics usually kill cells via inhibiting transcription and translation. In this study, we utilize cell-free protein synthesis (CFPS) platforms to produce colicin, a type of bacteriosin, which is toxic to *Escherichia coli* strains. CFPS systems are not constrained by cell-viability requirements and are therefore impervious to the cytotoxic nature of bacteriocins. We generated colicin variant library via error-prone PCR and synthesized using a CFPS platform. We will assess their efficacy of killing *E. coli* strains including pathogenic *E. coli*. We will discuss developing cell-free high-throughput screening methods for peptide-based drugs. The rapid design-build-test cycle of CFPS platforms provides a time and cost advantage in screening antimicrobial peptides and have phenomenal potential to open up a new area of pharmaceutical biotechnology.

Utilizing the Labcyte Echo® Liquid Handler and Gateway® Gene Assembly in the Generation and Transfection of a PTEN-GFP Construct into a PTEN Null PC-3 Cell Line

Lesnick, J.,

The study of protein function relies upon recombinant gene assembly for protein expression and functional assays. This strategy is particularly important when available information regarding function is limited. The vast amount of data generated from genome sequencing projects has paved the way for diverse functional genomic studies that involve cloning, modification and subsequent expression of target genes. This requires flexible and efficient procedures for generating vectors containing gene fusions, gene variants, genetic coding for protein tags and cellular targeting. Furthermore, efficient high throughput cloning and gene assembly procedures are essential for interrogating signaling pathways with multiple gene constructs. The resulting gene constructs can be analyzed in stable transgenic or in transiently expressing biological systems.

Utilizing the Labcyte Echo 525 liquid handler we demonstrate an efficient and robust gene assembly method to generate a fusion protein for functional cellular assays. A PTEN-GFP construct was generated using the Gateway chemistry. PTEN null PC-3 cells were transfected with this construct to generate a fluorescently labelled and functional PTEN cell line to interrogate the PI3K pathway. The assembled fusion protein now enables functional cellular assays including, but not limited to, viability and localization studies. The Echo 525 liquid handler enables flexibility and scalability for gene assembly while streamlining and automating the process. The acoustic, non-contact dispense eliminates tips and potential for cross contamination. The low 25 nL resolution enables reduction in reaction volume which results in reduced reagent consumption and reduced costs for the user.

A Scalable Epitope Tagging Approach for High Throughput ChIP-Seq Analysis

Xiong, X., University of Illinois at Urbana Champaign
Jain, S., University of Illinois at Urbana Champaign
Yan, J.,
Ren, B.,
Zhao, H., University of Illinois at Urbana-Champaign

Chromatin immunoprecipitation followed by deep sequencing (ChIP-seq) is a high throughput sequencing technique to detect the protein-DNA binding events and chemical modifications of histone proteins. Followed by recognition and direct sequencing, more comprehensive sequence information can be obtained in a high throughput fashion. The quality of any ChIP data is highly dependent on antibodies. Specific antibodies are required for enrichment and the increase of signal contrast over background. However, commercial antibodies for TFs in human genome are very limited. To overcome this major limitation, researchers have utilized an epitope tag to create a fusion protein through plasmid overexpression or bacterial artificial chromosome integration for tagged gene expression and allow the tracking of a TF with an antibody against the tag sequence. We have developed a new method to directly modify endogenous genomic sequence and enable fusion protein expression in the native context for ChIP-seq analysis. Specifically, we combine the clustered regularly interspaced short palindromic repeats (CRISPRs) technology with microhomology mediated end joining (MMEJ) to insert an 3×FLAG-tag with screening markers to the ends of TFs. Current strategies for chromosomal integration typically rely on CRISPR mediated integration vis homologous recombination (HR) or non-homologous end joining (NHEJ). Scaling HR mediated integration on human genome is less feasible; and the reported accurate NHEJ integration efficiency is low. MMEJ provides novel cellular repair mechanism with more precise integration over NHEJ. Using the CRISPR-MMEJ mediated tagging approach, we have enabled the ChIP-seq analysis towards both tagged monoclones and polyclones for three TFs including SP1, MYC and TCF7L2. Our method leverages the scalability of the knock-in process with easy donor construction in contrast to previously reported homologous recombination mediated strategy. We envision the power of our strategy to improve and broaden ChIP-seq technology on mapping the genomes of human beings and any other organisms on a genome-scale.

Consequences of Cas9 Cleavage in the Chromosome of Escherichia coli

CUI, L., Institut Pasteur
Bikard, D., Institut Pasteur

The RNA-guided Cas9 nuclease from CRISPR-Cas systems has emerged as a powerful biotechnological tool. The specificity of Cas9 can be reprogrammed to cleave desired sequences in a cell's chromosome simply by changing the sequence of a small guide RNA. Unlike in Eukaryotes, Cas9 cleavage in the chromosome of Bacteria has been reported to kill the cell. However, the mechanism of cell death remains to be investigated. Bacteria mainly rely on homologous recombination with sister chromosomes to repair double strand breaks (DSB). Here, we show that the simultaneous cleavage of all copies of the *Escherichia coli* chromosome at the same position cannot be repaired, leading to cell death. However, inefficient cleavage can be tolerated through continuous repair by the homologous recombination pathway. In order to kill cells reliably, homologous recombination can be blocked using the Mu phage Gam protein. We also found the efficiency of CRISPR spacer processing and cleavage varies from spacer to spacer in *E. coli*. Finally, the introduction of the Non-Homologous End Joining (NHEJ) pathway from *M. tuberculosis* was not able to rescue the cells from Cas9-mediated killing, but did introduce small deletions at a low frequency. This work provides a better understanding of the consequences of Cas9 cleavage in Bacterial chromosomes which will be instrumental in the development of future CRISPR tools.

Telling Tales to Recode Zebrafish Genome: Chemical Reprogramming of Zebrafish for High-Efficiency Genome Editing

Aksoy, Y. A.,
Mann, G., Defence Science & Technology Organisation
Goldys, E., Macquarie University
Hesselson, D.,
Cole, N.,

As next-generation sequencing costs continue to fall, whole-genome and exome-genome sequencing and genome-wide association studies are generating more candidate human disease genes than ever. However, the function and role of these genes in the human disease remain to be discovered. Model organisms are crucial for efficient and robust investigation of candidate human disease genes in vivo. Due to its short reproductive cycle, ease of maintenance and physiological similarity to humans, zebrafish provides a powerful model system for not only investigating these genes but also for rapid large-scale drug screening. However, inability to alter the genome with ease and accuracy has been a major challenge. Genome editing tools such as CRISPR-Cas9 and TALENs have recently become available allowing introduction of heritable and precise sequence modifications in the genome. Both of these tools induce highly precise, locus-specific double strand breaks in the genome to generate multiple target-specific gene knock-ins and knock-outs in vivo. These exciting genome editing tools redefine the boundaries of the medical research by allowing scientists to generate various human disease models in model organisms much more quickly than before. We can now, investigate the functional role of individual genes much faster and readily alter multiple genes in cells at once to reveal their interactions. In this study we report a novel way to reprogram zebrafish embryos for high efficiency homology-directed genome editing. Our work further presents the application of these powerful tools to generate human disease models in zebrafish.

Design a Tunable Expression and Display System on Bacterial Spore Surface

Drufva, E., University of New Hampshire
Wu, K., University of New Hampshire

Bacterial spores have been explored to display proteins of interest as it offers potential advantages compared to soluble systems or other displaying platforms. First, spores are robust structures that are able to survive extreme temperatures, desiccation, and exposure to solvents and noxious chemicals. A few proteins displayed on the surface of *Bacillus* spores have shown enhanced stability. Second, the displayed proteins are part of the spore and they are synthesized and assembled during the sporulation process. Therefore, no protein purification is needed other than collecting the spores, which will significantly reduce the production costs of proteins. Third, enzymes displayed on the spore surface are reusable and recyclable. Spore serves as an immobilization platform and allows for multiple rounds of reactions. In addition, these spores are viable and able to germinate into cells. So spores with decreased enzyme activity can be easily regenerated through a germination-sporulation cycle. We have been working on displaying multiple proteins on the spore surface to create miniature reactors with complex functions. One challenge is that the expression level of the spore surface proteins is tightly regulated and no inducible expression system is available in response to external signals. In this work, we designed a few hybrid promoters by incorporating operators for commonly used repressors into the promoters for a few spore surface proteins. The resulted promoters have the natural regulatory elements allowing for the timely expression during the sporulation process and also the operators controlling the expression level through the level of the inducers. The designed tunable expression and display system will greatly facilitate the study and engineering of multiply protein complexes on the spore surface.

Mining Sugar Transporters from Gut Fungi for Improved Fuel & Energy Production

Solomon, K. V., University of California, Santa Barbara
Henske, J., University of California, Santa Barbara
Gilmore, S. P., University of California, Santa Barbara
Seppala, S., University of California, Santa Barbara
O'Malley, M. A., University of California, Santa Barbara

Membrane transporters are cellular gatekeepers that control the uptake of needed substrates and the secretion of valuable fuel products. These transporters are substrate-specific and help determine cell phenotypes. For example, industrial workhorses such as *Saccharomyces* lack the needed transporters to fully use the sugars from renewable plant biomass and natively secrete toxic fuel compounds. In contrast, anaerobic fungi from the guts of large herbivores thrive on an array of sugars resulting from lignocellulose hydrolysis due to a number of sugar uptake pathways. Here, we mine this diverse phylum for novel sugar transporters using an integrated transcriptomics and phenotyping approach to identify new tools for improved bioprocessing. Three new gut fungal species from distinct genera were sequenced to generate *de novo* transcriptomes containing hundreds of putative transporters. These transporters were classified into Transporter Classification Database (TCDB) families on the basis of homology and assigned a putative substrate. Substrate predictions were subsequently validated by comparing phenotype against metabolic bottlenecks and the expression profile of predicted transporters to identify transporter candidates responsible for sugar uptake. Transcriptomic sequencing also revealed a number of multidrug transporter candidates which may further increase productivity by increasing fuel tolerance via product secretion.

Metabolic Design Principles for Engineering Synthetic Microbial Communities

Soyer, O. S., Warwick Centre for Integrative Synthetic Biology
Zerfass, C., University of Warwick

Biotechnological solutions to some of the key economic and societal challenges will require the engineering of multi-species systems, that can fully exploit the biochemical potential of the microbial world, and that can allow novel applications such as artificial gut, synthetic soil, waste-to-value conversion, and ecosystem functions in space missions. It is unlikely that the current focus on engineering genetic circuits in model organisms on its own can achieve these high level applications of "microbial engines" composed of multiple microbial species that are biochemically and industrially relevant. To successfully carry out this kind of complex engineering, synthetic biology will need to establish the design principles governing microbial interactions and develop the experimental tools for their characterisation and manipulation.

Here, I will describe our ongoing efforts in discovering the reasons for metabolic interactions among microbes. I will argue that these metabolic interactions can be understood as an emergent property of the biochemical environment and associated thermodynamics within a multi-species system. Particularly, results from our recent research has shown that the thermodynamic basis of microbial growth can lead to co-existence, even under conditions that are kinetically predicted to lead to competition-driven exclusion. This thermodynamic inhibition can have significant consequences for the engineering of multi-species system. I will illustrate these with examples from our ongoing experimental work in establishing metabolic interaction motifs among biochemically and biotechnologically relevant microbes.

Metabolic Design Principles for Engineering Synthetic microbial Communities

Christian Zerfaß, Xue Jiang, Orkun S Soyer

University of Warwick, Warwick Integrative Synthetic Biology

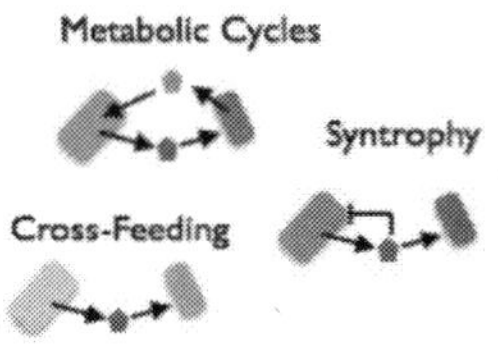

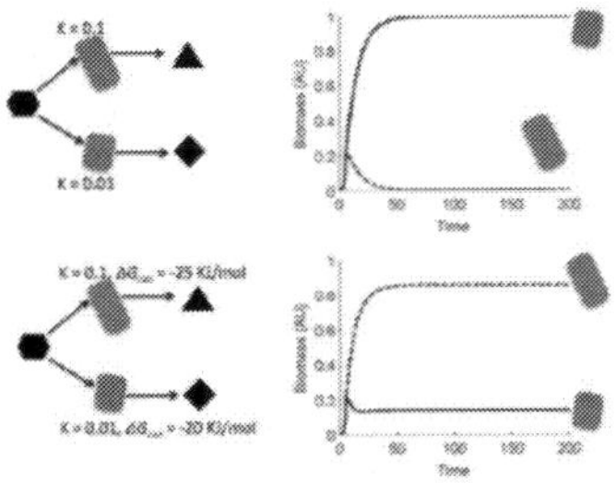

Kinetics dominated $v = \frac{v_{max} \cdot [S]}{K + [S]}$

Thermodynamics dominated $v = \frac{v_{max} \cdot [S] \cdot (1 - \exp(\Delta G_{rxn}))}{K + [S] \cdot (1 + k_r \cdot \exp(\Delta G_{rxn}))}$

Methanogenic microbial community

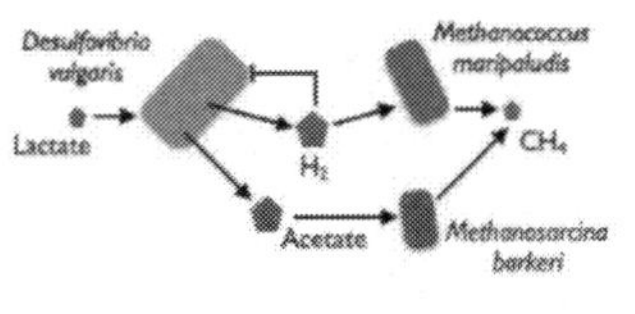

Plant-microbial (rhizosphere) community

Self-sufficient microbial community

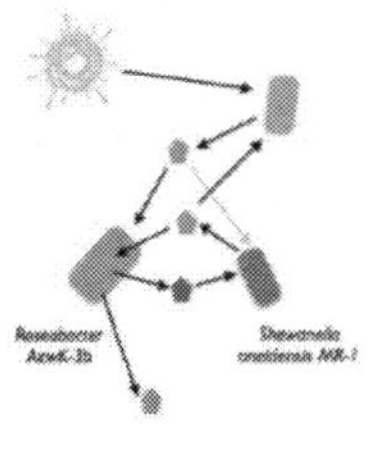

The Role of Single Occupancy Effects on Integrase Dynamics in a Cell-Free System

Artavanis, G.,
Hsiao, V., California Institute of Technology
Hayes, C. A., California Institute of Technology
Murray, R. M., California Institute of Technology

Phage integrase-based circuits are an alternative approach to relying on transcriptional and translational repression for biomolecular circuits. Previous research has shown that circuits based on integrases can perform a variety of functions, including counters, Boolean logic operators, memory modules and temporal event detectors. It is therefore essential to develop a principled theoretical and experimental framework for the design, implementation and study of such circuits. One of the fundamental questions that such a framework should address concerns the functionality limitations and temporal dynamics of the integrases as regulatory elements. We have tested the functionality of several large serine type integrases from a recently published library in a cell-free transcription-translation (TX-TL) platform. In addition, we have explored experimentally and through mathematical modelling and simulations how integrase dynamics depends on the concentration of integrase and that of its binding sites.
We report that sequestration of integrase molecules, either in the form of monomers or dimers, by the integrase's own binding sites dominates integrase dynamics, and that the delay in the activation of the reporter is negatively correlated with integrase plasmid concentration. We have validated our sequestration hypothesis by building a model with MATLAB's SimBiology toolbox, and running simulations with various integrase and binding sites concentrations. The simulation results qualitatively match the experimental results, and offer further insights into the system.

The role of single occupancy effects on integrase dynamics in a cell-free system

Georgios Artavanis[1], Victoria Hsiao[1], Clarmyra A. Hayes[1] and Richard M. Murray[1]

[1]Division of Biology and Bioengineering
California Institute of Technology, Pasadena, CA 91125

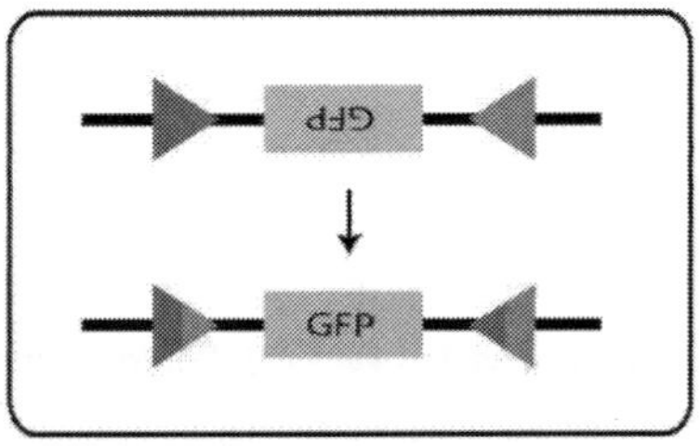

High Resolution Expression Profiling of Bacterial Genomes
Scholz, S., University of Michigan

High Resolution Expression Profiling of Bacterial Genomes

Scott Scholz, Elayne Fivenson and Xiaoxia (Nina) Lin

Cellular and Molecular Biology PhD Program and Department of Chemical Engineering, University of Michigan, Ann Arbor, MI

Synthetic genetic circuits are typically encoded on plasmids. However, to make them more robust and deployable at large scales, these systems will have to be integrated into host chromosomes. Resources relating a gene's position on the chromosome to its expression level, therefore, are highly desirable, yet currently not established. There have been limited studies examining the model bacterium *E. coli*, but previous results were not able to fully elucidate the effects of genome position on gene expression. In one work, examination of reporter gene expression level at 12 loci showed that expression is largely dependent on the copy number during the dynamic chromosomal replication process (i.e. distance from the origin). In another work, however, it was observed that expression of a reporter varied widely across 14 sites and the cause of this variation was not understood in most cases. Additionally, some chromosomal regions have very low expression of native and reporter genes and often have unique structural features.

To elucidate the effect of chromosomal position on gene expression at a high resolution, we have developed a multiplex strategy to construct and analyze several thousand genome-integrated reporters in a single mixed population of *E. coli*. We are generating normalized high resolution map of reporter transcription and protein expression variation across the bacterial genome. Libraries of mixed strains of *E. coli* are constructed through millions of simultaneous unique reporter integration events. Each integrated reporter transcribes a unique barcode which specifies the genome integration location. Using these mixed reporter strain libraries, we are mapping transcription and protein expression variation under different conditions at very high resolution onto the genome. These maps will serve as valuable resources for researchers seeking to identify ideal integration sites. In addition, expression variation maps generated across different conditions will advance our fundamental understanding of gene expression variation and regulation.

Repurposing Ribosomes for Synthetic Biology: Ribo-T v2.0

Carlson, E. D., Northwestern University
d'Aquino, A., Northwestern University
Fulk, E. M., Northwestern University
Kim, D. S., Northwestern University
Szal, T., University of Illinois at Chicago
Mankin, A., University of Illinois at Chicago
Jewett, M. C., Northwestern University

Our recent advancement engineering a ribosome with tethered subunits, Ribo-T, represents a paradigm shift in the ability to engineer the key catalyst in gene expression, the ribosome (Nature, 524, 119-124, 2015; ACS Synth. Biol., 4, 1173-1175, 2015). Across all kingdoms of life, the ribosome is composed of two separate, completely dissociable subunits. Extensive ribosome engineering *in vivo* was limited by the requirement for ribosomes to maintain life. By mutating the mRNA ribosome binding site and the corresponding site on the small subunit, specialized orthogonal small subunits are created that operate in parallel to the wild-type pool; since the cell is not dependent on these orthogonal small subunits for life, their function can be altered in extraordinary ways. However, the large subunit, which contains the key engineering targets of the peptidyl transferase center and the nascent polypeptide exit tunnel, was not orthogonal due to promiscuous exchange between the wild-type and orthogonal small subunit pools. We were the first to solve this problem by tethering the large subunit to the small subunit via two short RNA linkers to create Ribo-T. While this first version of Ribo-T represents a step-change in our ability to engineer the ribosome and gives great promise for advancing basic science and synthetic biology, several limitations need to be overcome to develop it into a versatile tool with broader impact. Here, we use knowledge-driven and evolution-based approaches to improve Ribo-T properties, including orthogonal function. The new Ribo-T version 2.0 will be more versatile and will serve as a better tool for many biotechnological, engineering and basic science applications.

Selective Nitrogen Delivery to Regulate and Optimize Synthetic Microbial Consortia for Biochemical Production

Scholz, S., University of Michigan

Selective Nitrogen Delivery to Regulate and Optimize Synthetic Microbial Consortia for Biochemical Production

Scott Scholz, Ian Graves and Xiaoxia (Nina) Lin

Cellular and Molecular Biology PhD Program and Department of Chemical Engineering, University of Michigan, Ann Arbor, MI

Many chemicals, including platform molecules used as building blocks for various final products, can potentially be produced using bio-based systems as sustainable alternatives. Lignocellulosic biomass is a particularly desirable feedstock due to its widespread availability. However, it is highly resistant to degradation. One approach to improve the efficiency of its conversion into desired products has been to combine different biochemical steps, from cellulase enzyme production, enzymatic hydrolysis of lignocellulose, to conversion of sugars into desired products, into a single reaction vessel, termed consolidated bioprocessing (CBP). In this project we employ a microbial consortia-based approach to enable CBP. Although consortia-based microbial production systems offer many advantages, such as modular design and high efficiency of individual specialized organisms, they present new challenges in population regulation.

We have developed a new method to dynamically regulate the growth and production performance of synthetic microbial consortia for the production of commodity chemicals from lignocellulosic biomass. Specifically, we have developed a selective nutrient delivery platform that can enable tuning of consortia member performance without making additional genetic modifications. By selectively delivering nitrogen, we can tune the composition of organisms within consortia over a wide range. We have used this method to control the performance of three different synthetic consortia for the production of two commodity chemicals from cellulosic feedstocks.

The Operon Calculator: Automated Design of Synthetic Operon Sequences Using 15 Models and Design Rules

Cetnar, D., Penn State Univ
Salis, H. M., Pennsylvania State University

Single and multi-cistronic operons are the central architectural unit of all natural and engineered genetic systems in bacteria, including multi-regulator genetic circuits and multi-enzyme pathways. An operon's sequence ultimately determines the expression levels of its RNAs and proteins, though the sequence-to-function rules that dictate optimal operon design remain to be elucidated or fully tested. Currently, operons are designed by appending several pre-existing promoters, ribosome binding sites, coding sequences, and terminators. As a consequence, many engineered operons are full of overlapping, context-dependent, and undesired genetic elements that confound rational design and will inevitably break the operon's function, such as repetitive DNA sequences, internal promoters and terminators, and highly translated internal start codons.

Here, we present a radically new algorithm, the Operon Calculator, that automatically designs *whole operon sequences* to yield maximum tunable control over its encoded RNA and protein expression levels, while eliminating the presence of all cryptic, undesired genetic elements that might break the operon's function. We combine 15 biophysical models and design rules using multi-objective genetic algorithm optimization to: (i) control transcription rates using models of transcriptional regulation; (ii) quantify transcriptional elongation efficiencies; (iii) eliminate RNAse binding sites; (iv) remove internal promoters and (v) transcriptional terminators, both intrinsic and rho-dependent; (vi) control translation and (vii) translational coupling rates using models of translation; (viii) harmonize synonymous codon usage to prevent translation elongation pile-ups; (ix) remove ribosomal pause sites, unless desired; (x) remove highly translated internal start codons to prevent expression of truncated proteins; remove genetic instability elements, including (xi) repetitive sequences, (xii) recombinase sites, and (xiii) transposon insertion sites; (xiv) remove undesired restriction sites; and (xv) calculate system-level RNAP and ribosomal loads. Using systematic experimentation, we present quantitative sequence-function relationships controlling mRNA stability and the expression of truncated proteins, two key factors that are often responsible for operon failure. Finally, we demonstrate an illustrative case study where we use the Operon Calculator to redesign the Central Metabolic network of the *E. coli MG1655*.

MYtxtl® an E. coli Based Cell-Free Transcription-Translation (TXTL) Kit for Bio-Manufacturing

Eggenstein, E., MYcroarray
Steil, B. P., MYcroarray
Noireaux, V., University of Minnesota
Rouillard, J. M., University of Michigan

At present, hybrid bacteriophage-*E. coli* TXTL systems are popular and widely applied cell-free TXTL platforms in synthetic biology and bioengineering applications. By combining a strong and robust bacteriophage RNA polymerase (often from T7) for gene transcription with a cytoplasmic *E. coli* cell extract providing the translation machinery, this technology has proven useful for protein design and engineering, especially in high-throughput setups. Their major disadvantage is the relatively small repertoire of transcriptional regulators, thus considerably limiting the complexity of *in vitro*dynamical systems that can be constructed, such as gene circuits, for synthetic biology applications.

Our MYtxtl® kit – entirely based on the endogenous TXTL machinery of *E. coli* – allows the implementation of a plethora of transcriptional activation units, thus providing a unique and unrivaled degree of regulatory freedom for dynamical systems in gene circuitries. We make use of the primary *E. coli* transcription machinery consisting of the core RNA polymerase and sigma factor 70 (s^{70}) as basic transcriptional regulator and complement it with other stable or degradable endogenous transcription factors for setting up multi-stage activation cascades. In addition, gene expression under the control of the bacteriophage T7 promoter/operator system is still feasible by using a two-plasmid system: one for the expression of T7 RNA polymerase under transcriptional control of the *E. coli* P_{70}promoter, the other comprising the gene of interest downstream of the T7 promoter.

Another promising, high potential field of application for cell-free TXTL systems is the large-scale protein synthesis (bio-manufacturing) for e.g. the production of biopharmaceuticals. Currently, major hurdles to enter this market are that (i) existing methods for cell extract preparation are developed for standard shake flask cultures only providing merely insufficient amounts of bacterial cell pellet as well as bear the risk of batch-batch variation and (ii) protein expression in > 20 µL TXTL reaction volume is limited most likely due to poor and inconsistent supply of oxygen which decelerates the TXTL machinery.

To overcome the first limitation, we investigated the preparation of *E. coli*cell extracts using a large fed-batch fermenter by permanently adjusting pH, temperature and supply of nutrients. This approach yielded in considerably higher cell densities during log-phase cell growth as compared to flask-production, giving TXTL reactions with proportionate protein production ability.

Secondly, we developed scaled-up TXTL reactions by evaluating the effect of several oxygenation systems and incubation methods on their protein production efficiency. Thereby, we demonstrated that high-yield production of correctly folded and biochemically active protein in TXTL milliliter-scale reactions is feasible.

Thus, by combining the regulatory flexibility of the *E. coli* TXTL machinery with effective oxygenation methods, we developed a cell-free TXTL systems applicable for both synthetic biology and high-mass protein production.

RNA Information Processors for Complex In Vivo Logic Computation

Green, A. A., Arizona State University
Kim, J.,
Ma, D.,
Silver, P. A., Harvard Medical School
Collins, J. J., Broad Institute of Harvard and MIT
Yin, P., Harvard Medical School

Synthetic biology aims to create functional devices, systems, and organisms with novel and useful functions taking advantage of engineering principles applied to biology. Starting from the first demonstration of synthetic circuits in bacteria more than a decade ago, synthetic biology has substantially advanced strategies for biomedical applications. Ultimately, sophisticated therapeutic sensor-effector devices that connect diagnostic input with therapeutic output may provide platforms for future gene- and cell-based therapies. Despite great progress over the last decade, an underlying problem in synthetic biology remains the limited number of high-performance, modular, composable parts. A potential route to solve parts bottleneck problem in synthetic biology utilizes the programmability of nucleic acids inspired by molecular programming approaches that have demonstrated complex biomolecular circuits evaluating logic expressions in test tubes. Central to our current work is a novel class of in vivo RNA-based devices called toehold switches that activate gene expression in bacteria only when they detect a cognate trigger RNA. The toehold switches take trigger RNAs without any sequence constraints overcoming the limitations of conventional synthetic riboregulators. Further, these toehold switches can be forward-engineered to modulate gene expression by several hundred-fold rivaling the best protein-based regulators and provide extremely low levels of crosstalk across large library of components. Using a toolkit of orthogonal devices and modular composability, we demonstrate how toehold switches can be incorporated into decision-making RNA networks to rapidly evaluate complex logic in living cells. We have successfully demonstrated a 4-input AND gate, a 6-input OR gate, and a 10-input expression in disjunctive normal form, surpassing the complexity of previous works. The compact encoding of RNA information processing system using a library of modular parts is amenable to aggressive scale-up towards complex control of in vivo circuitry towards autonomous behaviors and biomedical applications.

Ribocomputing devices for sophisticated in vivo logic computation

Jongmin Kim, Wyss Institute, Harvard University

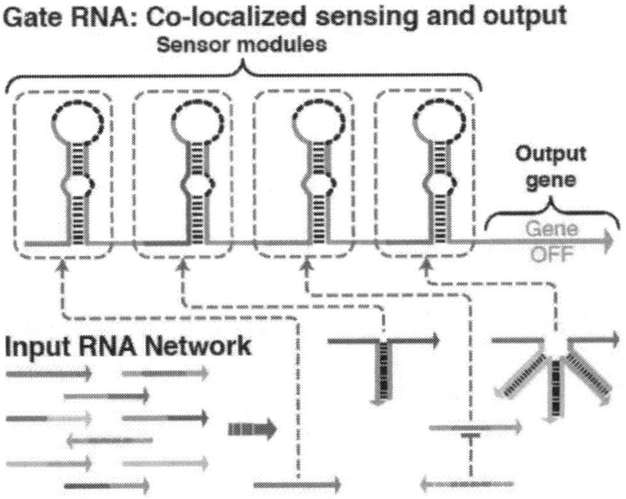

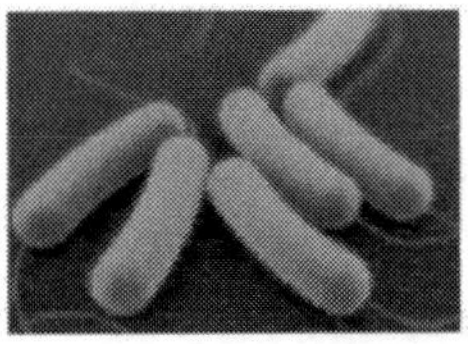

A Cas9-Based Toolkit to Program Gene Expression in Saccharomyces Cerevisiae

d'Espaux, L., Lawrence Berkeley National Lab
Reider Apel, A., Lawrence Berkeley National Lab
Wehrs, M., Lawrence Berkeley National Lab
Sachs, D., Lawrence Berkeley National Lab
Garber, M., Lawrence Berkeley National Lab
Mukhopadhyay, A., Berkeley National Lab
Keasling, J. D., Lawrence Berkeley National Laboratory

From bread, beer and wine to synthetic biology, *Saccharomyces cerevisiae* has long been our go-to eukaryotic host. More recently, CRISPR/Cas9-directed homologous recombination has allowed for rapid and efficient chromosomal integration of heterologous genes and pathways. What remains relatively unexplored is how best to choose gene expression parts—integration locus, promoter, etc.—for a gene of interest. Here we develop a toolkit of well-characterized parts to program gene expression. Our toolkit consists of (1) two dozen genomic integration loci and (2) four dozen native promoters characterized at different growth phases and in different media; (3) modular protein tags to control protein expression level, solubility, and sub-cellular localization; and (4) a web tool to use these data to design donor DNA cassettes. These integration cassettes can be generated in 1-2 rounds of PCR and integrated with high efficiency using pre-cloned Cas9-gRNA plasmids targeting our integration sites. We demonstrate using our toolkit to program various expression contexts for genes of interest as a way to develop and optimize engineered strains.

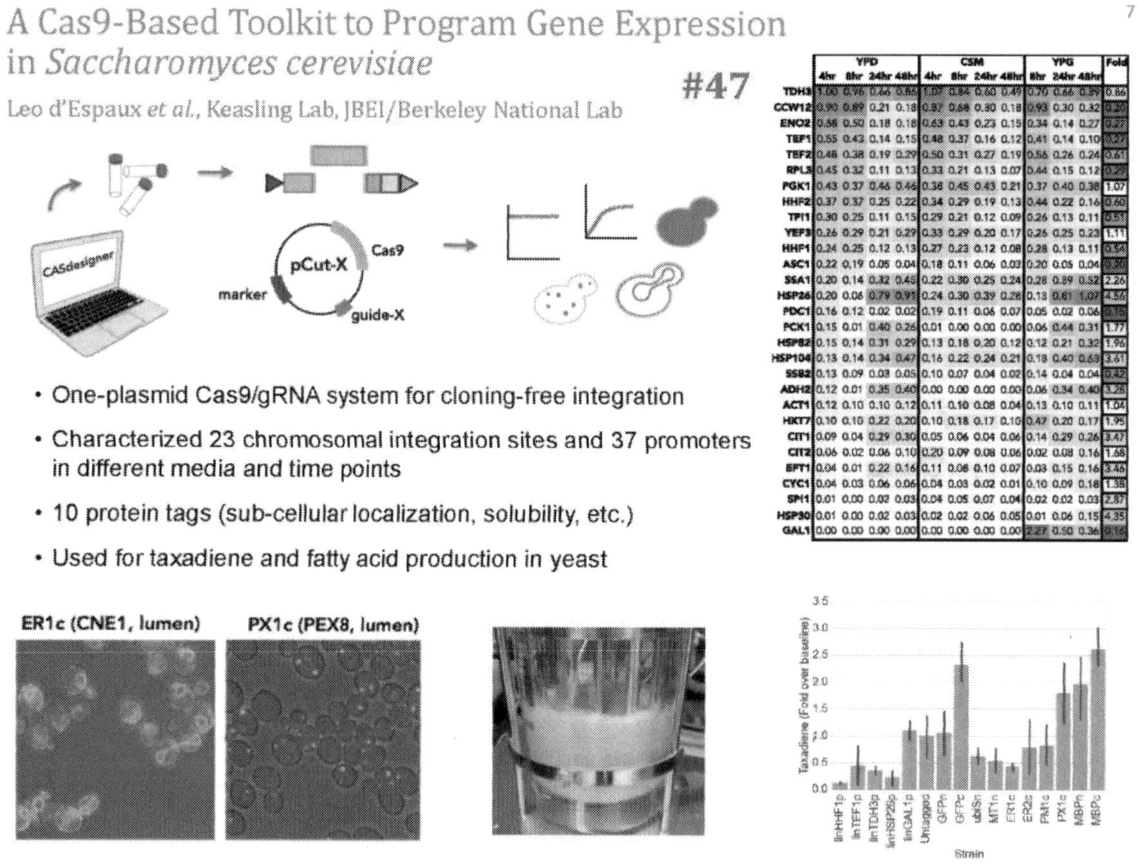

Bioengineering Magnetic Nanoparticles in Magnetotactic Bacteria

Furubayashi, M., Chiba Univ.
Voigt, C. A., University of California, San Francisco

Magnetic nanoparticles (MNPs) are in great demand in diverse modern technologies such as high-density data storage, electromagnetic shielding, and various applications for biomedicine including targeted drug delivery or hyperthermia. Compared to bulk magnets, MNPs shows unique and interesting properties, which largely dependent on their size, shape or crystal composition. The strong interest in the filed is to produce uniform-sized MNPs with desired properties required for each technological application. However, chemical synthesis of structurally perfect MNPs with narrow size distributions still remains a challenge. In contrast, magnetic nanoparticles biomineralized by natural organisms are known to show narrow size distributions, high crystal perfections that could be biosynthesized under environment-friendly reaction conditions. Our goal is to design and produce MNPs with various properties by genetically engineering Magnetospirillum magneticum AMB-1, a bacteria that can biosynthesize chains of magnetic nanoparticles (called magnetosomes). Magnetosomes are composed of magnetite ($Fe3O4$) nanoparticles with remarkably narrow size/shape distribution. By modifying bacterial MNPs from bottom-up and retaining its natural ability to produce them with narrow size distributions, we envision the production of "designer magnetic nanoparticles" which is otherwise difficult to achieve. We have engineered and characterized basic genetic parts, such as stable plasmids, promoters or reporters for M. magneticum. Using these genetic parts, we have changed the expression level of magnetosome genes to tune the size/shape of magnetosome nanoparticles. In the presentation, we will also discuss our attempts to decorate the surface or modify the chemical composition of MNPs.

A Novel Class of Programmable Hormone-Responsive Transcription Factors to Re-Engineer Plant-Environment Interactions

Khakhar, A., University of Washington
Nemhauser, J.,
Klavins, E.,

Human society is built on a plethora plant products from essentials such as food and feed stocks to luxuries such as pharmaceuticals and textiles. The indiscriminant use of pesticides and fertilizers coupled with the escalating effects of climate change has created a rapidly changing agricultural environment that gravely threatens the sustained productivity of crops. To address this challenge there is a need for tunable biological control systems that can connect the plants native perception of its environment to appropriate natural or synthetically engineered responses. In this work we present a novel class of hormone-responsive CRISPR-based transcription factors and demonstrate their function in *Arabidopsis thaliana*. We explore the modular nature of the transcription factors by testing a range of different transcriptional repressor domains and hormone responsive degrons. Using this approach we are exploring how hormone fluxes triggered by the environment can be coupled in a tunable fashion to the expression of a gene of choice. Specifically we are interested in coupling the perception of the plant hormone jasmonate isoleucine, which is released upon insect predation, to the expression of insect defense mechanisms. We are also interested in rewiring developmentally triggered fluxes of the plant growth hormone auxin to the expression of novel genes, to re-engineer plant development. Using this new class of transcription factors we hope to design biological control systems that will make crops more robust to a rapidly changing environment.

Programming plant development

Arjun Khakhar

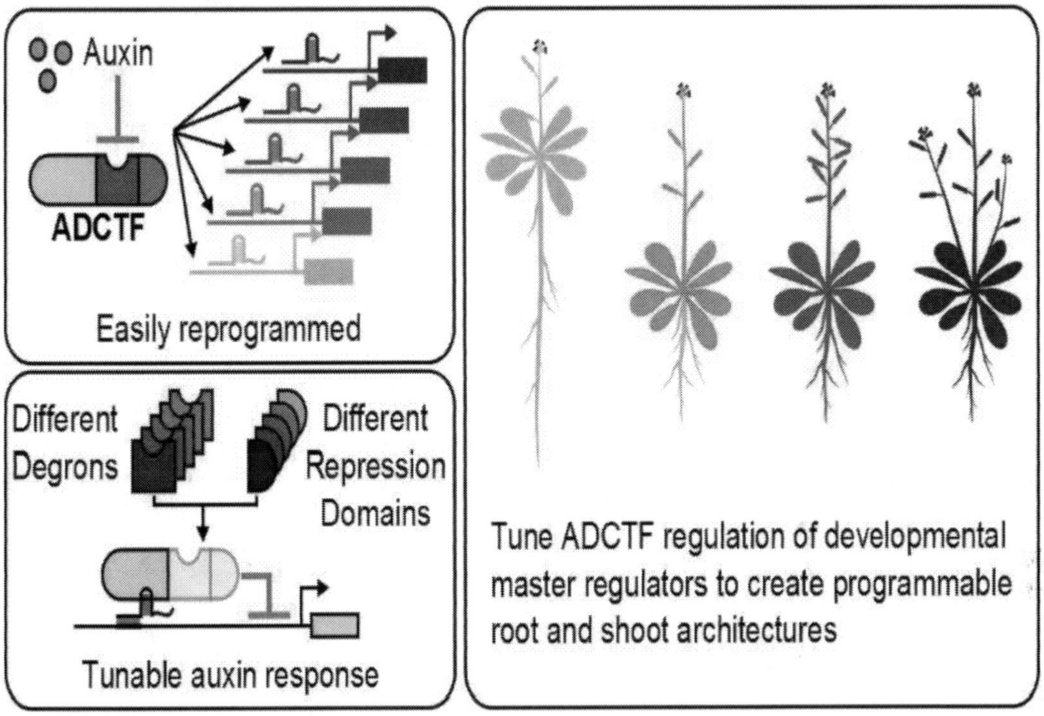

Expression Mutant Screening for Improved Benzylisoquinoline Production in Yeast

Savitskaya, J.,
Grewal, P., Genentech Inc.
Arkin, A. P., University of California at Berkeley
Dueber, J.,

Benzylisoquinoline alkaloids (BIAs) comprise a pharmaceutically valuable class of natural molecules. To relieve issues with chemical and plant-based production of BIAs, several metabolic engineering efforts have reconstituted natural BIA pathways in Saccharomyces cerevisiae. Our lab has previously used a fluorescent biosensor to identify an enzyme necessary to complete the pathway from glucose to reticuline, a key intermediate in the production of a variety of BIAs. Current yields of reticuline are approximately 5 orders of magnitude below theoretical yields, one of the main bottlenecks being the low and tightly regulated supply of free tyrosine, which is required for BIA synthesis. Using the same fluorescent biosensor, we screened collections of expression mutants to enhance our strain background for the increased production of reticuline.

Role of Plasmid Copy Number in the Dynamics of a Single Gene Oscillator

Swaminathan, A., California Institute of Technology
Gomez, M. M., UC Berkeley
Shis, D. L., MIT
Bennett, M. R., Rice University
Murray, R. M., California Institute of Technology

Although understanding all dynamics in the cell may be elusive, we can aim to understand the impact of design choices under our control. We consider a single gene oscillator as a suitable case study to better understand the influence of various parameters on resulting dynamics. With such a simple circuit the number of variables is greatly reduced while maintaining non-trivial dynamics.

As an initial study we consider the effects of repressor choice and plasmid copy number. Here, we create a new single gene oscillator similar to the published lacI oscillator in which lacI represses its own transcription. The repressor we use is the gene treRL, a chimeric repressor with a lacI DNA binding domain and a ligand binding domain that responds to trehalose as an inducer. This slightly modified system produces more regular oscillations than the original lacI oscillator. With a suitable oscillator we investigate the role of plasmid copy number.

Plasmid copy number is an important variable to consider, as plasmids are typical platforms on which synthetic circuits are designed. It is well known that it is difficult to ascertain exact plasmid copy numbers. Furthermore, the copy number varies dynamically with time. However, there are methods to vary relative plasmid copy numbers. Here, we study the influence of gene copy number on the dynamics of the self-repressing single gene oscillator. We compare the treRL oscillator at three different copy numbers: the medium copy origin p15a, the low copy origin psc101, and a single copy integrated into the genome. The reporter is a fast folding fluorescent protein on a separate plasmid with a colE1 origin.

We find that the period and amplitude of oscillations increases as the copy number decreases. Through analysis of a one-dimensional delay differential equation we find that we cannot explain a change in period without changing delays or degradation rates. We also find that the correlation in fluorescence between daughter cells after cell division falls off faster in the oscillator on the p15a origin compared to the psc101 and single copy variants. Similarly, we find that the phase difference between daughter cells grows more rapidly after cell division on the p15a oscillator compared to the psc101 and single copy variants. These results suggest that lower copy number variants of our single gene oscillator lead to more synchronized and regular oscillations.

Massively-Parallel Laboratory Evolution of Bacteriophage T7 Reveals Differential Evolutionary Utility of Expanded Genetic Codes

Hammerling, M. J.,
Barrick, J. E., University of Texas at Austin
Deatherage, D. E., The University of Texas at Austin

The development of the first synthetic organisms with truly reassigned codons has spurred research projects to determine the potential evolutionary utility of genetically-incorporated nonstandard amino acids in proteins and whole organisms. Initial results have been promising, and have included enzymes with greater activity on novel substrates as well as organisms with improved fitness due to nsAA incorporation in laboratory evolution experiments. We sought a more rigorous quantification of the evolutionary utility of a variety of expanded genetic codes in the context of an evolving biological system. To this end, we evolved 96 independent lines of bacteriophage T7 each on *Escherichia coli* hosts with six different expanded genetic codes, for a total of 576 independently evolved populations. Each independently-evolved population was deep-sequenced, and the frequency of nsAA incorporation in bacteriophage proteins were compared to the 20 canonical amino acids and each other. The nsAAs were found to be incorporated throughout the proteome of the bacteriophage, including instances of strong signals of positive selection. We also developed a substitution matrix for each nsAA to determine which adjacent canonical amino acids it most readily substituted for in the bacteriophage proteome. Finally, bacteriophage protein variants found to incorporate nsAAs at high frequency, such as T7 RNA polymerase, were assayed to assess the impact of the nsAA on protein function. Our results offer several lessons for the productive use of expanded genetic codes in complex evolving biological systems and biotechnology applications.

Highly-parallel bacteriophage evolution reveals differential evolutionary utility of expanded genetic codes

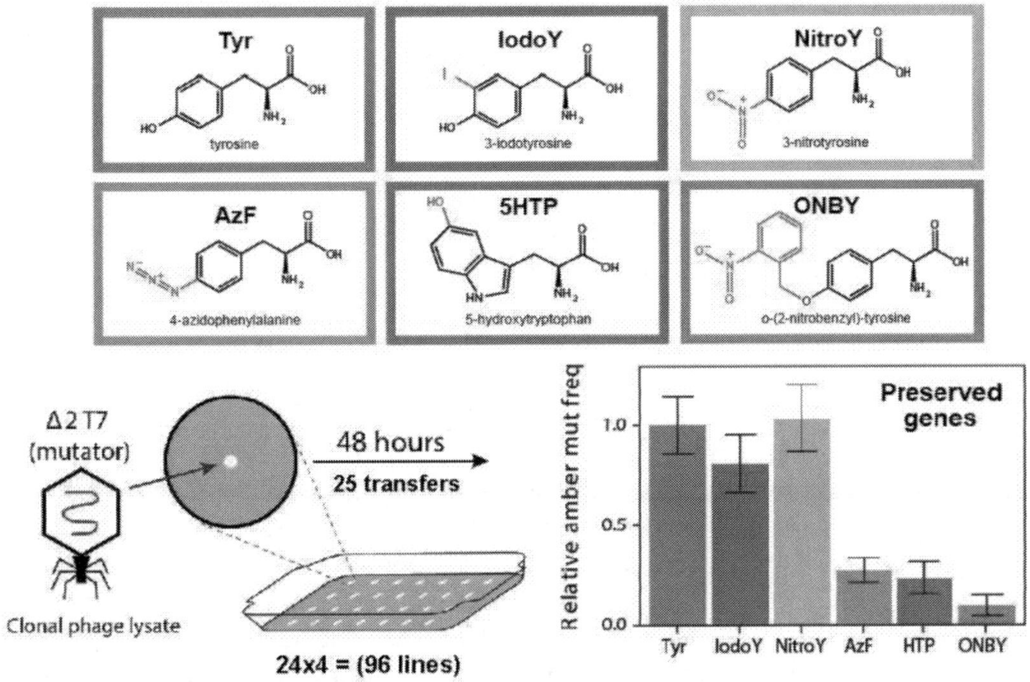

Poster #74

Michael J Hammerling, Barrick Lab - UT Austin

Gene Co-Expression Highlights Coordination of Microbial Community Behavior

McClure, R. S., Pacific Northwest National Laboratory
Overall, C. C., Pacific Northwest National Laboratory
McDermott, J.,
Hill, E., Pacific Northwest National Laboratory
Bernstein, H. C., Pacific Northwest National Laboratory
Beliaev, A. S., Pacific Northwest National Laboratory

Association networks built from parallel transcriptomic datasets can answer fundamental questions regarding gene control and species interactions in microbial systems. Here, we combine several RNA-seq datasets examining a model consortium comprised of a cyanobacterial phototroph (*Thermosynechococcus elongatus* BP1) and a heterotrophic partner (*Meiothermus ruber* strain A. During coculture, *M. ruber* is completely dependent upon *T. elongatus* for sources of energy, carbohydrates and organic nitrogen sources. Concurrently, *M. ruber* removes potentially toxic molecular oxygen from the system and provides key micro/macronutrients to *T. elongatus*. Because of these dependencies there is a large amount of interaction, metabolite exchange and crosstalk between the two organisms. To understand and identify these interactions at the systems level we utilized RNA-seq in conjunction with the Context Likelihood of Relatedness (CLR) program to build a network of genes (nodes) from both species and identify correlations of expression (edges) both within and between species of this consortium. Resulting networks were then parsed to identify possible points of interaction and the genes and processes that show the greatest number of links between species. Networks comprised of each species individually showed structures similar to what has been previously observed with genes involved in central functions included photosynthesis, oxidative phosphorylation, cell motility, nitrogen and amino acid metabolism, carbohydrate metabolism and translation grouped separately into tight modules of highly co-expressed genes. When examining both species networks became much larger (~10x more edges with only ~2x more genes). When focusing on edges between species, *M. ruber* genes with the highest number of interspecies links include those involved in amino acid metabolism, cell adhesion and B12 metabolism. *T. elongatus* genes with multiple interspecies links include those involved with photosynthesis, carbohydrate metabolism, nitrogen metabolism and terpenoid metabolism. Through a view of species interactions at the network level these studies identify putative points of interaction in phototroph-heterotroph communities. These include both central and expected processes (carbon, nitrogen and energy metabolism) as well as additional areas of interaction (adhesion, terpenoid metabolism and B12 metabolism). Both the developed methodology and conclusions derived from this work are widely applicable to microbial communities for identification of interactions between species and characterization of community functioning as a whole. Future efforts will be focused on examining these putative interactions further and in constructing biological networks to examine interactions of microbial communities comprised of several heterotrophs supported by a single photoautotroph.

RNA Roadblocks: Hairpins for Protein Level Tuning in S. Cerevisiae.

Ellis, T., Imperial College London
Weenink, T., Imperial College London

To modulate protein levels in eukaryotes, gene expression is normally modified at the transcriptional level by exchanging or mutating the promoter of a gene of interest. Unfortunately this approach often also alters the regulation of a protein as promoter sequences encode information both on the strength of expression and when expression occurs. To enable control over protein levels without modifying promoters in *S. cerevisiae*, we introduce here a novel method for tuning the efficiency of protein translation. By placing strong hairpins in the 5' UTR of expressed mRNAs we show here that we can down-tune translation. We use the predicted free energy of folding of the 5'UTR to estimate the final expression level of the protein in question.

We designed and constructed a library of secondary structures, calculating the free energy for every possible member, optimizing the design to obtain a wide spread of folding energies with only minor changes to 5'UTR sequences. We characterized several libraries by expression of green fluorescent protein and went on to investigate the further effect of 5'UTR length and the location of the hairpin within the 5'UTR. We then applied this approach to modulate expression from 4 different constitutive promoters and from the GAL1 promoter, without having to disrupt the tight regulation of the native promoter. This work thus provides a new class of modular part for manipulating eukaryotic gene expression, and by following RNA design principles stemming from free energy calculations, allows predictable tuning of protein production.

A Cell-Free Transcription-Translation Kit (MYtxtl®) for Rapid Prototyping of Transcription Units Assembled from Error-Free Modular Parts

Steil, B. P., MYcroarray
Eggenstein, E., MYcroarray
Noireaux, V., University of Minnesota
Rouillard, J. M., University of Michigan

Rapid prototyping of genetic circuitry requires the ability to assemble and test different combinations of modular, error-free DNA parts in a high throughput manner. We describe here a modular parts system for rapidly assembling and testing linear DNA transcription units in an in vitro cell-free transcription-translation system (MYtxtl®) derived from *E. coli*. This modular parts system consists of a collection of separate promoters, open reading frames and terminators that can be combined on demand into transcription units that can be assembled and tested in the same day.

We synthesized and sequence-verified several plasmids that each contains one modular part. As a demonstration of rapid prototyping, we assembled multiple transcription units in parallel on a microplate using one plasmid each of a promoter, open reading frame, and terminator. These plasmids were digested and assembled in a directional Golden Gate cloning reaction into linear transcription units. This linear template was then PCR amplified, purified and added to the MYtxtl® reaction. Linear templates were incubated for as short as 4 hours followed by measurement of mRNA synthesis and/or protein production. This allowed us to compare the expression of a particular protein under the control of promoters of different strengths in one work day.

One limitation of using linear DNA molecules as templates in MYtxtl® reactions is their susceptibility to rapid nuclease degradation by the RecBCD enzyme and other nucleases present in the MYtxtl® extract. We therefore used direct modifications of the linear DNA templates and nuclease inhibitors added to the reaction to increase protection of templates from degradation. These highly efficient and cost-effective methods increased protection of linear DNA templates from degradation by several fold relative to reactions using templates without modifications or nuclease inhibitors. This allowed for protein production from linear templates that approached that of circular plasmids, eliminating the need to spend time cloning genetic circuits.

Starting from a plasmid library of modular parts, we successfully assembled and tested multiple transcription units within an 8 to 24 hour period. These reactions can be used to analyze transcription, translation, gene circuitry and protein function (with the additional step of protein purification). This rapid assembly and testing of modular parts in the MYtxtl® system allows for high-throughput prototyping of a diverse number of transcription units and proteins, making the MYtxtl® system a valuable tool to synthetic biologists.

Engineering RNA Aptamer-Based Gene Circuits for Control of Microbial Interactions

McClure, R. S., Pacific Northwest National Laboratory
Carothers, J.,
Beliaev, A. S., Pacific Northwest National Laboratory

The ability to precisely and accurately control genetic circuits has far reaching implications for the engineering of robust and safe microbial systems with desired outputs. To achieve that goal, we propose to develop a scalable and generalizable pipeline for the development of functional genetic circuits that respond to specific target ligands using RNA-based devices, specifically riboswitches, which induce downstream gene expression as a function of target ligand concentration. Here, previously designed RNA sequences that bind to two small molecule ligands, *p*-aminophenylalanine (pAF) and theophylline (THP) were used as starting material. These sequences, termed aptamers, were cloned upstream of 15 nucleotide random sequences driving expression of a tetracycline resistance (*tetA*) readout gene. In a subset of these random sequences presence of the ligand causes a structural change in the aptamer such that the ribosome binding site of *tetA* is available for binding. Duplicate cultures of such aptamer/switch libraries were grown either in the presence or absence of the ligand (pAF, or THP). Functional aptamer/switch pairs, riboswitches, were therefore enriched only in cultures containing the ligand. Sequencing of the resulting cultures identified several hundred switch sequences that were > 2-fold more abundant in response to the presence or absence of the ligand. Such changes in abundance represent putative successful riboswitches. These studies are currently being expanded and refined so that enrichment and selection of functional riboswitches can take place within a single community of continuously cultured bacteria. Additional experiments have been aimed at increasing the modularity if riboswitches through the inclusion of fusion proteins containing the gene the riboswitch was originally evolved with. This has resulted in a fusion riboswitch evolved to induce expression of beta-galactosidase modified into a riboswitch designed to induce the expression of *tetA*. This approach of rapidly and efficiently screening putative riboswitches, as well as increasing their modularity, has several applications as aptamers for new ligands are developed and synthetic genetic circuits become cornerstones of engineered microbial consortia of bio-industrial importance.

Engineering Red Blood Cell-Based Biosensors for Physiological Monitoring

Dolberg, T.,
Schwarz, K. A., Chemical and Biological Engineering, Chemistry of Life Processes Institute, R.H. Lurie Comprehensive Cancer Center, Northwestern University
Leonard, J. N., Northwestern University

Cell-based therapies have a wide range of applications ranging from cancer immunotherapy to regenerative medicine. A promising emerging frontier of this field is the development of engineered red blood cells (eRBCs) for therapeutic and diagnostic applications. RBCs have exceptionally long circulation times (around 40 days – far longer than synthetic vehicles), lack DNA (and thus are safe), and can be loaded with drugs, proteins, or other cargo. Technologies that enable one to engineer RBCs to perform specific functions *in vivo* could serve unmet diagnostic and therapeutic needs. In particular, new technologies are required for non-invasive, routine monitoring for pathogen exposure (e.g., in the context of first responders) and for "actionable" analytes (e.g., markers of inflammation post-surgery).

In this project, we sought to develop eRBC biosensors than detect highly toxic agents, with the long-term goal of enabling one to detect exposure to these agents prior to the onset of physical symptoms. As a first step towards this goal, we designed and evaluated a novel biosensor strategy that is suitable to achieving biosensing in eRBCs, which lack DNA and thus require a readout other than gene expression. Towards this end, we engineered a novel cell-surface receptor protein in which ligand binding induces receptor dimerization, which then facilitates reconstitution of an intracellular split fluorescent protein. Ultimately, eRBC fluorescence may be monitored non-invasively using established technologies for fluorescent imaging of the retina. Importantly, our strategy involves modification of RBC-resident proteins, since retention of membrane proteins during RBC maturation is a tightly regulated and an incompletely understood process. In this study, we comparatively evaluated a range of biosensor architectures that implement the proposed mechanism, identified design biosensor features that successfully conferred significant ligand-induced generation of fluorescent output, and investigated strategies for improving biosensor performance (e.g., minimization of background fluorescence and enhancing fold-induction upon exposure to ligand). This crucial proof-of-principle demonstration establishes a foundation for developing eRBC biosensors that could ultimately address an unmet need for non-invasive monitoring of physiological signals for a range of diagnostic applications.

Arrowland: A Browser-Based Environment for Omics Mapping and Simulation

Garcia-Martin, H., Joint Bioenergy Institute
Birkel, G., Lawrence Berkeley National Laboratory
Morrell, W., Sandia National Laboratory
Hillson, N. J., Joint BioEnergy Institute
Keasling, J. D., Lawrence Berkeley National Laboratory

Today, the biological sciences are at the center of a rapidly expanding universe, of new kinds of data. We've only just begun to curate, explore, and interpret it. While pursuing this at JBEI we've developed a software tool called Arrowland: a combined omics visualization, showing metabolomics, fluxomics, proteomics, and transcriptomics together on multi-scale, customizable maps, with a unique method of navigation. The fundamental design goal of Arrowland is to make it easy to get omics data in, and easy to explore.

Arrowland runs in a web browser, as well as on tablets and phones. It's fully searchable, and integrated with a modeling library called the Quantitative Metabolic Modeling library, and a data curation tool called the Experiment Data Depot. All are open-source and free for academic use. JBEI is seeking collaborators, testers, and co-developers to help us improve Arrowland.

Functional Reconstitution of a Pyruvate Dehydrogenase in the Cytosol of Saccharomyces Cerevisiae through Lipoylation Machinery Engineering

Lian, J., University of Illinois at Urbana Champaign
Zhao, H., University of Illinois at Urbana-Champaign

Acetyl-CoA is a key precursor for the biosynthesis of a wide range of fuels, chemicals, and value-added compounds, whose biosynthesis in *Saccharomyces cerevisiae* involves acetyl-CoA synthetase (ACS) and is energy intensive. Previous studies have demonstrated that functional expression of a pyruvate dehydrogenase (PDH) could fully replace the endogenous ACS-dependent pathway for cytosolic acetyl-CoA biosynthesis in an ATP-independent manner. However, the requirement for lipoic acid supplementation hinders its wide industrial applications. In the present study, we focus on the engineering of a *de novo* synthetic lipoylation machinery for reconstitution of a functional PDH in the cytosol of yeast. First, a lipoic acid auxotrophic yeast strain was constructed through the expression of the *Escherichia coli* PDH structural genes and a lipoate-protein ligase gene in an ACS deficient (*acs1*Δ *acs2*Δ) strain, based on which an *in vivo* acetyl-CoA reporter was developed for following studies. Then the *de novo* lipoylation pathway was reconstituted in the cytosol of yeast by co-expressing the yeast mitochondrial lipoylation machinery genes and the *E. coli* type II fatty acid synthase (FAS) genes. Alternatively, an unnatural *de novo* synthetic lipoylation pathway was constructed by combining the reversed β-oxidation pathway with an acyl-ACP synthetase gene. To the best of our knowledge, reconstitution of natural and unnatural *de novo* synthetic lipoylation pathways for functional expression of a PDH in the cytosol of yeast has never been reported. Our study has laid a solid foundation for the construction and further optimization of acetyl-CoA overproducing yeast strains.

Engineering Insulation Between Synthetic Bacterial Signaling Pathways

McClune, C.,
Laub, M.,
Voigt, C. A., University of California, San Francisco

Two component systems (TCS) are the primary family of sensors in prokaryotes and are popular tools for synthetic biology. In response to environmental stimuli, a sensor histidine kinase (HK) phosphorylates a downstream response regulator (RR), eliciting changes in gene expression. A given bacterium uses 10-200 different HK/RR pairs to sense the various elements of its environment. TCSs are attractive sensors for engineered systems because of their natural abundance, their diverse inputs, and domain modularity that permits protein engineering. However, because the HK/RR interaction is mediated by highly conserved domains, engineered TCS sensors must avoid interactions with tens or hundreds of endogenous TCS proteins. To generate new orthogonal TCS pairs and learn design principles of TCS insulation, we mutated the specificity-determining regions of HKs and RRs and built large mutant libraries. We are leveraging cell sorting and deep-sequencing to quantify the compatibility of mutant HKs and RRs in E. coli, which has 32 endogenous TCS. Furthermore, we're submitting these libraries to in vitro selections to test how the function of these pathways is affected by the presence or absence of paralogous pathways. In addition to generating toolkits of insulated sensors, this work will provide tools for integrating or propagating signals, interfacing with host physiology and monitoring cell states.

Engineered, Evolved, and Re-Engineered Yeast for Producing Cellulosic Biofuels

Jin, Y. S., University of Illinois at Urbana-Champaign

While there are anticipated benefits from using cellulosic biomass as a feedstock to produce biofuels and chemicals, commercial production of cellulosic biofuel and chemicals has been hampered by technical difficulties, such as inefficient fermentation of cellulosic sugars, and the toxicity of acetic acid which are abundant in plant cell wall hydrolysates. As an effort to overcome the problems, we constructed a highly engineered *Saccharomyces cerevisiae* strain capable of converting mixed carbon components (cellobiose, xylose, and acetate) into ethanol simultaneously.

As *S. cerevisiae* cannot ferment xylose, we first constructed a rapid and efficient xylose fermenting *S. cerevisiae* (SR8) through a rational design that allows strong and balanced expression levels of the xylose metabolic genes (*XYL1*, *XYL2*, and *XYL3*), and evolution of the engineered strain for rapid xylose fermentation (1). Through systems-level characterization of the evolved strain, we learned that deletion of *PHO13* is a beneficial genetic perturbation eliciting up-regulation of the pentose phosphate pathway (2). Second, we introduced a cellobiose metabolic pathway into the efficient xylose fermenting strain (SR8) for co-fermentation of cellobiose and xylose (3). Again, a rational design for integrating genes (*cdt-1* and *gh1-1*) coding for cellobiose transporter and intracellular β-glucosidase, and evolution of the engineered strain were combined to isolate an engineered strain (EJ4) co-fermenting cellobiose and xylose rapidly. After measuring the copy numbers of *cdt-1* and *gh1-1* in the evolved strain, we learned that cellobiose fermentation can be improved by amplification of *cdt-1* and *gh1-1*. Finally, we introduced an acetate reduction pathway, which can enhance xylose fermentation through redox coupling (4), into the xylose and cellobiose co-fermenting strain (EJ4) for efficient co-utilization of cellobiose, xylose, and acetate. The resulting strain was able to produce ethanol from a mixture of cellobiose, xylose and acetate with a substantially higher yield and productivity than the control strains, demonstrating the synergistic effects from integration of multiple metabolic pathways (5).

The Design, Build, Test and Learn (DBTL) cycle has been formulated as a standard practice of synthetic biology to construct optimal microbial strains with desired traits. Our strain improvement efforts exemplifies the DBTL cycle.

References

1. Kim, S. R., Skerker, J. M., Kang, W., Lesmana, A., Wei, N., Arkin, A. P., and Jin, Y. S. (2013) *PloS ONE* 8, e57048
2. Kim, S.R., Xu, H., Lesmana, A., Kuzmanovic, U., Au, M., Florencia, C., Oh, E.J., Zhang, G.C., Kim, K. H., Jin, Y.S. (2015). *Applied and Enviromental Microbiology* 81,1601-1609
3. Ha, S. J., Galazka, J. M., Kim, S. R., Choi, J. H., Yang, X., Seo, J. H., Glass, N. L., Cate, J. H., and Jin, Y. S. (2011). *Proceedings of the National Academy of Sciences of the United States of America* 108, 504-509
4. Wei, N., Quarterman, J., Kim, S. R., Cate, J. H., and Jin, Y. S. (2013). *Nature*

Communications 4: 2805
5. Wei, N., Oh, E.J., Million, G., Cate, J.H., Jin, Y.S (2015). *ACS Synthetic Biology* 4, 707-13

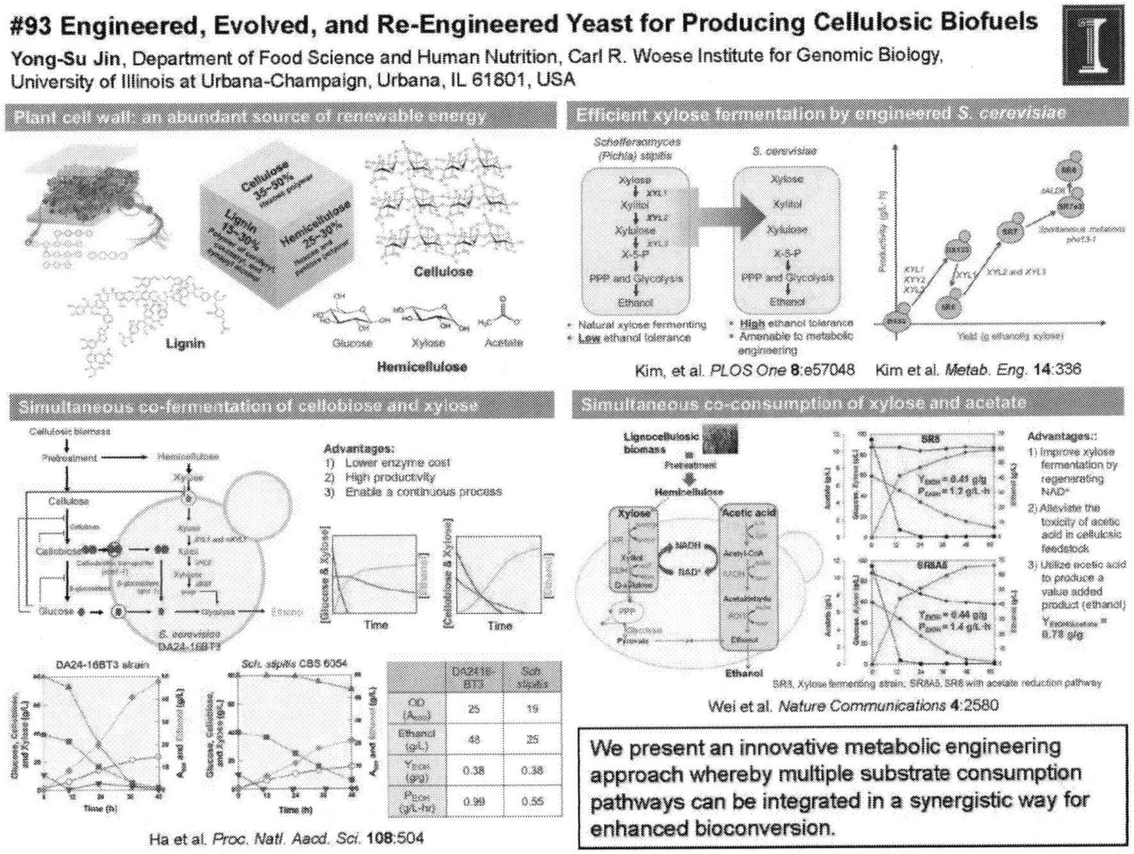

	DA2416-BT3	Sch. stipitis
OD (A_{600})	25	19
Ethanol (g/L)	48	25
Y_{EtOH} (g/g)	0.38	0.38
P_{EtOH} (g/L-hr)	0.99	0.55

Evolutionarily Identified Biological Design Allowing Simultaneous Sugar Utilization in Yeast

Lane, S., University of Illinois at Urbana-Champaign
Xu, H., University of Illinois at Urbana Champaign
Jin, Y. S., University of Illinois at Urbana-Champaign

Diauxic growth on mixtures of glucose and other sugars is a common phenotype throughout microorganisms. Glucose is generally consumed first followed by slower growth and consumption of remaining carbon sources. Considering that hydrolysates of renewable biomass are generally comprised of mixtures of glucose and other sugars, diauxic growth patterns are problematic for enabling efficient renewable bioconversion. Additionally, while techniques have been developed to enable simultaneous consumption of specific mixtures of sugars, a widely-applicable technique is desirable. In this study we report a generalizable sugar cofermentation strategy in the yeast *Saccharomyces cerevisiae*. Through evolutionary engineering with the glucose analog 2-deoxyglucose, we isolate a mutant strain capable of simultaneously consuming glucose and xylose at nearly equivalent rates and identify three mutations in the yeast hexokinases sufficient to confer the phenotype. Through analysis of deletion mutants, we rule out the role of specific yeast transporters and identify the overall cause behind the simultaneous glucose and xylose utilization: a reduced rate of glucose phosphorylation. We then recreate the simultaneous utilization phenotype with wild-type hexokinases by controlling their expression via the yeast doxycycline induction system. These results demonstrate that despite significant recent work in engineering xylose-specific sugar transporters, sugar mixtures can indeed be simultaneously consumed without modifications to the endogenous yeast transport machinery. We ultimately show that this technique is also effective at enabling co-consumption of mixtures of glucose and galactose. In the case of both glucose/xylose and glucose/galactose mixtures, tuning the rate of glucose phosphorylation allowed adjusting the relative sugar consumption rates as desired. This is to our knowledge the first report of a broadly-applicable cofermentation strategy that works across a range of metabolic pathways and is effective without regard to competitive transport inhibition.

A Cell-Free Framework for Rapid Biosynthetic Pathway Prototyping and Enzyme Discovery

Karim, A., The University of Texas at Austin
Jewett, M.,

Speeding up design-build-test (DBT) cycles is a fundamental challenge facing biochemical engineering. To address this challenge, we report a new cell-free protein synthesis driven metabolic engineering (CFPS-ME) framework for rapid biosynthetic pathway prototyping. In our framework, cell-free cocktails for synthesizing target small molecules are assembled in a mix-and-match fashion from crude cell lysates either containing selectively enriched pathway enzymes from heterologous overexpression or directly producing pathway enzymes in lysates by CFPS. As a model, we apply our approach to *n*-butanol biosynthesis showing that *Escherichia coli* lysates support a highly active 17-step CoA-dependent *n*-butanol pathway *in vitro*. Further, we leverage robotic liquid-handling systems to access combinatorial design space exceeding typical pipelines pursued in cells. The elevated degree of flexibility in the cell-free environment allows us to manipulate physiochemical conditions, access enzymatic nodes, discover new enzymes, and prototype enzyme sets with linear DNA templates to study pathway performance. We anticipate that CFPS-ME will facilitate efforts to define, manipulate, and understand metabolic pathways for accelerated DBT cycles without the need to reengineer organisms.

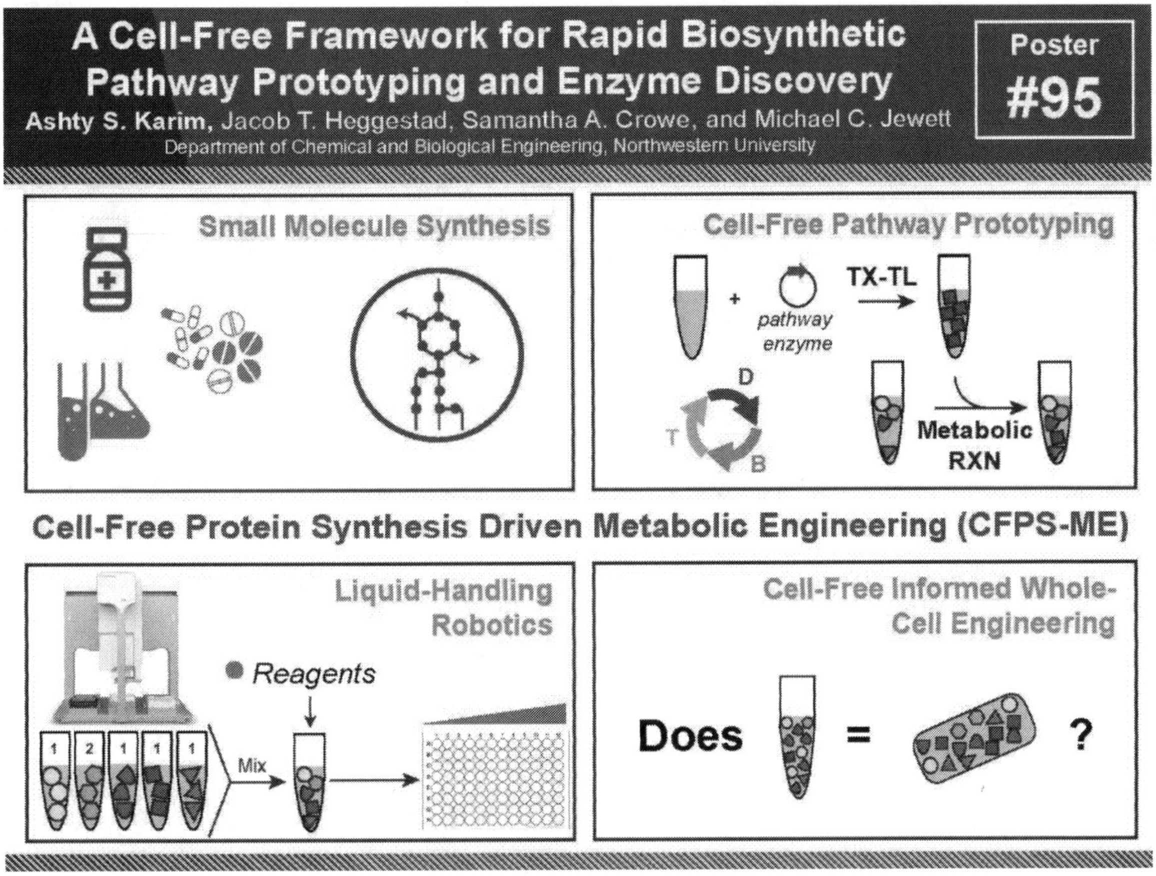

Easyclone-Markerfree: A Vector Toolkit for Marker-Less Integration of Genes into Saccharomyces Cerevisiae

jessop Fabre, M., DTU Biosustain
Jakociunas, T., The Novo Nordisk Foundation Center for Biosustainability
Stovicek, V., Technical University of Denmark
Dai, Z., Chalmers University of Technology
Jensen, M. K., Technical University of Denmark
Keasling, J. D., Joint Bioenergy Institute
Borodina, I., Technical University of Denmark

Saccharomyces cerevisiae is an established industrial host for production of recombinant proteins, fuels and chemicals. To enable stable integration of multiple marker-free overexpression cassettes in the genome of *S. cerevisiae*, we have developed a vector toolkit EasyClone-MarkerFree. The integration of vectors into previously described loci is facilitated by Cas9 endonuclease introducing double strand break at the particular sites in the genome. Cas9 is recruited to the chromosomal location by specific gRNA molecules expressed from a set of gRNA helper vectors. Single and triple insertions are obtained with 90-100% and 60-70% targeting efficiency correspondingly. We demonstrate application of the vector toolkit by constructing a haploid laboratory (CEN.PK113-7D) and a diploid industrial strain (Ethanol Red) for production of 3-hydroxypropionic acid, where we tested three different acetyl-CoA supply strategies, requiring overexpression of three to six genes each.

Towards an Engineered, Self-Sustained Microbial Ecosystem

Zerfass, C., University of Warwick
Christie-Oleza, J.,
Soyer, O. S., Warwick Centre for Integrative Synthetic Biology

Life on earth does not need any input apart from the energy provided through sunlight, and maintains itself by the cycling of nutrients. The sunlight is converted into chemical energy by photosynthetic plants and algae, which in turn provide food to other, non-photosynthetic organisms, i.e. heterotrophs. We are currently engineering a similarly self-sufficient and light-driven microbial ecosystem with tailored capabilities for chemical conversions. This will serve a dual purpose; the resulting engineered ecosystem will serve as a model system to better understand complex natural microbial communities, and at the same time, it is a novel low-cost system for the biotechnological production of value-added chemicals.

Communities of algae and bacteria are widespread in nature and are promising for applications as wastewater treatment and production of value added chemicals such as biofuels.[1] They are advantageous over single bacterial strains since they can work in cooperative ways. Multiple steps of a chemical synthesis can be compartmentalised among different bacteria, which reduces the energy-investment (and hence metabolic burden) for each individual cell.[2,3] Furthermore the community members support each other's growth by clearing up toxic metabolic side-products[3–5] and providing growth factors such as vitamins.[6] As an example, the superiority of a synthetic community was demonstrated by the efficient degradation of chemical contaminants associated with oil spills, while the individual organisms were less or not capable to degrade these compounds.[7,8]

To exploit microbial communities, it is necessary to gain a fundamental understanding of how and why microbes interact, and how such interactions react to disturbances.[9] Answering these questions in natural systems is almost impossible due to their sheer complexity and limitations on performing controlled experiments. Thus, we are engineering a minimalist self-sustained microbial ecosystem, which can maintain itself over extended time-scales, hence facilitating the investigation of interactions between the community members. The applicability of this ecosystem to chemical synthesis will be examined by creating a biomineralisation-cycle within the community, which allows the production of manganese oxide, a reactive mineral with significant potential in biodegradation and conversion of organic wastes into high-value chemicals.[10] This knowledge will support the bottom-up design of light-driven bacterial communities with desired synthetic capabilities, but also give new insights into natural microbial communities with a defined engineered model-ecosystem.

References

(1) Ramanan, R. et al. *Biotechnol. Adv.* **2015**.

(2) Zhou, K. et al. *Nat. Biotechnol.* **2015**, *33* (4), 377–383.

(3) Kouzuma, A.; Kato, S.; Watanabe, K. *Front. Microbiol.* **2015**, *6*, 477.

(4) Christie-Oleza, J. A.; Scanlan, D. J.; Armengaud, J. *Proteomics* **2015**, *15* (20), 3454–3462.

(5) Großkopf, T.; Soyer, O. S. *Curr. Opin. Microbiol.* **2014**, *18*, 72–77.

(6) Amin, S. A.; Parker, M. S.; Armbrust, E. V. *Microbiol. Mol. Biol. Rev.* **2012**, *76* (3), 667–684.

(7) Tang, X. et al. *J. Hazard. Mater.* **2010**, *181* (1-3), 1158–1162.

(8) Abed, R. M. M.; Köster, J. *Int. Biodeterior. Biodegradation* **2005**, *55* (1), 29–37.

(9) Goers, L.; Freemont, P.; Polizzi, K. M. *J. R. Soc. Interface* **2014**, *11* (96), 20140065.

(10) Remucal, C. K.; Ginder-Vogel, M. *Environ. Sci. Process. Impacts* **2014**, *16* (6), 1247–1266.

Front Propagation in an Array of Artificial Cells

Tayar, A., Weizmann institute

Alexandra M. Tayar, Eyal Karzbrun, Vincent Noireaux, Roy H. Bar-Ziv

Living systems employ front propagation and spatiotemporal patterns, encoded in nonlinear biochemical reactions, for communication, self-organization, and computation Emulating such dynamics in minimal systems is important for understanding physical principles in living cells and *in vitro*. Here, we report a 1D array of DNA compartments in a silicon chip as an artificial multicellular system, offering means to implement reaction-diffusion dynamics by integrated genetic circuits and chip geometry. Using a bistable circuit we programmed a sharp front of protein synthesis propagating in the array as a cascade of signal amplification and short-range diffusion. The front velocity is maximal at a saddle-node bifurcation from a bistable regime with traveling fronts to a monostable one that is spatially homogenous. Near the bifurcation the system exhibits large variability between compartments, providing a possible mechanism for population diversity not due to stochastic fluctuations. This puts forth a man-made biological system with programmable information processing from genes to cells and up to the multicellular scale.

Nature Physics 11, 1037–1041 (2015) doi:10.1038/nphys3469

Rational Design of Large Crispr/dCas9 Transcriptional Logic Circuits in Eukaryotic Cells

Vrana, J.,
Gander, M. W., University of Washington
Voje, W. Jr., University of Washington
Carothers, J.,
Klavins, E.,

Natural cells perform incredible feats of computation through a beautiful orchestra of inter-connected and interacting molecules. A grand challenge in synthetic biology is to emulate the complexities of natural cellular systems to create new sophisticated cellular behaviors through the introduction of synthetic molecular functions. The creation of new molecular functions that are orthogonal, inter-connectable, and scalable is essential for engineering complex molecular networks. However, many natural molecular functions often exhibit unpredictable behaviors when used outside of their natural genetic context, impeding the engineering of grand molecular networks *in vivo*. To address these issues, we engineered a set of un-natural transcriptional components in *S. cervisiae* that can be interconnected and assembled into large genetic circuits. Our system is comprised of a large set of orthogonal synthetic transcriptional CRISPR/dCas9 'NOR' gates. We leveraged a RNA-guided synthetic chromatin remodeler, a dCas9-mxi1 fusion, to substantially improve dCas9 guided transcriptional repression and used computer-aided design to create a set of orthogonally interacting guide-RNAs and promoter components. Since all of the components are orthogonal to the native *S. cervisiae* host, our transcriptional system exhibits exceptional predictability and scalability with minimal cell loading. We demonstrate the power of our synthetic NOR gates by assembling them into large circuits that perform basic combinatorial computation (NOT, OR, AND, NAND, XOR, XNOR functions) and long delay cascading circuits (up to seven interconnected NOR gates). Complementing this system, we also constructed a biochemical model that predicts synthetic network behavior through global parameter optimization and Monte Carlo simulation. The model allows for predictable forward-engineering of synthetic gene circuits through dynamic behavior prediction and examination of parameter sensitivity estimation. Our approach has enabled the construction of among the largest eukaryotic gene circuits to date. We envision future iterations of the system will form the basis for large, synthetic, decision-making systems in living cells.

Development of a Multiplexed Cadmium Biosensor for the Detection of Heavy Metals in the Environment

Bereza-Malcolm, L., La Trobe University
Aracic, S.,
Mann, G., Defence Science & Technology Organisation
Franks, A.,

Continuous environmental monitoring is necessary for rapid identification of dangerous contaminants present in our ecosystems. Heavy metals are of particular concern with anthropogenic activities resulting in an increase in levels present in soil and water. Current methods for detection of heavy metals are reliant on the transport of samples to off-site facilities, resulting in a time delay between sample collection and analysis. This delay can result in accidental consumption of contaminated substances, consequently leading to cellular accumulation and health related problems, such as cancer. Engineering bacteria to act as on-site microbial biosensors is one approach which provides the ability to rapidly detect heavy metals in the environment.

We are developing and testing the abilities of multiplexed cadmium responsive biosensors. The biosensing construct consists of an operator/promoter sequence regulated by the transcriptional regulator, CadR. CadR binds to the operator sequence, inhibiting transcription of downstream genes. Typically, these genes are responsible for cadmium uptake, reduction and export. The transcriptional regulator, *cadR*, together with the cognate operator/promoter was cloned upstream of a promoterless *gfp*. The biosensing construct was sub-cloned into a broad-host range vector, along with a constitutively expressed *rfp*. In this system an uptake of cadmium chloride into the bacterial cell results in GFP expression which increases over time, allowing qualitative and quantitative analyses.

The biosensing construct was transferred into commonly used laboratory bacteria; *Escherichia coli*, *Shewanella oneidensis* and *Pseudomonas aeruginosa*, alongside wild-type *Enterobacter* spp. originally isolated from cadmium contaminated soil. Fluorescence microscopy and cadmium slope agar plates were used as an initial assessment of each biosensors functionality. The limit of detection of the microbial biosensors in response to cadmium chloride was found to range between 0.005-0.5 $\mu g\ ml^{-1}$ with the response time ranging from 5 minutes to 24 hours. These results highlight the microbial biosensors effectiveness at detecting cadmium chloride at levels deemed dangerous to human health by the World Health Organisation. The specificity of the microbial biosensors was also tested and it was found that exposure to alternate heavy metals, particularly mercury chloride, results in a detectable output. This revealed that the biosensing construct can produce a multiplexed response to multiple inputs. The cadmium biosensors are currently being applied for the detection of cadmium in environmental samples including ground and ocean water, along with soil. Alongside these assays the facultative anaerobe biosensors are currently being tested under anaerobic conditions, opening the potential for the incorporation of alternate outputs.

Future research on the capabilities of the multiplexed cadmium biosensors will add knowledge to how bacteria can be applied as long-term environmental monitoring devices. This PhD project has been financially supported by the Human Protection and Performance Program of the Defence Science Institute, Australia.

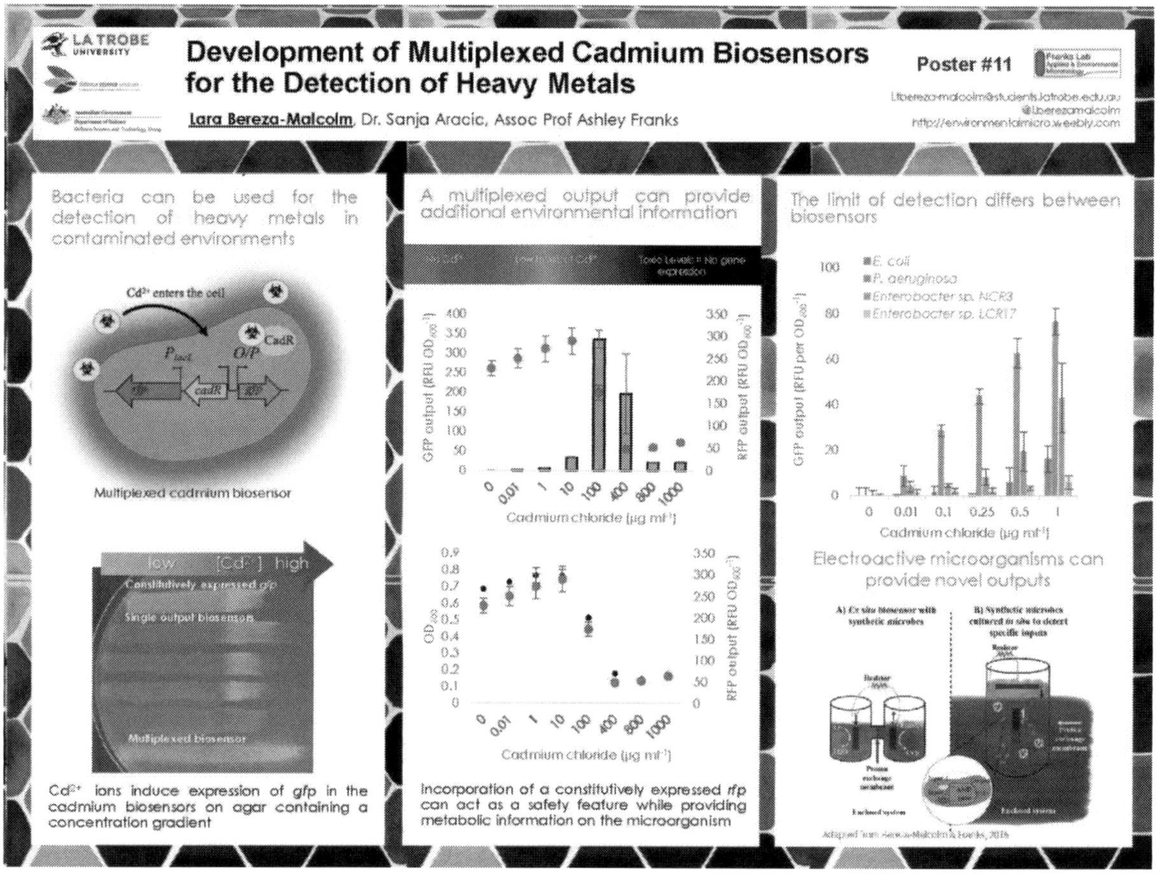

Rewriting the Genome Novel Tools in Synthetic Biology

Clore, A., Integrated DNA Technologies

The Synthetic Biology revolution has allowed unprecedented advances in the areas of disease treatment, vaccine development, and small molecule discovery. Two advances have facilitated this growth. The inexpensive, fast and reliable synthesis of large, accurate fragments of DNA, and the ability to rapidly and accurately modify genomes using CRISPR technology. This presentation will discuss the development, methodology, and implementation of these technologies drawing from IDTs in house research and recent publications.

Rational Design of High-Number DNA Fragments for Gene Construction

Birla, B.,

Gene synthesis is frequently used in modern molecular biology research either to create novel genes or to obtain natural genes when the synthesis approach is more flexible and reliable than cloning. DNA chemical synthesis has limits on both its length and yield, thus full-length genes have to be hierarchically constructed from synthesized DNA fragments. Gibson Assembly and its derivatives are the simplest methods to assemble multiple double-stranded DNA fragments. Currently, up to 12 dsDNA fragments can be assembled at once with Gibson Assembly according to its vendor. In practice, the number of dsDNA fragments that can be assembled in a single reaction are much lower. We have developed a rational design method for gene construction that allows high-number dsDNA fragments to be assembled into full-length genes in a single reaction. Using this new design method and a modified version of the Gibson Assembly protocol, we have assembled 3 different genes from up to 45 dsDNA fragments at once. Our design method uses the thermodynamic analysis software Picky that identifies all unique junctions in a gene where consecutive DNA fragments are specifically made to connect to each other. Our novel method is generally applicable to most gene sequences, and can improve both the efficiency and cost of gene assembly.

Quality and Quantity: Protein Production and Folding Using a Bacterial Protein Secretion System

Metcalf, K. J., University of California, Berkeley
Rosales, S., University of California, Berkeley
Bevington, J.,
Valdivia, E., University of California, Berkeley
Tullman-Ercek, D., University of California, Berkeley

Two process metrics must be considered when producing a heterologous protein: product titer (i.e., quantity); and activity and folding of the protein (i.e., quality). Protein secretion can increase product titer by simplifying purification steps and enabling a continuous process. We studied the effect of secretion on protein folding to improve product quality. In Gram-negative bacteria, proteins can be secreted in one-step from the cytosol to the extracellular space using the type III secretion system. Recent structural data shows that the protein is unfolded and linearized as it is passes through the secretion needle, which is 2 nm in diameter and 50 nm long. Proteins must thus refold in the extracellular space after secretion for this approach to be viable. Indeed, functional assays show enzymatic activity or antigen binding after secretion of two model enzymes and a single chain variable fragment of an antibody, suggesting that these proteins do fold after secretion. Genetic and chemical methods show these proteins form disulfide bonds and multimerize in the extracellular space after secretion. Finally, we find the fraction of folding is dependent on both the protein of interest and media composition. The ability to recover folded heterologous protein in the extracellular space demonstrates a strength of the use of a protein secretion chassis.

Quality and Quantity: Protein Production and Folding Using a Bacterial Protein Secretion System

Kevin Metcalf, Sandy Rosales, James Bevington, Elias Valdivia, and Danielle Tullman-Ercek

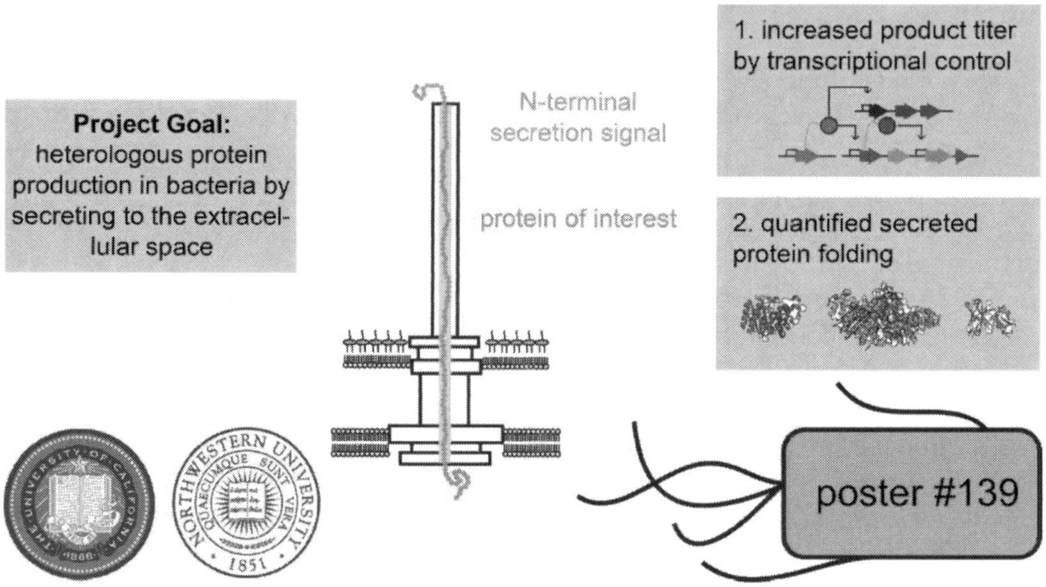

RNA-Programmed Gene Circuit Engineering Using Crispr-Nickases

Standage-Beier, K., Arizona State University
Wang, X.,
Zhang, Q., Arizona State University

There is a need for programmable tools to remodel genomes. Such tools have been integral for synthetic biologists to engineer organisms and implement synthetic gene circuitry. CRISPR-Cas systems provide an easily programmable tool for targeting many of these precise manipulations. Here we demonstrate the ability of CRISPR-Cas derived nickases to direct targeted-recombination of large genomic regions flanked by repetitive elements in *E.coli*. While CRISPR directed double-stranded DNA breaks are highly lethal in many bacteria, we show that CRISPR-guided nickase systems can be programmed to make precise, non-lethal, single-stranded incisions in targeted genomic regions. We demonstrate that dual-targeted nicking induces recombination events resulting in large-scale deletion. Multiplex targeting enables deletion of 133 Kb, approximately 3% of the entire *E.coli* genome. This provides a framework for methods to manipulate bacterial genomes using CRISPR-nickase systems. We envision this system working synergistically with preexisting bacterial genome engineering methods with further extension enabling targeted and multiplex integration of synthetic gene circuits. Future development of CRISPR based methods may synergistically incorporate aspects of protein engineering and multi-targeted sgRNA design to accelerate genome engineering.

Transhydrogenase Promotes the Robustness and Evolvability of Metabolic Networks

Chou, H. H.,
Marx, C. J., University of Idaho
Sauer, U., ETH Zurich

Homeostasis of redox cofactors NAD(P)H is crucial for metabolism but inevitably disturbed when microbes switch growth substrates, evolve biochemical pathways, or are engineered to produce valuable chemicals. To illuminate how metabolic systems endure NADPH perturbations brought instantly by pathway modification and restore homeostasis in the long run, we studied laboratory evolution of two pathway-engineered bacteria, *Escherichia coli* and *Methylobacterium extorquens*, which underproduced NADPH during growth on glucose and on methanol, respectively. Literature suggests multiple metabolic strategies to restore NADPH homeostasis. Surprisingly, genetic dissection of isolates from twelve populations of *E. coli* evolved on glucose and eight populations of *M. extorquens* evolved on methanol revealed merely two solutions: (1) enhancing the NADH/NADPH exchange reaction by boosting the expression of membrane-bound transhydrogenase (mTH) in every population of *E. coli* and *M. extorquens*; (2) simultaneously consuming glucose with acetate, an disfavored byproduct normally excreted during glucose catabolism, in two *E. coli* sub-populations. Notably, mTH is spontaneously upregulated upon pathway modification in both bacteria and distributes broadly among prokaryotes and eukaryotes. Convergent evolution of two phylogenetically and metabolically distinct bacterial species suggests mTH as a conserved buffering mechanism that promotes the robustness and evolvability of metabolism. Moreover, adaptive diversification of *E. coli* via evolving dual substrate consumption highlights the physiological flexibility to exploit ecological opportunities and points to a bioengineering alternative to fuel the redox currency production.

Cysteinyl-DOPA Crosslinking and the Road to an Expanded Metabolism

Thyer, R., UT Austin
Cotham, V., University of Texas at Austin
Gollihar, J., UT Austin
Tack, D. S., University of Texas at Austin
Maranhao, A. C., University of Texas at Austin
Brodbelt, J. S., UT Austin
Ellington, A. D., University of Texas at Austin

Across all domains of life, the proteome is confined to 20 standard amino acids, with just two rare exceptions; selenocysteine and pyrrolysine. Expansion of the genetic code with a growing number of non-canonical amino acids (ncAAs) bearing new amino acid chemistries, has enabled the synthesis of proteins with diverse new functionality. 3,4-dihydroxy-L-phenylalanine (L-DOPA), is a naturally occurring amino acid containing a catechol moiety, a new redox chemistry not occurring in the 20 standard amino acids. Although found widely in nature, 3,4-dihydroxy-L-phenylalanine only occurs in proteins as a post-translational modification of tyrosine (or as a damaging oxidation product). The presence of 3,4-dihydroxy-L-phenylalanine in proteins confers unique properties such as redox dependent crosslinking with nearby nucleophiles or other 3,4-dihydroxy-L-phenylalanine residues, as well as coordination of metal ions. In addition to its use in nature, catechol containing polymers are under investigation as self healing, programmable biomaterials. The ability to harness these properties for protein engineering is of great interest, as proteins are a scalable and programmable scaffold. Although an orthogonal aminoacyl-tRNA synthetase (aaRS):tRNA pair for the incorporation of L-DOPA has previously been reported, it, like many other early aaRS:tRNA pairs, suffers from poor activity and a lack of discrimination between its new substrate and tyrosine. Here we describe the evolution of an aaRS for highly efficient site-specific incorporation of 3,4-dihydroxy-L-phenylalanine, the development of cysteinyl-DOPA crosslinking motifs and our progress towards an expanded metabolism in *E. coli*. This new aaRS enabled the production of 3,4-dihydroxy-L-phenylalanine containing proteins with yields exceeding 20 $mg.L^{-1}$ at over 90% incorporation efficiency. Using a combination of rational design, and refinement by computational modelling, several engineered pairs of 3,4-dihydroxy-L-phenylalanine and cysteine residues were shown to form covalent cysteinyl-DOPA crosslinks in different proteins. This new bioorthogonal crosslinking chemistry and the catechol moiety in general are valuable and highly enabling additions to the protein engineering toolkit.

Synthetic Evolutionary Origin of a Proofreading Reverse Transcriptase

Ellefson, J., University of Texas at Austin
Ellington, A., University of Texas at Austin

While proteins involved in fundamental biological processes are under extraordinary functional constraints, recent work has shown that core components in the central dogma can be engineered to have radically altered properties, such as the creation of expanded genetic alphabets, tethered ribosomes, and rewiring of the genetic code. One of the last discovered components of the central dogma, reverse transcriptase (RT), has for the most part been the province of a single family of enzymes of ancient evolutionary origin and are inherently error prone due to their lack of a proofreading (3'- 5' exonuclease) domain. To determine if the lack of proofreading is a historical coincidence or a functional limitation of reverse transcription, we attempted to evolve a high fidelity, thermostable DNA polymerase to efficiently use RNA templates. The Reverse Transcription Xenopolymerase (RTX) is a new, distinct evolutionary lineage that maintains the DNA proofreading activity of its parent with a RNA template, which was shown to greatly improve RT fidelity. Mutations in highly conserved molecular checkpoints render RTX capable of polymerizing either DNA or RNA templates with similar efficiencies, enabling applications such as single enzyme RT-PCR and direct RNA sequencing without cDNA isolation. The creation of RTX confirms proofreading is compatible with reverse transcription and reemphasizes that core molecular machinery that has otherwise been conserved over billions of years can be engineered to attain novel functionality.

RNA- Programmed DNA Methylation

Xiong, T., Johns Hopkins University
Meister, G., Dana-Farber Cancer Institute
Kato, N., Johns Hopkins University
Timp, W., Johns Hopkins University
Novina, C., Dana-Farber Cancer Institute
Ostermeier, M., Johns Hopkins University

Appropriate epigenetic modification of DNA via cytosine methylation has been shown to be vital to the proper functioning of cellular processes in eukaryotes. High levels of methylation in promoter regions are associated with transcriptional silencing. The ability to precisely target methylation to specific sites in promoters could be used as a tool for gene silencing and for understanding the relationship between DNA methylation and cell phenotype. In addition, since abnormal hypomethylation is associated with diseases such as leukemia, atherosclerosis and, Alzheimer's disease, precise targeting of methylation to a single promoter is a potential therapeutic strategy. The typical strategy for targeting DNA methylation involves end-to-end fusion of a DNA binding domain (DBD), such as zinc fingers and TALE, to a cytosine DNA methyltransferases (MTase), typically bacterial enzymes or human DNMT3a. The two main drawbacks of this approach are (1) methylation occurs at non-target sites since the MTase domain remains active to methylate these sites when the DBD is not interacting with DNA and (2) new sequence recognition requires DBD redesign. Our solution to the first problem is to use the target site for methylation as a template for assembling the active form of an otherwise ineffective split methyltransferase (sMTase). This strategy has been shown to result in very specific targeting in bacteria (~80% methylation at the target site and ≤0.8% methylation at other CpG sites) using zinc fingers as the DBD. The second problem can be addressed by the CRISPR-Cas9 genome engineering tool, which has not yet been applied to target DNA methylation. A key feature of nuclease- null Cas9 (dCas9) as a DBD is its ability to bind specific sites defined by the single-stranded guide RNA (sgRNA) sequence. A carefully selected sgRNA allows for highly specific targeting without the requirement for DBD redesign. An ideal dCas9-sMTase would have many exemplary features: the precision of true single-site targeting of methylation, ease of targeting via simple design of sgRNA, and multiplexed targeting of several sites via multiple sgRNAs. The later advantage addresses the concern that single site methylation might not be sufficient to have a desired phenotypic effect. To this end, we designed fusion proteins of dCas9 and a sMTase. The dCas9-sMTase provides high precision and control over CpG methylation in bacteria and human cells.

Understanding Cell Reprogramming By Adaptive Regulatory Mutations As a Guide for Cell Design

Utrilla, J., University of California, San Diego
O'Brien, E.,
Feist, A. M., University of California, San Diego
Palsson, B. O., University of California, San Diego

Understanding cell reprogramming by adaptive regulatory mutations as a guide for cell design

Adaptive regulatory mutations are common genetic variants in evolutionary experiments. These mutations shift gene expression at a global scale to switch functions between different cellular states. In this work Escherichia coli RNA Polymerase mutants were analyzed to understand the implications of such evolutionary mechanisms in synthetic biology . Using multi-omic data, key environmental controls and a genome scale model of metabolism and gene expression it was possible to unravel the system reprogramming by single amino acid changes in the RNAP. The regulatory effect of the mutations results in a perturbed cellular state that increase growth and cell yield by reducing non-growth related protein expression. Transcriptional regulatory network analysis show the mechanism of cellular function shift. Econometric analysis reveal the resource allocation constraints that organisms face. Detailed phenotypic assays show consistent fitness benefits and detriments in specific environments. Our results advance in the understanding of the multi-scale relationship between a genetic variation and the mechanism of phenotypic change. Thus they should be able to guide synthetic biology efforts and provide a basis for the design of synthetic cell phenotypes.

Conversion of Glucose to 2,3-Butanediol in Escherichia coli Cell-Free Lysates: A Model System

Kay, J. E.,
Jewett, M. C., Northwestern University

In recent decades we are seeing more and more examples of whole pathways being activated in crude lysates to generate a desired product *in vitro*. For instance, inexpensive substrates glucose or maltose have been used to regenerate ATP to power cell-free protein synthesis. Buffers and reaction conditions designed to mimic the cytoplasm have allowed a remarkable number of enzymes to be coordinately activated and to remain stable for up to about a week. It begs the question, of whether cytoplasmic mimicry will benefit other endeavors such as rapid enzyme screening or *in vitro* multi-enzyme catalysis.

Here, as a model study, we sought to determine if the conversion of glucose to 2,3-butanediol (2,3-BD), a medium level commodity chemical with many industrial applications, could be achieved in an *in vitro* system designed to have high cofactor regeneration. Extracts of *Escherichia coli* expressing three heterologous enzymes are able to convert glucose to m2,3-BD at surprisingly high rates and concentrations. With no strain optimization, we have observed a maximal synthesis rate of m2,3-BD of 11.3 ± 0.1 g/L/h with a theoretical yield of 71% (0.36 g m2,3-BD / g glucose) and concentrations of 82 ± 8 g/L m2,3-BD in batch reactions. We have found the system to be robust to working concentrations of antibiotics and antifoam, and other compounds that are toxic to cell growth but do not denature or inhibit relevant enzymes. Current work includes tuning enzyme ratios to try to understand what is necessary and limiting in the system, and testing potential enzyme screening protocols. These results highlight the ability for high-level co-factor regeneration in cell-free lysates and suggest exciting opportunities to use lysate-based systems to rapidly prototype metabolic pathways and carry out molecular transformations when bioconversion yields (g product/L), productivities (g product/L/h), or cellular toxicity limit commercial feasibility of whole-cell fermentation.

A Highly Productive One-Pot System for the Incorporation of Non-Standard Amino Acids into Cell-Free Synthesized Proteins

Des Soye, B. J., Northwestern University
Jewett, M. C., Northwestern University

The site-specific installation of non-standard amino acids (NSAAs) into proteins via amber suppression enables the synthesis of peptides and biopolymers with new chemical properties and functions. However, in most systems competition between release factor 1 (RF1) and aminoacylated orthogonal tRNAs results in the formation of a significant quantity of undesired truncated products. Recent efforts have addressed this limitation through the generation of a high-yielding cell-free protein synthesis (CFPS) platform using extracts derived from a recoded organism lacking RF1 that has been optimized for CFPS. These extracts, while highly productive, are dependent upon supplementation with purified T7 RNA polymerase (T7RNAP) to catalyze orthogonal transcription *in vitro*. Here, we describe the development of a truly one-pot CFPS system for synthesis of NSAA-containing proteins at high yields. As the aforementioned CFPS-optimized strain was incapable of synthesizing T7RNAP, we applied lambda Red recombination to site-specifically integrate the T7RNAP gene into the genome of the strain. We investigated different combinations of insertion loci and promoter strength to identify the combination that best balances strain health and T7RNAP overexpression levels. Further analysis of these strains revealed that T7RNAP produced was being cleaved by the membrane-bound periplasmic protease OmpT, potentially reducing the activity of the enzyme. To address this, we leveraged multiplex advanced genome engineering (MAGE) in the highest-performing T7RNAP-expressing strain, mutating the coding region of the T7RNAP gene to eliminate OmpT recognition and cleavage sites. This mutant T7RNAP remains highly productive, and is highly resistant to OmpT-mediated degradation. Our most highly productive cell extracts featuring this mutant T7RNAP achieve more than 2.0 g/L of active superfolder green fluorescent protein (sfGFP) in the presence of supplemental T7RNAP, and ~1.6 g/L of sfGFP in the *absence* of supplemental T7RNAP. We show that these extracts enable significantly higher levels of NSAA incorporation than current state-of-the-art one-pot CFPS systems produced from BL21 (DE3) and its derivatives. To our knowledge, this is the highest-yielding one-pot CFPS system developed to date. Our work has implications for using recombineering approaches for CFPS strain development, expanding the chemistry of biological systems, and cell-free synthetic biology.

One-Step in Vitro Large Gene Cluster Cloning By Crispr-Cas9 with Gibson Assembly

Lou, C.,

We describe a method that allows the targeted cloning of near-arbitrary, long genomic sequences (50-150 kb) from bacteria to accomplish in a single step within 1-2 days. Target DNA is excised from bacterial chromosomes *in vitro* by the RNA-guided Cas9 nuclease at two designated loci, and ligated to the cloning vector by Gibson assembly. This method can be a useful molecular engineering tool for efficiently cloning large gene clusters that are difficult to obtain directly by traditional PCR or restriction-enzyme-based methods.

Cas9-Assisted Targeting of CHromosome segments (CATCH) Method

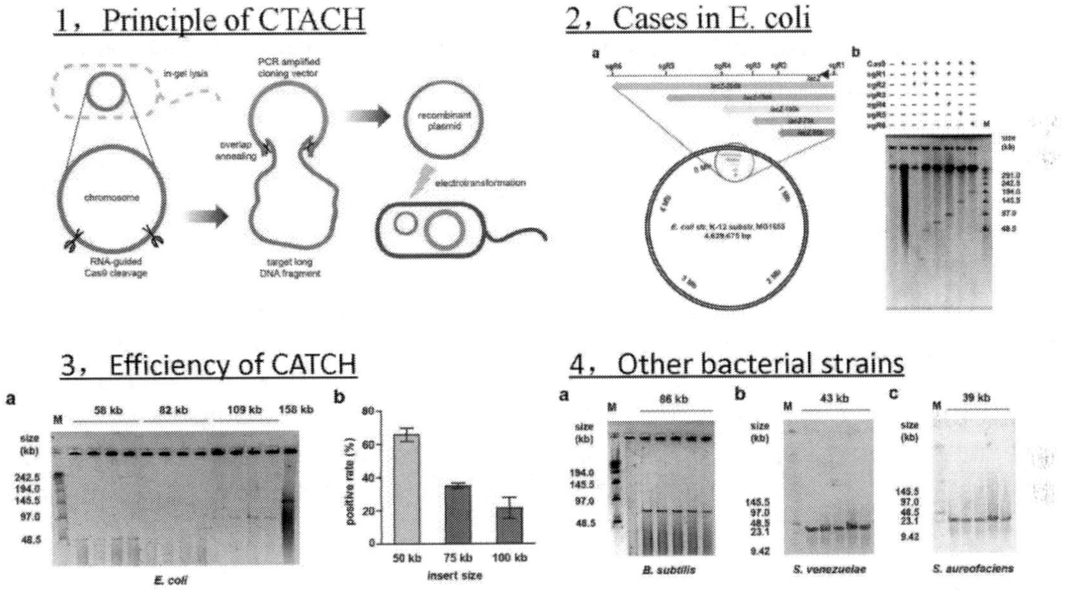

Jiang, W., Zhao, X., Gabrieli, T., Lou, C*., Ebenstein, Y*., & Zhu, T. F*. (2015). *Nature communications*, *6*.

Engineering Small Transcription Activating RNAs (STARs)

Chappell, J., Cornell University
Takahashi, M. K., Cornell University
Lucks, J. B., Cornell University

Efficient engineering of synthetic cellular functions is ultimately dependent on our ability to control gene expression in a precise, predictable and scalable manor. In bacteria, small RNA (sRNA) regulators of gene expression are an increasingly attractive substrate for this. As sRNAs regulate gene expression by formation of RNA structures, they are highly designable and characterizable – with a wealth of computational and experimental tools that can rapidly unlock functional design principles. Moreover, sRNAs are compact and place low-burden on cellular resources, meaning they are potentially highly scalable for multiplexing regulation in a single cell. Finally, sRNA transcriptional regulators can be utilized for higher-order regulation such as logic gates and networks, that process signals rapidly due to fast degradation rates of RNAs.

However, a fundamental gap in both natural and synthetic sRNA regulatory capabilities has been the lack of transcriptional activators – critically limiting their application for higher-order regulation. Recently we have addressed this gap by creating the first Small Transcription Activating RNAs, or STARs. The STAR system involves placing an RNA transcriptional terminator upstream of a target gene that acts as a blockade and prevents the cell from transcribing the gene. When the STAR is present, it removes this blockade, turning ON the downstream gene by allowing transcription to take place. The simplicity of this strategy enabled libraries of highly orthogonal STARs to be created that can be used to control expression of diverse genes, combinatorially combined with libraries of other regulators to allow for tunable and precise control of gene expression and combined to construct RNA-only logic gates. Finally, we present recent work harnessing computational RNA design tools to develop second-generation libraries of highly functional STARs with near digital dynamic ranges that can further expand the utility of this novel class of regulator.

Engineering Bacteria for Biofuel Production

Macklyne, H.,
Paget, P.,

Our research into biofuels addresses the need for environmentally and socially responsible sources of energy. Biofuels, made from organic matter, have recently become a viable alternative to petroleum-based fossil fuel. Sugar and starch make up the majority of feedstock used in biofuel production as it is easily digested. However, the use of these feedstocks is problematic as it consumes resources with negative environmental and social implications. By using a bacterium able to utilise five and six carbon sugars, such as the thermophile *Geobacillus thermoglucosidasius,* organic lignocellulosic waste material can be used as a feedstock.

The aim of this project is to investigate and utilise key genetic regulators of fermentation in *G.thermoglucosidasius* and to construct genetic engineering tools that enable strain development, for second generation biofuel production. We have focused on the redox-sensing transcriptional regulator Rex, widespread in Gram positive bacteria, which controls the major fermentation pathways in response to changes in the cellular NAD+/NADH ratio. Having identified several members of the Rex regulon, we are now building novel synthetic regulatory switches that can be used to dynamically control gene expression in response to changes in cellular redox state. In addition, new approaches and tools for the efficient genetic engineering of *G.thermoglucosidasius* will be discussed including an optimized *E. coli* - *Geobacillus* conjugation method for plasmid transfer and gene disruption.

Road Rules for Traffic on DNA: Systematic Analysis of Gene Regulation By Encounters

Hao, N., The University of Adelaide
Shearwin, K. E., The University of Adelaide
Dodd, I. B.,

Textbook depictions of the transcription of a gene show RNA polymerase binding to the DNA at the promoter, initiating transcription and then transcribing uneventfully across the gene until reaching a terminator sequence and falling off the DNA. However, this picture of the DNA as a freeway for RNA polymerase does not apply within a cell, where the DNA is more like a crowded one-lane two-way street. As an RNA polymerase makes its way along the DNA, its progress can be affected by a myriad of proteins bound to the DNA, by difficult DNA sequences that make it pause, and also by other traffic - RNA polymerases moving either in the opposite direction or more slowly in the same direction. The outcomes of these collisions are crucial in gene regulation, yet the factors that determine the 'winner' – whether the DNA-bound protein blocks transcription by RNA polymerase or whether the polymerase dislodges the protein and interferes with its function – are unknown.

Here we integrate *in vivo* experiments in *E. coli*, using well-defined genetic components, with mathematical modelling to study the factors which influence the outcome of these interactions. Construction of a modular, chromosomally integrated single-copy chassis allows us to systematically vary parameters such as RNA polymerase density, promoter-roadblock spacing, and roadblock binding affinity. This combined approach allows us to evaluate the significance of RNA polymerase - roadblock encounters, the factors that affect them and may provide new tools for the manipulation of gene expression.

Creating Libraries of Orthogonal RNA Toehold Repressors

Carlson, P. D., Cornell University
Lucks, J. B., Cornell University

RNA regulators of gene expression are becoming increasingly important components of the synthetic biology toolbox. RNA is a versatile and designable molecule, and is one of the most powerful substrates at our disposal for engineering gene expression. The ability to propagate information directly via RNA-RNA interactions creates the basis for the bottom-up design of genetic networks to implement logic or carry out complex patterns of gene expression. An important class of such regulators is RNA-mediated repressors, which repress downstream gene expression in response to a *trans*-acting antisense RNA. Rational mutagenesis and sequence mining have been successful in creating libraries of orthogonal (independently-acting) RNA repressors that function on the transcriptional and translational levels. However, the construction of large networks will require additional regulators with enhanced dynamic range and reduced crosstalk. Inspired by work demonstrating the successful *de novo* design of RNA-mediated translational activators called toehold switches (Cell, 159, 925-939, 2014), we sought to use a similar approach to create *de novo* designed RNA translational repressors. In this setup, the default structural state frees the ribosome binding site (RBS) and start codon, allowing translation to proceed in an ON state. Binding of a designed *trans*-acting antisense RNA leads to structural rearrangement and occlusion of the RBS and start codon, preventing translation in the OFF state. Using two different approaches, we successfully designed toehold repressors with the Nucleic Acid Package (NUPACK) capable of attenuating gene expression up to 98% (50-fold repression). We further demonstrate the ability to construct libraries of orthogonal repressors with potential applications ranging from nucleic acid-based diagnostics to metabolic engineering.

Creating Libraries of Orthogonal RNA Toehold Repressors

Paul Carlson and Julius Lucks

Robert Frederick Smith School of Chemical and Biomolecular Engineering, Cornell University

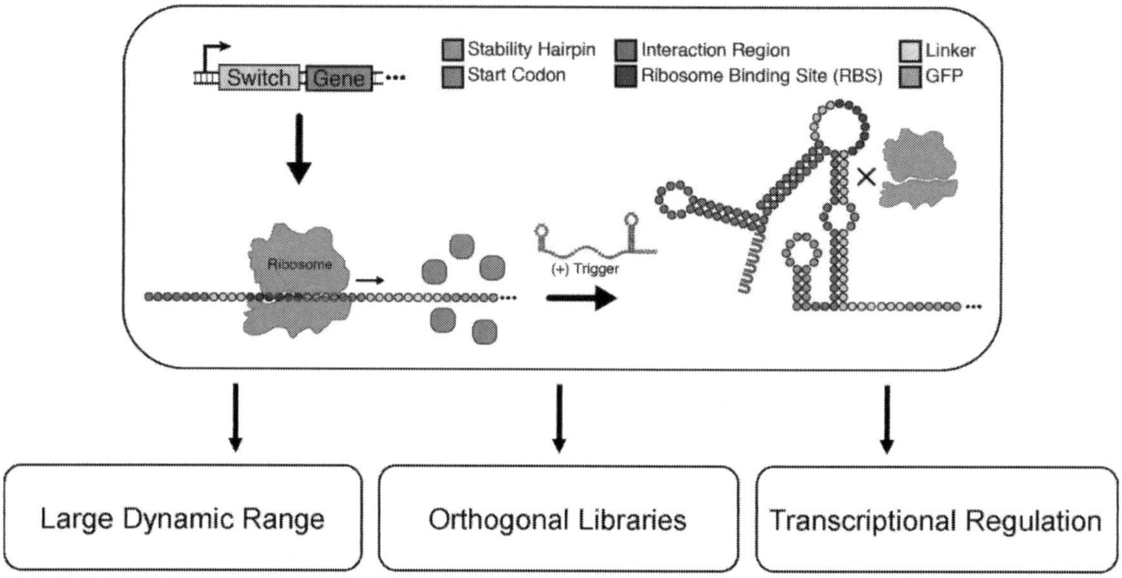

Programming the Timing of Gene Expression in Synthetic Circuits

Tei, M., UC Berkeley
Venturelli, O. S., UC Berkeley
Arkin, A. P., University of California at Berkeley

The control of gene expression timing in synthetic circuits has many potential applications for performing a target function using genetically modified cells including biocontainment by triggering cell death following a defined delay and controlling the timing of cargo release for drug delivery. Previous studies have demonstrated that bistable switches or transcriptional cascades can control the timing of gene expression. However, tuning the delay time requires sequence modifications and the length of the delay is correlated with output noise. To identify a robust and tunable delay generating circuit, we systematically enumerated a broad range of circuit topologies and scored circuit performance. We identified a molecular sequestration circuit that exhibits tunable control over the duration of the delay and does not significantly amplify output noise. This circuit establishes an adjustable threshold using a repressor to stoichiometrically inhibit an activator. The delay is determined by the time for the activator to accumulate to the level of the repressor. We experimentally characterized the range of delays for this circuit using heterologous extracytoplasmic function (ECF) σ and anti-σ factors in E.coli. This delay generator circuit can be used as a building block for complex synthetic circuits including developmental programs and pulse generators.

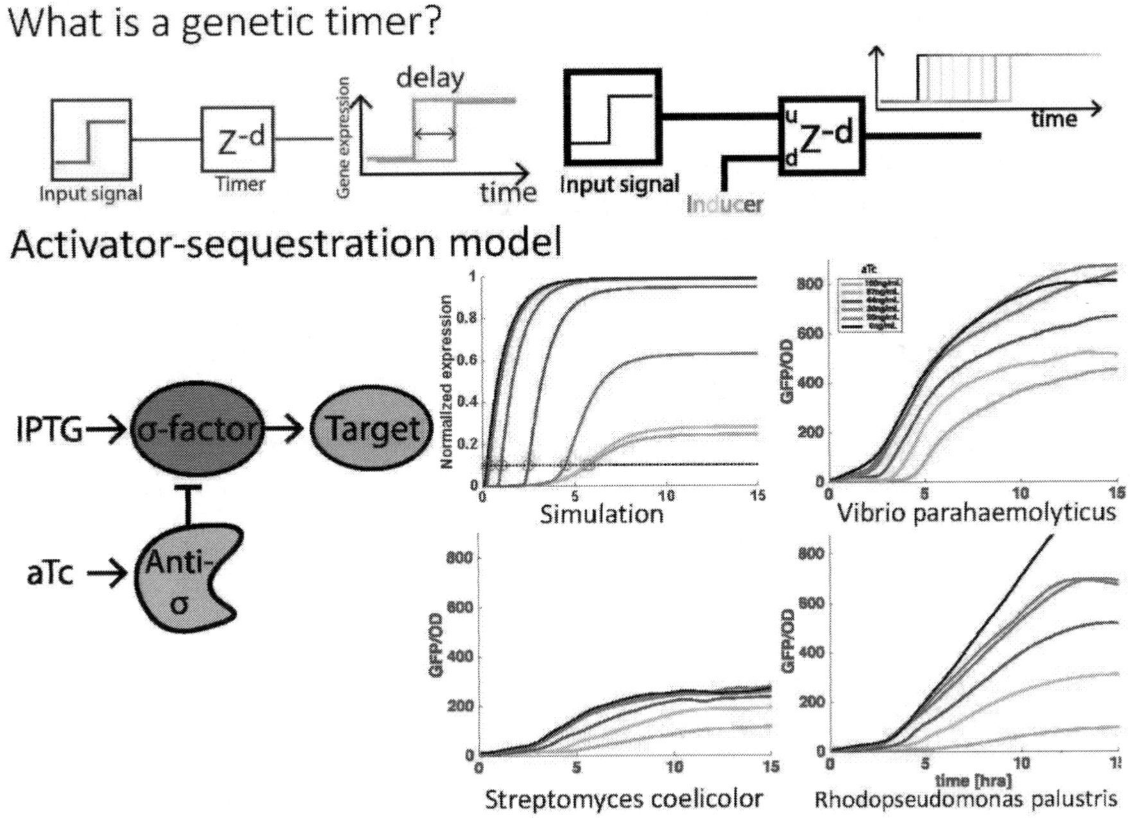

Metabolic Engineering of Escherichia coli for Efficient Eukaryotic Glycan Biosynthesis

Glasscock, C. J., Cornell University
Lucks, J. B., Cornell University
DeLisa, M. P., Cornell University

One of the most important modifications for many proteins is *N*-linked glycosylation, which modulates an array of therapeutic properties from pharmacokinetic activity to immunogenicity. Large-scale production of therapeutic glycoproteins is currently accomplished using eukaryotic cell culture platforms; however, recent advances in *Escherichia coli*-based glycoprotein synthesis could complement existing platforms with high yields and better control over glycoform. These *E. coli*-based platforms involve heterologous, multi-enzyme pathways for glycan synthesis that require optimization for production-scale yields. The aim of this study is to optimize a previously reported pathway for eukaryotic core mannose$_3$-*N*-acetylglucosamine$_2$ ($Man_3GlcNAc_2$) *N*-linked glycosylation in *E. coli* with synthetic small regulatory RNAs (sRNAs), gene overexpression, and rational enzyme selection. We first constructed a panel of synthetic sRNAs to repress the endogenous expression of *E. coli* genes that were either rationally-selected or predicted by a genome-scale flux balance model to reduce sugar precursor levels. We then screened our sRNA panel for improved glycan production with a fluorescence-based assay of cell-surface displayed glycans and identified *E. coli* strains with over 2-fold increases in fluorescence intensity. Further, we overexpressed two *E. coli* genes involved in GDP-mannose biosynthesis to achieve an ~2.5-fold improvement glycan production, which increased to ~3.5-fold when combined with our best sRNAs. Finally, we expected that translocation of lipid-linked oligosaccharides (LLOs) into the periplasm was a bottleneck in our pathway. To overcome this, we selected an LLO flippase from *Campyobacter jejuni* that is known to exhibit relaxed substrate specificity and expressed this enzyme alongside our optimized system. This strategy resulted in an ~10-fold improvement in translocated glycan production. Future efforts will attempt to show that optimized glycan production significantly increases $Man_3GlcNAc_2$ protein glycosylation in *E. coli*.

Intracellular Spectral Recompositioning of Light: A Novel Strategy for Improving Photosynthetic Efficiency of Diatoms

Fu, W., New York University Abu Dhabi
Chaiboonchoe, A., New York University Abu Dhabi
Sultana, M., New York University Abu Dhabi
Salehi-Ashtiani, K., New York University Abu Dhabi

Intracellular Spectral Recompositioning of Light: A Novel Strategy for Improving Photosynthetic Efficiency of Diatoms

Weiqi Fu[1,2,3], Amphun Chaiboonchoe[1,2], Mehar Sultana[2], Kourosh Salehi-Ashtiani[1,2]

[1]Laboratory of Algal, Systems, and Synthetic Biology, Division of Science and Math, and [2]Center for Genomics and Systems Biology (CGSB), New York University Abu Dhabi, P.O. Box 129188, Abu Dhabi, UAE

[3]Center for Systems Biology and Faculty of Industrial Engineering, Mechanical Engineering and Computer Science, School of Engineering and Natural Sciences, University of Iceland, Reykjavik 101, Iceland

Abstract

In order to concurrently address global resource scarcity and the impending climate change, the use of photosynthetic diatoms can provide an exciting solution for sustainable production of biofuels and bioactive compounds. However, photosynthetic efficiency in diatoms needs to be optimized to reduce the energy costs for sustainable biomass production. We have developed and implemented a strategy herein referred to as Intracellular Spectral Recompositioning of light (or *ISR*), which can increase the quantum yield of light if the otherwise wasted portion of blue light is shifted to green, which diatoms have evolved to harvest through accessory pigments. We demonstrate that *ISR* can be employed chemically or biogenically to improve photosynthesis in the diatom *Phaeodactylum tricornutum* in photobioreactors. Genetically engineered diatom cells with enhanced green fluorescent protein (EGFP) achieved desired spectrum recompositioning with high photo-stability in batch culture. These engineered strains exhibited higher efficiency of photosynthesis than their wild type counterpart by 30%. This increase can be attributed to an increase of quantum yields in photosystem II in diatom cells under light stress conditions. Long-term experiments demonstrated the stability of EGFP transformants and the robustness of GFP expression upon the presence of nitrate during cultivation. Pond simulator experiments also observed the GFP transformants could outperform their wild type counterpart. In addition, genome-scale transcriptome and gene set enrichment analysis of the obtained strains will be presented in comparison to the wild type to characterize these strains at the system level. The *ISR* approach is expected to be broadly applicable toward improving energy efficiency for cultivation of microalgae at industrial production scales.

Cell Free Synthesis of Protein Microcompartments

Garcia, D., ORNL
Doktycz, M. J., Oak Ridge National Laboratory

Cell Free Synthesis of Protein Microcompartments

David C. Garcia & Mitchel J. Doktycz

Cell free synthesis (CFS) has developed into a technique with diverse uses. One emerging application utilizes CFS to design and build microscale structures with potential uses in biotechnology. For example, bacterial microcompartments (BMCs) are self-assembling bioreactors that house metabolic functions and regulate access of substrates and metabolites. The use of BMCs to contain non-native enzymes and substrates could develop into a technique for creating chemicals and pharmaceuticals. Cell free synthesis approaches can be valuable for engineering the targeting and assembly of enzymes into synthetic microcompartments. Here, we report on the production of Salmonella enterica ethanolamine utilization (eut) bacterial microcompartment components in an E. coli based cell free environment. Recombinant expression of the EutS microcompartment proteins are being generated to form polyhedral compartments. We have designed a system to co-express EutS and sfGFP tagged with an encapsulation peptide which can facilitate the targeting of sfGFP to the protein shell. These experiments will demonstrate a proof of concept for using bacterial microcompartments as nano-bioreactors for diverse applications.

Engineering an Escherichia coli Strain to Characterize Heme Metabolism

Couture, M., Université Laval
Gagné, M.,
Rey, S.,
Labrie, G.,

Iron is necessary for growth of bacteria. In this context, iron restriction is one of the mechanisms used in humans to fight bacterial infection. Iron restriction also includes limited access to the host heme that may constitute a potential abundant source of iron for invading bacteria. Most bacteria including laboratory *Escherichia coli* do not have the genes necessary to import heme from the environment and to degrade this porphyrin and release iron. However, several human pathogens possess a set of genes (chu) that allow the import and processing of heme, including pathogenic *E. coli* strains such as O157:H7 that causes severe food infection. The *chu* gene cluster is presumably involved in capturing, internalizing and metabolizing heme to release iron which would be used for growth under iron restricted condition (1). The *chu* gene cluster contains 8 genes some of which have no clearly defined function.

To facilitate the characterization of the function of the individual chu genes, all 8 genes (*chu*ASTUVWXY) were introduced in a single copy the genome of a non-pathogenic *E. coli* strain in which the *ent*E gene encoding the siderophore enterobactin is inactivated to exacerbate the effect of iron restriction on growth. The *chu* genes were first cloned as sub-clusters in two pOSIP vectors (2) that allow the easy integration in the *E. coli* genome from the plasmid-encoded integrases. The *chu* genes were then integrated in the genome of the *E. coli* JW0586-1 (D*ent*E) strain of the Keio collection (3). Gene expression of the chu genes was regulated from their natural promoters. *In vivo* assays revealed that this engineered *E. coli*/chuASTUVWXY grows well on minimal medium depleted of iron when supplemented with various sources of heme (free heme, myoglobin and hemoglobin). As expected, growth was strictly dependant on the inclusion of the gene encoding the outer membrane heme receptor chuA (4). On the other hand, several *chu* genes including those encoding the periplasmic transporters appear to contribute to the optimal growth of the engineered strain depending of the growth conditions. This work reports the building of the *E. coli*/*chu*ASTUVWXY engineered strain that is an important tool to fully characterize the function of the *chu* genes under various growth conditions, environmental stresses and the presence of competing strains form the community.

1- Ouellet, Y. H. et al. 2016. *J. Inorg. Biochem.* 154 : 103-113

2- St-Pierre, F. et al. 2014. *ACS Synt. Biol.* 2 : 537-541

3- Baba, T. et al. 2006. *Mol. Syst. Biol.* 2006.0008

4- Hagan, E. C. and Mobley, H. L. 2009. *Mol. Microbiol.* 71: 79-91

Enhancing Orthogonal Behavior and Amber Suppression Activity for Heterologous tRNA and tRNA Synthetase Variants

Maranhao, A., University of Texas at Austin
Thyer, R., UT Austin
Ellington, A., University of Texas at Austin

There have been considerable advancements in the translational incorporation of non-canonical or unnatural amino acids (UAA) into proteins over the last two decades. As the most widely used method for site-specific UAA incorporation, amber stop codon suppression employs an orthogonal translation system (OTS) consisting of a heterologous tRNA synthetase:tRNA pair, which function to augment or expand an organism's genetic code. More recent advances such as "Amberless" *E. coli* as well as engineering UAA-dependence or "addiction" build on earlier work by expanding the utility of OTSs. However, heterologous tRNA synthetase:tRNA pairs remain critical to establishing efficient OTSs. To this end, we have conducted a complete overhaul of the heterologous tRNA from *Methanocaldococcus janaschii* (*Mj*) tyrosyl pair. Using compartmentalized partnered replication (CPR) and liquid-based pheS negative selection, we evolved a new set of orthogonal *Mj*-tRNA variants with enhanced amber suppression activity for incorporation of tyrosine as well as three 3-halo-tyrosine UAAs (3-Cl-Y, 3-Br-Y, and 3-I-Y) in *E. coli*. These tRNAs show up to three-fold reduction in promiscuous aminoacylation by endogenous tRNA synthetases while increasing amber suppression activity by almost 40%. With respect to host translation machinery, these new *Mj*-tRNA variants also display enhanced orthogonality, thus eliminating organismal fitness costs associated with tRNA mischarging. Our study identified a tRNA variant displaying tyrosine-dependent suppression activity, which demonstrates the capacity to engineer suppressor tRNA variants with amino acid specificity. Additionally, we successfully employed our CPR selection to enhance the activity of the *Methanocaldococcus janaschii* (*Mj*) tyrosyl tRNA synthetase with some mutant variants demonstrating over fifteen-fold increase in amber suppression activity.

Noncanonical Amino Acid Addictions in Biological Systems

Tack, D. S., University of Texas at Austin
Thyer, R., UT Austin
Ellefson, J., University of Texas at Austin
Ellington, A. D., University of Texas at Austin

Engineered orthogonal translation systems (OTSs) for the incorporation of non-canonical amino acids (NCAAs) allow for the addition of new chemistries to genetic code. OTSs are generally detrimental to cell fitness, and are quickly lost in natural bacterial systems. We have demonstrated that a single, essential enzyme can be engineered to require a NCAA, 3-nitrotyrosine, for activity. Furthermore, that enzyme can be used to leverage a cellular addiction to a NCAA for hundreds of generations without escape from NCAA dependence. A β-lactamase engineered for NCAA addiction is sufficient for organismal addiction to the NCAA, and this addiction operon, consisting of the engineered β-lactamase and an engineered OTS, provided addiction in *E. coli* for over 500 generations without detectable escape ($<10^{-9}$ escape frequency). At the conclusion of 500 generations, not only do the *E. coli* become more tolerant to the NCAA, but a portion of the population appears to be significantly hindered in growth in the absence of the NCAA, suggesting that the NCAA is not only required for lactam resistance, but incorporated as a part of the bacterial genome. Additionally, the addiction operon can be quickly transferred to a wide variety of bacterial orders, including *Pseudomonadales*, *Vibrionales*, *Pasteurellales*, and *Enterbacteriales*, by changing lactamase specificity from penicillins to cephalosporins, demonstrating for the first time that the *Methanocaldococcus jannaschii* tyrosine OTS functions in a diverse array of bacteria. Our work paves the way for the use of OTSs, and their many chemistries, in a variety of bacteria, and may lay a foundation for the use of NCAAs as a facile means for biocontainment using single engineered essential enzymes.

Commercial Expression Levels of a Potent Recombinant Antiviral Therapeutic Protein in Sugarcane

Padilla, C. S., Texas A&M AgriLife Research
Damaj, M. B., Texas A&M AgriLife Research
Molina, J.,
Mirkov, T. E., Texas A&M University

Expression of recombinant proteins in plants represent a promising system to produce therapeutic proteins since plants can be grown in large quantities at low cost. In addition to those benefits, plant-produced proteins are free of human pathogen contaminants, reducing the risk of potential adverse effects. However, to be a reliable system, plant-synthesized recombinant proteins should be expressed in large-enough quantities to be commercially extracted. Sugarcane (*Saccharum* spp. hybrids) has been proposed as an ideal crop for recombinant protein production because of its rapid growth rate and high light and water to biomass conversion efficiency. Moreover, sugarcane is an excellent biological gene containment crop due to its vegetative propagation system. This project aims to develop an efficient expression system in sugarcane for the production of a proprietary recombinant protein with antiviral and antitumor activities. Initial work generated several transgenic sugarcane varieties that overexpress the therapeutic protein under the control of the constitutive *maize ubiquitin 1* (*Ubi1*) promoter, and we have recovered protein expression levels up to 0.1% of total soluble protein (TSP). In order to increase recombinant protein yield, sugarcane varieties were transformed with five new constructs carrying the proprietary therapeutic protein under the control of sugarcane stem-regulated promoters from *DIRIGENT5-1 (SHDIR5-1), Sugarcane bacilliform virus (SCBV21), DIRIGENT16 (SHDIR16), PROLINE RICH PROTEIN (SHPRP), and ELONGATION FACTOR1a (SHEF1a)*. A 35S-NOS double terminator was used to stabilize transcription. Two sugarcane varieties (cv. CP72-1210 and CP89-2143) were transformed using a biolistic method with single, or up-to a five-construct combination to stack stem specific promoters driving the expression of the recombinant protein. Newly generated T_0 staked lines show yields of up to 0.4% TSP equivalent to 4mg/Kg fresh weight of stalk tissue. This represents a four-fold yield increase from previously generated transgenic lines.

Developing Sugarcane and Energy Cane As Highly Productive Biofactory Platforms

Damaj, M. B., Texas A&M AgriLife Research
Padilla, C. S., Texas A&M AgriLife Research
Beyene, G.,
Jifon, J. L.,
Enciso, J.,
Mirkov, T. E., Texas A&M University

Sugarcane and energy cane, *Saccharum* spp., are key players in the expanding bioeconomy for sugars, bioenergy, and platform for production of native and recombinant proteins. *Saccharum* species are ideal for such applications due to their high resource use efficiencies, rapid growth, and efficient photosynthetic pathway, and high biomass production capacity with potential yields of up to 45 Mg of dry biomass per hectare per annum. Our technology platform integrates molecular genetics, genomics and biotechnology to develop new traits through optimized transgene delivery. We have applied this approach in three fundamental research areas: biofuel development, recombinant protein expression and stress tolerance trait development. One major area focuses on the enhancement of biomass yield, fiber content, and water and nutrient use efficiencies for biofuel production. Our current platform focuses on the down regulation of a set of isolated sugarcane genes involved in hormonal biosynthesis, resulting in up to an 81% increase in biomass, and 26% and 95% in total fiber and lignin content, respectively. A second major area is the cost-effective and large-scale commercial recombinant production of high value proteins, achieving up to 10.0 mg recombinant protein/kg of stalk weight or 1.4% of total soluble protein. High protein expression levels in the stalk-based production system are achieved by using our promoter staking expression system and our "in-house" toolbox of sugarcane tissue-regulated promoters for targeted gene expression to vascular tissues. A third major area consists of the development of enhanced traits for tolerance to stresses such as cold, salinity and drought to help broaden sugarcane production in marginal growth environments. Current method development has resulted in enhanced stress tolerance of up to 63% survival after 5 days of cold (10°C), 80% survival at high salt (1 M sodium chloride), and 90% survival and an 11% decrease in water usage after 3 weeks of drought. We are now focusing on optimizing existing methodologies and developing new technologies such as gene editing and targeted mutagenesis to enhance our currently used transgenic technologies to improve sugarcane and energy cane.

Construction and Characterization of Biological Circuits to Create a Synthetic Bacterial Consortium to Divide Genetic Functions and Reduce the Metabolic Burden

Zuñiga, A.,
Goñi, A., Centro Nacional de Biotecnología
de Lorenzo, V., Centro Nacional de Biotecnología (CNB-CSIC)
Ruz, G.,
Gonzalez, B.,

Bacterial quorum sensing (QS) signal synthases, receptors, and cognate promoter elements are important components of a wide variety of engineered biological devices. We employed QS components to construct a bidirectional cell to cell communication network using three compartmentalized circuits to create a synthetic bacterial consortium. This strategy allows us to obtain a better control of gene expression, divide functions of these genes and has the potential to become a powerful biotechnology for various applications. We constructed a synthetic bacterial consortium, in three different *Cupriavidus pinatubonensis* JMP134 bacterial cell, creating a genetic variant monospecie consortium. *luxI* gene present in circuits of bacteria A and C and *rhlI* gene present in circuit of bacterium B, catalyze the synthesis of the acyl homoserine lactone (HSL), 3-oxododecanoyl-HSL and butanoil-HSL, respectively. When HSLs signal accumulate at high enough concentrations these molecules diffuse through the cell membrane and binds to their cognate transcription factor (LuxR in A and C and RhlR in B) and activate respectively promoter. We characterized the promoter activity of each circuit in presence or not of each HSL and in different combinations by flow cytometry. The results suggest that genetic variants strains of *C. pinatubonensis*, expressing the constructed circuits, seem to have a predictable behavior. The data obtained can be used to model and simulate the consortium dynamics and predict the complex behaviors of bacteria in response to specific environment.

Proyecto FONDECYT postdoctorado Nº 3140031

Seeing the Invisible: Exploting Crosstalk to Measure Genetic Constructs In Vivo

Hecht, A., National Institute of Standards and Technology
Endy, D., Stanford University
Salit, M., National Institute of Standards and Technology
Munson, M. S., National Institute of Standards and Technology

The emerging bioeconomy is in need of a metrology infrastructure. One area where infrastructure is needed is for the measurement of the activity of genetic constructs that do not produce fluorescent proteins. This abstract describes a competitive assay for measuring genetic construct activity in *Escherichia coli.*

In cells, transcriptional and translational resources are limited, resulting in a limited capacity for protein expression. It has been shown that genetic constructs compete for resources, with expression of one construct coming at the expense of other constructs. Our assay leverages this competition to measure the activity of spectroscopically invisible genetic constructs. Several new results from previously presented work are highlighted here:

1) *In vivo* there is a monotonic relationship between the co-expression of two constructs, where increases in expression of one lead to decreases in express of the other.

2) We exploit this relationship to develop a calibration method where we can infer the expression level of an undetectable protein, by observing the reduction in expression of a fluorescent reference protein.

3) We show the impact of reference construct copy number and expression strain on the performance of the assay.

These data show that the competitive assay can be used to measure the expression level of spectroscopically undetectable constructs, and can be useful for enabling forward engineering in synthetic biology.

Seeing the Invisible: Exploiting Metabolic Resource Coupling to Measure Gene Expression

Ariel Hecht, Drew Endy, Marc Salit and Matthew S. Munson

Poster #224

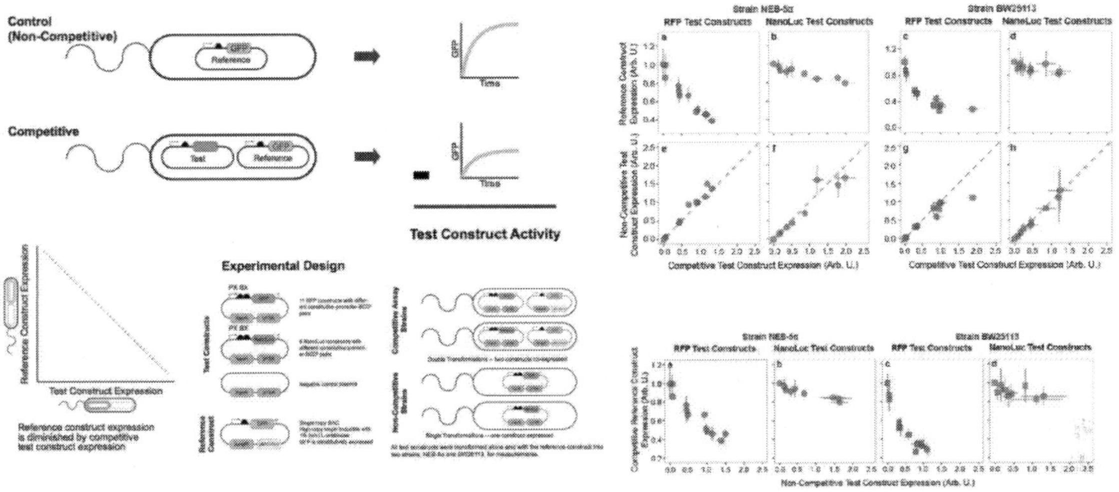

NIST National Institute of Standards and Technology U.S. Department of Commerce

JIMB The Joint Initiative for Metrology in Biology

Stanford University

When Wavelengths Collide: Bias in Cell Abundance Measurements Due to Expressed Fluorescent Proteins

Hecht, A., National Institute of Standards and Technology
Endy, D., Stanford University
Salit, M., National Institute of Standards and Technology
Munson, M. S., National Institute of Standards and Technology

The abundance of bacteria in liquid culture is commonly inferred by measuring optical density (OD) at 600 nm. In doing so, it is commonly assumed that the dominant source of the attenuation of incident light is scattering by cells. If cells express fluorescent proteins that absorb light at the wavelength used for measuring OD, then this absorption can also attenuate incident light. Red fluorescent proteins (RFP) can strongly absorb light at 600 nm. Increasing RFP expression can falsely inflate apparent cell density and lead to underestimations of mean per-cell fluorescence by up to 10%. A literature search shows that this error is prevalent through the synthetic biology community.

We will present data showing that this underestimation of mean per-cell fluorescence is linear with the amount of expressed RFP. We will also present data showing that measuring OD at 700 nm would allow estimation of cell abundance unaffected by the presence of nearly all fluorescent proteins.

As the practice of engineering cells becomes ever more precise there is a corresponding need for better characterization of the impact of genetic elements on gene expression. We believe that raising awareness of this issue within the synthetic biology community, and promoting adoption of the simple change that we propose, will improve the quality of data generated throughout this community.

Universal Expression System for Eukaryotic Microorganisms

Rantasalo, A., VTT Technical Research Centre of Finland
Landowski, C., VTT – Technical Research Centre of Finland
Kuivanen, J., VTT Technical Research Centre of Finland
Jantti, J., VTT Technical Research Centre of Finland
Mojzita, D., VTT Technical Research Centre of Finland

This work describes the development of an orthogonal expression system which is functional (transferable) in a large spectrum of eukaryotic microorganisms. The expression system is based on a synthetic transcription factor (sTF), which regulates expression of underlying target gene(s) containing the sTF-dependent promoter(s). The expression of the sTF is achieved by a specific (universal) core promoter, instead of a full promoter, which provides a low, but sufficient expression level of the sTF in a broad spectrum of host organisms. This expression is constitutive, therefore the sTF is present in the cells in all growth conditions, and in all developmental and growth stages. Also the sTF-dependent promoter, regulating the expression of the target gene(s), contains a similar type of universal core promoter functional in diverse species. The combination of the sTF-binding sites and the core promoter in the sTF-dependent promoter enables a wide range of potential target gene expression levels, which can be modulated by the choice of these components from very low to very high. Therefore, the developed expression system provides robust, stable, and tuneable expression levels of the target genes in a broad spectrum of host organisms. This could simplify and focus the genetic tools needed for constructing new expression hosts, including those with undeveloped know-how, and it could enable the use of these (often very promising) hosts in novel industrial applications. The method for selecting the universal core promoters, construction of the expression systems, and demonstration of their performance in diverse hosts, including yeasts and filamentous fungi, will be presented.

Spatial Ratiometric Analysis for Intrinsic Plant Promoter Activity

Pollak, B., University of Cambridge
Federici, F.,
Haseloff, J., University of Cambridge

Rational design of synthetic gene networks requires accurate characterisation of the genetic elements controlling them. In eukaryotic multicellular systems, gene activity can either be restricted spatially to particular tissues or have ubiquitous expression, a trait which is in part controlled by the promoter which drives that gene. Furthermore, transcriptional activity can be affected by the genetic context in which the circuit is located as well as by extrinsic factors such as metabolic state.

Here we present a method for studying promoter activities in plants through ratiometric normalisation spatially and show its application in plants. We use measured promoter activity over a reference promoter *in vivo* through dual-nuclear fluorescent protein expression in *Marchantia polymorpha* gemmae, vegetative reproductive buds with stable cellular arrangements, which have fluorescently labelled membranes for spatially determining promoter activity. Spatial ratiometric analysis allows extruding measures for intrinsic properties of promoters and map their activities to cells and tissues. This enables comparing promoters in a range of different contexts, therefore improving our description of tissue-specific promoter activity in multicellular systems.

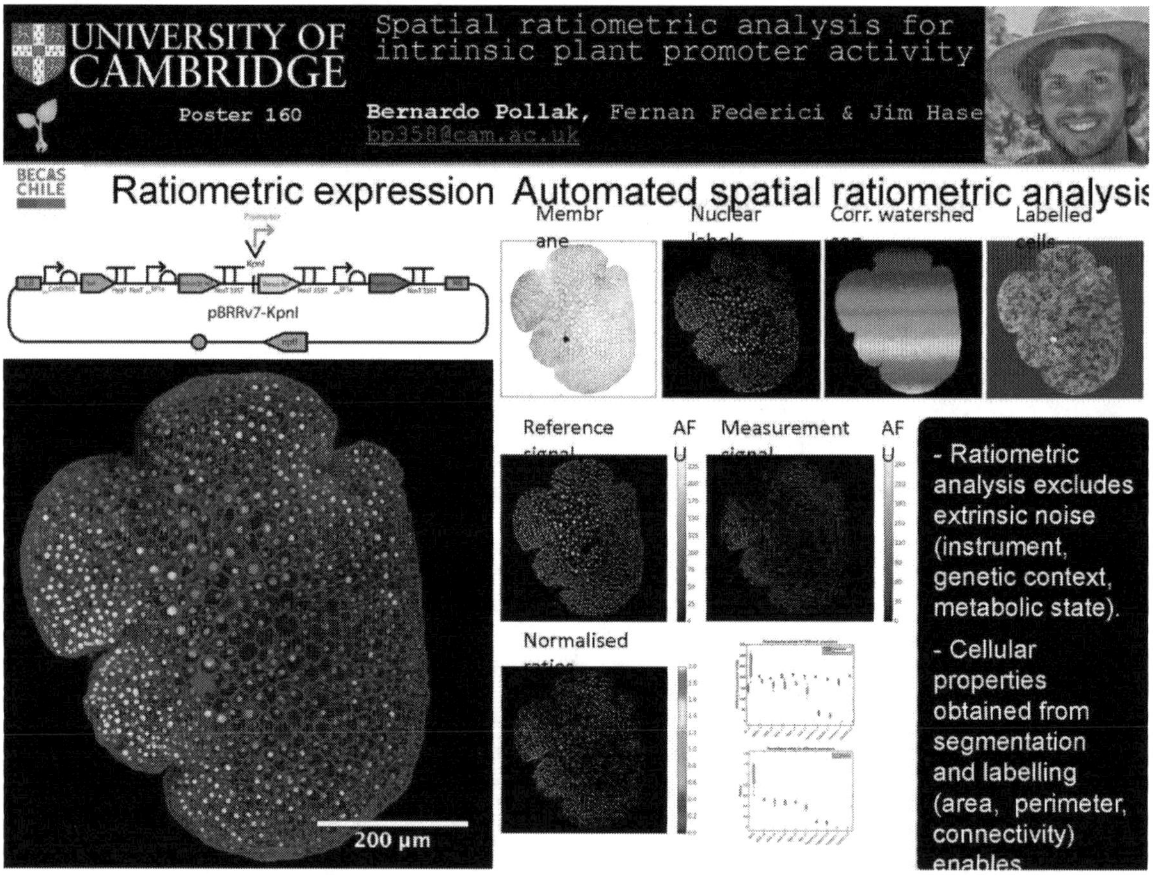

Characterization of Co-Transcriptional RNA Structure Evolution Using Shape-Seq

Strobel, E. J., Cornell University
Watters, K. E., Cornell University
Yu, A. M., Cornell University, Ithaca, New York, Weill Cornell Medical College, New York, New York, Memorial Sloan-Kettering Cancer Center, New York, New York
Lucks, J. B., Cornell University

RNA is a versatile molecule that is capable of both storing information and adopting structures that serve specific biological functions. Our understanding of the relationship between RNA structure and function is dependent on our ability to interrogate these structures, however, current approaches to RNA structure determination are limited because they focus on full-length RNA and do not provide information about intermediate folding events that are frequently critical to the function of the RNA. The ability to characterize RNA folding pathways is of particular importance because RNAs not only fold co-transcriptionally, but also frequently perform specific functions, such as termination, antitermination and splicing, co-transcriptionally. We have developed a novel technique that can map the dynamic folding pathways that RNAs undergo co-transcriptionally using SHAPE-seq, a high-throughput method that combines chemical probing of RNA structure with next generation sequencing to characterize the structures of thousands of RNAs in a mixture. Co-transcriptional SHAPE-seq allows us to observe the maturation of RNAs into a final structure by characterizing the structure of each length of RNA as it is transcribed. We have used co-transcriptional SHAPE-seq to uncover novel insights into the mechanism through which the *B. cereus crcB* fluoride riboswitch regulates transcription by observing the RNA folding events that lead to transcription termination or antitermination at nucleotide resolution.

Acoustic Reporter Genes for High-Resolution Ultrasound Imaging of Engineered Bacteria In Vivo

Bourdeau, R. W., California Institute of Technology
Lee-Gosselin, A., California Institute of Technology
Lakshmanan, A., California Institute of Technology
Kumar, S., California Institute of Technology
Farhadi, A., California Institute of Technology
Shapiro, M. G., California Institute of Technology

More than 90% of the cells in a typical human body are microbes, and there is a rapidly growing appreciation of the crucial roles these organisms play in human health and disease. Additionally, advances in synthetic biology have enabled the development of genetically engineered bacterial therapies and diagnostics that are starting to advance into the clinic. These diverse activities of natural and engineered microbes depend critically on their anatomical location within host organisms, such as the various regions of the mammalian gastrointestinal tract. However, no effective methods are currently available to image microbes inside mammalian hosts with high spatial and temporal resolution. The development of such methods would transform the study of microbial-host interactions and facilitate the development of an entirely new class of bacterial diagnostics.

We address this challenge by developing acoustic reporter genes (ARGs) to enable the imaging of microbes with ultrasound. Unlike optical techniques, which have poor spatial resolution in vivo due to the scattering of light by tissue and are not commonly used in the clinic, ultrasound is capable of very high spatial resolution in vivo (tens of microns in mice) and is used widely in medicine. We recently discovered a unique class of protein nanostructures encoded in the genomes of certain photosynthetic microbes that can serve as imaging agents for ultrasound[1]. These nanostructures, called gas vesicles (GVs), comprise all-protein shells that are ~ 250 nm in size and are hollow on the inside. We showed that the hollow interior and unique mechanical properties of GVs allow them to scatter sound waves and thereby serve as sensitive *in vivo* imaging agents for ultrasound.

Now, we have succeeded in adapting the gene clusters encoding GV formation for heterologous expression as ARGs in *Escherichia coli* and *Salmonella typhimurium*, two important model organisms and chasses for synthetic biology. ARG expression allows us to image these bacteria with ultrasound in vitro and in vivo. Cells can be detected at concentrations below 10^8 cells/ml. In addition, engineering of the GV gene cluster has yielded variants that produce non-linear ultrasound contrast and can be "erased" with acoustic pressures above specific thresholds, thus enabling more sensitive detection within biological tissues. Furthermore, the engineering of ARGs with different thresholds for acoustic collapse enables multiplexed imaging of more than one cellular population.

ARGs will allow microbes to be imaged dynamically in vivo with unprecedented precision, unlocking a wide range of new biological research and the development of spatially informative bacterial diagnostics.

1. Shapiro, M.G. et al. Biogenic gas nanostructures as ultrasonic molecular reporters. *Nature nanotechnology* **9**, 311-316 (2014).

Enhancing Tolerance to Potato Zebra Chip Disease Using a Transgenic Approach

Chiong, K.,
De Hoyos, O.,
Mirkov, T. E., Texas A&M University

Zebra Chip (ZC) disease, associated with the phloem-limited bacterium *Candidatus* Liberibacter solanacearum (Lso) and vectored by the potato psyllid, *Bactericera cockerelli*, is a serious threat to the potato industry that leads to significant economic losses. To date, there are only a few registered insecticides that are effective against the potato psyllid and not much research has been done on controlling the causal bacterium. Recent focus has shifted from chemical control of Lso and the psyllid vector, to establishing genetic tolerance to Lso in potato through breeding and genetic engineering. Our current approach to study Lso-potato interactions involves an optimized transgenic delivery system of a plant-derived antimicrobial gene family into the commercially important chipping variety, 'Atlantic'. To efficiently target Lso, the use of co-delivery of antimicrobial genes and targeted transgene vascular expression were optimized. The first component consists of using one delivery vector to deliver two different gene expression cassettes and identifying the most effective combination of these genes. The second component is to target gene expression in phloem vascular tissues, where Lso is confined, by using phloem-specific promoters. The transformants were then screened for reduction of Lso titer and the loss of disease symptoms. Effectiveness of the introduced antimicrobial genes was evaluated by exposing transgenic lines to Lso positive potato psyllids in controlled cages. Data from plant health observations and semi-quantitative PCR analysis after 14 days and 24 days of exposure to the psyllids showed that several of the transgenic potato lines remained Lso free after 24 days of exposure to Lso positive psyllids. We are now investigating the use of Lso-potato system as a model for identification of potential transgenic for delivering resistance in citrus to the greening disease (Huanglongbing) caused by a closely related bacterium, *Candidatus* Liberibacter asiaticus. The Lso-potato model is ideal to study *Candidatus* Liberibacter-plant interactions, as it presents the ability to screen many different potentially tolerant genes in a short period of time due to the rapid growth, propagation, and ease of transformation of the potato. We hope that our Lso-potato model system can greatly reduce the use of chemical control and provide useful insights on using transgenic approaches to control other related bacteria in economically important crops, such as citrus, tomato, and bell pepper.

Morphogenetic Engineering of Bacterial Colonies.

Matute, T., Pontificia Universidad Catolica de Chile
Nuñez, I., Pontificia Universidad Catolica de Chile
del Valle, I.,
Kan, A.,
Haseloff, J., University of Cambridge
Rudge, T.,
Federici, F.,

We combine synthetic biology approaches, biophysical modeling and CRISPR/Cas for the engineering of morphogenetic mechanisms in bacterial colonies. We have created a system for prototyping morphogenesis starting from a single cell. We defined and implemented two elementary functions, symmetry-breaking and domain-specific cell regulation, and used characterized gene regulatory mechanisms to improve the design-build-test cycle.

This toolkit is enabling the study of morphogenetic principles and the engineering of artificial shapes. We are currently expanding this toolkit to include artificial cell-cell signaling, colony size controllers and metabolic cross-feeding mechanisms to explore the implementation of higher order designs from core structures. Our long term goal is to define computationally abstracted rules and elementary functions that make morphogenetic engineering more predictable, tractable and scalable.

Improvements to Nitrogen Production in a Model Bacterium for Biofertilizer Application

Barney, B. M., University of Minnesota

A large percentage of the nitrogen needed by society to sustain agricultural processes is accomplished through the industrial Haber-Bosch process. This industrial process is accomplished at a significant cost in energy. In the natural ecosystem, organisms have also had to evolve to overcome the limits of reduced nitrogen compound availability in the environment, even though a large pool of nitrogen is readily available in our atmosphere. Nature has solved this limitation through a select group of organisms known as diazotrophs, which evolved the ability to transform atmospheric nitrogen gas into fixed ammonia through the action of nitrogenase. Other species have evolved to work together with diazotrophs by evolving intricate symbiotic relationships based primarily on the exchange of nitrogen for energy rich compounds. As we look to the future for solutions and alternatives to energy demanding processes such as Haber-Bosch, it is important to understand the methods employed by biological systems that produce the same commodity chemicals, and look at examples where these systems can be altered to enhance nitrogen production in an effort to generate effective biofertilizer strains.

Our laboratory has been focusing on methods to induce and then screen for improved nitrogen production. In relation to this work, we reconstructed a strain which is a prolific ammonium producer. In this specific strain, the genes controlling nitrogen fixation are deregulated, through a modification of the *nifLA* operon. This modification results in a strain that yields extracellular ammonium levels in the 15-35 mM range, depending on the temperature in which it is cultured. Under these conditions, metabolism has been extensively reorganized to yield such high levels of ammonium. Using whole-genome RNA-Sequencing analysis, we have probed how the cell has altered gene regulation at the transcriptional level to achieve these high levels of nitrogen production. These results will be utilized to inform further biosynthetic strategies to construct stable strains that might be suitable for the production of ammonium and other nitrogen compounds. Recent work is aimed at the development of additional tools to probe alternative regulation at the transcriptional level that would result in elevated nitrogen production. These results are revealing an array of additional genes that play an important role in nitrogen regulation.

Our efforts to produce high levels of nitrogen have resulted in the undesirable evolution of cheaters during extended culture. Cheaters are single cells that have evolved within a community to take advantage of the elevated product levels generated within the culture, and then quit contributing toward the production of the desired product. Based on this evolution, these cells proliferate rapidly to become a dominant percentage of the culture and lower production yields. Our current efforts to direct fixed-nitrogen species toward a terminal product would help to eliminate the role that cheaters are able to play by making it exceedingly difficult to evolve a means to cheat. Our findings related to these efforts will be discussed.

Continuous Evolution of Proteases Using a T7 Bacteriophage System

Garry, D., UT Austin
Ellington, A., University of Texas at Austin

The directed evolution of proteases for enhanced activity or altered specificity is a difficult process on account of cell toxicity and difficulty to select for dynamic range of function. Here, we present a T7 bacteriophage-based continuous evolution scheme for protease activity with distinct advantages over other methods, chief among them ease-of-use. Our system works by adding the protease of interest to be evolved in place of the major capsid protein of the T7 bacteriophage (T7Δ10). T7Δ10 bacteriophages will not propagate on their native *E. coli* host without complementation with the major capsid protein in *trans*. We have created a fusion between the major capsid protein and the *E. coli* thioredoxin gene with a protease-specific cleavage site in between which does not allow for successful bacteriophage replication in the absence of protease cleavage. We initially show a proof of concept validation of the system whereas a known tobacco etch virus protease (TEV protease) mutant reverts back to the more active wild-type TEV protease after ten passages. We also show how simple genetic drift of a self-cleavage resistant TEV protease can mutate to lead to increased enzymatic speed in twenty passages. Additionally, we have drifted TEV protease to recognize four different single mutant cleavage sites away from the wild-type cleavage site (WT:ENLYFQ/S, EΔW: **W**NLYFQ/S, YΔH:ENL**H**FQ/S, YΔE:ENL**E**FQ/S, and QΔH:ENLYF**H**/S). As four distinction populations, these quasi-species evolved for 20-30 passages on the single mutant cleavage sites so that each population showed preference for its own versus any other single mutant cleavage sites in cell lysis assays. Although most lines to some degree lyse on other mutant cleavage sites, the YΔE and EΔW lines lyse on the QΔH cleavage site the slowest and, in turn, the QΔH line lyses slowest on the YΔE and EΔW cleavage sites, suggesting some degree of orthogonality. Preliminary results suggest other proteases can be used in the same system. In summary, we have developed a novel, T7-based continuous evolution system for evolving proteases which, in addition to evolving better protease activity and unique specificity, could be used to evolve proteases to cleave therapeutically relevant targets as well as rapidly characterize the emergence of resistant proteases in response to inhibitors.

Non-Standard Amino Acid Incorporation Using a Ribosome with Tethered Subunits

Kim, D. S., Northwestern University
Carlson, E. D., Northwestern University
Gan, R.,
Perez, J., Northwestern University
d'Aquino, A., Northwestern University
Jewett, M. C., Northwestern University

Previously, we developed an orthogonal ribosome with tethered subunits (oRibo-T) capable of translating specific orthogonal messenger RNAs (o-mRNA). oRibo-T allows researchers to investigate and engineer the ribosome's 50S subunit *in vivo* without perturbing cell viability. The ribosome's 50S subunit contains the peptidyltransferase center (PTC), which catalyzes peptide bond formation, and the exit tunnel, through which the nascent polypeptide is excreted. By decoupling the catalytic and secretion functions of the orthogonal ribosome from the pool of wild type ribosomes, oRibo-T represents a significant advance to previous works developing *in vivo* orthogonal translation machinery. Orthogonal translation systems have been applied in synthetic biology to incorporate non-standard amino acids (nsAAs) into proteins. Site-specific incorporation of nsAAs into proteins via amber suppression, where an nsAA is incorporated at a site corresponding to the amber codon (UAG) in the mRNA, has expanded our capacity for producing a variety of proteins for applications such as studying protein structure and function, mimicking posttranslational modifications, and producing improved protein therapeutics. Here, we apply oRibo-T for site-specific incorporation of nsAAs into proteins towards evolving oRibo-T's 50S subunit for incorporating nsAAs currently limited by the architecture of the wild type PTC. By combining oRibo-T with existing orthogonal translation systems developed for synthetic biology, we showcase its applicability and potential for future efforts in engineering the translation machinery for both engineering and fundamental science applications.

Design and Application of Stationary Phase Combinatorial Promoters

Hsiao, V., California Institute of Technology
Cheng, A., California Institute of Technology
Murray, R. M., California Institute of Technology

Current bacterial synthetic circuits rely on the fast dilution and high protein expression that occurs during exponential phase. However, constant exponential phase is both difficult to ensure in a lab environment and almost certainly impractical in any natural setting. Here, we characterize the performance of 13 *E. coli* native σ38 promoters, as well as a previously identified σ38 consensus promoter. We then make tetO combinatorial versions of the three strongest promoters to allow for inducible delayed expression. The design of these combinatorial promoters allows for design of circuits with inducible stationary phase activity that can be used for phase-dependent delays in dynamic circuits or spatial partitioning of biofilms.

Developing an in Vitro Platform for the Production of Structurally Homogeneous Glycoproteins Using Cell-Free Protein Synthesis and Samdi Mass Spectrometry

Kightlinger, W. K., Northwestern University
Techner, J. M., Northwestern University
Lin, L., Northwestern University
Mrksich, M.,
Jewett, M. C., Northwestern University

Glycosylation, the attachment of complex oligosaccharides (glycans) to proteins, is the most abundant polypeptide chain modification in nature and plays a pivotal role in protein folding, quality control, sorting, degradation, secretion and activity. Glycoproteins make up 70% of approved or preclinical protein therapeutics and a predicted two-thirds of human proteins. Changes in glycosylation are often a hallmark of cancer, inflammation, and Alzheimer's disease states and have been shown to profoundly affect therapeutically relevant protein properties including pharmacokinetics, immunogenicity, and biological activity. In fact, intentional engineering of a protein's glycosylation pattern has led to therapeutics with improved efficacy. However, the fundamental knowledge base and experimental techniques required to systematically assess which human glycoforms would be most effective at a given site are still lacking, as are scalable methods to site-specifically conjugate desired glycoforms to proteins. These knowledge and technology gaps stem from challenges in analysis of glycoforms and the complexity of glycan biosynthesis pathways. Glycan production is not template driven but is accomplished by the coordinated expression, regulation, and localization of glycosyltransferases (GTs) and oligosaccharyltransferases (OSTs) in various cellular compartments. The complex biosynthesis leads to a massive diversity of glycan structures. The complexity of this mixture has prohibited a systematic analysis of the effect of a particular glycan structure on glycoprotein function and has impeded efforts to engineer glycoprotein therapeutics with desired properties.

We will discuss our efforts to develop a cell-free expression platform where synthetic glycan biosynthesis pathways can be constructed and carefully controlled to yield structurally homogeneous glycoproteins. The open reaction environment of our *in vitro* glycosylation system allows us to apply a unit operations approach in which homogeneous glycans are installed on a target protein and elaborated in a step-wise fashion by rigorously controlling the reaction conditions, enzyme composition, and residence time in various compartments which mimic specific cellular organelles. In order to identify enzymes with desired specificities and kinetic profiles, we have expressed dozens of N-glycan modifying GT and OST enzymes from bacteria, human, mouse, and Chinese Hamster Ovary cell lines in *E. coli* based cell-free protein synthesis (CFPS) systems. By manipulating the redox environment and adding membrane mimics, we have achieved soluble titers of 150-700 μg/μL of these diverse enzymes that are often difficult to express *in vivo*. Unlike yeast or CHO systems, *E. coli* lysates lack native glycosylation machinery, providing a blank canvas for bottom-up glycoengineering.

In order to determine ideal enzyme and substrate combinations and reaction conditions for this *in vitro* platform, we are combining our CFPS system with a high-throughput, label-free screening tool known as SAMDI (Self-Assembled Monolayers for matrix-assisted Laser Desorption/Ionization) mass spectrometry. Here we report the synthesis and characterization of GTs and OSTs in crude lysates by screening their activity on immobilized glycosylation substrates.

By harnessing the versatility and throughput of cell-free systems, we hope that our glycoprotein synthesis platform will enable a deeper understanding of glycan structure and function and help to bring about a new generation of rationally designed glycoprotein therapeutics, vaccines, and diagnostics.

C1 Biorefinery By Designer Enzyme Complexes As Protein Synthesis Machinery on the Cell Surface of Ralstonia Eutropha

Hyeon, J. E., Korea University
Shin, S. K., Korea University
Joo, Y. C., Korea University
Jeong, D. W., Korea University
Han, S. O., Korea University

The designer microbes are a key biological technology that can be used for biorefinery. The use of complexed enzyme systems is one of the strategies for effective bioresources utilization and offers considerably greater potential through a highly ordered structural organization that enables enzyme proximity synergy. Carbon monoxide (CO) was successfully efficiently converted by functional complexes containing carbon monoxide dehydrogenase and carbon monoxide sensing heme protein with enhanced CO binding affinity. Carbon monoxide sensing heme protein acts as mediator for increasing of carbon monoxide dehydrogenase activity by substrate proximity and assembled complexes caused an efficiently conversion of CO gas with aid of Carbon monoxide sensing heme protein. An enzyme complex for biological conversion of CO to CO_2 was anchored on the cell surface of CO_2-utilizing *Ralstonia eutropha*and successfully showed 3.3-fold increased conversion efficiency. These results suggest that the complexed system may be a promising strategy for C1 biorefinery as biological tools for production of biochemicals such as bioplastics. The design of enzyme and microbe is a useful strategy as tools in whole-cell biocatalyst systems for C1 biorefinery and has drawn considerable attention as an attractive strategy for bioprocess applications.

Heuristic Design and Engineering of Synthetic Yeast Promoters

Rajkumar, A. S.,
Liu, G., Chalmers University of Technology
Jullesson, D. B., Chalmers University of Technology
Arsovska, D., Technical University of Denmark
Kristensen, M., Technical University of Denmark
Jensen, M. K., Technical University of Denmark
Siewers, V., Chalmers University of Technology
Nielsen, J., Technical University of Denmark
Keasling, J. D., Lawrence Berkeley National Laboratory

We present a general strategy to iteratively design and engineer synthetic promoters for Saccharomyces cerevisae inducible to environmental conditions or stresses of interest, exclusively employing yeast's native regulatory mechanisms. Starting from available experimental data from the relevant literature, focusing on transcriptome and transcription factor binding data, we identify transcription factor binding sites (TFBS) and other elements relevant to regulation by environmental conditions of interest. Where necessary, we generated our own transcriptome data. We then used these to modify the upstream activation sequences of yeast promoters and either create or improve their regulatory response to a condition of interest. We also discuss approaches to increase and stabilize basal output, and our results in this area. We demonstrate the validity and flexibility of this approach by creating sets of strong promoters inducible by either low pH or the ionic liquid ethylmethylimidazolium (EMIM) acetate, conditions chosen for their relevance to industry. The promoters are characterized at different levels of the inducing signals over different time points and phases of growth, and in the case of low pH induction ultimately used in the production industrially relevant chemical in a yeast cell factory, outperforming under the same conditions a native, constitutive promoter commonly used in metabolic engineering. In summary, our approach provides a working solution bridging true de novo synthesis espoused by synthetic biology and the use of endogenous yeast promoters in metabolic engineering applications irrespective of their suitability for the task.

Evolving Strand-Displacing Polymerases with Isothermal CSR

Milligan, J. N., University of Texas at Austin
Ellington, A. D., University of Texas at Austin

Compartmentalized self-replication (CSR) has proven useful for DNA polymerase evolution by creating a feedback loop in which a polymerase replicates its own DNA in an isolated emulsified compartment. This method has been used to evolve a number of enhanced phenotypes, including increased thermostability, inhibitor tolerance, and unnatural and modified base incorporation. However, CSR currently relies on thermal cycling and PCR. We hope to evolve polymerases that can function better in the context of isothermal amplification reactions, including polymerases with improved strand displacement activity. We therefore modified CSR for an isothermal amplification reaction, rolling circle amplification (RCA), in which the DNA polymerase must continuously displace downstream base-paired DNA in order to effectively replicate its own gene. Starting from a shuffled library of the DNA polymerases from *Thermus aquaticus* (Taq) and *Bacillus stearothermophilus* (Bst), 5 rounds of selection have yielded functional chimeric variants (65%), in which the polymerase best known for isothermal amplification, Bst, has acquired small Taq insertions. We observed Taq insertions at residues 530 to 535 in 71% of the population and at residues 494 to 497 in 29% of the population, indicating that positive selection is occurring. Some 16% of the sequenced population contained additional mutations known to increase thermostability or inhibitor resistance. Compared to Bst, the successful chimeras showed increased amplification rates, improved strand displacement activity, and greater thermostability in isothermal amplification assays. Surprisingly, they also showed increased activity in PCR reactions with difficult templates. These experiments represent the first directed evolution of a polymerase at a single temperature, opening up new avenues for improving amplification-based point-of-care diagnostics.

Self-Selection Using Artificial Networks: An Adaptive Tool for Directed Evolution

Dramé-Maigné, A., CNRS UMR 7083 ESPCI
Amikura, K.,
Kiga, D.,
Fujii, T., University of Tokyo
Rondelez, Y., CNRS UMR 7083 ESPCI

Self-selection using artificial networks: an adaptive tool for directed evolution

Selection-based strategy proved to be very efficient for directed evolution of enzymes but relies on the possibility of finding a case-specific selection tool that links the enzyme activity to its gene survival. Our self-selection process involves the use of a simple artificial molecular network. Introduced as a feedback loop it links directly the activity of the enzyme with the PCR amplification of its own gene. Molecular networks, such as the one using the DNA toolbox (1), are designed to produce short DNA strands interacting within each other's. They can then generate short oligonucleotides at the output. Assessing enzyme activity via the network allows to produce a correlated amount of primers and therefore will link enzyme activity to a PCR amplification yield.

In an approach inspired by the compartmentalized self-replication of Ghadessy and al. (2) we aim at using microdroplets as compartments to perform the self-selection process. Microfluidics allows us to generate a large amount of water-in-oil microdroplet. Up to 10^8 parallel tests on individual copy of the gene can be run simultaneously with few amount of space and reagents.

After generation of a mutant library, bacteria carrying and expressing the mutant are separated in individual droplets containing the molecular program. The program is next launched in the droplets before the PCR step. Then droplets are lysed to retrieve mutant gene copies and a new evolution cycle can be run. Promising results have been obtained with the nicking enzyme Nt.Bst.NBI used as a first model with this adaptive selection tool for directed evolution.

(1) K. Montagne, R. Plasson, Y. Sakai, T. Fujii, and Y. Rondelez, "Programming an in vitro DNA oscillator using a molecular networking strategy," Mol. Syst. Biol., vol. 7, p. 466, Feb. 2011.

(2) F. J. Ghadessy, J. L. Ong, and P. Holliger, "Directed evolution of polymerase function by compartmentalized self-replication," Proc. Natl. Acad. Sci. U. S. A., vol. 98, no. 8, pp. 4552–4557, Apr. 2001.

Universal Expression Tools to Improve Nutrient Acquisition of Energy Crops

Mansoori, N., Joint BioEnergy Institute- Lawrence Berkeley National Lab
Hernandez-Garcia, C., Joint BioEnergy Institute- Lawrence Berkeley National Lab
Masson, F., Joint BioEnergy Institute- Lawrence Berkeley National Lab
Shih, P., Joint BioEnergy Institute, Lawrence Berkeley National Laboratory
Vuu, K., Joint BioEnergy Institute, Lawrence Berkeley National Laboratory
Goyal, G.,
Nath, S., Lawrence Berkeley National Laboratory
Hillson, N. J., Joint BioEnergy Institute
Deutsch, S., Lawrence Berkeley National Laboratory
Loqué, D., Joint BioEnergy Institute, Lawrence Berkeley National Laboratory

Plant growth and development relies on roots as a means to anchor the plant body as well as to absorb water and nutrients such as N, P and K. The current development of growing crops on marginal land (low water content, low nutrient supply, vulnerability to erosion and heavy-metal pollution) for bioenergy will reduce competition with food crops and the pressure on high-quality arable lands utilization. In order to compensate for the lack of nutrients in the soil, root systems in energy crops can be engineered through synthetic biology to generate complex metabolic pathways to improve nutrient acquisition/accumulation in plants.

The introduction of complex metabolic pathways into plants opens new opportunities to utilize plant-based platforms for production of molecules of interest. However the tools for such characterization and modifications specifically in plants are lacking. Designing "universal" expression tools for plant root engineering functional across diverse plant species would dramatically advance our ability to modify crops beyond traditional methods. Our aim is to generate "universal" promoter libraries which can ultimately be used for tissue specific metabolic pathway engineering in bioenergy crops. In house transcriptome analysis of various bioenergy crops will identify root specific genes that are constitutively expressed, induced or repressed through conserved metabolic responses to N, P and Fe starvation in hydroponic conditions across a widely diverse selection of plant species. Promoters from the identified genes are isolated and characterized for spatiotemporal expression patterns and expression levels across different taxonomic classes using composite plants and transgenic approaches. An efficient method for DNA assembly into a suite of plant transformation vectors leveraging *in vivo* yeast homologous recombination has been developed to this aim. Thus the identified promoters along with a number of genes could be used for complex metabolic pathways in bioenergy crops in a tissue specific manner.

Increasing Extracellular Nitrogen Production By Azotobacter Vinelandii dj

Knutson, C. M.,
Barney, B. M., University of Minnesota

In photosynthetic organisms, nitrogen is the most expensive and limiting nutrient input. Traditionally, nitrogen fertilizer has been produced by energy intensive methods such as the Haber-Bosch process. Our research aims to provide alternative means of producing the required nitrogen for photosynthetic crops. *Azotobacter vinelandii* DJ is a model diazotrophic and aerobic bacterium which can be genetically manipulated to increase its extracellular nitrogen production. Here we have performed insertional transposon mutagenesis on *A. vinelandii* DJ in tandem with two facilitated screens to identify the nitrogen producing phenotype. Facilitated screens involved the use of a novel biosensor strain developed by our lab and the co-culture with various algal species. Here we present the mutants identified using the methods described as well as the characterization and quantification of the nitrogen products resultant of each mutation.

Delineating the Mechanisms of Quorum Sensing Cross-Talks By Coupling Activation and Repression

Kim, S., Harvard University
Lips, D., Harvard Medical School
Silver, P. A., Harvard Medical School

Many Gram-negative bacterial species use acyl homoserine lactone (AHL)-mediated quorum sensing to detect local population density or diffusion rates and change gene expressions accordingly. In synthetic biology, components of quorum sensing systems are often adopted to build synthetic circuits that require cell to cell communication using diffusible signals. Because different species of bacteria synthesize different AHLs and their corresponding sensor/responder proteins, quorum sensing is commonly referred to as a species-specific communication method bacteria adopted. Ironically, it is also well noted that there are considerable cross-talks between AHLs and their non-cognate receptors, making the extent of species-specificity questionable. In synthetic circuits, cross-talks between non-cognate AHLs and receptors can disrupt desired functions that are based on quorum sensing components, especially when they were employed to mimic species-specific communication effect. Such observation led to studies that examined signal cross-talks and promoter cross-talks by using GFP reporters that are activated when an AHL/receptor complex activates transcription. While these studies provided valuable information on the presence of activating cross-talks, there has been a lack of a systematic approach to easily detect and quantify the nature of repressive cross-talks. Here, we report the synthetic reporters we built to test both activating and repressive cross-talks across 10 quorum sensing systems. Our study allows us to further delineate the mechanism behind cross-talks, by distinguishing the following three aspects: AHL binding to the receptor, AHL/receptor complex binding to the DNA, and the viable interaction with the RNA polymerase. Finally, our reporter design also features a Memory element, allowing us to test the presence or the absence of AHLs in the mouse gut using engineered AHL sensor E.coli strains.

DNA Origami Sensor That Controls Gene Expression

Zadegan, R. M., Boise State University
Hughes, W. L., Boise State University

Self-assembled nucleic acid structures such as DNA origami are examples of the integration of materials engineering and synthetic biology. These nanostructures are being used as nanomachines to perform various roles in drug delivery,[1] biology,[2] photonics,[3] semiconductors,[4] and molecular computation.[5,6] In this study, we designed and synthesized a self-assembled DNA complex that detects microRNA biomarkers (miR146, miR151 miR193 and miR212) that are specific to lung cancer and produces a fluorescence signal and a proof of concept RNA output. Two fluorophores (a donor and acceptor pair) were placed at the DNA origami with an optimum inter-distance of ~2 nm and FRET (Förster resonance energy transfer) was used to monitor the activity of the nanomachine (Fig. 1a). In the closed state, the donor and acceptor dyes are in close proximity, leading to a high FRET value (state 1, Fig. 1b). Addition of microRNA mimics (i.e. DNA strands with identical genetic content), triggers strand displacement of the DNA locks on the closed origami. As a result the DNA origami conformation switches to an open state and thus the donor and acceptor dyes become distant, leading to a low FRET value (state 0, fig 1b). Once opened by the aforementioned keys, transcription of a target DNA molecule showed multifold increase in the RNA production (Fig. 1b). This study successfully demonstrates that DNA origami can be programed to control transcription, an essential step in gene expression, providing a tool for controlling biomolecular reactions for nanomedicine, synthetic biology, biotechnology and molecular computating.

References

(1) Zadegan *et al.*, Construction of a 4 Zeptoliters Switchable 3D DNA Box Origami, ACS Nano 2012.

(2) Reif & LaBean, DNA Nanotechnology and its Biological Applications, in Bio-Inspired and Nanoscale Integrated Computing 2009.

(3) Klein *et al.*, Multi-Scaffold DNA Origami Nanoparticle Waveguides, Nano Letters 2013.

(4) Zhirnov, *et al.*, Nucleic Acid Memory, Nature Materials 2016.

(5) Zadegan *et al.*, Construction of a Fuzzy and Boolean Logic Gates Based on DNA, Small, 2015.

(6) Yurke *et al.*, A DNA-fuelled molecular machine made of DNA, Nature 2000.

Poster # 212; DNA Origami Sensor That Controls Gene Expression, By Reza Zadegan, Boise State Uni.

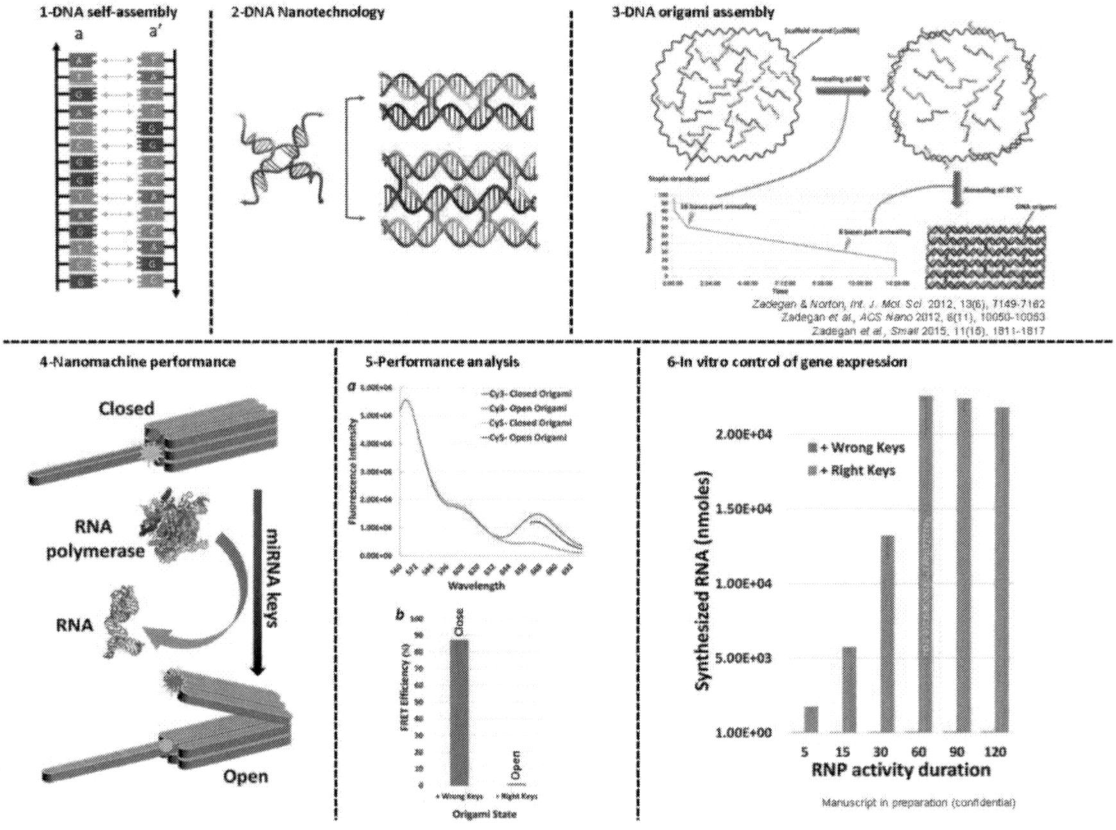

Design of a Minimized pT181 Attenuator Using Shape-Seq

Watters, K. E., Cornell University
Settle, A.,
Zhuang, R.,
Liao, J.,
Lucks, J. B., Cornell University

In an era where targeted gene therapy is slowly becoming a reality, the ability to effectively deliver stable gene circuits remains a difficult challenge. One issue is that nucleic acid delivery becomes increasingly inefficient as the size of the delivered gene circuit gets larger. Thus, there is a need to be able to design smaller gene circuitry. In this work, we describe our efforts to minimize a synthetic RNA regulator, informed with structural insights. The regulator we chose for this study is the pT181 attenuator, which controls transcription by alternatively forming a terminated or anti-terminated structure in response to an antisense RNA. We use our newly updated SHAPE-Seq (Selective 2'-Hydroxyl Acylation analyzed by Primer Extension Sequencing) technique to examine the structural features of the attenuator RNA to find regions that are non-essential. By sequential deletion and mutation, we were able to greatly reduce the size of the pT181 attenuator and show that it still functions in a similar fashion to the original design.

Pooled Assembly, Genotyping and Scoring of Diverse Genomic Libraries

Egbert, R.,
Yu, E.,
Arkin, A., UC Berkeley

Precision engineering of complex behaviors in living systems is complicated by our limited understanding of multiple underlying contexts of gene expression and our limited ability to manipulate the genetic code in individual cells at scale. Iterative build, design and test cycles using indexed strains thus often sample only a small region of the explorable parameter space for a given gene network. Further, the relative ease of optimizing gene networks on multi-copy plasmids for bacterial expression often is not predictive of system performance when transferred to the chromosome for deployment in complex environments ranging from the bioreactor to the mammalian gut.
We have developed a scheme for the pooled assembly, genotyping and single-variant fitness scoring of barcoded genomic libraries coupled with CRIPSRi-mediated retrieval of individual genotypes from the diverse pool. To validate the workflow we generated a genomic library of over one million barcoded variants of the violacein biosynthetic pathway vioABEDC in E. coli, sampling from a genotype space of over 260,000 combinations.
We employed Lambda Red recombination to serially integrate a diversified DNA fragment - here, a gene with a degenerate ribosome binding site - fused to a selectionfluorescence
marker with genome homology. Successive “inchworm” integration stages replaced the selection-fluorescence marker from the previous stage, cycling among combinations of two fluorescence markers and three antibiotic resistance markers. We used fluorescence activated cell sorting of pooled integrants to enrich for the expected fluorescence phenotype, screening out spontaneous resistance mutants and off-target integration events manifest as dual fluorescence phenotypes.
We genotyped the genomic library by associating individual barcodes with ribosome binding site variants of each vio gene by deep sequencing amplicon fusion libraries generated by emulsion PCR. Using time-series measurements of barcode abundance we quantified differential fitness scores for each genotype under pathway induction conditions and in resource competition with violacein-sensitive Bacillus subtilis to identify variants that optimally balance the benefit of antimicrobial production against the cost of gene expression.
To isolate individual genotypes from the pooled library the final integration stage included a degenerate (N)20 barcode embedded in the untranslated region of a gfp gene. We validated the extraction of single genotypes in an indexed library of eight clones with diverse GFP expression levels transformed with CRISPRi plasmids that coexpresses
catalytically dead Cas9 with a guide RNA that target each barcode. We also demonstrated the retrieval of low-abundance genotypes from the pooled genomic library by transforming the library with CRISPRi plasmids encoding guides that target rare barcodes.
We believe our approach to constructing pooled genomic libraries by comprehensively sampling the parameter space of a gene network will expand understanding of

sequence to function relationships and enable engineered phenotypes that are currently unreachable. We anticipate this genomic assembly, screening and isolation platform will advance synthetic biology efforts to optimize cryptic biosynthetic gene clusters for natural product discovery and other engineered cellular behaviors.

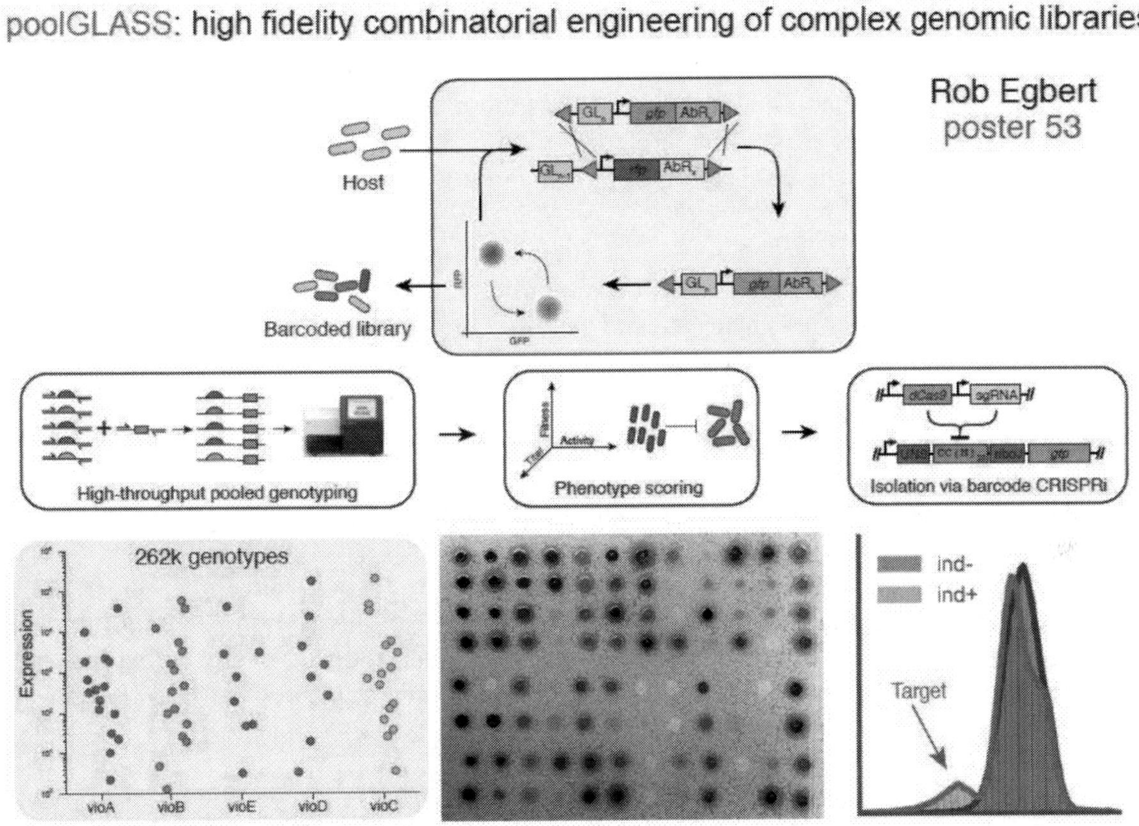

Using bacteria to make improved, nacre-inspired materials

Schmieden, D. T.,
Aubin-Tam, M. E., Delft University of Technology
Meyer, A. S.,

New materials inspired by nature could find applications as coatings in the civil and aeronautical/aerospace industries. Mollusks have developed hard shells to protect themselves from predators. One of the main components, nacre, is a composite consisting of calcium carbonate platelets which are connected by an organic matrix. Nacre has caught the attention of researchers due to its outstanding material properties such as very high compressive and tensile strength.

We use prokaryotes to generate new, nacre-inspired composite materials, consisting of alternating layers of calcium carbonate and poly-gamma-glutamate (PGA). Calcium carbonate is precipitated by the action of Sporosarcina pasteurii, which hydrolyzes urea, increasing the pH of its growth medium and thus making calcium carbonate fall out of solution.

PGA, an anionic polymer produced by many Bacillus species, serves to mimic the complex organic layers found in natural nacre. Due to its unusual gamma linkage, it is resistant to proteases and is used for industrial applications, as it is non-toxic, non-immunogenic, biodegradable, edible, and water soluble. Anionic polymers are good substrates for calcium carbonate precipitation, as their negative side chains allow binding of calcium ions. We alternate calcium carbonate precipitation and PGA application to produce our final product.

Scanning electron microscopy of our samples reveals large regions of layered calcium carbonate structures, which resemble the ones found in nacre. In the negative control (without PGA), layered structures are much less prevalent and limited to spherical crystals.

Both the ability to crystallize calcium carbonate and the ability to produce PGA can be transferred to other bacteria. We will combine the corresponding genes in E. coli to generate an organism that produces nacre-inspired materials in a continuous and tunable production process, replacing the current time-consuming workflow.

This will be one of the first examples of the generation of high-performance composite materials produced by synthetic biology, potentially providing more environmentally-friendly and sustainable materials with applications in e. g. construction or medicine.

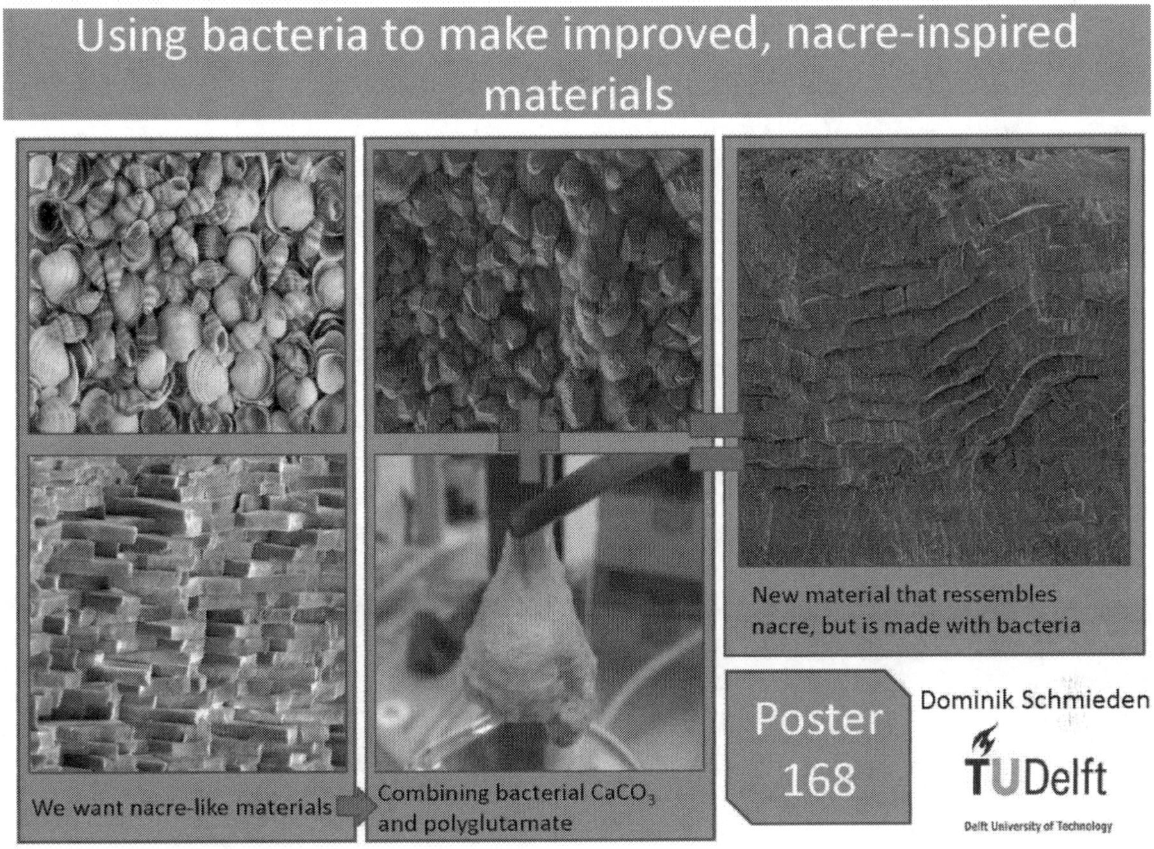
Using bacteria to make improved, nacre-inspired materials
We want nacre-like materials
Combining bacterial $CaCO_3$ and polyglutamate
New material that ressembles nacre, but is made with bacteria
Poster 168
Dominik Schmieden
TUDelft
Delft University of Technology

Leveraging Chemical Insights in Computational Identification of Interesting Biosynthetic Gene Clusters

de los Santos, E. L. C., California Institute of Technology
Bates, D., University of Warwick
Corre, C., University of Warwick
Challis, G. L.,

Natural products have been shown to have diverse applications in fields such as agriculture and medicine. The diminishing returns from screens of traditional chemical libraries coupled with the reduced costs of genome sequencing have fuelled a renewed interest in mining genetic sequences for biosynthetic gene clusters (BGCs). This has led to the development of methods to analyze genome sequences to identify potential BGCs. These have been successful at identifying gene clusters, particularly those which contain polyketide synthase or non-ribosomal peptide synthetase modules. However, the success of these algorithms in finding potential gene clusters has necessitated the need for further refinement and classification: the amount of putative gene clusters exceeds our ability to experimentally characterize them. While advances in DNA synthesis, cloning, and heterologous expression will increase our ability to characterize a potential gene cluster, methods to select which clusters to prioritize are still needed.

We present a framework that does this by examining the mechanistic steps involved in the biosynthesis of specific chemical moieties. By focusing on the chemical steps and the corresponding enzymes required to produce a specific structure, we can quickly screen *in silico* large genomic databases to a small, experimentally tractable list of potential BGCs that are likely to produce the targeted chemical moiety. Furthermore, by focusing on the relatively smaller space of biologically available starting products and chemical reactions, this framework can uncover novel, uncharacterized classes of BGCs complementing the discovery and annotation provided by existing methods.

Yeast ribosome display for high-throughput selection of cap-independent translation regulatory elements impacting both in vitro and in vivo protein expression

Gan, R.,
Jewett, M. C., Northwestern University

Here we report a yeast ribosome display method. This *in vitro* selection and evolution method, based on a cell-free protein synthesis (CFPS) system derived from *Saccharomyces cerevisiae* extracts, is able to enrich a target gene from a model library by 100-fold per round of selection. To demonstrate utility of the technology, and in order to improve protein synthesis in the *S. cerevisiae* CFPS system, we successfully evolved a 5' end untranslated region (5-UTR) of a reporter protein containing 15 randomized nucleotides in front of the start codon. After 4-rounds of selection, the most beneficial mutant showed a 13-fold improvement in protein synthesis as compared to the starting library, which is mainly due to enhanced cap-independent translation initiation (CITI) efficiency of the selected 5-UTR sequence. By repeating the beneficial short sequences in tandem, protein yield was further improved by ~3-fold in CFPS. Additionally, the sequence information of 24 5-UTR mutants with various CITI activities was collected and the determinants of activity were analyzed. We found that the CITI activity is partially dependent on the A/T content ($R^2 = 0.50$) of the 5-UTR sequence. Finally, twelve 5-UTR mutants were cloned into a yeast expression vector and tested for protein yield *in vivo*. These CITI sequences were able to tune protein expression *in vivo* across a 20-fold range in expression, similar to *in vitro* results, though due to an unclear mechanism. Thus, the yeast ribosome display method was used to identify reliable genetic elements *in vitro* prior to putting them into a host. Looking forward, the yeast ribosome display method holds promise for evolving libraries of proteins (*e.g.*, antibodies) as well as DNA regulatory parts (*e.g.*, promoter function, terminator strength) for synthetic biology.

Long-term adaptive evolution of highly recoded E. coli strains

Wannier, T. M., Harvard Medical School
Kunjapur, A. M., MIT
Church, G. M., Harvard Medical School

Advances in genome engineering and DNA synthesis technologies have facilitated the large-scale recoding of bacterial genomes, whose objectives include expanded amino acid incorporation, multi-virus resistance, and biocontainment. Our current understanding, however, of the consequences of specific codon substitutions or the deletion of translation elements, such as release factor 1 (RF1), is limited. Furthermore, highly recoded strains tend to have significantly decreased growth rates. To shed greater light on genome design and to concurrently improve strain fitness, we have performed adaptive evolution of several recoded E. coli strains for more than 1,000 generations in independent replicate populations grown in glucose minimal media. Among these strains is C321.ΔA, in which all instances of TAG stop codons and RF1 have been substituted and deleted, respectively. The growth rate of C321.ΔA is nearly half of the growth rate of its parent strain (ECNR2) in minimal media. However, the average growth rates of evolved C321.ΔA populations are more than four-fold the growth rate of the ancestral C321.ΔA strain and two-fold the growth rate of ECNR2. Additionally, the growth rates of some independent subpopulations of evolved recoded strains in minimal media approach the growth rate of the parent strain in rich media. Preliminary analysis of next-generation sequencing results on dominant clones from independent generation ~1,100 populations is ongoing. Results from this study will guide current and future recoding efforts in the lab while improving utility of existing strains for downstream applications.

Genome-scale regulatory landscape of Escherichia coli in response to various environmental stresses

Seo, S. W., Seoul National University
Palsson, B. O., University of California - San Diego

Environmental stresses such as metal iron depletion, oxidative stress, low pH stress, and osmotic stress greatly affect the cellular performance during fermentation. Understanding the genome-wide regulatory network in response to these stresses is important to rationally design microbes with improved tolerance towards them. In this study, we comprehensively reconstructed transcriptional regulatory networks of several transcription factors related to iron depletion, oxidative stress, low pH stress, and osmotic stress in *Escherichia coli.* Integrative data analysis from genome-wide measurements found that each transcription factor (TF) exhibits complex role in many fundamental cellular processes to coordinately overall respond of cells towards theses stresses. Functional analysis of newly found regulon members provide evidence of importance of those genes for survival under stress condition. These types of data can also provide more accurate information of regulation allowing us to design genomes for synthetic cells in the future.

Reference

Seo et al. Nat. Commun. 5:4910. 2014

Seo et al. Nat. Commun. 6:7970. 2015

Seo et al. Cell Rep. 12(8):1289-99. 2015

Novel luminescent reporters from marine environmental bacteria for prokaryotic and eukaryotic systems

Kan, A.,
Pollak, B., University of Cambridge
Haseloff, J., University of Cambridge

Bioluminescent reporters are commonly used as reporters in molecular biology. Luciferases from various sources have been cloned and exploited as reporters due to their fast time dynamics and easily detectable and quantifiable outputs. While all eukaryotic luciferases require the addition of an exogenous substrate to produce light, bacterial bioluminescence is generated by a single operon and has the benefit of autonomous luminescence. Yet, compared to fluorescent protein reporters, little engineering has been applied to improve and optimize the lux operon on a protein level. In this work, we have sequenced the genomes of 19 luminescent marine bacteria from the genus *Vibrio*, *Aliivibrio* and *Photobacterium* isolated from shore seawater around the world. After identifying the *lux* operons, we cloned both the full operons and *LuxAB* genes separately into an *E. coli* for characterization in a common background, where they exhibited a wide range of luminescence. Using this pool of homologous luminescent *lux* operon sequences from diverse sources, we used a DNA-shuffling directed evolution technique to evolve *luxAB* proteins specific to an exogenous decanal substrate 1.5 times brighter than the best wild-type construct. The whole operon was also restructured into a reporter component and substrate pathway component, which were co-evolved to generate an autonomous luminescent construct 25 times brighter than the best wild-type construct. The evolved luciferase proteins were further fused into a single protein and also fused with a yellow fluorescent protein to create a spectrally distinct luminescent reporter. Furthermore, the evolved reporter proteins and substrate pathway were refactored for use in the plant *Marchantia polymorpha*, where they exhibit autonomous luminescence.

Engineering DNA polymerase delta to make its activity Ca2+ dependent

Bhan, N., Rensselear Polytechnic Institute

We are developing a DNA polymerase-based "molecular ticker-tape" for generating spatially and temporally resolved calcium ion concentration recordings in neurons. Calcium ion concentration changes are used as a proxy for neural firing. Such a "ticker-tape", with its DNA synthesis error rate dependent on Ca^{2+} ion concentrations, would write the temporal trace of neural activity as error patterns onto DNA molecules making it possible to uncover neural interactions with single neuron precision. Two approaches are being attempted; direct tweaking of DNA polymerases to make their activity Ca^{2+} dependent and a modular approach where a DNA polymerase accessory protein is being made Ca^{2+} dependent. Proliferating cell nuclear antigen (PCNA) is a ring-shaped homotrimeric accessory protein that slides onto linear DNA and interacts with DNA polymerases, increasing their processivity at the cost of fidelity. *S. cerevisiae* polymerase delta (Pol delta) suffers approximately 2 to 4 fold loss in fidelity *in vitro* upon associating with its PCNA. Since a nontrimeric PCNA cannot interact with DNA polymerases we have made its trimerization Ca^{2+}-dependent, by rationally fusing it with calmodulin (CaM). We are now studying the effect of Ca^{2+} on PCNA-CaM fusion and Pol delta interactions. Using the same rational approach we plan to create several Pol delta and calmodulin fusions. These fusions will be screened through an established *S. cerevisiae* plasmid-shuffling scheme, where only a functional DNA polymerase fusion will result in cell survival.

De novo design of a genetically encoded simple viral-like coat protein for artificial viruses

Hernandez-Garcia, A., Wageningen University
Kraft, D.,
van der Schoot, P.,
de Vries, R.,

Viruses are naturally evolved non-living entities and key to understand the emergence of biological complexity and life. Viruses have been implemented for gene therapy and construction of nanomaterials due to their remarkable properties such as nano-scale organization, precise self-assembly and efficient nucleic acid delivery into cells. However, their engineering and application have been severely limited by complex and not fully understood structure-function relations, safety risks, immunogenicity, intricate production, etc.

In order to overcome these limitations, we have designed a polypeptide that mimics fundamental properties of viral capsid proteins [1, 2]. The polypeptide consists of three independent blocks made up of simple amino acid sequences that provide colloidal stability, self-assembly and DNA binding. The viral-mimetic polypeptide self-assembles around a DNA template into precise and monodisperse rod-like nanostructures, similarly to the Tobacco Mosaic virus. Furthermore, the artificial rod-like viruses have high stability in physiological conditions and against enzymatic attack, and deliver pDNA and mRNA into cells with high efficiency and low toxicity. The viral-like coat polypeptide is produced monodisperse and secrete with high yield into the medium using a genetically modified P. pastoris yeast strain. Its genetically encoded multi-block design allows to program additional functionalities such as targeted delivery, self-replication and others.

References

[1] Coating of Single DNA Molecules by Genetically Engineered Protein Diblock Copolymers. Armando Hernandez-Garcia, et. al., Small 8, 3491–3501 (2012)

[2] Design and self-assembly of simple coat proteins for artificial viruses. Armando Hernandez-Garcia, et. al., Nature Nanotechnology 9, 698–702 (2014)

Computationally Reconstructing RNA Folding Pathways with Co-Transcriptional SHAPE-Seq

Yu, A. M., Cornell University, Ithaca, New York, Weill Cornell Medical College, New York, New York, Memorial Sloan-Kettering Cancer Center, New York, New York
Watters, K. E., Cornell University
Strobel, E. J., Cornell University
Lucks, J. B., Cornell University

Co-transcriptional folding of RNAs happens every time an RNA is synthesized in the cell, making it an important process that underlies many conserved RNA functions. Formation of RNA structure happens during RNA transcription, and the final folded structure can depend greatly on the folding pathway. However, our understanding of RNA co-transcriptional folding is hampered by a lack of computational methods that can accurately determine these intricate folding pathways. Existing RNA folding algorithms often do not consider co-transcriptional folding. Even those that do consider co-transcriptional effects still require improvements in accuracy and have limitations such as being unable to take experimental data as input. Here we present a computational method that uses co-transcriptional SHAPE-Seq data, which can uncover nucleotide-resolution structural information for each intermediate length of an RNA during synthesis, to reconstruct the folding pathway of RNAs.

An RNA-Based Negative Autoregulation Circuit

Zhang, Y.,
Hu, C., Cornell University
Takahashi, M. K., Cornell University
Lucks, J. B., Cornell University

The central focus of synthetic biology is to develop programmable genetic regulatory systems to control gene expression. Recently, regulatory RNAs have emerged as powerful components of the synthetic biology toolbox for constructing genetic circuits. RNA-only circuits have the advantage of fast signal propagation that is set by the rapid degradation rate of RNA molecules. Negative autoregulation (NAR) as a network motif is an essential part of several designs of genetic oscillators. NARs can also decrease the response time of genetic circuits, and promote the robustness of steady state expression levels. Being able to engineer an RNA-only negative autoregulatory circuit expands the possibilities to engineer more sophisticated RNA-only circuitry. In this work, we used a model-guided design to construct an RNA negative autoregulatory circuit with transcriptional sRNA repressors. We showed that the RNA-only transcriptional negative autoregulator speeds circuit response time both in vivo and in ell free transcription and translation systems (TX-TL). We also used an effective model to describe the negative autoregulator's dynamics in TX-TL, and parameterized unknown parameters to predict future designs. This work presents the first known RNA-only negative autoregulatory circuit, and broadens our ability to construct increasingly sophisticated RNA genetic circuitry.

Benchmarking RNA Switches for Diagnostic Applications

Verosloff, M.,
Chappell, J., Cornell University
Lucks, J. B., Cornell University

The current golden standards employed for pathogen detection are generally too expensive and/or require technical expertise that becomes prohibitive for their widespread use in developing nations. Recent developments in RNA design and engineering have led to an explosion in the potential of this biomolecule for the constructing of sophisticated sensors and bio-circuitry aimed at a plethora of problem solving applications. Centrally to this idea is the plethora of natural, and more importantly, laboratory designed RNA regulatory switches. Upon the binding of nucleic acid or small ligands, these switches can be tailored to act upon both the transcriptional and translational landscape. We are currently working to characterize and further improve upon RNA-based bio-circuitry in order to ascertain what is needed to convert input signals, such as pathogen DNA/RNA, into rapid and comprehensible visual outputs. Specifically we will present work on the characterization of several reported RNA switches. Using fluorescent and colorimetric outputs, we have begun to establish a benchmark characterization to compare the performance of switches in terms of speed of response, dynamic range and limit of detection. This also enables us to explore additional important factors such as constraints for naked eye visualization. In harnessing the synergism of cell free systems, synthetic bio-circuitry can strive towards improving the standard of living of many of the world's impoverished.

A Cell-Free Approach to Fine Chemical Pathway Engineering

Moore, S., Imperial College London
Polizzi, K., Imperial College London
Freemont, P., Imperial College London

Pathway engineering studies often neglect a key requirement to balance the supply and demand of enzyme loading at the molecular level. Furthermore, toxic intermediates can also hinder growth and promote genetic instability, so fine-tuning of pathway gene expression is essential to hit the metabolic sweet spot of pathway design[1].

Our aim is to draw a link between cell-free *in vitro* kinetics and pathway design *in vivo* in living cells [2,3]. Using the biosynthesis of raspberry ketone (Type III polyketide) as a 5-enzyme model pathway, we have designed a synthetic *in vitro* pathway to study the effects of enzyme loading and cofactor regeneration on pathway productivity. We have also discovered a reductase enzyme that fluoresces upon binding of a late pathway intermediate, therefore serving as a reporter by measuring product accumulation in real time. Using this fluorescence reporter, we have studied the pathway under massively parallel (384- to 1536-well plate format) experimental conditions with acoustic liquid handling robotics. In summary, the pathway enzyme benzalacetone synthase represents a rate-limiting step in the pathway, whilst low productivity occurs through imbalances in coenzyme A utilisation and cofactor regeneration.

Using this cell-free insight, we have implemented an *in vivo* pathway using combinatorial pathway refactoring[4]and natural riboswitch insulators to improve genetic stability and increase benzalacetone synthase levels. We are currently quantifying pathway designs with enzyme-GFP fusions and whole-cell extract LC-MS/MS analysis to correlate enzyme loading to pathway productivity. This will enable us to provide a fingerprint of an optimal enzyme pathway for production of Type III polyketides.

1. Jeschek, M., Gerngross, D. & Panke, S. Rationally reduced libraries for combinatorial pathway optimization minimizing experimental effort. *Nat. Commun.*7, 11163 (2016).

2. Weaver, L. J. *et al.* A kinetic-based approach to understanding heterologous mevalonate pathway function in *E. coli. Biotechnol. Bioeng.*112, 111â??9 (2015).

3. Opgenorth, P. H., Korman, T. P. & Bowie, J. U. A synthetic biochemistry module for production of bio-based chemicals from glucose. *Nat. Chem. Biol.*(2016). doi:10.1038/nchembio.2062

4. Moore, S. J. *et al.* EcoFlex - A Multifunctional MoClo Kit for *E. coli* Synthetic Biology. *ACS Synth. Biol.* (2016). doi:10.1021/acssynbio.6b00031

Orchestrating an Enzyme Ensemble Towards Novel Antibiotics

Lai, H. E., Imperial College London
Moore, S., Imperial College London
Polizzi, K., Imperial College London
Freemont, P., Imperial College London

Violacein is a violet pigment first isolated from Chromobacterium violaceum and is part of the indolocarbazole biosynthetic family that utilises L-tryptophan as the starting substrate. The violacein pathway is encoded within a conserved operon of 5 genes (vioABCDE), whose gene-products catalyse a 14-electron oxidative biosynthesis pathway[1]. The indolocarbazole biosynthetic pathways have attracted immense interests because of their therapeutic potential for medical applications, including antimicrobial, antiviral, trypanocidal and anti-tumorigenic properties[2].

The violacein pathway is a favourable model pathway in synthetic biology for combinatorial biosynthesis because of its relative simplicity and production of coloured intermediates, which allows high-throughput screening. Recent study into the effect of promoter strength[3] on violacein yield have indicated that heterologous production of violacein has the potential to surpass natural supply. In my project, I have implemented a combinatorial library screening method for violacein production using EcoFlex, a Golden Gate DNA assembly kit developed by our lab (in review). The library variants, which differ by their promoter, RBS and terminator sequences at each of the three selected violacein pathway genes, exhibited a wide range of violacein yields with the highest producing strain at 66 mg L^{-1}. It was also found that low-copy plasmid number was better suited for library construction for toxic pathways like violacein, as observed by the decrease in frequency of mutation in the gene regulatory parts. In the presence of strong selective pressure (toxic pathway), the cells will likely to mutate the gene regulatory components to increase survivability.

As a separate story, due to the homology of enzymes involved in the indolocarbazole biosynthetic pathways, some researchers have attempted to create novel metabolites using a combination of enzymes involved in different pathways[4]. It is envisaged that the diversification of metabolites will lead to discovery of novel compounds, which may exhibit therapeutic effect better than violacein upon screening against a wide range of disease biomarkers. Potentially synthetic biology can be used to incorporate site-specific labelling enzymes for halogenation, methylation and hydroxylation to create synthetic analogues of violacein, possibly via homologous enzymes from related indolocarbazole pathways or through rational structure-based engineering. To begin, we have crystallised and solved the structure of a tryptophan-bound VioA, FAD-dependent L-tryptophan oxidase, showing the close proximity of the tryptophan substrate with FAD cofactor in a 2.6 Å crystal structure. Using the structure of VioA, we will attempt to rationally engineer VioA to accept tryptophan derivative as a substrate, thus paving the way towards biosynthesis of violacein analogues.

References

1. Balibar, C. J. & Walsh, C. T. In Vitro Biosynthesis of Violacein from L -Tryptophan

by the Enzymes VioA - E from Chromobacterium V iolaceum †. 15444–15457 (2006).
2. Durán, M. et al. Potential applications of violacein: a microbial pigment. Med. Chem. Res. 21, 1524–1532 (2011).
3. Jones, J. A. et al. ePathOptimize: A Combinatorial Approach for Transcriptional Balancing of Metabolic Pathways. Sci. Rep. 5, 11301 (2015).
4. Du, Y.-L. & Ryan, K. S. Expansion of Bisindole Biosynthetic Pathways by Combinatorial Construction. ACS Synth. Biol. 150102122128000 (2015). doi:10.1021/sb5003218

In vitro glycosylation by cell-free protein synthesis-derived oligosaccharyltransferases

Hershewe, J. M., Northwestern University
Schoborg, J. A., Iowa State University
Jewett, M. C., Northwestern University

Asparagine (N)-linked glycosylation is an important protein modification that is ubiquitous in nature and therapeutically relevant. The crucial cellular and molecular processes affected by N-glycosylation are far reaching, including protein folding, immunogenicity, activity, half-life, and regulation of signaling cascades. Despite its vast importance, our ability to study functional and structural consequences of site-specific glycosylation remains limited. For these reasons, developing tools to study individual glycosylation components expands our ability to understand and engineer glycosylation for developing novel therapeutics. One such component is the large, membrane-bound enzyme, termed an oligosaccharyltransferase (OST). The OST is responsible for catalyzing the formation of glycosidic bonds between a target protein bearing an asparagine residue within a consensus sequence and lipid-linked oligosaccharides. Understanding and studying OSTs is extremely important since they dictate *where* a glycan is attached, and *which* olicosaccharide is attached. It has been shown elsewhere that full heterologous glycosylation pathways can be expressed in *E. coli,* conferring the strain with the ability to glycosylate protein. This method, however, does not allow for user-defined tuning of glycosylation components, and relies heavily on the viability of cells. In particular, the over-expression of membrane proteins- such as OSTs- can be highly toxic to cells and impede expression. In order to address these issues, we have leveraged the *E. coli* cell-free protein synthesis (CFPS) platform to rapidly express and study six bacterial OSTs. By optimizing CFPS for use with lipid-protein nanodiscs, we are able to synthesize properly-folded OSTs whose transmembrane stretches are supported within the lipid bilayer of the nanodiscs. This method results in up to 750 ug/mL of soluble enzyme in six hours. We show that CFPS-derived OSTs can then be directly introduced into *in vitro* glycosylation (IVG) reactions without the need for lengthy purifications or overexpression. These CFPS-derived OSTs are active on multiple protein substrates, and site-specifically glycosylate target proteins comparably to *in vivo* produced OSTs. In addition, the chemical environments of IVG reaction conditions can be easily optimized for higher glycosylation efficiencies, and can be tuned to any OST. By simply optimizing buffer concentration, supplementation of cofactors, and addition of macromolecular crowding reagents, we have improved efficiency of IVG reactions by over 50%. We anticipate this broadly applicable method for quickly manufacturing and studying OSTs will provide new opportunities for studying glycosylation components, thereby bringing us closer to applying this knowledge to develop novel glycoprotein therapeutics.

In Vitro Expression and Assembly of Nonribosomal Peptide Synthetase for Biosynthesizing D-Phe-L-Pro Diketopiperazine

Li, J.,
Goering, A. W.,
Kelleher, N. L., Northwestern University
Jewett, M. C., Northwestern University

Nonribosomal peptides (NRPs) are of great interest because they possess significant bioactivities such as antiviral, antimicrobial and anticancer activities. These compounds are natively biosynthesized by large multimodular enzymes called nonribosomal peptide synthetases (NRPSs). Current metabolic engineering and synthetic biology studies have tried to utilize surrogate microbes such as *E. coli* for the heterologous production of pharmaceutical molecules (e.g. NRPs) by reconstitution of their entire gene clusters in the host cells. Although those *in vivo* surrogate hosts may seem promising to produce complex NRP compounds, it still remains problematic to obtain high yields, which are mainly due to metabolic burden inhibiting cell growth, unavailability of necessary precursors and toxic products. Alternatively, in this work we demonstrated a novel approach to synthesize natural products based on a completely *in vitro*, cell-free, platform consisting of enzyme expression and product formation. As a model system, we utilized two NRPSs that are involved in the first step of gramicidin S biosynthesis, namely, GrsA (126 kDa) and GrsB1 (121 kDa, the first module of GrsB). GrsA and GrsB1 can function in together to catalyze the formation of the D-Phe-L-Pro diketopiperazine (DKP). We initially expressed GrsA and GrsB1, respectively, using an *E. coli*-based cell-free protein expression (CFPS) system. The results indicated that both enzymes were expressed with yields more than 100 μg/mL. Then, we aimed to activate the newly expressed *apo* enzymes to be functional *holo* enzymes by addition of the purified phosphopantetheinyl transferase (PPTase, for example, Sfp from *Bacillus subtilis*). The post-translational modification of GrsA and GrsB1 was confirmed by labeling with a fluorescent dye (BODIPY-CoA) after addition of the PPTase Sfp. We next sought to synthesize D-Phe-L-Pro DKP using the *in vitro* platform. Instead expression of GrsA and GrsB1 separately, two enzymes were coexpressed in one-pot mixture allowing reconstitution of the full NRPS assembly line for product formation. The results indicated that our *in vitro* system can produce 12 mg/L of D-Phe-L-Pro DKP, which is higher than the reported *in vivo* yield at 9 mg/L. Our studies notably demonstrate that *in vitro*, cell-free, platforms are robust to produce “difficult-to-express” proteins like the very large NRPS enzymes, and more importantly, complex natural products that could be biosynthesized via the *in situ* CFPS expressed enzymes. We anticipate that our approach could be generalized for the construction of highly efficient cell-free based artificial biosynthetic factories to enable target product discovery and synthesis.

Guiding Receptor Directed Evolution Using a Physiochemical Specificity Model

Stainbrook, S.,
Yu, J., Northwestern University
Reddick, M.,
Bagheri, N., Northwestern University
Tyo, K. E. J., Northwestern University

We recently evolved of a yeast G-protein coupled receptor (GPCR) to detect a peptide biomarker for chronic kidney disease. The capability to evolve a receptor for any peptide biomarker enables the production of yeast-based biosensor (YBB) diagnostics for nearly any disease. YBBs use yeast to detect biomarkers in readily accessible patient samples such as urine and produce a colorimetric output discernable to the naked eye. Because YBBs use active-dry yeast, they are inexpensive to produce and have a long shelf life at room temperature. These features make YBBs ideal for point-of-care diagnostics and patient self-test applications, especially in resource-limited settings.

A major bottleneck to production of YBBs remains the difficulty of predicting an optimal path for directed evolution of the yeast GPCR, Ste2. Of two peptide targets for which we attempted to evolve receptors, one receptor took approximately 11 months to develop. A receptor for the other peptide target was never evolved, likely due to an infeasible directed evolution strategy. Through our work, it became clear that each receptor has a characteristic evolutionary step size, which is the maximum difference in the ligand for which a responsive receptor can be found in a single round of directed evolution. We also found that increased receptor promiscuity is a hallmark of intermediates along successful directed evolution pathways. However, the mechanisms underlying receptor promiscuity are poorly understood, preventing predictive design of receptor libraries. A better understanding of mechanisms for receptor promiscuity would enable the design of libraries that are more likely to contain successful variants, and of experiments that are more likely to isolate them.

The promiscuity of a receptor can be selected directly. We use fluorescence-activated cell sorting (FACS) to evolve receptors of varying promiscuity. The receptors were then characterized against a library of single amino acid variants of α-factor. Partial least squares regression (PLSR) was applied to model the responses of each receptor to the peptide library. For each peptide, 8 physiochemical properties of each amino acid were provided as predictors. This modeling approach accurately predicts the response of a receptor to a peptide, and can be used to estimate the ligand range of a receptor. This provides a way to estimate the allowable evolutionary step size for any receptor and enables design of a more efficient directed evolution strategy. Finally, examination of the subset of highly predictive physiochemical properties provides insight into the mechanisms that produce promiscuity.

Single-Chain Variable Fragment Antibodies Targeting Amyloid β Oligomers: Alzheimer's Disease Toxins

Cline, E.,
Popova, I.,
Sebollela, A.,
Kamel, J.,
Qin, A.,
Klein, W.,

Amyloid beta oligomers (AβOs) accumulate early in Alzheimer's disease and experimentally cause memory dysfunction and the major cellular pathologies associated with AD (*e.g.*, tau abnormalities, synapse loss, oxidative damage, *etc.*). However, the structures of the AβO species most germane to AD pathogenesis are ill-defined. This uncertainty regarding the pathophysiologically relevant AβO structures has diminished the perceived therapeutic value of targeting Aβ-derived toxins. Our long-term research goals are to identify AβO species germane to AD onset and to determine the structural characteristics of these AβOs that contribute to their role in the pathogenesis of AD. To help achieve this goal, we have identified multiple single-chain variable fragment (scFv) antibodies with high specificity for AβOs (and minimal affinity for Aβ monomers and fibrils) by panning phage-displayed human scFv libraries. We have determined that at least one of these scFvs, NUsc1, is specific for a small sub-population of synapse-binding AβOs. Furthermore, we have demonstrated that NUsc1 retains its AβO binding activity when it is expressed in soluble form, not attached to phage. The specificity of NUsc1, and the other AβO-specific scFvs, makes them promising tools for (1) determining the role of individual AβO conformations in AD pathogenesis; and (2) application as brain imaging probes for AD diagnostics (*e.g.*, Viola *et al.*, 2015, Nature Nanotechnology).

Applying Evolutionary Principles to Stabilize Genetic Circuits

Yi, X.,
Travisano, M.,
Kazlauskas, R. J., University of Minnesota

A fundamental difference between an electric circuit and a genetic one is that while the former is stable, the latter is subject to spontaneous mutations and ensuing instability caused by adaptation. This fragility is pronounced when the human-desired function that is engineered to the genetic circuit conflicts with fitness — the ability to reproduce — of the organisms that express the circuit. This problem is present in virtually all engineered biological systems. Works in synthetic biology from the recent decade have chartered the basic framework for the functionality of genetic circuits, such as logic gates, feedback loops, noise control; they were all done as proof of concept at the small scale in laboratory, thereby risking serious fragility. For instance, the synthetic gene oscillator, an early breakthrough in the field, could only oscillate for no more than a few cycles before breaking down. Hence, the robustness of synthetic systems is becoming a bottleneck to delivering discoveries from laboratories to more complex and demanding settings of industry and medicine. Here, we propose a novel approach to address robustness where knowledge about how organisms evolve and how competition occurs is integrated into the design of synthetic biological systems. Indeed, we have identified several evolution-aware principles of design that, according to our modeling, should improve robustness by several to effectively infinite folds, depending on the situation. Consider that a mutant carries two mutations and has a high fitness, and mutants that carry either of the two mutations have worse fitness than wildtype that carries no mutations. Evolution is short-sighted: only mutants with increased fitness are selected and capable of sweeping the population. Hence, assuming mutations occur one at a time, this short-sightedness dictates that the system is trapped in the local fitness optimum represented by the wildtype and that the double mutant, despite its large fitness, is unlikely to invade wildtype. Therefore, engineering a bacterial system that expresses a human-desired function to be a local fitness optimum should improve its stability. We believe our novel approach constitutes a missing link that is needed to potentiate the exciting findings emerging from synthetic biology to the real world.

Deciphering and Exploitation of the Adjacent Transcriptional Regulators for Synthetic Biology

Lebar, T.,
Jerala, R.,

Designable DNA binding domains are an extremely useful tool for the transcriptional regulation of the selected endogenous genes, as well as for the construction of synthetic genetic circuits. Natural transcriptional regulators sometimes act synergistically and can affect binding of other transcriptional regulators; however this phenomenon usually depends on each combination. Here we report that we can affect binding of transcription activator-like effectors (TALEs) by other TALEs with an overlapping or an adjacent recognition site. This occurs in a polarized manner, by preferential binding of the 5â?? TALE, which could be explained by the stronger binding of the RVDs at the TALE N-terminal domain of TALEs. This phenomenon allows introduction of negative regulation by displacing 3` bound TALE-based activators and construction of logic functions with kinetics superior to the traditionally used KRAB domains. We demonstrate the function of this mechanism on up to 5 consecutive TALE target sites and construct all 16 two-input logic gates and a half-adder circuit. Displacement by the adjacent TALEs can also be used to displace other DNA binding domains, such as Cas9-sgRNA.

Design, Synthesis and Testing Towards a 57-Codon Genome

Ostrov, N.,
Landon, M. M., Harvard University
Guell, M., Harvard Medical School
Kuznetsov, G., Harvard Medical School
Teramoto, J., Harvard Medical School
Norville, J. E., Harvard Medical School
Cervantes, N.,
Zhou, M.,
Napolitano, M. G., Harvard Medical School
Moosburner, M.,
Pruitt, B. W.,
Conway, N.,
Goodman, D. B., Harvard Medical School
Singh, K.,
Gardner, C. L.,
Wanner, B. L., Harvard Medical School
Lajoie, M. J., Harvard Medical School
Church, G. M., Wyss Institute, Harvard University

Recoding, the re-purposing of genetic codons, is a powerful strategy for enhancing genomes with functions not commonly found in nature. We report the computational design, synthesis, and progress towards assembly of a 3.97 MB, 57-codon *Escherichia coli* genome in which all 62,214 instances of seven codons were replaced with synonymous alternatives across all protein coding genes. Thus far, we have validated 63% of recoded genes by individually testing 55 segments of 50-kb each. Importantly, 90% of tested essential genes retained functionality with limited fitness effect. We demonstrate identification and correction of lethal design exceptions, only 13 of which were found in 2229 genes tested. This work underscores the feasibility of rewriting genomes and establishes a framework for large-scale design, assembly, troubleshooting and phenotypic analysis of synthetic organisms.

Towards a 57-Codon Genome

Matthieu Landon[1,2,3*†], Nili Ostrov[1,†], Marc Guell[1,4†], Gleb Kuznetsov[1,5†], Jun Teramoto[1,4], Natalie Cervantes[1], Minerva Zhou[7], Kerry Singh[7], Michael G. Napolitano[1,6], Mark Moosburner[1], Ellen Shrock[1], Benjamin W. Pruitt[8], Nicholas Conway[9], Daniel B. Goodman[1,5], Cameron L. Gardner[1], Gary Tyree[1], Alexandra Gonzales[1], Barry L. Wanner[1,8], Julie E. Norville[1], Marc J. Lajoie[1*], George M. Church[1*]

WHY?

Phage Resistance

Theoretical resistance to all phages.

Expanded chemistry

4 non-standard amino acids could be supported at once.

Biocontainment

$6x10^{-11}$ escape frequency of our best biocontained strain is ~150x lower than NIH standard.

WHAT?

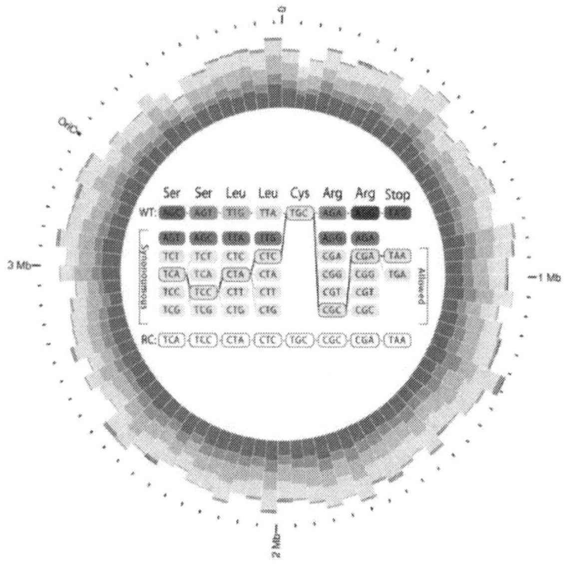

HOW?

Unprecedented full-genome synthesis and assembly effort:

- 7 codons removed
- 3.98Mb synthesized
- 148,955 bp changed

Robust Pipeline for Assembly:

- 63% tested genes
- 99.3% viable tested genes

Troubleshooting efforts:

- Effects on fitness
- Lethal recoding

MORE ON POSTER 110!

Engineered Probiotics for Cholera Treatment and Diagnosis

Mao, N., Boston University
Cameron, D. E., Massachusetts Institute of Technology
Collins, J. J., Massachusetts Institute of Technology

New synthetic biology techniques have enabled us to engineer the human microbiota for therapeutic purposes. In this project, we have engineered the lactic acid bacterium *Lactococcus lactis* to detect *Vibrio cholerae*, the causative agent of the infectious disease cholera, and may initiate therapeutic and diagnostic functions *in situ*. For the pathogen detection module, we used the CqsS sensor domain from the *V. cholerae* CAI-1 quorum-sensing system to create a hybrid two component system in *L. lactis* that can directly detect the *V. cholerae* CAI-1 signal. Incorporation of a TetR-based signal inverter into the circuit allowed us to couple CAI-1 recognition to activation of the output module. For the output modules, we are developing functions for therapeutic and diagnostic applications. We identified two antimicrobial peptides (AMPs) that inhibit *V. cholerae* growth, and are working to develop an efficient AMP release mechanism in *L. lactis*. When coupled to the *V. cholerae* detection module, we will be able to produce a detect-and-kill genetic circuit. In parallel, we have developed a colorimetric output module, based on Ã?-lactamase production, that can serve as a diagnostic function in *L. lactis*. We have tested the circuit in an in vitro co-culture system and will conduct in vivo tests in a *V. cholerae* mouse infection model. Such engineered â??smartâ?□ probiotics will provide a cost-effective, point-of-care solution to the prophylaxis, diagnosis and treatment of cholera in the low-resource settings of a cholera endemic area.

Towards Mechanically-Activated Artificial Cell

Ho, K. K. Y.,
Lee, L. M., The University of Arizona
Liu, A. P., University of Michigan

Artificial cells are minimal and well-defined systems that can be used to decipher the design principles behind biological complexity. One current platform for building artificial cells involve the encapsulation of *in vitro* transcription-translation system in compartmentalized structures such as single or double emulsions or lipid vesicles. Information is encoded in artificial cells through the use of gene circuits and the control of artificial cells has largely been limited to the addition of membrane-permeable small molecule inducers. To expand the capability of information processing by artificial cells, we seek to construct artificial cells that can respond to mechanical forces, which are arguably the most primitive sensory information cells can perceive. As a model of artificial cells, we generated double emulsion droplets with surfactant-stabilized oil in the middle phase. We developed a multilayer microfluidic device capable of trapping individual double emulsion droplet, and apply controlled aspiration and compression to the trapped droplet to simulate different mechanical loading on an artificial cell. We demonstrated the salient features of this single droplet manipulation device and showed that aspiration of the middle phase oil can lead to thinning of the oil. By encapsulating mammalian cell free expression system encoding a calcium biosensor, we show that thinning of oil combined with hypoosmotic shock can lead to calcium influx into the artificial cell. The transport of calcium ions through otherwise impermeable oil represents an advance in artificial cell research and opens up a new dimension in artificial cell activation using a mechanical stimulus.

Engineering a Synthetic Microbial Consortium for Efficient Production of Water Retaining Biopolymers via in-situ Precursor Production

Minty, J., University of Michigan
Dunham, D., Ecovia Renewables Inc.
Boyer, D.,
Kohley, S.,
Singer, M., University of Michigan
Lin, X. N., University of Michigan - Ann Arbor

Synergistic microbial communities are ubiquitous in nature and exhibit appealing features, such as sophisticated metabolic capabilities and robustness. This has inspired fast-growing interest in engineering synthetic microbial consortia for bioprocessing and biotechnology development. However, despite the potential advantages offered by microbial consortia, there are relatively few reports of their use in real-world applications. In this work, we have constructed a synthetic fungal-bacterial consortium for the efficient production of water retaining biopolymers with diverse commercial applications from low-cost biomass resources. Conventional fermentation methods for producing these biopolymers generally require exogenously supplied metabolic precursors, contributing to high feedstock costs. We have designed a binary consortium in which biopolymer precursors are produced *in-situ* from low-cost biomass materials, eliminating the need for exogenous addition. In this system, one microbial species converts biomass feedstocks into biopolymer precursors, which are subsequently co-metabolized to the final biopolymer product by a second species. Biopolymer production with the consortium has been demonstrated using biodiesel waste glycerol, a low-cost biomass resource, as the primary feedstock. Feedstock flexibility can be achieved by utilizing pre-cursor producing species with different substrate preferences, and work is currently underway to adapt the system to carbohydrate-rich biomass feedstocks.

An Engineered Open Cell Free Synthesis Platform for the Production of Non-natural Amino Acid Based Antibody-Drug Conjugates

Groff, D.,

We describe an open cell free synthesis (OCFS) platform that facilitates the high-level production of IgGâ??s that contain non-natural amino acid (nnAA) substitutions for the site-specific conjugation of small molecule warheads.Â We demonstrate that targeted proteolytic inactivation can be used as a strategy to destroy the E.coli protein Release Factor 1 (RF1), boosting Amber codon mediated nnAA incorporation into a nascent polypeptide, due to lack of competition between a truncated RF1 and an orthogonal, CUA-encoded, nnAA-charged tRNA.Â Several strains were generated that contain variants in the switch loop of RF1 which are susceptible to cleavage by the E.coli protease OmpT.Â During the cell lysis step of extract production for OCFS, OmpT, which is normally compartmentalized in the E.coli outer membrane as an integral membrane protein, is mixed with the E.coli cytoplasmic fraction, and targeted proteolysis of the variant RF1 occurs on demand. Using an extract from this designed strain for OCFS, we demonstrate that itâ??s possible to incorporate nnAAâ??s into previously unobtainable sites in an IgG, producing antibodies at titers of several hundred milligrams per liter in an overnight reaction. These variants were then site specifically conjugated to a small molecule warhead and the corresponding ADCâ??s effectively killed cells in a highly efficient manner.

Retargeting a Dual-Acting sRNA for Multiple mRNA Transcript Regulation

Lahiry, A., The Ohio State University
Stimple, S. D., The Ohio State University
Wood, D. W., Ohio State University
Lease, R. A., Ohio State University

We are interested in regulating multiple mRNAs within a metabolic pathway for balancing flux of metabolites during fermentation. Bacterial small regulatory sRNAs act by specific "antisense" sRNA:mRNA complementary base pairing interactions with one or more mRNAs. The sRNAs vary the efficiency of translation of mRNA into proteins by perturbing mRNA structures, leading to targeted mRNA turnover or enhanced translation *in vivo*. Alterations of sRNA antisense sequence frequently diminish sRNA activity or stability. We hypothesized that distinctive stem-loop antisense structural motifs of the *E. coli* DsrA sRNA were critical to its robust formation of antisense sRNA:mRNA interactions. Two separate DsrA sequences within stem-loops each specify an sRNA:mRNA interaction that targets translation of a native mRNA. We built an *E. coli* genetic system to quantify perturbation of mRNA translation *in vivo* by DsrA sRNA and its derivatives. In this system, levels of DsrA and its two mRNA reporters are precisely controlled by orthogonal, small molecule-induced transcription control. Native mRNA target sequences were each fused to distinct (mCherry, GFP) reporter gene fluorescent proteins. The activity of DsrA at these reporter genes was validated during cell growth in a 96-well format for high-throughput screens at different steady-state transcript levels of DsrA and mRNAs. We then rationally designed retargeted DsrA stem-loop antisense regions by editing complementary sequences to make them pair with desired mRNA targets, while conserving the DsrA native-like stem-loop structure. We retargeted each antisense stem-loop of DsrA sRNA to act singly at one of two non-native mRNA reporter gene fusion targets from the *n*-butanol fermentation pathway of *Clostridium acetobutylicum*. Most of our singly-retargeted sRNAs functioned as anticipated by tuning down target mRNA translation (~2-9 fold), validating our approach for metabolic engineering by mRNA target regulation. Further, we recombined single-retargeted stem-loops to create dual-retargeted DsrA variants that can act at two non-native mRNA targets in *E. coli*. This sRNA stem-loop antisense motif is therefore a robust structural element that can be retargeted to govern non-native mRNAs without loss of sRNA function, enabling tailored targeting to desired mRNA sequences. The universality of sRNAs as a genetic control mechanism in bacteria suggests that these sRNA derivatives can be validated in *E. coli* and then implemented *in situ* in industrially relevant host organisms such as *Clostridium*.

Engineering Thermostable Proteins to Selectively Extract Strategic Metals from Low Temperature Geothermal Brines

Charrier, M.,
Ajo-Franklin, C.,

Geothermal brines can contain comparatively high concentrations of rare earth elements (REEs), high-value metals, and/or critical metals that offer the opportunity to add value to geothermal energy production. However, existing technologies are either insufficiently selective or too expensive to feasibly extract these elements. Here we test whether engineered thermostable cell surface proteins can unite the affinity, selectivity, and robustness needed to act as a low-cost, selective, and reusable metal adsorbent. Using heterologously expressed S-layer proteins with a C-terminal metal-binding moiety purified by affinity chromatography from *E. coli*, we show that our engineered variant can bind Zn^{2+} more tightly at 40°C and more selectively than its native counterpart via ICP-MS. We also show that two other variants can bind Gd^{3+} more tightly at 40°C. Thus, the results presented here suggest that using thermophilic microorganisms to display engineered metal-binding proteins is an extremely promising approach for strategic metal extraction in low-temperature geothermal arena.

Non-Canonical Amino Acid Incorporation via Sense Codon Reassignment in E. coli

Biddle, W., Colorado State University
Schmitt, M. A., Colorado State University
Fisk, J. D., Colorado State University

Suppression of nonsense codons leading to the incorporation non-canonical amino acids (ncAAs) using orthogonal tRNA and aminoacyl tRNA synthetase (aaRS) pairs has been the most commonly employed technology for the expansion of the genetic code. Nonsense suppression is limited due to competition with the termination machinery in the number of ncAAs that can be incorporated into a given protein and in the yield of protein containing ncAAs that can be obtained. Additionally, the number of nonsense codons is limited, presently enabling expansion to 21 amino acid genetic codes. We describe investigations of *sense codon reassignment* in *E. coli*: the incorporation of ncAAs into proteins in response to codons normally directing the incorporation of natural amino acids, as a means to expand the number of substitutions and yield of modified protein produced.

Variants of the orthogonal tyrosine tRNA/aaRS pair from *Methanocaldococcus jannaschii* have been evolved to aminoacylate nearly 100 different ncAAs onto forms of the *M. jannaschii* tyrosine tRNA directed to nonsense codons. In order to expand the utility of the suite of ncAA incorporating *M. jannaschii* aaRS variants presently available, we investigated the ability of already-evolved ncAA incorporating pairs to reassign sense codons *in vivo*. A goal of this work is to produce 21 and 22 amino acid genetics codes without eliminating any of the 20 canonical amino acids. By modifying the anticodon on the tRNA to decode a sense codon typically read through a wobble pairing interaction and directed evolution of improved variants, we have been able to reassign histidine, lysine and arginine sense codons to incorporate tyrosine with efficiencies between 10 and 70%. Using our improved lysine system the ncAA O-methyl-L-tyrosine can be incorporated in response to AAG codons with lower efficiencies. Further, we have also performed directed evolution experiments to increase this efficiency as well as investigate the transferability of modifications made between ncAA and sense codon reassignment systems, *e.g.* Do improvements to the Tyr-incorporating aaRS also increase the efficiency of para-aminophenylalanine incorporation?

Interactions between the orthogonal aaRSs and tRNAs are one of several factors that affect ncAA incorporation efficiencies. In attempts to reassign one of the two histidine codons in *E. coli*, the unintended modification of the orthogonal tRNA by an endogenous *E. coli* enzyme, tadA was observed. TadAâ??s essential function is to modify the A34 in the anticodon of the *E. coli* arginine tRNA. We observed the modification of A34 of the histidine-targeted *M. jannaschii* tRNA to inosine which produced a tRNA variant relatively incapable of discriminating between the two histidine codons. We identified variants of both the tRNA and synthetase that simultaneously increased the sense codon reassigning efficiency of the pair and eliminated post transcriptional modification of the *M. jannaschii* Tyr-tRNA bearing an AUG anticodon via directed evolution. The engineered variant is capable of reassigning the His CAU codon with 2.5-fold increased efficiency and has near perfect discrimination between the two histidine codons.

Utilizing Sense Codon Reassignment to Understand the Plasticity of the E. coli Genetic Code

Schmitt, M. A., Colorado State University
Biddle, W., Colorado State University
Fisk, J. D., Colorado State University

The genetic code was set and has remained largely unchanged since the evolution of the last universal common ancestor of life on earth. Several lines of evidence suggest that this probable â??frozen accidentâ?□ is potentially fairly plastic. Although the meaning of the 64 codons has not changed in 2 billion years of evolution, huge variation in the use of synonymous codons exists. The components that are employed to translate the code are also highly variable across species. The orthogonality of certain tRNA/aminoacyl tRNA synthetase pairs between species has been exploited to expand the genetic code, most commonly by incorporating non-canonical amino acids in response to the amber stop codon. Unfortunately, the number of stop codons is limited, and the method of nonsense codon suppression is not readily extendable to additional codons. In contrast, sense codon reassignment should be a broadly extendable, general method for genetic code expansion. The ability to expand the number of available genetic codes from the 100s presently available (employing 21 amino acids) to 10,000s (22 amino acids) to 1,000,000s (23 amino acids) would open vast new vistas of protein sequence and functional space for exploration and exploitation.

Engineering the genetic code requires a detailed understanding of the molecular and systems level interactions governing the entire process of translation. To evaluate the potential for sense codon reassignment in *E. coli*, we have evaluated 20 of the 21 codons read via a wobble interaction. Our approach evaluates the ability of a tyrosine-charging *M. jannaschii* aminoacyl tRNA synthetase to recognize *M. jannaschii* tRNAs with altered anticodons and the extent to which Tyr-aminoacylated *M. jannaschii* tRNAs can compete with endogenous *E. coli* tRNAs to decode and reassign the meaning of sense codons. Variants of the *M. jannaschii* tyrosine pair have been evolved to incorporate ~100 different non-canonical amino acids into proteins in response to the amber stop codon.

Our work augments studies of natural translational systems and examines the plasticity and fitness costs imposed by redefining the *E. coli* genetic code. We show that to some extent *every* wobble codon in *E. coli* can be reassigned to another amino acid. We examine sense codon reassignment efficiency in the context of the prevalence of each codon in the *E. coli* genetic code and the predicted effect on tRNA aminoacylation as a result of altering the *M. jannaschii* tRNA anticodon to decode a given sense codon. We discuss observed growth defects of the *E. coli* strains as a result of reassignment, which are not always correlated with increased reassignment. We evaluate the ability of orthogonal tRNAs to decode only the desired sense codon and show that consideration of possible tRNA modifications by endogenous *E. coli* enzymes is of high importance for maintaining the fidelity of orthogonal tRNAs.

A Tunable Bacterial Selection System for Engineering Condition-Dependent Protein Function

Brechun, K. E.,
Woolley, A.,
Arndt, K.,

Protein engineering towards the development of protein switches requires evolution in opposing directions, producing proteins that are highly active in one state and inactive in the other. Numerous assays exist to aid in evolving protein function, however these assays are of limited use in the development of condition-dependent protein function, as they do not control for absence of function in the off-state. To enable bi-directional evolution of conditional function, we have engineered an *in vivo* assay with a selection parameter that can be tuned, allowing for enrichment of the on- or off-state as required by external conditions. This assay combines principles from two previously established *in vivo* assays: the â??hitchhikerâ?? assay[1], a selection assay for protein-protein interactions based on the production of beta-lactamase; and the â??band-passâ?? assay[2], an externally tuneable bacterial selection system for beta-lactamase activity. The resulting genetic circuit enables one to select for cells with specific levels of beta-lactamase activity, which correspond to different protein-protein interaction strengths. We demonstrated the function of this assay using designed leucine zipper peptides with a range of interaction strengths, comparing the results with data from *in vitro* characterization of the peptides. We will use this assay in the development of a synthetic photo-controlled protein, using the ability of the assay to select for cells with the protein switch in the active or inactive state, depending on the experimental set up. It is anticipated that this assay could be used to facilitate the directed evolution of a wide variety of protein switches involving protein-protein interactions

AUTHOR INDEX

AUTHOR INDEX

AUTHOR INDEX

AUTHOR INDEX

AUTHOR INDEX

AUTHOR INDEX

AUTHOR INDEX